D1607635

GRASSLAND ECOPHYSIOLOGY AND GRAZING ECOLOGY

Grassland Ecophysiology and Grazing Ecology

Edited by

G. Lemaire
INRA, France

J. Hodgson
Massey University, New Zealand

A. de Moraes
Universidade Federal do Paraná, Brazil

C. Nabinger
Universidade Federal de Rio Grande do Sul, Brazil

and

P.C. de F. Carvalho
Universidade Federal de Rio Grande do Sul, Brazil

CABI *Publishing*

CABI *Publishing* is a division of CAB *International*

CABI Publishing
CAB International
Wallingford
Oxon OX10 8DE
UK

Tel: +44 (0)1491 832111
Fax: +44 (0)1491 833508
Email: cabi@cabi.org
Web site: http://www.cabi.org

CABI Publishing
10 E 40th Street
Suite 3203
New York, NY 10016
USA

Tel: +1 212 481 7018
Fax: +1 212 686 7993
Email: cabi-nao@cabi.org

A catalogue record for this book is available from the British Library, London, UK.

Library of Congress Cataloging-in-Publication Data
Grassland ecophysiology and grazing ecology / edited by G. Lemaire ... [*et al.*].
p. cm.
Edited and revised versions of papers presented at an international conference held in Curitiba, Parana, Brazil, Aug. 24-26, 1999.
Includes bibliographical references.
ISBN 0-85199-452-0 (alk. paper)
1. Rangelands--Management--Congresses. 2. Grasses--Ecophysiology--Congresses. 3. Range ecology--Congresses. 4. Rangelands--South America--Management--Congresses. 5. Grasses--Ecophysiology--South America--Congresses. 6. Range ecology--South America--Congresses. I. Lemaire, Gilles, 1945-

SF84.84 G68 2000
633.2902--dc21 00-028916

ISBN 0 85199 452 0

Typeset in 10/12pt Goudy by Columns Design Ltd, Reading.
Printed and bound in the UK by the University Press, Cambridge.

Contents

Part II: Morphogenesis of Pasture Species and Adaptation to Defoliation

Part III: Plant–Animal Interactions

Part IV: Sustainable Grazing Management of Natural Pastures

Part V: Problems of Animal Production Related to Pastures in Subtropical and Temperate Regions of South America

Contributors

M. Agnusdei, INTA-EEA, Balcarce 7620, Provincia de Buenos Aires, Argentina
S.G. Assuero, Universidad Nacional de Mar del Plata, CC 276, 760 Balcarce, Argentina
U. Bausenwein, Plant Science Group, Macaulay Land Use Research Institute, Craigiebuckler, Aberdeen AB15 8QH, UK
E.J. Berretta, Instituto Nacional de Investigación Agropecuaria (INIA), Estación Experimental del Norte, Ruta 5, km 386, 45000 Tacuarembó, Uruguay
C.K. Black, Williams and Kettle Ltd, PO Box 501, Feilding, New Zealand
M. Boval, INRA, Station de Recherches Zootechniques, BP 515, 97165 Pointe-à-Pitre, France
P. Carrère, INRA, Unité d'Agronomie, 234 Av. Du Brézet, 69039 Clermont-Ferrand, Cedex 2, France
P.C. de F. Carvalho, Facultade de Agronomia, Universidade Federal do Rio Grande do Sul (UFRGS), Av. Paulo Gama, 110, 90046–900 Porto Alegre, Brazil
P. Cruz, INRA, Station d'Agronomie, BP 27, 31326 Castanet-Tolosan, France
S.C. Da Silva, Departamento de Zootecnia, ESALQ/USP, Universidade de São Paulo, Campus 'Luis Queroz', Av. Padua Dias 11, Cx Postal 9, 13418–900 Piracicaba, S.P., Brazil
L.A. Dawson, Plant Science Group, Macaulay Land Use Research Institute, Craigiebuckler, Aberdeen AB15 8QH, UK
A. de Moraes, Departemendo de Fitotecnia et Fitossanitarismo, Universidade Federal do Paraná, Rua dos Funcionarios 1540, CP 2959, 80.001–970 Curitiba, Brazil
V.A. Deregibus, Faculty of Agronomy, University of Buenos Aires, 1417 Buenos Aires, Argentina

R. de Visser, Research Institute of Agrobiology and Soil Fertility (AB-DLO), PO Box 14, NL-6700 AA, Wageningen, The Netherlands

J.-L. Durand, Unité d'Ecophysiologie des Plantes Fourragères, INRA-UEPF, 86600 Lusignan, France

F. Gastal, Unité d'Ecophysiologie des Plantes Fourragères, INRA-UEPF, 86600 Lusignan, France

I.J. Gordon, Macaulay Land Use Research Institute, Craigiebuckler, Aberdeen AB15 8QH, UK

S.J. Grayston, Plant Science Group, Macaulay Land Use Research Institute, Craigiebuckler, Aberdeen AB15 8QH, UK

J. Hodgson, Institute of Natural Resources, Massey University, Private Bag 11222, Palmerston North, New Zealand

S.C. Jarvis, Institute of Grassland and Environmental Research, North Wyke Research Station, Okehampton, Devon EX20 2SB, UK

E.A. Laca, Department of Agronomy and Range Science, University of California, One Shields Avenue, Davis, California 95616–8515, USA

C.E. Lascano, CIAT, AA 67-13, Cali, Colombia

G. Lemaire, Unité d'Ecophysiologie, INRA, Lusignan 86600, France

G.E. Maraschin, Universidade Federal do Rio Grande do Sul, Porto Alegre, Brazil

C. Matthew, Institute of Natural Resources, Massey University, Private Bag 11-222, Palmerston North, New Zealand

P. Millard, Plant Science Group, Macaulay Land Use Research Institute, Craigiebuckler, Aberdeen AB15 8QH, UK

F. Montossi, Instituto Nacional de Investigación Agropecuaria (INIA), Estación Experimental del Norte, Ruta 5, km 386, 45000 Tacuarembó, Uruguay

C. Nabinger, Facultade de Agronomia, Universidade Federal do Rio Grande do Sul, Av. Benro Goncalves, 7712, 91501-970 Porto Alegre, Brazil

C.J. Nelson, Department of Agronomy, University of Missouri, Columbia, MO 65211, USA

A. Oliveira Machado, Departamentato de Fiotecnia e Fitossanitarismo, Universidade Federal do Paraná, Rua dos Funcionários 1540, CEP 80.035–050, Curitiba, Andrea, Brazil

A.J. Parsons, AgResearch Grasslands, Private Bag 11008, Palmerston North, New Zealand

E. Paterson, Plant Science Group, Macaulay Land Use Research Institute, Craigiebuckler, Aberdeen AB15 8QH, UK

G. Pigurina, Instituto Nacional de Investigación Agropecuaria (INIA), Estación Experimental del Norte, Ruta 5, km 386, 45000 Tacuarembó, Uruguay

D.F. Risso, Instituto Nacional de Investigación Agropecuaria (INIA), Estación Experimental del Norte, Ruta 5, km 386, 45000 Tacuarembó, Uruguay

N.R. Sackville Hamilton, Institute of Grassland and Environmental Research, Plas Gogerddan, Aberystwyth SY23 3EB, UK

R. Schäufele, Department of Grassland Science, Technische Universität München, 85350 Freising-Weihenstephan, Germany

H. Schnyder, Department of Grassland Science, Technische Universität München, 85350 Freising-Weihenstephan, Germany
S. Schwinning, Department of Biology, University of Utah, Salt Lake City, UT 84112, USA
J.F. Soussana, INRA, Unité d'Agronomie, 234 Av. du Brézet, F-63100 Clermont-Ferrand, France
J. Stuth, Department of Rangeland Ecology and Management, Texas A&M University, College Station, TX 77843–2126, USA
B. Thornton, Plant Science Group, Macaulay Land Use Research Institute, Craigiebuckler, Aberdeen AB15 8QH, UK
M.H. Wade, Departamento de Produccion Animal, Facultad de Ciencias Veterinarias, Universidad Nacional del Centro de la Provincia de Buenos Aires (UNCPBA), 7000 Tandil, Argentina

Preface

This book consists of edited and revised versions of the invited papers presented at an International Symposium on Grassland Ecophysiology and Grazing Ecology held in Curitiba, Parana, Brazil, 24–26 August 1999. This Symposium, organized by the Federal University of Parana (Curitiba) and the Federal University of Rio Grande do Sul (Porto Alegre) in Brazil had as its main objective to assemble the leading scientists in the world in the research area of plant ecophysiology and ecology focused on the management and sustainability of natural grassland ecosystems. Sixteen international speakers coming from different parts of the world were invited to deliver review papers on different scientific topics concerning plant responses to the environment, plant adaptation to defoliation, plant–animal interactions and the sustainability of the grazing management of natural pastures.

The reasons for the organization of such a symposium, with a large international audience, in South America were as follows:

1. The need to focus all the scientific disciplines dealing with plant ecophysiology and ecology on a more global approach to the functioning and the evolution of grazing ecosystems in natural grassland areas.
2. The need for increasing integrated research from plant through to ecosystem levels for understanding the contribution of grasslands to terrestrial ecology and global changes.
3. The recognition that the large areas of natural grassland of the subtropical and temperate regions of South America (i.e. *pampas* in Argentina and *campos* in Uruguay and south Brazil) represent a very important example on a world scale of the necessity to combine agronomical and ecological approaches. This is essential in order to allow sustainable development of rural areas with respect for natural resources (water and soil quality, biodiversity) and the improvement of the rural economy.

4. The wish to create and develop links between the different research teams in the world and those in these regions in order to have a rapid transfer of scientific concepts and methods, allowing the emergence of integrated research projects and enhancing multilateral collaboration.

A poster session with 45 contributions was organized for the presentation of the main research activities which are in progress in the different research teams from South America. All the invited speakers were impressed by the high quality of these contributions showing that the up-to-date concepts, approaches and methods in ecophysiology and in grazing ecology were being implemented and well used by the majority of the young researchers and PhD students. This gave an excellent view of the high quality of the human resources involved in grassland science in the different countries of the region.

Scientific Committee

Presidents

Prof. G.E. Maraschin (Federal University of Rio Grande do Sul, Porto Alegre, Brazil), jorge@netmarket.com.br
Dr G. Lemaire (INRA, Lusignan, France), lemaire@lusignan.inra.fr

Organizing Committee

President

Prof. Anibal de Moraes (Federal University of Parana, Curitiba, Brazil), anibalm@agrarias.ufpr.br

1

Sustainability of Grazing Systems: Goals, Concepts and Methods

John Hodgson[1] and S.C. Da Silva[2]

[1]*Institute of Natural Resources, Massey University, New Zealand;* [2]*Departamento de Produção Animal, Universidade de São Paulo, Brazil*

Introduction

The theme of this symposium focuses on the principles that underlie the behaviour of populations of plants and animals in grazing systems and their influence on system productivity and sustainability. We see it as our remit to provide an overview of the context for the four technical sessions and attempt to define a set of goals against which the outcomes of research on the ecology of grazing systems can be judged. Interest in the behaviour of grazing systems and our ability to control them is explicit only in the final technical chapter (Stuth and Maraschin, Chapter 17, this volume), though it is implicit in many of the other chapters. This makes the choice of title for the introductory chapter particularly challenging.

It would be difficult to deny the importance of 'sustainability' in grazing systems, though there is room to argue about the specific meaning of the term. We can define sustainability in a broad sense, taking as a model the definition of the World Commission on Environment and Development: 'Sustainable development ... meets the needs of the present without compromising the ability of future generations to meet their own needs' (WCED, 1987). This is probably too broad a definition to provide useful guidance for the participants in this symposium, and more useful for our purposes would be a narrow-sense definition of sustainability that focuses on the maintenance of productivity and stability in the soil, plant and animal components of pastoral systems.

Stability can itself be a difficult term to define objectively (see, for example, Illius and Hodgson, 1996). Pimm (1984) defines an ecosystem as stable 'if all variables return to equilibrium values following a perturbation away from

 Grassland Ecophysiology and Grazing Ecology (eds G. Lemaire, J. Hodgson, A. de Moraes, C. Nabinger and P.C. de F. Carvalho)

equilibrium'. In a sense, this begs the question, given that many grasslands exist in forest climax zones, so that any concept of stability must take into account the management controls necessary to maintain a subclimax state. There are also interesting implications for concepts of system stability in the contrasts between classical successional theory (Clements, 1916) and 'state-and-transition' theory (Westoby *et al.*, 1989) as applied to pastoral systems. Hadley (1993) and Tainton *et al.* (1996) present challenging analyses of the links between complexity and stability. In this context the practical definition of stability by Conway (1987) – 'the constancy of productivity in the face of small disturbing forces arising from the natural fluctuations and cycles in the surrounding environment' – is preferred, though with due regard to the strictures of Hadley (1993) and Tainton *et al.* (1996) about the need for flexibility of management against a background of environmental variability.

Goals and Concepts

Goals for research can be defined in relatively objective and practical terms. The concepts on which research programmes are built are more intuitive and even philosophical in nature, and their testing is often dependent upon the availability or development of appropriate methodology. In the context of a narrow-sense definition of sustainability, we suggest that the goals for research on the ecology of grazing systems should be as follows:

1. To enhance the productivity and stability of production in grazing systems.
2. To enhance the stability and predictability of grass/legume balance in such systems.

There is nothing new in these suggestions (see Hadley, 1993; Riveros, 1993), but they are particularly appropriate for this symposium. It could be argued that goal 1 is not strictly attainable without support from goal 2, but for convenience it is appropriate to deal with them as complementary rather than mutually dependent entities.

This symposium provides an excellent opportunity to compare and contrast goals and concepts of sustainability for pastoral systems in temperate and tropical zones, and it is essential that we make the most of the opportunity provided. In particular, we need to consider whether the generally greater range of climatic variation within and between years in tropical than in temperate countries results inevitably in greater variability in pasture production, and whether the greater contrasts in plant morphology and phenology in tropical than in temperate plants can be harnessed to our advantage rather than treated simply as contributors to instability (Toledo and Formoso, 1993). We shall return to this theme throughout the chapter, because it has a major impact on the conceptual framework within which much of our pastoral research is conducted. Specifically, we can ask the question: 'To what extent is it feasible to develop concepts and paradigms that can be generalized across a range of plant and animal systems (and, by inference, across climatic zones), without losing the focus that is needed to define appropriate management strategies for specific systems?'

Concepts and Methods

Plants

There is now a strong factual basis for understanding the dynamics of plant growth and development and their relevance to concepts of herbage accumulation and utilization in grazed pastures. This understanding is based on knowledge of shoot structure and phenology, shoot and leaf dynamics, light interception and carbon balance, size/density compensation, root structure and dynamics, tissue and nutrient fluxes, and their response to variations in climatic and edaphic variables and management strategies (e.g. Chapman and Lemaire, 1993; Matthew *et al.*, 1995; Bullock, 1996; Lemaire, 1997; Thornton *et al.*, Chapter 5, this volume; Nelson, Chapter 6, this volume; Lemaire and Agnusdei, Chapter 14, this volume). Linkage of concepts of tissue turnover in plants (Lemaire, 1997) with evidence on solute movement and nutrient cycling in plant tissues (Schnyder *et al.*, Chapter 3, this volume; Thornton *et al.*, Chapter 5, this volume; Cruz and Boval, Chapter 8, this volume) provides a powerful tool for considering the nutrient and carbon economy of plant communities (Gastal and Durand, Chapter 2, this volume) and the scope for controlling these by sward manipulation. Growing evidence on root dynamics (Dawson *et al.*, Chapter 4, this volume) and the synchrony of root and shoot development on tiller axes (Matthew *et al.*, 1998) will add substantially to knowledge on the control of nutrient uptake and utilization (Jarvis, Chapter 16, this volume).

Briske (1996), working with rangeland communities, draws a distinction between the morphological characteristics and adaptations that contribute to 'avoidance of grazing' and the short-term physiological responses that contribute to regrowth capability or 'tolerance of grazing'. Lemaire and Chapman (1996) make much the same distinction for temperate pastures between relatively short-term adjustments to C and N fluxes and longer-term adjustments in plant morphology. Conceptually, this is a useful distinction to draw, though there are likely to be overlaps at the boundary between the two sets of attributes (Richards, 1993). However, Briske (1996) also comments that:

> although a reasonable understanding of the major plant attributes that confer grazing avoidance and grazing tolerance has been achieved, the concepts are largely based on empirical evidence with a limited theoretical base … and progress towards a greater functional interpretation will be required to address critical questions associated with plant–herbivore interactions.

In this context, it is important to recognize that a comprehensive information base really exists for only two temperate grasses (*Lolium perenne* and *Festuca arundinacea*) and one temperate legume (*Trifolium repens*) (Lemaire and

Chapman, 1996). There is complementary evidence for other temperate species, offering some contrast in plant size and morphology, and for an increasing range of tropical species (Richards, 1993; Lemaire and Chapman, 1996; Gomide, 1997; Lemaire, 1997). It seems unlikely that there will ever be an opportunity in the future to repeat the detailed analytical work originally carried out on *L. perenne* and *T. repens*, so it will be important to maximize the opportunity to examine the extent to which the detailed evidence from these species can be generalized to provide effective management frameworks for alternative species. It will be particularly important to augment the limited information on tropical species (e.g. Gomide, 1997), where contrasts in pathways of C metabolism, as well as greater ranges of variation in plant size and structure, will make generalization more difficult. We suggest that this should be a major priority for future research in tropical centres, and we are pleased to note the emphasis on aspects of the ecophysiology of tropical species in the papers offered to this symposium.

Bullock (1996) argues that the principles of the dynamics of plant populations apply to all communities and all grazed systems, but evidence in patterns of tissue flow and tiller population dynamics for tropical species is only now becoming available (e.g. Carnevalli and Da Silva, 1999) for comparison with evidence for perennial ryegrass/white clover pastures (Lemaire and Chapman, 1996). Chapman and Lemaire (1993) emphasize the need for more research on phenotypic plasticity and the limits to flexibility in size/density adjustment as a means of defining the range of management tolerances for alternative species, and suggest that limited plasticity may be an important reason for the relatively poor persistence of tropical and subtropical species. A conceptual framework is thus already in place, but contrasts in plant size and morphology will demand ingenuity in the development of experimental methodology for tropical pastures.

In our view, one of the strengths of the current understanding of the behaviour of pasture plants has to do with the complementary nature of information at different levels of aggregation (Lemaire and Chapman, 1996). This is by no means a universal view, however. In reviewing a session at the Seventeenth International Grassland Congress concerned with plant growth, Parsons (1993) commented that 'leaves are a difficult currency by which to understand plant growth'. Reporting on a discussion on plant communities at the same congress, Harris (1993) commented on the compensating adjustments in tiller size and density, and asked 'do we need to concern ourselves about tiller dynamics?' – suggesting rather that a more pragmatic approach would be to focus attention on levels of community biomass. These are not necessarily contradictory views, but they serve to emphasize the need for critical evaluation of the objectives of specific research programmes and the methods adopted to meet these objectives.

The concepts of plant and plant population behaviour outlined above probably provide a reasonable basis for defining management strategies to help maintain stability of production in a range of pasture species. There is much less certainty, however, about the strategies required to control species balance in mixed swards, with particular reference to the balance between grasses and legumes. The issue is complicated by climatic variability, which influences the balance between competitive ability and resource use of alternative species (e.g. Harris, 1987), and by structural and biochemical contrasts between species, which may affect preferential grazing (Gordon and Lascano, 1993; Launchbaugh, 1996). Even the supposed stability of grass/legume balance in long-established temperate pastures dominated by *L. perenne* and *T. repens* now appears to be an illusion based on the self-compensating effects of rapid spatiotemporal changes in species balance on individual sward patches (Schwinning and Parsons, 1996).

Instability in grass/legume balance appears to be a particular problem in tropical pastures, and is certainly one of the major factors limiting levels of animal output relative to temperate pasture conditions (Cameron *et al.*, 1993; Maraschin and Jacques, 1993; Toledo and Formoso, 1993; Lascano, 1994). The definition of plant characteristics influencing grazing avoidance and grazing tolerance mechanisms (Briske, 1996) provides the basis for explaining grazing impact on species balance (Briske and Silvertown, 1993). However, the latter authors comment that 'an evaluation of the mechanisms potentially capable of regulating tiller populations to maintain grassland stability indicates that relatively little is known about the mode of operation or ecological significance of these mechanisms'.

Against this background, the search for effective combinations of tropical grasses and legumes has been essentially pragmatic in nature, rather than driven by conceptual models of compatibility. There are encouraging indications of the potential for stability in the combination of *Brachiaria* spp. with *Desmodium ovalifolium* (Toledo and Formoso, 1993) or with *Arachis pintoi* (Fisher *et al.*, 1996), where grass and legume components have compatible sward-forming habits. However, attempts to develop stable legume associations in tall-grass communities have been less successful, except perhaps in the case of shrubby legumes, such as *Stylosanthes* spp. (Cameron *et al.*, 1993). The evidence of Clements (1989) serves to illustrate the difficulties of maintaining stability in mixed communities where legume growing points are susceptible to grazing damage. It is perhaps reasonable to point out that there are equivalent problems in maintaining species balance in combinations of alternative temperate legume species and companion grasses, which are much less tolerant of variations in grazing management than are conventional *L. perenne*/*T. repens* combinations.

Animals

Studies on the ingestive behaviour of grazing animals at the plant or patch level have made rapid progress, with particularly profitable links between interests in natural ecology and in managed systems (Gordon and Lascano, 1993). This has provided a firm basis for the development of a conceptual framework linking the structural and biochemical characteristics of forage plants and sward canopies to the components of ingestive behaviour and to the rate of herbage intake and diet composition (Gordon and Lascano, 1993; Ungar, 1996; Cosgrove, 1997). However, effective models of the control of forage intake and diet selection in free-grazing animals have been slower to develop (Wade and Carvalho, Chapter

12, this volume). This is partly because of the difficulties of monitoring variations in animal behaviour in space and time (Illius and Hodgson, 1996) and partly because of the shortage of information linking external (behavioural) controls and internal (digestive) controls of foraging behaviour (Demment *et al.*, 1995; Dove, 1996). There is a real need for progress in both of these areas of research. Useful conceptual models that link foraging strategy to spatially variable vegetation characteristics are now being developed (e.g. Laca and Demment, 1996; Laca, Chapter 11, this volume), and this should help to stimulate the development of concepts from which to build more spatially explicit models of foraging strategy.

As in the examples of plant behaviour considered earlier, more progress has been made in conceptualizing the effects of sward canopy characteristics on the mechanics of the grazing process in temperate than in tropical conditions (Gordon and Lascano, 1993), despite the fact that some of the pioneer work in this subject was carried out on tropical species (Stobbs, 1973a, b). This is a consequence of both a more limited database and the difficulty of generalizing concepts across very wide ranges of variation in size and structure of tropical plants. Comparative evidence from temperate and tropical pastures would be particularly helpful in defining the relative importance and range of influence of alternative canopy characteristics on ingestive behaviour.

Much of the early work on selective grazing behaviour was too descriptive to provide an effective basis for understanding foraging strategy, but current concepts facilitate research on the ways in which the physical and biochemical characteristics of plants influence preferential behaviour (Laca and Demment, 1996; Launchbaugh, 1996; Gordon, Chapter 10, this volume). One of the major problems has been, and remains, a description of the vegetation resource that is meaningful in terms of the cues to which animals respond and the decisions they make as they move through the resource. There has been useful synergy in the development of theoretical concepts of foraging strategy across a very wide range of animal species with contrasting dietary resources (e.g. Crawley, 1983; Hodgson *et al.*, 1997). However, there is still room for argument about the extent to which classical foraging theory (Pyke, 1984) can be applied to the specific case of grazing ruminants (O'Reagain and Schwartz, 1995; Hodgson *et al.*, 1997). In their selective strategy, grazing animals must strike a balance between maximizing potential intake rate and protecting biochemical defences (Illius and Hodgson, 1996), and they are capable of adjusting selective behaviour in response to changes in the spatial or temporal variability of their food environment (O'Reagain and Schwartz, 1995).

It seems axiomatic that selection is seldom absolute, in the sense of either complete rejection of one plant species or complete concentration on another, and this would be consistent with the trade-offs between alternative plant characteristics noted above. Indeed, the concept of partial preference is now firmly established (Parsons *et al.*, 1994), though the reasons for this pattern of behaviour and its place within the suite of factors influencing choice at a particular locus and point in time are not clear. Concepts of partial preference (Parsons *et al.*, 1994), dietary trade-off between the physical and biochemical characteristics of food sources (Illius and Hodgson, 1996) and sampling behaviour (Hodgson *et al.*, 1997) may all be implicated.

Given the difficulties noted above in defining objectively either the selective strategy of grazing animals or the effects of differential defoliation on the competitive strategy of alternative plant species, it would be difficult and probably dangerous to attempt to make general predictions about the impact of selective grazing on species balance in plant communities. This serves to illustrate the limited predictive ability of current concepts of differential defoliation by grazing animals and differential resistance to grazing by forage species (see also Briske, 1996).

Scale and heterogeneity

Hierarchies of scale, ranging from individual plants or plant parts to landscape or even regional units, can be defined in studies of both plant and animal behaviour (Stuth, 1991; Tainton *et al.*, 1996). There has been criticism of the essentially reductionist approach adopted in studies of the ingestive behaviour of grazing animals, on the grounds that it does not help the understanding of foraging strategy at the practical level of paddock-scale or landscape-scale behaviour (Taylor, 1993). This is a reasonable argument, though it tends to ignore both the need for understanding at several levels of complexity (Ungar, 1996) and the potential advantages of concepts of animal behaviour that function at similar scales to the concepts of plant behaviour outlined earlier. Here, too, there has been something of a dichotomy of interest between small-scale, detailed studies at the level of the individual plant or shoot (Lemaire and Chapman, 1996) and broad-scale studies at the plant-community level (Archer, 1996; Briske, 1996), and similar concerns have been expressed about the difficulties of scaling up from the plant to the community or landscape level (e.g. Harris, 1993; Briske, 1996). It is, however, noteworthy that the plant-orientated papers at this symposium concentrate primarily towards the lower end of the hierarchy of scale outlined above, whereas the animal-orientated papers focus primarily at the upper end.

Hadley (1993), quoting Chambers (1990), comments perceptively on the recognition in both agricultural and social sciences that complexity and diversity are underperceived and consequently undervalued. Tainton *et al.* (1996) emphasize the links between scale of observation and perception of complexity, arguing that the degree of complexity observed will be scale-dependent, and they comment on the difficulty of viewing complexity across scales of observation. Scale in investigation and prediction is likely to be more important in the relatively extensive and often spatially heterogeneous conditions typical of tropical and subtropical pastoral systems than in the relatively small-scale and homogeneous conditions of temperate systems. However, it should be noted that the preceding analysis suggests that a weakness in dealing with tropical and subtropical grazing systems is often the lack of detailed information on specific aspects of plant or animal behaviour, rather than shortage of broad-scale information.

Increasing scale of operation tends to increase the importance of heterogeneity in the vegetation resource and in resource use, the issue to which Taylor (1993) draws attention. It has been argued that foraging behaviour at a large scale still involves a series of small-scale decisions as animals move on a foraging path (Hodgson *et al.*, 1997). However, a conceptual basis for visualizing the hierarchy of decision scales in a heterogeneous environment has been slow to develop, not least because of the limitations of procedures for describing and quantifying heterogeneity in vegetation distribution and animal movement. Diversity and heterogeneity may be an integral part of the sustainability of non-equilibrium systems, and should be exploited rather than attempting to reduce them (Tainton *et al.*, 1996).

The question of scale is also implicit in work on patch dynamics in plant communities and its relevance to long-term community stability (Schwinning and Parsons, 1996). In this context, the evidence suggests that 'stability' is hardly a useful concept at the scale of the individual plant or shoot, and the general impression is one of mobility also at the patch level. Rather, its value depends upon the scale of measurement relative to the hierarchy of 'grain' size inherent in specific plant communities (Tainton *et al.*, 1996). In this regard, contrasts in 'grain' size and plant size in temperate and tropical communities may be largely self-cancelling, but there are likely to be substantial differences in appropriate observational procedures in these contrasting conditions, which may confuse interpretation of the underlying patterns of behaviour.

Integration

Substantial progress has been made over the last 25 years in developing understanding of the behaviour of plant and animal populations in grazing systems, but there has been less progress in conceptualizing the outcomes of interactions between these populations. As a consequence, we are still reliant on a set of subjective and often rather vague generalizations in defining the management requirements for a specified outcome in terms of pasture or animal performance. Control of stability of production or sward composition is no more predictable than control of level of output in this regard, in both cases dictating the need for regular monitoring and iterative responses in order to achieve a specified performance target. This is not to deny the value of the detailed information on plant or animal behaviour, but rather to argue the need for a more objective and mechanistic conceptualization of the outcomes of their interactions.

Solutions to these problems will ultimately be dependent upon the development of mathematical models, which provide a way of quantifying the principles implicit in the conceptual models discussed earlier in this chapter. However, systems models are unlikely to be successful unless they take into account the realities of plant and animal behaviour which these concepts embrace. For example, conventional models based on estimates of net herbage accumulation do not deal adequately with feedback mechanisms affecting either herbage growth or herbage consumption. Models based on the concept of tissue dynamics (Woodward, 1998) provide much greater flexibility in this regard. Similarly, it seems likely that the effectiveness of interactive pasture models will always be circumscribed unless it is possible to incorporate into them concepts of the control of shoot size and population density, the influence of tiller morphology and phenology on canopy structure and their impacts on tissue dynamics and herbage intake.

A factor in the development and refinement of pasture management procedures in temperate regions has been the concept of sward state as a major driving variable influencing herbage production and utilization, on the one hand, and herbage intake and animal performance, on the other (Hodgson, 1985). Its value lies in the ability to relate information on plant and animal processes to a common set of sward indicators and to use these as link factors in mathematical models (Woodward, 1998) or to define a set of target indices for pasture management (Sheath and Clark, 1996). Concepts of leaf mass and area, meristem populations, herbage mass and canopy structure may all be involved, but in practical terms these indicators are often condensed for convenience to a single index, such as sward height (Hodgson, 1990) or herbage mass (Matthews, 1995). This approach has so far had little impact on research in tropical and subtropical environments (Fisher *et al.*, 1996), though in principle it should have just as much to offer in these conditions. The wide variation in the size and morphology of tropical plant species may make for greater difficulty in establishing single indices for use in the management of tropical pastures. However, the approach has been used successfully in the management of temperate tussock-grass communities, which encompass substantial contrasts in species characteristics (Armstrong *et al.*, 1997). In these circumstances, the requirement may be to investigate the use of indicators that are more species-specific.

Conclusion

In this introductory chapter, it has only been possible to deal briefly with the range of concepts and levels of understanding in a complex area of research. Given the complexity, there can be no doubt about the valuable contribution made by the development and use of conceptual models in promoting interchange between the constituent science disciplines, in achieving progress in understanding the principles of the behaviour of grazing systems and in communicating these principles to the pastoral industries. The organizers of this symposium are to be congratulated on the development of a programme that should maximize the opportunity for such interchange, and it will be up to the participants to make the maximum possible use of this opportunity. In doing so, we should bear in mind the criticisms of the 'grassland profession' made by Nores and Vera (1993) in their keynote address to the Seventeenth International Congress. Comments on the need for greater engagement at the level of community problem solving have been taken as the main message from this address, but the authors also identified the needs for greater emphasis on interdisciplinary research and for greater breadth in academic and technical training.

References

Archer, S. (1996) Assessing and interpreting grass–woody plant dynamics. In: Hodgson, J. and Illius, A.W. (eds) *The Ecology and Management of Grazing Systems*. CAB International, Wallingford, pp. 101–134.

Armstrong, R.H.A., Grant, S.A., Common, T.G. and Beattie, M.M. (1997) Controlled grazing studies on *Nardus* grassland: effects of between-tussock sward height and species of grazer on diet selection and intake. *Grass and Forage Science* 52, 219–231.

Briske, D.D. (1996) Strategies of plant survival in grazed systems: a functional interpretation. In: Hodgson, J. and Illius, A.W. (eds) *The Ecology and Management of Grazing Systems*. CAB International, Wallingford, pp. 37–67.

Briske, D.D. and Silvertown, J.W. (1993) Plant demography and grassland community balance: the contribution of population regulation mechanism. In: *Proceedings of the 17th International Grassland Congress*. SIR Publishing, Wellington, New Zealand, pp. 291–298.

Bullock, J.M. (1996) Plant competition and population dynamics. In: Hodgson, J. and Illius, A.W. (eds) *The Ecology and Management of Grazing Systems*. CAB International, Wallingford, pp. 69–100.

Cameron, D.F., Miller, C.P., Edye, L.A. and Miles, J.W. (1993) Advances in research and development with *Stylosanthes* and other tropical pasture legumes. In: *Proceedings of the 17th International Grassland Congress*. SIR Publishing, Wellington, New Zealand, pp. 2109–2114.

Carnevalli, R.C.A. and Da Silva, S.C. (1999) Validacao de tecnicas experimentais para avaliacao de caracteristicas agronomicas e ecologicas de pastagens de *Cynodon dactylon* cv. 'Coastcross 1'. *Scientia Agricola* 56, 489–499.

Chambers, R. (1990) *Microenvironments Unobserved*. IEED Gatekeeper Series 22, International Institute for Environment and Development, London.

Chapman, D.F. and Lemaire, G. (1993) Morphogenetic and structural determinants of plant growth after defoliation. In: *Proceedings of the 17th International Grassland Congress*. SIR Publishing, Wellington, New Zealand, pp. 95–104.

Clements, F.E. (1916) *Plant Succession: An Analysis of the Development of Vegetation*. Publication 242, Carnegie Institute, Washington, 512 pp.

Clements, R.J. (1989) Rates of destruction of growing points of pasture legumes by grazing cattle. In: *Proceedings of the 16th International Grassland Congress*. AFPF, Versailles, France, pp. 1027–1028.

Conway, G.R. (1987) The properties of agroecosystems. *Agricultural Systems* 24, 95–117.

Cosgrove, G.P. (1997) Animal grazing behaviour and forage intake. In: Gomide, J.A. (ed.) *Proceedings of the International Symposium on Animal Production Under Grazing, Vicosa, Minas Gerais, Brazil, November 1997*, pp. 59–80.

Crawley, M.J. (1983) *Herbivory, the Dynamics of Animal Plant Interactions*. University of California Press, Berkeley, 437 pp.

Demment, M.W., Peyraud, J.-L. and Laca, E.A. (1995) Herbage intake at grazing: a modelling approach. In: Journet, M., Grenet, E., Farce, M.-H., Theriez, M. and Demarquilly, C. (eds) *Recent Developments in the Nutrition of Ruminants*. INRA, Paris, pp. 121–141.

Dove, H. (1996) The ruminant, the rumen and the pasture resource: nutrient interactions in the grazing animal. In: Hodgson, J. and Illius, A.W. (eds) *The Ecology and Management of Grazing Systems*. CAB International, Wallingford, pp. 219–246.

Fisher, M.J., Rao, I.M., Thomas, R.J. and Lascano, C.E. (1996) Grasslands in the well-watered tropical lowlands. In: Hodgson, J. and Illius, A.W. (eds) *The Ecology and Management of Grazing Systems*. CAB International, Wallingford, pp. 393–425.

Gomide, J.A. (1997) Morphogenesis and growth analysis of tropical grasses. In: Gomide, J.A. (ed.) *Proceedings of the International Symposium on Animal Production Under Grazing, Vicosa, Minas Gerais, Brazil, November 1997*, pp. 97–115.

Gordon, I. and Lascano, C. (1993) Foraging strategies of ruminant livestock on intensively managed grasslands – potential and constraints. In: *Proceedings of the 17th International Grassland Congress*. SIR Publishing, Wellington, New Zealand, pp. 681–689.

Hadley, M. (1993) Grasslands for sustainable ecosystems. In: *Proceedings of the 17th International Grassland Congress*. SIR Publishing, Wellington, New Zealand, pp. 21–27.

Harris, W. (1987) Population dynamics and competition. In: Baker, M.J. and Williams, W.M. (eds) *White Clover*. CAB International, Wallingford, UK, pp. 203–297.

Harris, W. (1993) Chairperson's summary paper. Session 11: plant communities. In: *Proceedings of the 17th International Grassland Congress*. SIR Publishing, Wellington, New Zealand, pp. 373–375.

Hodgson, J. (1985) The significance of sward characteristics in the management of temperate sown pastures. In: *Proceedings of the 15th International Grassland Congress*. The Science Council of Japan, Nishi Nasuno, Japan, pp. 63–66.

Hodgson, J. (1990) *Grazing Management: Science into Practice*. Longman Scientific and Technical, Harlow, UK, 203 pp.

Hodgson, J., Cosgrove, G.P. and Woodward, S.J.R. (1997) Research on foraging behaviour: progress and priorities. In: *Proceedings of the 18th International Grassland Congress, Association Management Centre, Calagary, Canada*, Vol. III, pp. 109–118.

Illius, A.W. and Hodgson, J. (1996) Progress in understanding the ecology and management of grazing systems. In: Hodgson, J. and Illius, A.W. (eds) *The Ecology and Management of Grazing Systems*. CAB International, Wallingford, pp. 429–457.

Laca, E.A. and Demment, M.W. (1996) Foraging strategies of grazing animals. In: Hodgson, J. and Illius, A.W. (eds) *The Ecology and Management of Grazing Systems*. CAB International, Wallingford, pp. 137–158.

Lascano, C.E. (1994) Nutritive value and animal production of forage *Arachis*. In: Kerridge, P.C. and Hardy, B. (eds) *Biology and Agronomy of Forage* Arachis. CIAT, Cali, Colombia, pp. 109–121.

Launchbaugh, K.L. (1996) Biochemical aspects of grazing behaviour. In: Hodgson, J. and Illius, A.W. (eds) *The Ecology and Management of Grazing Systems*. CAB International, Wallingford, pp. 159–184.

Lemaire, G. (1997) The physiology of grass growth under grazing: tissue turn-over. In: Gomide, J.A. (ed.) *Proceedings of the International Symposium on Animal Production Under Grazing, Vicosa, Minas Gerais, Brazil, November 1997*. Departamento de Zootechinica, Universidad Federal de Viçosa, Minas Gerais, Brazil, pp. 117–144.

Lemaire, G. and Chapman, D. (1996) Tissue flows in grazed plant communities. In: Hodgson, J. and Illius, A.W. (eds) *The Ecology and Management of Grazing Systems*. CAB International, Wallingford, pp. 3–36.

Maraschin, G.E. and Jacques, A.V.A. (1993) Grassland opportunities in the subtropical region of South America. In: *Proceedings of the 17th International Grassland Congress*. SIR Publishing, Wellington, New Zealand, pp. 1977–1981.

Matthew, C., Lemaire, G., Sackville Hamilton, N.R. and Hernandez-Garay, A. (1995) A modified self-thinning equation to describe size/density relationships for defoliated swards. *Annals of Botany* 76, 579–587.

Matthew, C., Yang, J.Z. and Potter, J.F. (1998) Determination of tiller and root appearance in perennial ryegrass (*Lolium perenne*) swards by observation of the tiller axis, and potential application in mechanistic modelling. *New Zealand Journal of Agricultural Research* 41, 1–10.

Matthews, P.N.P. (1995) Grazing management principles and targets: a case study. *Dairyfarming Annual* 47, 171–174.

Nores, G.A. and Vera, R.R. (1993) Science and information for our grasslands. In: *Proceedings of the 17th International Grassland Congress*. SIR Publishing, Wellington, New Zealand, pp. 33–37.

O'Reagain, P.J. and Schwartz, J. (1995) Dietary selection and foraging strategies of animals on rangeland: coping with spatial and temporal variability. In: Journet, M., Grenet, E., Farce, M.-H., Theriez, M. and Demarquilly, C. (eds) *Recent Developments in the Nutrition of Herbivores*. INRA, Paris, pp. 407–423.

Parsons, A.J. (1993) Chairperson's summary paper. Session 8: plant growth. In: *Proceedings of the 17th International Grassland Congress*. SIR Publishing, Wellington, New Zealand, pp. 176–178.

Parsons, A.J., Newman, A.J., Penning, P.D., Harvey, A. and Orr, R.J. (1994) Diet preference of sheep: effects of recent diet, physiological state and species abundance. *Journal of Animal Ecology* 63, 465–478.

Pimm, S.L. (1984) The complexity and stability of ecosystems. *Nature* 307, 321–326.

Pyke, G.H. (1984) Optimal foraging theory: a critical review. *Annual Review of Ecological Systems* 15, 523–575.

Richards, J.H. (1993) Physiology of plants recovering from defoliation. In: *Proceedings of the 17th International Grassland Congress*. SIR Publishing, Wellington, New Zealand, pp. 85–93.

Riveros, F. (1993) Grasslands for our world. In: *Proceedings of the 17th International Grassland Congress*. SIR Publishing, Wellington, New Zealand, pp. 15–20.

Schwinning, S. and Parsons, A.J. (1996) A spatially explicit population model of stoloniferous N-fixing legumes in mixed pasture with grass. *Journal of Ecology* 84, 815–826.

Sheath, G.W. and Clark, D.A. (1996) Management of grazing systems: temperate pastures. In: Hodgson, J. and Illius, A.W. (eds) *The Ecology and Management of Grazing Systems*. CAB International, Wallingford, pp. 301–323.

Stobbs, J.H. (1973a) The effects of plant structure on the intake of tropical pastures. I. Variation in the bite size of cattle. *Australian Journal of Agricultural Research* 24, 809–819.

Stobbs, J.H. (1973b) The effects of plant structure on the intake of tropical pastures. II. Differences in sward structure, nutritive value, and bite size of animals grazing *Setaria anceps* and *Chloris gayana* at various stages of growth. *Australian Journal of Agricultural Research* 24, 821–829.

Stuth, J.W. (1991) Foraging behaviour. In: Heitschmidt, K. and Stuth, J.W. (eds) *Grazing Management: An Ecological Perspective*. Timber Press, Oregon, pp. 65–83.

Tainton, N.M., Morris, C.D. and Hardy, M.B. (1996) Complexity and stability in grazing systems. In: Hodgson, J. and Illius, A.W. (eds) *The Ecology and Management of Grazing Systems*. CAB International, Wallingford, pp. 275–299.

Taylor, J.A. (1993) Chairperson's summary paper. Session 19: Foraging strategy. In: *Proceedings of the 17th International Grassland Congress*. SIR Publishing, Wellington, New Zealand, pp. 739–740.

Toledo, J.M. and Formoso, D. (1993) Sustainability of sown pastures in the tropics and subtropics. In: *Proceedings of the 17th International Grassland Congress*. SIR Publishing, Wellington, New Zealand, pp. 1891–1896.

Ungar, E.D. (1996) Ingestive behaviour. In: Hodgson, J. and Illius, A.W. (eds) *The Ecology and Management of Grazing Systems*. CAB International, Wallingford, pp. 185–218.

WCED (World Commission on Environment and Development) (1987) *Our Common Future*. Oxford University Press, Oxford, 383 pp.

Westoby, M., Walker, B.H. and Noy-Meir, I. (1989) Opportunistic management for rangelands not at equilibrium. *Journal of Range Management* 42, 266–274.

Woodward, S.J.R. (1998) Quantifying different causes of leaf and tiller death in grazed perennial ryegrass swards. *New Zealand Journal of Agricultural Research* 41, 149–159.

2

Effects of Nitrogen and Water Supply on N and C Fluxes and Partitioning in Defoliated Swards

François Gastal and Jean-Louis Durand

Unité d'Ecophysiologie des Plantes Fourragères, INRA-UEPF, 86600 Lusignan, France

Introduction

In most of the world, pasture growth and productivity is limited by soil water and nutrient availability. Even in the areas of Europe where fertilizer inputs are particularly high, pastures often undergo periods of water deficit and possible associated mineral deficiency. In addition, pasture management needs new rules in the many areas of the world where classical models, adapted for intensive practices, are inadequate. Therefore growth, functioning and simulation of pastures under water and nitrogen limitations require special attention. In this chapter, a general overview of current knowledge on sward responses to water and nitrogen supply is presented and illustrated for a limited number of examples, mostly related to grasses.

It is convenient to consider the responses of swards to environmental variables like water and nitrogen supply through two basic aspects: resource acquisition and resource utilization. These two aspects may be interrelated through variables representing sward status with respect to water and nitrogen from the soil and with respect to sward physiology. These variables may also be used as an experimental tool to evaluate sward status in field situations.

Carbon and N fluxes in swards depend on plant integrative processes and cannot be explained solely at the sward level, particularly in the case of pastures, which are often plurispecific. Understanding the responses of swards to environment and to management strategies requires integration of the responses of the plants at the sward level. Given the number of processes involved, it is important to develop a modelling approach. One example of this approach will be presented. The current agenda of research will also be presented, as we are far from being

 Grassland Ecophysiology and Grazing Ecology (eds G. Lemaire, J. Hodgson, A. de Moraes, C. Nabinger and P.C. de F. Carvalho)

able to predict with confidence the long-term trends in grassland productivity under new management practices or the performance of newly introduced genotypes.

Resource Acquisition by Plants

Water and nitrogen uptake by plants are determined by three fundamental and interactive processes: (i) water and nitrogen distribution in the soil; (ii) root distribution (depth and density); and (iii) uptake efficiency of the roots. Points (ii) and (iii) directly depend on the functioning of plants and swards. Aspect (i) results from water and nitrogen balance in the soil–plant–atmosphere (s.p.a.) system and is not considered explicitly in this chapter. Soil nitrogen mostly refers to mineral N (nitrate and ammonium). However, a number of studies have suggested that organic N compounds may be taken up by plants (Hodge *et al.*, 1998).

Availability of water and nitrogen for plants and swards

The soil volume actually investigated by roots is limited by physical characteristics, particularly soil resistance to root penetration. This factor is probably the most widespread factor limiting root depth in pastures. Rooting depth also varies between species. Among grasses commonly grown in northern Europe, water extraction is deeper in tall fescue and perennial ryegrass than in Italian ryegrass and cocksfoot (Lemaire and Denoix, 1987; Durand *et al.*, 1997). In sown grasslands, rooting depth increases during the phase of sward installation. Later, rooting depth is more stable and results from the previously mentioned species and soil characteristics. Vertical root distribution is important for water acquisition, through its influence on the amount of water available for plants or sward. The relevance of rooting depth for N uptake is less clear, considering that nitrogen mineralization mostly occurs in the top horizons of the soil. In mixed communities, the rooting depth of each species needs to be considered. In this case, considerable seasonal fluctuations may occur in the rooting activity of individual species, in particular through the regeneration of the rooting system of annual species. In perennial species, large changes in the explored soil volume may be brought about by seasonal growth patterns and by competition, especially for light, between species.

Root density (mass or area of root per volume of soil) is highly variable between species. Similarly, root diameter and branching frequency differ to a large extent between species and ecotypes. Root density may also depend on pasture management. It is generally considered that intensive management leads to a decrease in root density (Troughton, 1957; Evans, 1973). However, Deinum (1985) reported a higher root density under grazing than cutting, due to larger tiller density in grazed swards. Seasonal variations in root biomass may be larger than changes due to grazing intensity (Matthew *et al.*, 1991). Root density is partly determined by turnover of the rooting system, which results from root emission and root longevity, and which also varies significantly between species (see Dawson *et al.*, Chapter 4, this volume).

Water flux in the soil–plant–atmosphere continuum

The concept of an s.p.a. continuum was developed from the representation of water flow in the liquid phase as a catenary process (Van den Honert, 1948). Water demand is determined by evaporation at the leaf surface. It depends on radiation, air temperature, air humidity, wind and surface properties (cuticular and stomatal resistances, rugosity, leaf shape). Useful models of potential evapotranspiration (PET) have been validated in the past 40 years. However, actual transpiration differs from PET. First, only the absorbed energy may be converted into evaporated water by the foliage. Secondly, the soil can only provide the liquid water that is reached by roots. The resistance to liquid water flow depends on the soil/root interface, root endodermis and plant hydraulic network. These various resistances limit the actual transpiration of the plant. They also generate a pressure drop between the soil solution, which is generally close to 0 MPa, the xylem elements and the evaporating leaf surfaces. The pressure drop within the xylem elements leads to a drop in the water potential of the different plant organs below the soil water potential. In a simple form, the s.p.a. continuity equation describing a conservative flux is:

$$T \approx A = \frac{\psi_{soil} - \psi_{plant}}{R_{soil-plant}} \quad (1)$$

where T is transpiration, A water absorption, $R_{soil-plant}$ the sum of resistances on the water flow path and ψ_{soil} and ψ_{plant} the soil and plant water potential, respectively. This equation provides a tool for linking the water balance of a plant or any crop to its water status (ψ_{plant}).

Nitrogen uptake

Nitrate and ammonium uptake relies on specific molecular transport systems. It is currently considered that several transport systems, having high and low affinity for NO_3^- and NH_4^+, operate simultaneously and allow uptake under low and high external N concentrations, respectively (Le Bot *et al.*, 1998). The apparent K_m of high-affinity carriers ranges between 0.01 and 0.1 mM, depending on the ion and the species. The combination of different transport systems in roots would allow efficient mineral N uptake under a large range of concentrations in the soil solution.

In addition to the role of external N concentration, nitrogen uptake is regulated by the physiological status of the whole plant (Touraine *et al.*, 1994), as suggested by the following experimental evidence: limited range of variation in N content of whole plant under ample N supply; large and rapid adaptation of N uptake to growth rate of the whole plant (Gastal and Saugier, 1989); and low uptake potential of plants supplied with abundant NO_3^- compared with the uptake potential of plants grown under low NO_3^- supply (Larsson, 1994), indicating feedback mechanisms and regulation. Nitrate concentration of root tissues is likely to regulate NO_3^- uptake. However, this mechanism alone is insufficient to explain the overall regulation of N uptake at the whole-plant level, since NO_3^-, which is reduced in shoots in many herbaceous species, is not recirculated to roots through the phloem. Thus, it seems likely that the concentration in free amino acids also regulates (negatively) root N uptake, as suggested by manipulation of external and internal concentration in amino acids (Muller and Touraine, 1992). Root carbohydrate concentration might also alter N uptake, but this still remains to be clearly demonstrated.

In low-input grassland systems, N fixation is the major source of N. The dependence of N fixation on the whole-plant C/N relationships has been demonstrated. Nitrogen fixation clearly depends on the carbon status of the plant. It is also known that free amino acid concentrations reduce nitrogenase activity (Parsons *et al.*, 1993). N fixation requires high oxygen fluxes and therefore is restricted to the top horizons of the soil. Oxygen diffusion into the nodules is limited by variable resistance, which responds to the environment and plant status (Sheehy *et al.*, 1985). Water deficit or excess and low temperatures increase O_2 diffusion resistance. High amino acid concentrations and low carbon supply also increase the resistance to O_2 diffusion, reducing N fixation (Joy, 1998).

Water deficit may largely limit N uptake and N fixation. In grass swards experiencing water shortage, it is commonly observed that N concentration decreases significantly despite a high N supply (Onillon *et al.*, 1995). This might be due to a reduction in N availability at the sites of root uptake, related to a limitation of soil water fluxes. In legumes, N fixation responds strongly to water deficits. The direct effect of water status on nitrogenase activity via the increase of O_2 diffusion resistance has been demonstrated (Durand *et al.*, 1987; Serraj *et al.*, 1999). Therefore, water availability and nitrogen nutrition interact strongly.

Plant and Sward Status

Water and nitrogen status result from water and N balance in the soil and from water and nitrogen uptake of plants or sward. Both affect their functioning. Identifying relevant variables with respect to plant growth and productivity offers the opportunity to experimentally evaluate resource accessibility to plants and sward.

Plant water status

The water flux transpired out of the plant does not necessarily match the flux of water uptake into the plant. The relationship between water content and water flux varies between species and environmental conditions. This relationship is determined by the osmotic pressure (π) of the cells and the elasticity (ε) of the cell walls, ϵ representing the ratio between a change in tissue turgor (P) and a relative change in water volume (dV/V):

$$\frac{dP}{\left(dV/V\right)} = \varepsilon \tag{2}$$

The osmotic pressure (π) also depends on the volume of water in the plant :

$$\pi = \pi_0 \, V_0/V \tag{3}$$

where π_0 and V_0 are the osmotic pressure and water volume at full turgor. The water potential (ψ) is equal to the difference between the turgor pressure and the osmotic pressure.

$$\psi = P - \pi \tag{4}$$

The mass conservation equation relates the volume change (dV/dt) to absorption (A) and transpiration (T).

$$\frac{dV}{dt} = A - T \tag{5}$$

Hence, the water balance ($A-T$) can be related to the plant water status, combining equations 1, 2, 3, 4 and 5. During the day, ψ_{plant} varies according to transpiration. It is highest at dawn (predawn water potential) and minimal at the beginning of the afternoon (Fig. 2.1).

However, the relevant variable with respect to plant growth is P. Throughout the day, the variations in P depend on adjustment of both transpiration (T) and osmotic potential (π). The predawn water potential represents an integrated soil water potential. Its value is therefore a good estimate of water availability.

Nitrogen status

It has long been known that the N concentration of plants or sward increases with N supply. This increase in total N concentration is brought about by the conjugated increase in concentration of free amino acids, proteins and nitrate (under situations of excess). A linear relationship between relative growth rate (RGR) and total N concentration has been shown during early development of

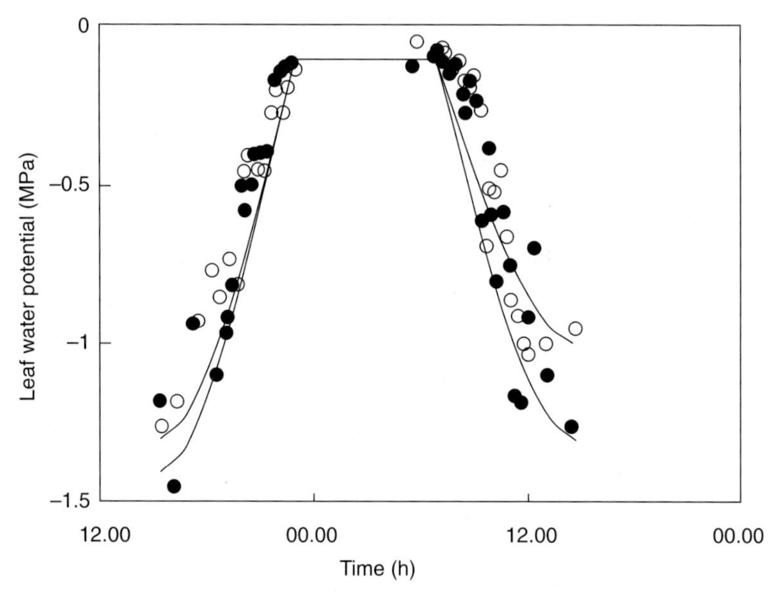

Fig. 2.1. Typical hourly variations of the leaf water potential of an irrigated tall fescue sward in summer. Low (●) and high (○) nitrogen supply (J.-L. Durand, unpublished data).

seedlings (Agren, 1985). However, in dense swards, growth is generally not exponential and, under such circumstances, N concentration content declines as RGR decreases (Greenwood *et al.*, 1991). This has led to the definition of critical N concentration (N_c) of swards as the lowest N concentration at which maximum growth rate is achieved. By definition, N_c determines the limit between situations of suboptimal and supraoptimal N nutrition. N_c can be evaluated graphically by representing N concentration as a function of biomass (Fig. 2.2).

Using a large set of experimental conditions in N supply and in climate, N_c has been extensively evaluated. It has been shown that N_c decreases during the growth of the crop (Greenwood *et al.*, 1990). Furthermore, the relationship between N_c and crop biomass does not significantly depend on environmental conditions other than N supply. An allometric function has been established empirically:

$$N_c = \alpha W^{\beta} \quad (6)$$

where W is the standing biomass and α and β are parameters.

It has been shown that α and β are similar for species of the same metabolic group but differ between C_3 and C_4 species (Fig. 2.3). At any stage of biomass accumulation by the crop, a nitrogen nutrition index (NNI) can be defined as the ratio between actual N concentration and N_c. This ratio provides a precise estimate of the crop nitrogen status (Lemaire and Gastal, 1997).

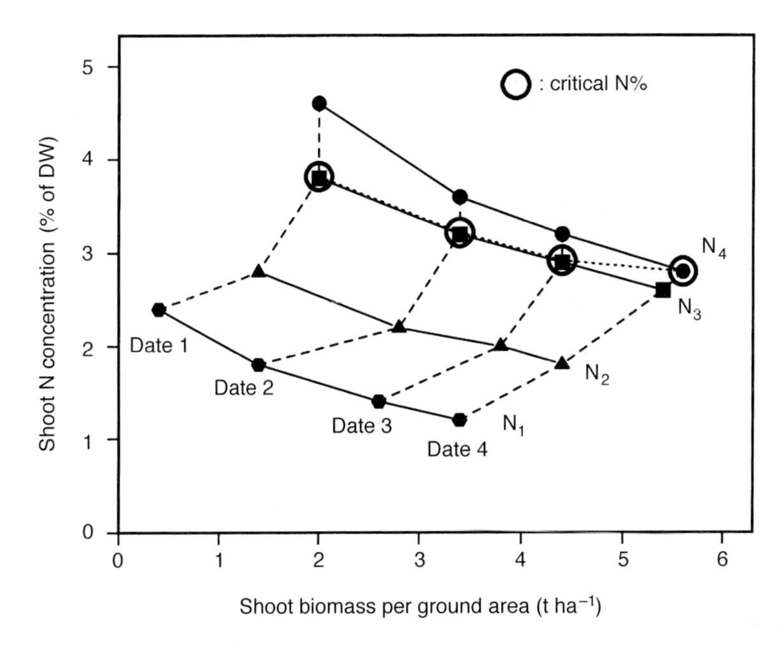

Fig. 2.2. Relationship between biomass accumulation (four measurement dates along a growth period) and total N concentration in the canopy of a tall fescue sward grown at four N fertilization rates (N_1 to N_4). DW, dry weight.

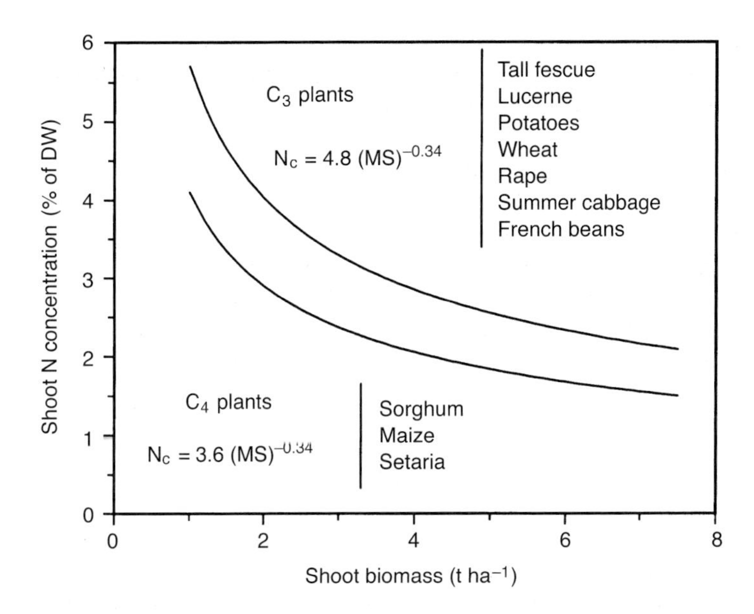

Fig. 2.3. Critical nitrogen concentration (N_c) for several C_3 and C_4 species. Each reference curve is fitted to an allometric relationship where MS stands for shoot biomass (t ha^{-1}). DW, dry weight.

Plant Productivity

Conceptual framework

Plant growth and therefore sward productivity result from the interaction between the genome and the environment (Durand *et al.*, 1991; Fig. 2.4). Morphogenesis depends on a morphogenetic programme (genome) and local climatic and edaphic conditions (light quality and quantity, organ temperature, water and nutrient status, etc.). The rate of volume expansion of each organ is associated with energy and structural costs and determines a carbon demand. Carbon supply is determined by the size of the non-structural carbohydrate pool, which is fed by photosynthesis and depolymerization of reserves. At sward level, in several instances the expansion rate of organs is not limited by carbon supply. At plant level, a different picture emerges, due to possible competition for light between individuals.

There are important feedbacks between these two classes of processes. First, canopy photosynthesis depends on leaf area, which results from leaf expansion. Secondly, the conversion efficiency of absorbed light by leaves (i.e. photosynthesis per unit of leaf area) can in several instances be regulated in response to the

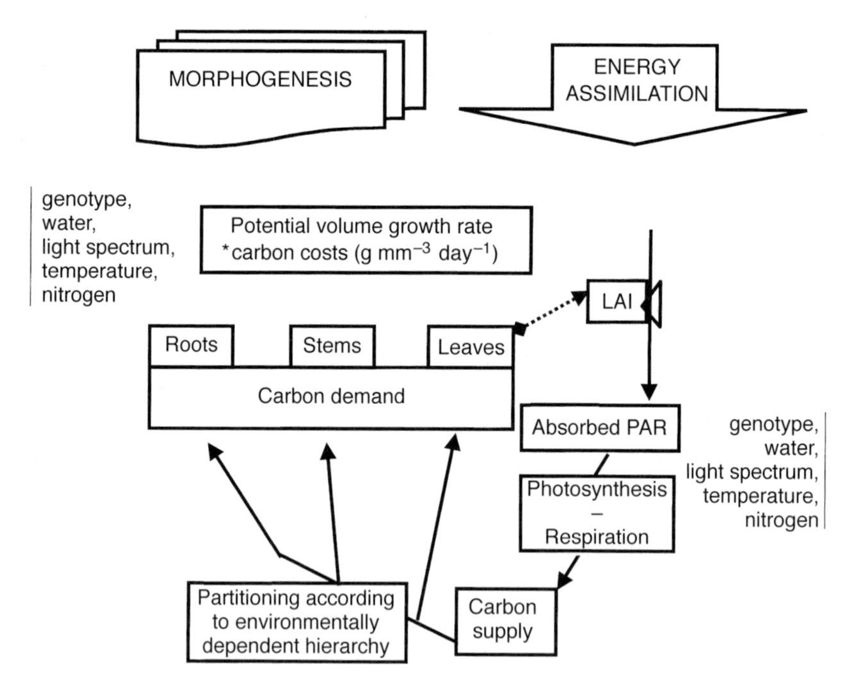

Fig. 2.4. Conceptual framework of plant growth analysis in relation to nitrogen and water supply. Solid arrows: carbon fluxes (after Durand *et al.*, 1991). LAI, leaf area index; PAR, photosynthetically active radiation.

export and utilization of photoassimilates (sink activity). In forage plants, however, where restoration of leaf has to proceed following defoliation, the first aspect predominates.

Partitioning of photoassimilates may be viewed as part of the morphogenetic programme of the plant, and may be interpreted as following a hierarchy of carbon allocation between different plant organs: growing leaves, stems, new axes and roots, in a decreasing order of allocation priority (Durand *et al.*, 1991). For instance, the increase in the root/shoot partitioning ratio under drought is not a change in priority but results from it: under water shortage, leaf volume expansion is more sensitive than root elongation, fewer assimilates are used for leaf growth and hence more photoassimilates are available for root growth. A major task of current research in crop science is to analyse the differential responses of the plant organs to environmental constraints. Another important task is to analyse the response of photosynthesis to the same conditions. Modelling allows one to integrate these different features and predict the overall functioning of the sward, and therefore its growth and productivity.

Carbon assimilation

CO_2 assimilation and losses through photosynthesis and respiration are largely dependent on water and nitrogen supply. The response of photosynthesis to water and nitrogen supply is related to leaf photosynthetic activity and to spatial distribution of green leaf area. Green leaf area results from the balance between leaf growth and leaf senescence. The physiological processes underlying leaf photosynthesis, leaf growth and leaf senescence fundamentally differ; therefore, regulation of these processes by water and N supply also differs.

Light interception

At levels of leaf area index (LAI) below 3 to 4, nitrogen and water affect canopy photosynthesis mainly through leaf area expansion and hence PAR interception. In contrast, under higher LAI, the effect of water or nitrogen is mostly due to their effect on photosynthetic capacity. For a given LAI, the light absorption efficiency depends on leaf optical properties and geometry. Water or nitrogen deficit hardly alters the former in the visible range (PAR). However, for certain species, leaf geometry may drastically change due to low water potential, through rolling or inclination. In tall fescue, leaves roll under water stress. This may lead to a significant reduction in the actual leaf area and light interception, to less than 50% of the maximum unrolled area under very severe water deficit (Tournebize, 1990). Thus, in situations of LAI below 3 to 4, which are commonly encountered under drought conditions, significant reductions of carbon assimilation due to leaf rolling may occur before any effect of water deficit on leaf photosynthetic capacity.

Leaf photosynthesis

A correlation has been observed between leaf N concentration and leaf photosynthetic capacity (P_{max}) under ambient CO_2 concentration and saturating incident irradiance (Evans, 1983; Field and Mooney, 1986). In C_3 plants, chloroplast proteins account for about 75% of total nitrogen of a well-illuminated leaf, and mostly comprise N invested in Rubisco and in membrane proteins associated with light harvesting complexes (Chapin *et al.*, 1987). This probably explains the correlation observed between P_{max} and N concentration. In C_4 plants, this correlation is also observed. In this case, changes in leaf N concentration occur through changes in Rubisco and PEPcarboxylase concentrations (Sage *et al.*, 1987).

In contrast to the correlation found between P_{max} and N concentration, it is generally observed that the dependence of leaf photosynthesis on N concentration is very limited at low irradiance (Fig. 2.5).

Although water deficit may affect non-stomatal conductance, the stomatal response is the first and most pronounced process explaining the effect of water deficit on leaf photosynthetic capacity. The determinism of stomatal closure under water deficit is still a matter of discussion (Tardieu *et al.*, 1992; Monteith, 1995). A detailed review on this topic is outside the scope of this chapter. Twenty years ago, ψ_1 or its turgor component was considered to be the determining variable in stomatal response to water deficit. Experimental evidence for direct control of

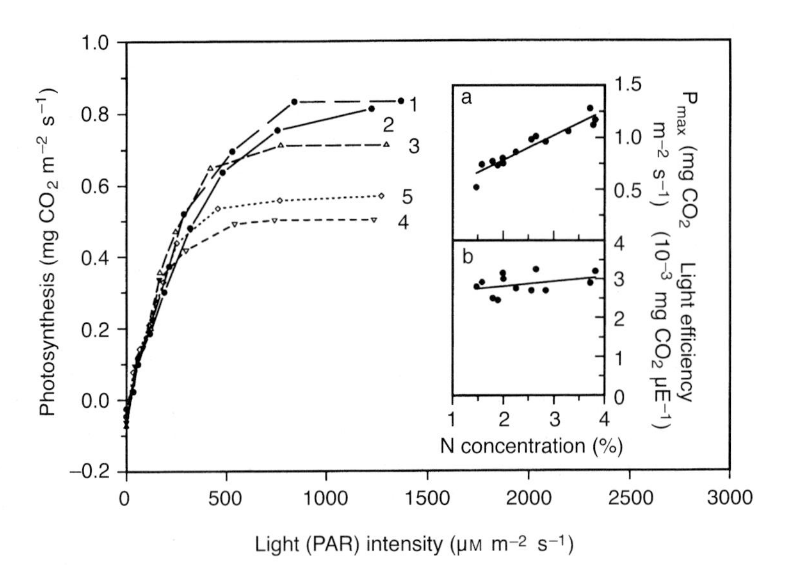

Fig. 2.5. Effect of N supply on the response of photosynthesis of a tall fescue leaf to light, under ambient CO_2. 1 to 5: plants grown under increasing N deficiency. Inset a: P_{max}: leaf photosynthesis at saturating light. Inset b: Light efficiency: initial slope of the photosynthesis/light relationship.

stomatal conductance by roots has been provided in several species, under controlled as well as under field conditions (Gollan *et al.*, 1986). Chemical messengers appeared to play a key role in the control mechanism. There are obvious large differences between species. Lucerne, for instance, exhibits poor stomatal control, in contrast to maize or tall fescue. In lucerne, ψ_l can decline to low values (−3 MPa) without turgor losses, due to continuing carbon assimilation and potassium assimilation both bringing about an efficient maintenance of osmotic pressure (J.-L. Durand, unpublished data). New models of stomatal regulation are still being produced. A reinterpretation of the relationship between air relative humidity, soil water content and stomatal resistance has been published (Dewar, 1995; Monteith, 1995).

In addition to mechanisms at the plant level, the density of the canopy strongly influences stomatal behaviour and leaf water relations. In dense swards, the aerodynamic resistance of the crop is high, the air humidity in the canopy tends to remain high and stomata open. At high irradiance, transpiration is large, but, where the tiller density is high, the transpiration per leaf can be lower than in sparser crops. The leaf ψ_l can therefore be higher under high irradiance in dense canopies than in sparse canopies. In such conditions, ψ_l remains a good estimate of plant water status. The response of leaf P_{max} to ψ_l does not significantly depend on N nutrition (Ghashghaie and Saugier, 1989).

Canopy photosynthesis

Following the integration of leaf to canopy proposed by Thornley and Johnson (1990), a simple model of canopy photosynthesis and its response to nitrogen and water status was developed for tall fescue swards. The response of leaf photosynthesis to light (PAR) was described by a non-equilateral hyperbola. P_{max} was linearly related to leaf nitrogen concentration (see Fig. 2.5). In addition, P_{max} was also linearly related to ψ_l (Fig. 2.6b), according to data from Ghashghaie and Saugier (1989). The reduction in effective leaf area due to leaf rolling was taken into account (Fig. 2.6a) through a relationship derived from Tournebize (1990) and Viratelle (1992).

The carbon dioxide exchange rate (CER) was measured on swards grown under different N and irrigation regimes (Gastal and Bélanger, 1993; B. Onillon and J.-L. Durand, unpublished data). According to the model output, it appears that the simple relationships used between P_{max}, leaf N concentration and leaf water potential provide a reasonable estimation of canopy photosynthesis and its response to nitrogen and water supply (Fig. 2.6c). In most observed situations, the main effect of nitrogen and water supply was to alter LAI. The model also showed that, in addition to the reduction in unrolled LAI, leaf rolling and stomatal opening contributed equally to the reduction in photosynthesis under an LAI of 0.9.

Recovery from N or water deficits is not fully understood. In particular, lasting effects of water limitations are not yet taken into account.

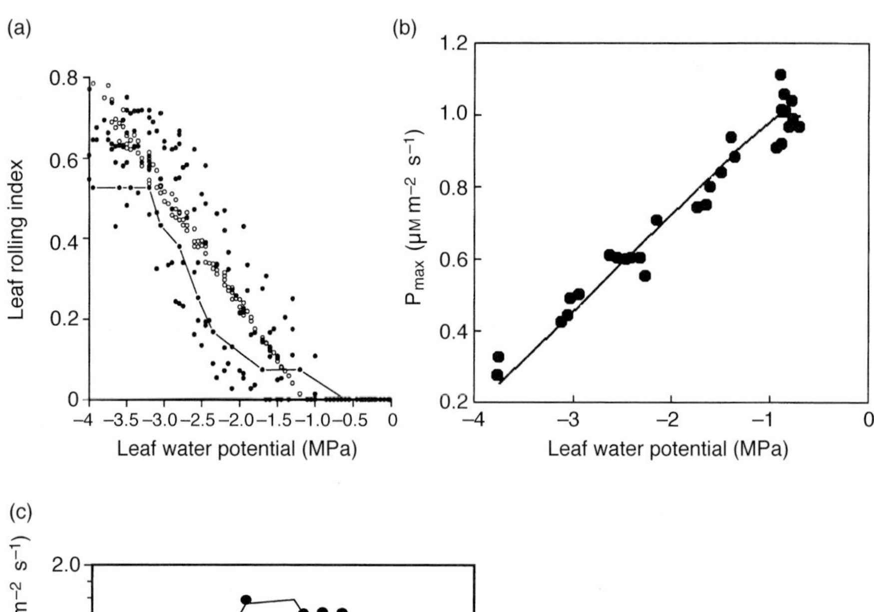

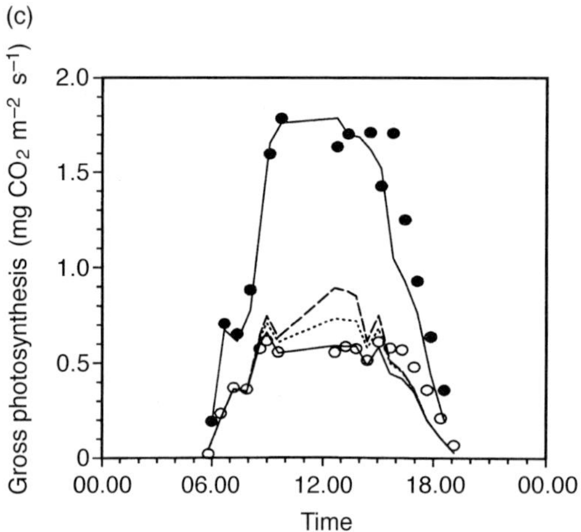

Fig. 2.6. (a) Rolling index of tall fescue leaves versus water potential (J.-L. Durand, unpublished data). (b) P_{max} versus leaf water potential in tall fescue (after Ghashghaie and Saugier, 1989). (c) Comparison of simulated (lines) and observed (points) canopy photosynthesis during a typical sunny day at high N supply in a tall fescue sward. ●: irrigated, LAI = 3; ○: rain-fed, LAI = 0.9. Continuous line: full simulation. Dashed line: simulation without effect of ψ_1 on P_{max} or on leaf rolling. Dotted line: effect of ψ_1 on P_{max} without effect on leaf rolling potential (J.-L. Durand, unpublished data).

Carbon–nitrogen partitioning and morphogenesis

Following the framework developed earlier (see Fig. 2.4), C and N partitioning represents the second aspect that needs to be considered in growth studies. In grasslands, and particularly in perennial plants or swards, C and N partitioning is often difficult to deduce from measurement of biomass and C and N concentrations. This is due to senescence and decay of organs, which may lead to a substantial underestimation of actual growth, particularly for underground organs. Therefore, proper growth studies need to be conducted under conditions where root senescence is minimized (short-term experiments) or, alternatively, when isotopes can be utilized.

In artificial cultures, a larger partitioning of C to shoots relative to roots has repeatedly been shown when N supply increases (Brouwer, 1962). The use of C isotope on swards of tall fescue has shown that this change in C partitioning in response to N supply systematically occurs under field conditions, at any season and even under water shortage (Bélanger *et al.*, 1992; Onillon *et al.*, 1995).

There is evidence showing that leaf growth is more responsive to nitrogen or water supply than root growth (Robson and Parsons, 1978; Gastal and Saugier, 1986; Onillon *et al.*, 1995). For nitrogen supply, this can be illustrated by the progressive but substantial decline in the leaf elongation rate of tall fescue plants transferred from ample to limited N nutrient solutions, whereas root growth is slightly stimulated (Fig. 2.7). This response can be considered as a response of sink activity to N or water supply, independent of carbohydrate supply, for a number of reasons suggesting that carbon is not the primary limiting event. First, it is generally observed under conditions of moderate nitrogen or water limitation that the concentration of non-structural carbohydrates increases in both source and sink tissues (Volenec and Nelson, 1984; Spollen and Nelson, 1994). Secondly, it is often reported that a moderate nitrogen or water shortage affects C partitioning between shoots and roots substantially without significant effect on biomass accumulation of whole plants (Gastal and Saugier, 1986).

Sink activity is spatially distributed within the plant (Fig. 2.8). In the case of plants that are regularly defoliated, it is convenient to consider the localization of the growth zones (i.e. zones of organs where primary growth occurs). The possible physical suppression of part of a growth zone or whole growth zones through defoliation may substantially affect plant responses to water or nitrogen. This consequence of defoliation needs to be considered in addition to other aspects, such as the use of C and N reserves. In grasses, the leaf growth zone is located at the base of the leaf, within the sheath, and therefore is entirely or at least partly protected from defoliation in the absence of internode elongation. In contrast, for dicot species which undergo stem elongation during most of the developmental period, many of leaf and stem growth zones or meristems (axillary buds) are removed by defoliation. This implies that meristems and growth zones need to be regenerated after defoliation, in order for the plant to recover its initial growth rate. Lucerne is an example where defoliation removes a large number of active growth zones or quiescent meristems. Their regeneration takes time, and this explains the relatively large lag period commonly observed for restoration of leaf area index and the overall low tolerance of this species to defoliation.

Within growth zones, growth proceeds through cell division, cell expansion and cell maturation. Current knowledge is quite inadequate to give a general overview of the underlying regulations. However, in a number of situations, the

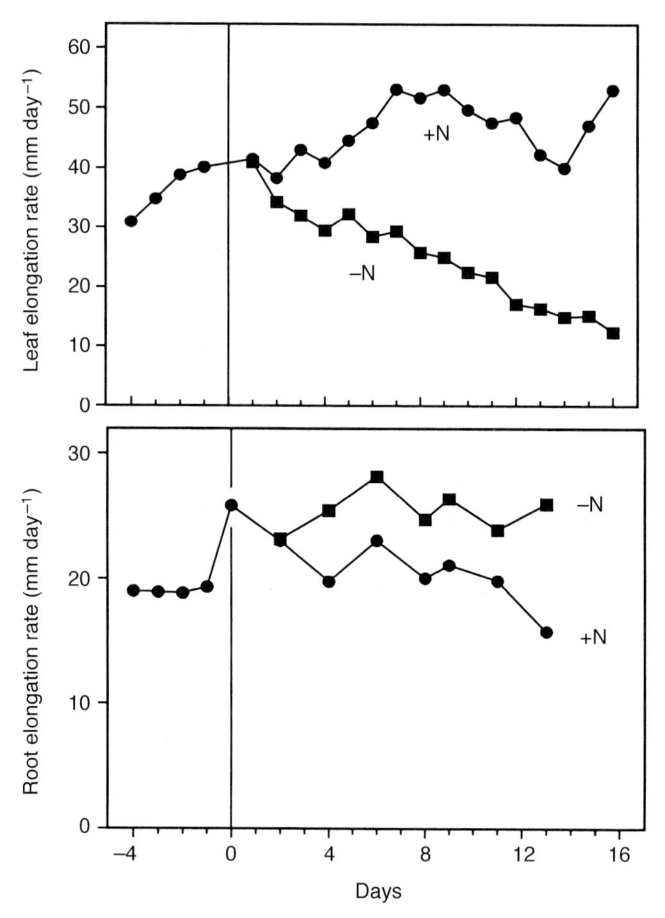

Fig. 2.7. Response of leaf and root elongation rates to N supply in tall fescue. Plants grown with ample (+N) or limited (−N) N supply. (After Gastal *et al.*, 1992.)

role of nitrogen and water in cellular dynamics has been identified. It has been shown that nitrogen affects leaf elongation of grasses mostly by increasing the rate of epidermal cell division, whereas the elongation rate of individual epidermal cells is changed to a much lesser extent by N nutrition (MacAdam *et al.*, 1989; Gastal and Nelson, 1994; Fricke *et al.*, 1997). Thus, in grasses it appears that the stimulation of leaf growth by nitrogen is mostly determined by a larger number of cells simultaneously undergoing expansion at a rate that is altered by nitrogen to only a small extent. There is a substantial deposition of N in the zone of cell division of grass leaves (Gastal and Nelson, 1994). Part of this N seems to be internally recycled to support later chloroplast development and associated synthesis of Rubisco (Fig. 2.9). This large apparent requirement for N in the zone of cell division provides additional support for the view developed earlier that the response of leaf growth to N is not primarily determined by carbohydrate availability but rather by local N availability.

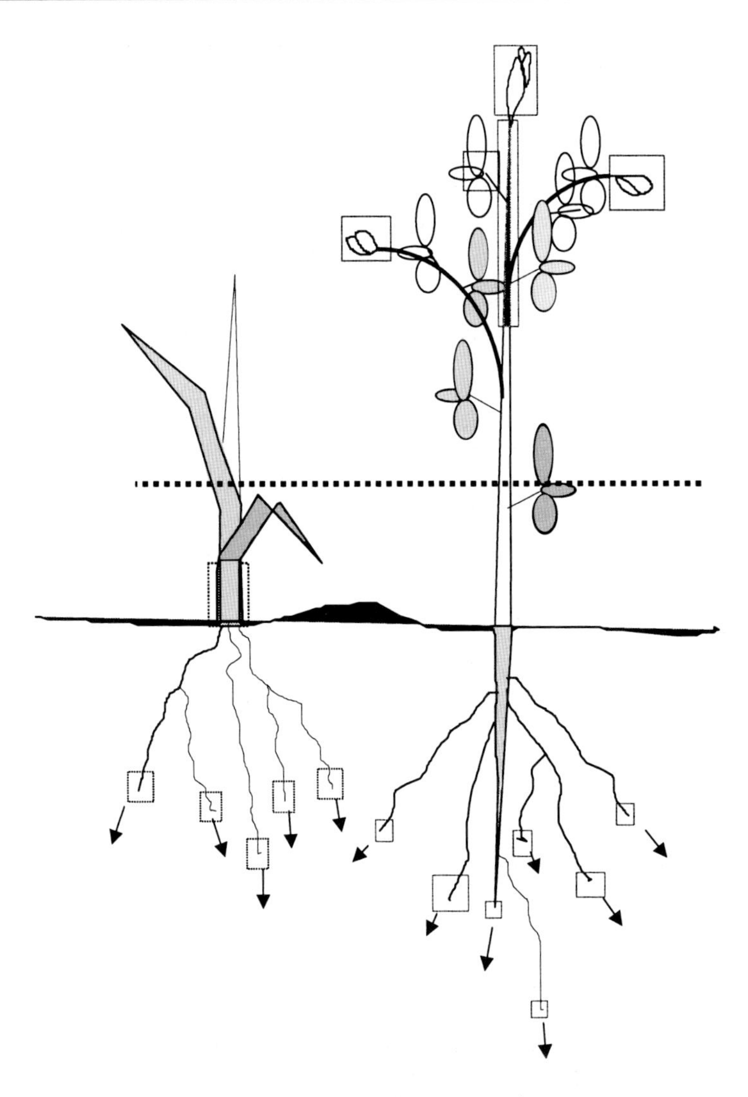

Fig. 2.8. Defoliation and removal of growth zones (within squares) in grasses and dicots.

In addition, split-root experiments on grasses have shown that leaf growth is not reduced when half of the root system is provided with ample N and the other half is kept in an N-free medium (in comparison to the control with all roots kept in full N solution), provided that N uptake by the whole plant is compensated by N uptake by the roots to which N is available (Gastal *et al.*, 1992). This suggests that a possible involvement of root-to-shoot signalling related to nitrogen starvation is unlikely to affect leaf growth negatively, in contrast to the hypothesis suggested by Clarkson and Touraine (1994) that abscisic acid (ABA) could be involved in the response of growth to nitrogen. In dicots, the effects of nitrogen

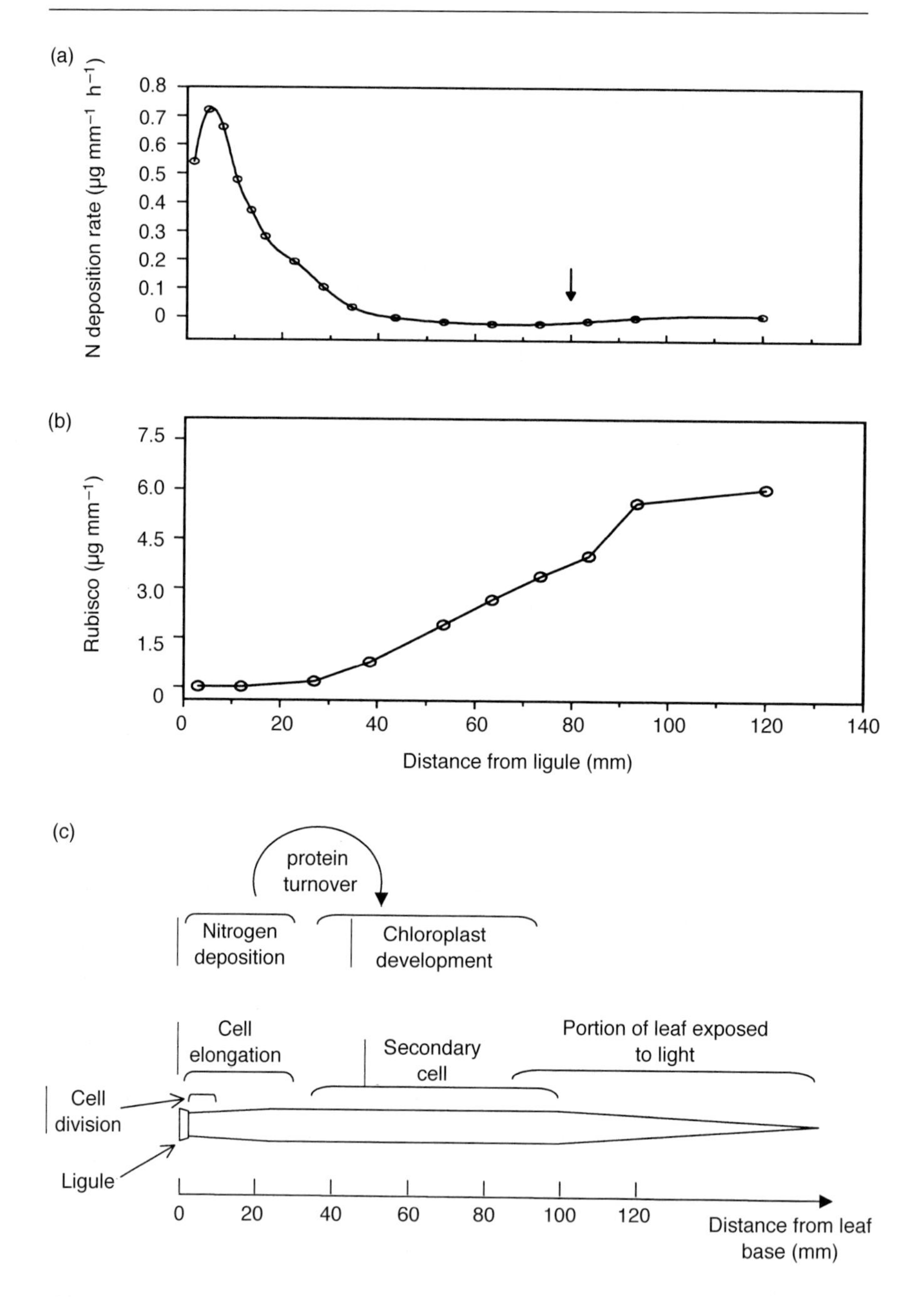

Fig. 2.9. (a) Deposition of N and (b) synthesis of Rubisco along the growth zone (c) of a tall fescue blade (after Gastal and Nelson, 1994).

on cell production and cell expansion appear to be more balanced. The effect of N on cell production, which is still large, seems relatively more important for cell elongation than it is in the case of grass leaf elongation (Palmer *et al.*, 1996).

In the growth zone of grass leaves, water deposition closely corresponds to the volumetric expansion of tissues (Schnyder and Nelson, 1988; Schnyder *et al.*, Chapter 3, this volume). The amount of water required for volume expansion is quantitatively close to the calculated volume of phloem sap necessary to sustain the carbohydrate requirements of the leaf growth zone. However, there is no definitive evidence that phloem is the only local source of water for growth. It is doubtful that water deficit limits growth through a lack of water for volumetric growth. It is rather the water status which is relevant (Martre *et al.*, 1999). There has been much discussion in the past 20 years about the signals limiting growth rates. It has been shown that root signals probably reduce leaf expansion independently from growth-zone water status (Passioura, 1988; Davies and Zhang, 1991). On the other hand, any change in local water status (gradient of water potential between roots and shoot growth zone, or pressure in growing cells) would lead to a reduction of growth rate. During drought, the length of the growth zone, i.e. the potential expansion rate, is reduced. Thus, even if growing tissues were at the same water status as control plants, their growth would be reduced. The shorter growth zone observed in plants under restricted water supply might be induced by short periods of low water status in the growth zone itself. In contrast to nitrogen, water deficit greatly alters cell expansion rate. This inevitably results in also reducing the cell division rate, as is generally observed under long term water deficits (Durand *et al.*, 1995).

Beyond the general view previously developed that leaf and stem growth are more sensitive to nitrogen and water supply than growth of roots or storage organs, there is also experimental evidence indicating a differential sensitivity of growth to nitrogen and to water within shoots and within roots. Within shoots, it is generally observed that development of axillary buds (branching or tillering) is more sensitive to nitrogen supply than leaf growth. Under moderate nitrogen supply, tillering is almost suppressed in grass swards, whereas leaf elongation proceeds, although at a reduced rate (Bélanger, 1990; Nelson, Chapter 6, this volume). In addition, there is a large interspecific variability in the differential sensitivity of components of shoot morphogenesis to N. In many temperate grasses, the response of leaf growth to N is accompanied by a limited response in leaf appearance rate but a large response in final leaf size (Gastal and Lemaire, 1988). In contrast, in several C_4 grasses, the response of leaf appearance rate to N is large, whereas final leaf size varies to a lesser extent (Cruz and Boval, Chapter 8, this volume). Within roots, there is also a large spectrum of responses of the various morphogenetical components to nitrogen. Differentiation and growth of secondary roots respond differently to N from emission and growth of primary roots. Here again, there is a very large range of responses between species (Davidson, 1994). This large variability in root morphological responses to N supply has potentially large consequences under circumstances where distribution of N within the soil volume explored by roots is highly heterogeneous (Dawson *et al.*, Chapter 4, this volume).

The responses of the plant and hence of the sward to water and nitrogen in terms of C partitioning are considered to be largely determined by the response of morphogenesis to water and N supply, rather than by availability of non-structural carbohydrate. However, this view probably does not hold for the accumulation of non-structural carbohydrates in storage organs.

Our current knowledge about regulation of organ senescence in response to N or water supply is still very limited, despite the present understanding that tissue flows associated with growth and with senescence are regulated differently and that their relative importance is a strong determinant of grazing efficiency (Lemaire and Agnusdei, Chapter 14, this volume). The lack of knowledge is even larger for roots, although it is recognized that root longevity differs significantly between species (Troughton, 1981) and between situations of water and N availability.

Plant and Sward Modelling

Plants and swards represent a complex system where a large number of interactions and feedback mechanisms occur between resource distribution within the environment, mineral and water uptake, light interception and C assimilation and resource utilization. As discussed previously, plant and sward responses to the environment and particularly to water and N supply are numerous. Thus it is necessary to carefully select the major processes that will be incorporated into a model. Guidelines for such a modelling approach are outlined here, on the basis of the MecaNiCAL model developed by Tabourel-Tayot and Gastal (1998a, b).

When considering the plant responses to N and to water in terms of C and N acquisition and utilization, as stated in the previous sections, the plant may be described by a number of compartments: substrates, representing the plant resources that are assimilated (or recycled) and can be rapidly used for synthesis; and structures, representing the plant material that controls resource acquisition through a number of mechanisms, but which is not used in the short term as substrate for synthesis (Fig. 2.10). Substrate compartments include a pool of carbohydrates (W_C) and a pool of free amino acids (or organic-N substrate, W_O) and may include also a pool of nitrate (W_N) and longer-term reserves. Structure compartments include, as a minimum, shoots and roots separately. Within shoot and root structures, it is necessary to consider separately proteic and non-proteic structures, due to their respective role in physiological processes.

Being given a set of compartments, the next step is to define fluxes. The rate of change in the mass of the substrate compartments results from the fluxes of carbon and nitrogen associated with acquisition and utilization: *Pd* and A_N are fluxes of CO_2 assimilation and NO_3^- uptake, respectively. A simple model of CO_2 assimilation has been given earlier (see 'Carbon assimilation'). Nitrate uptake is represented as a function of NO_3^- concentration in the rooting medium and as a function of root mass, and is negatively affected by the concentration in organic-N substrate in the plant, in accordance with current knowledge, presented earlier,

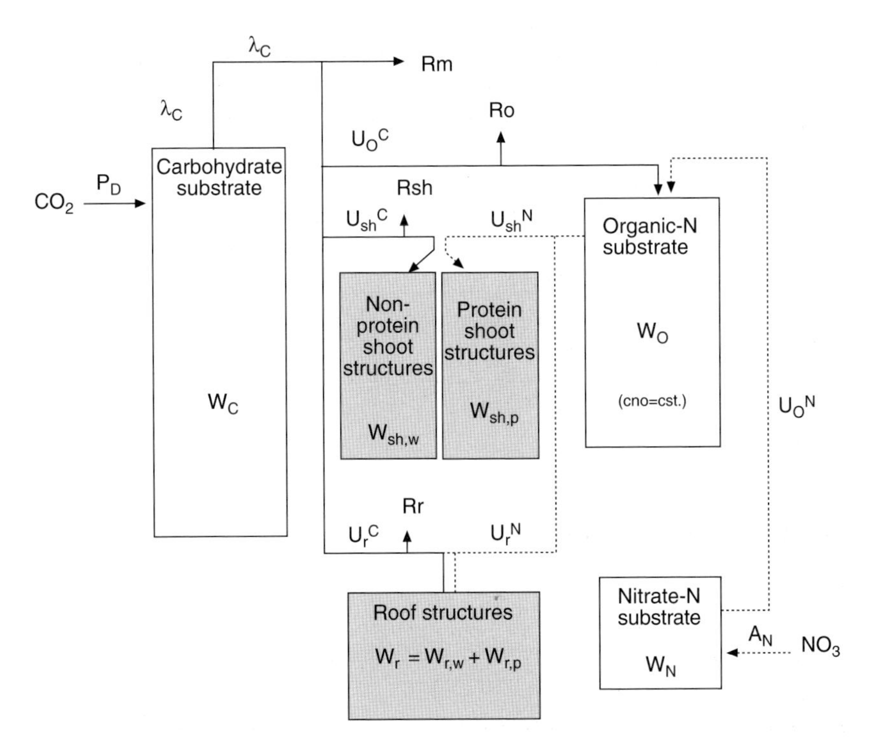

Fig. 2.10. Schematic representation of compartments and fluxes of C and N in the MecaNiCAL model (after Tabourel-Tayot and Gastal, 1998a).

about the regulation of N uptake. The carbohydrate substrate is used for maintenance of existing structures (*Rm*), for the reduction of NO_3^- (U_o^C), for the synthesis of protein and non-protein shoot and root structures (U_{sh}^C, U_r^C) and for respiration associated with structural synthesis and nitrate reduction (*Rsh*, *Rr*, *Ro*). Nitrate-N is reduced to the organic-N substrate, which in turn is used for the synthesis of protein and non-protein root and shoot structures (U_{sh}^N, U_r^N). Carbohydrate and nitrogen substrates could be partitioned separately within shoot and root structures. In this case, it would be necessary to consider translocation of C and N through mass flow, as a formalization of the Munch hypothesis (Sheehy *et al.*, 1996), or through diffusion (Thornley, 1972). Alternatively, carbohydrate and nitrogen substrates may be partitioned homogeneously within the structures of the whole plant. This approach, which has been followed in MecaNiCAL, allows one to avoid modelling translocation of the carbohydrate and nitrogen substrates, but requires additional hypotheses in the use of substrates for synthesis of shoot and root structures.

Synthesis of shoot and root structures depends on the balance between morphogenetic demand and availability in the carbohydrate substrate (see 'Carbon assimilation' and Fig. 2.4). Morphogenetic demand is determined by plant morphogenetic potential (the growth potential of leaves, stems, roots, tillers and

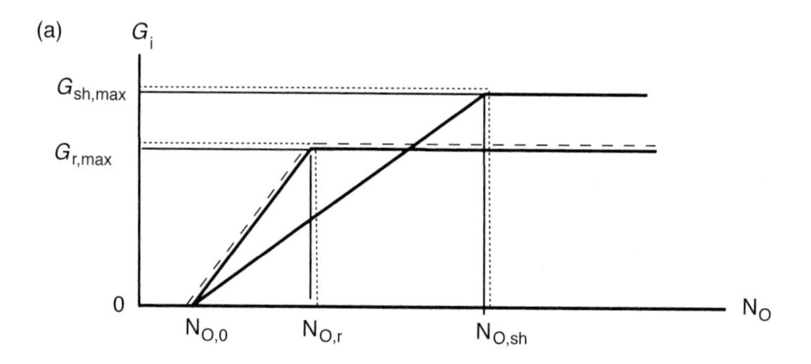

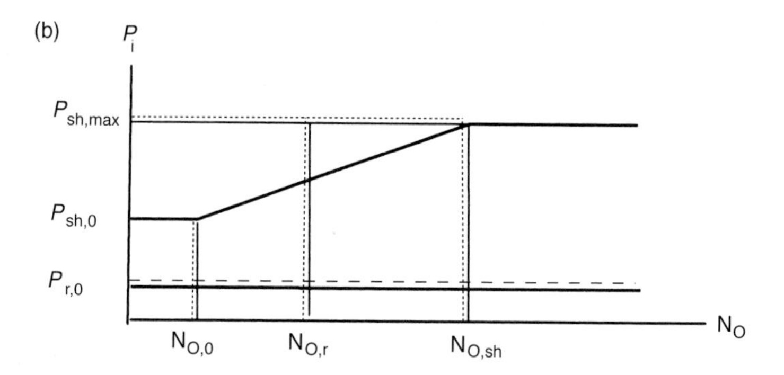

Fig. 2.11. Regulation of plant demands in C and N by concentration in organic-N substrate (N_o). (a) G_{sh}, G_r: potential growth rate of shoots and roots (solid and dashed line respectively). (b) P_{sh}, P_r: ratio between proteic and non-proteic structural mass (solid and dashed line respectively) (after Tabourel-Tayot and Gastal, 1998a).

reproductive organs), under the control of the genome and under regulation of its expression by environmental factors (temperature, light quality). Shoot and root demands also depend on the concentration of organic-N substrate: both the rate of structure synthesis (G) and the ratio between proteic and non-proteic structures (*P*) are linearly related to the concentration of the organic-N substrate above a minimum and below a maximum threshold, for shoots and roots separately (Fig. 2.11). Therefore, organic-N concentration plays a central role, since it regulates both the uptake of N (negatively) and structure synthesis (positively).

Finally, the sum of plant carbon demands is compared to carbon availability in the carbohydrate substrate pool. If total plant demand is higher than availability, individual organ demands are satisfied, following a strict priority for allocation to shoots versus roots.

Such a simple model in its principles allows one to simulate a number of responses of a grass (shoot and root growth rates, shoot and root composition) to major environmental conditions (temperature, light intensity, nitrogen availability) and to defoliation. In addition, it allows one to parameterize morphological plasticity through several components (growth demand of the various organs). Even if it has to be recognized that the number of physiological responses considered explicitly in such a model is limited, even relative to the physiological responses reviewed in the previous sections, it can be seen as a starting-point where additional complexity can be introduced progressively later on. This approach represents an alternative to models where a large number of processes are considered simultaneously, leading to major difficulties in calibration and evaluation procedures. It also offers alternatives to the use of major teleonomic hypotheses on which the carbon and nitrogen partitioning processes are based (functional equilibrium).

MecaNiCAL deals with a simplified plant, which is considered as a number of similar tillers. However, intra- and interspecific competition for light and nutrients generates a distribution of sizes and growth rates among tillers. This is of limited relevance for growth in a relatively short period of time. In the longer term, the perenniality of the grass sward obviously depends on tiller dynamics (appearance and disappearance). Therefore, tillering or, more generally, branching processes need to be explicitly introduced in current models.

Conclusion

In the context of pasture swards, our understanding of crop physiology in general and our understanding of responses to nutrient and water supply in particular have improved significantly. Integration of processes through modelling has also improved but still requires effort. However, in the context of pastures, there are important aspects where our knowledge needs to be improved. First, long-term sward functioning needs to be investigated more specifically, particularly through tiller (branch) and root dynamics and turnover. Secondly, responses of plants to defoliation and interactions with nutrient and water supply are not fully understood, particularly the adaptations of shoot morphogenesis, which lead to alterations of sward structure in response to grazing management. Thirdly, our knowledge about plant-to-sward integration (the dynamics of individuals within the sward) is limited, particularly in the context of pasture swards, despite the obvious importance of intra- and interspecific plant competition.

References

Agren, G.I. (1985) Theory for growth of plants derived from the nitrogen productivity concept. *Physiologia Plantarum* 64, 17–28.

Bélanger, G. (1990) Incidence de la nutrition azotée et de la saison sur la croissance, l'assimilation et la répartition du carbone dans un couvert de fétuque élevée en conditions naturelles. Thèse Doctorat en Sciences, Université Paris XI, Orsay, 169 pp.

Bélanger, G., Gastal, F. and Warembourg, F.R. (1992) The effects of nitrogen fertilization and the growing season on carbon partitioning in a sward of tall fescue (*Festuca arundinacea* Schreb.). *Annals of Botany* 70, 239–244.

Brouwer, R. (1962) Distribution of dry matter in the plant. *Netherlands Journal of Agricultural Science* 10, 361–376.
Chapin, F.S., Bloom, A.J., Field, C.B. and Waring, R.H. (1987) Plant response to multiple environmental factors. *BioScience* 37(1), 49–57.
Clarkson, D.T. and Touraine, B. (1994) Morphological responses of plants to nitrate deprivation: a role for abscissic acid? In: Roy, J. and Garnier, E. (eds) *A Whole Plant Perspective on Carbon–Nitrogen Interactions*. SPB Academic Publishing, The Hague, the Netherlands, pp. 187–196.
Davidson, I. (1994) The responses of plants to non-uniform supplies of nutrients. *New Phytologist* 127, 635–674.
Davies, W.J. and Zhang, J. (1991) Root signals and the regulation of growth and development of plants in drying soil. *Annual Review of Plant Physiology and Molecular Biology* 42, 55–76.
Deinum, B. (1985) Root mass of grass swards in different grazing systems. *Netherlands Journal of Agricultural Science* 33, 377–384.
Dewar, R.C. (1995) Interpretation of an empirical model for stomatal conductance in terms of guard cell function. *Plant, Cell Environment* 18, 365–372.
Durand, J.-L., Sheehy, J.E. and Minchin, F. (1987) Nitrogenase activity, photosynthesis and nodule water potential of soybean plants experiencing water deprivation. *Journal of Experimental Botany* 38, 311–321.
Durand, J.-L., Varlet-Grancher, C., Lemaire, G., Gastal, F. and Moulia, B. (1991) Carbon partitioning in forage crops. *Acta Biotheoretica* 39, 213–224.
Durand, J.-L., Onillon, B., Schnyder, H. and Rademacher, I. (1995) Drought effects on cellular and spatial parameters of leaf growth in tall fescue. *Journal of Experimental Botany* 46(290), 1147–1155.
Durand, J.-L., Gastal, F., Etchebest, S., Bonnet, A.C. and Ghesquière, M. (1997) Interspecific variability of plant water status and leaf morphogenesis in temperate forage grasses under summer water deficit. *European Journal of Agronomy* 7, 99–107.
Evans, P.S. (1973) The effect of repeated defoliation to three different levels on root growth of five pasture species. *New Zealand Journal of Agriculture* 16, 31–34.
Evans, J.R. (1983) Nitrogen and photosynthesis in the flag leaf of wheat. *Plant Physiology* 72, 297–302.
Field, C. and Mooney, H.A. (1986) The photosynthesis–nitrogen relationship in wild plants. In: Givnish, T. (ed.) *On the Economy of Plant Form and Function*. Cambridge University Press, Cambridge, pp. 25–55.
Fricke, W., McDonald, A.J.S. and Mattson-Djos, L. (1997) Why do leaves and leaf cells of N-limited barley elongate at reduced rates? *Planta* 202, 522–530.
Gastal, F. and Bélanger, G. (1993) The effect of the nitrogen fertilisation and the growing season on photosynthesis of field-grown tall fescue canopies. *Annals of Botany* 72, 401–408.
Gastal, F. and Lemaire, G. (1988) Study of a tall fescue sward grown under nitrogen deficiency conditions. In: *Proceedings of the XIIth Meeting of the European Grassland Federation, Dublin, Ireland, 4–7 July 1988*. Irish Grassland Association, Belclare, Ireland.
Gastal, F. and Nelson, C.J. (1994) Nitrogen use within the growing leaf blade of tall fescue. *Plant Physiology* 105, 191–197.
Gastal, F. and Saugier, B. (1986) Alimentation azotée et croissance de la fétuque élevée. I – Assimilation du carbone et répartition entre organes. *Agronomie* 6(2), 157–166.

Gastal, F. and Saugier, B. (1989) Relationships between nitrogen uptake and carbon assimilation in whole plants of tall fescue. *Plant, Cell and Environment* 12, 407–418.
Gastal, F., Nelson, C.J. and Coutts, J.M. (1992) Role of N on leaf growth of grasses: assessment of a root signal hypothesis. In: *Proceedings of the Annual Meeting of the American Society of Agronomy, Minneapolis, USA*. American Society of Agronomy, p. 125.
Ghashghaie, J. and Saugier, B. (1989) Effects of nitrogen deficiency on leaf photosynthetic response of tall fescue to water deficit. *Plant, Cell and Environment* 12, 261–271.
Gollan, T., Passioura, J.B. and Munns, R. (1986) Soil water status affects the stomatal conductance of fully turgid wheat and sunflower leaves. *Australian Journal of Plant Physiology* 13, 459–464.
Greenwood, D.J., Lemaire, G., Gosse, G., Cruz, P., Draycott, A. and Neeteson, J.J. (1990) Decline in percentage N of C3 and C4 crops with increasing plant mass. *Annals of Botany* 66, 425–436.
Greenwood, D.J., Gastal, F., Lemaire, G., Draycott, A., Millard, P. and Neeteson, J.J. (1991) Growth rate and N% of field grown crops: theory and experiments. *Annals of Botany* 67, 181–190.
Hodge, A., Stewart, J., Robinson, D., Griffiths, B.S. and Fitter, A.H. (1998) Root proliferation, soil fauna and plant nitrogen capture from nutrient-rich patches in soil. *New Phytologist* 139, 479–494.
Joy, K.W. (1988) Ammonia, glutamine and asparagine: a carbon–nitrogen interface. *Canadian Journal of Botany* 66, 2103–2109.
Larsson, C.M. (1994) Responses of the nitrate uptake system to external nitrate availability: a whole plant perspective. In: Roy, J. and Garnier, E. (eds) *A Whole Plant Perspective on Carbon–Nitrogen Interactions*. SPB Academic Publishing, The Hague, pp. 31–45.
Le Bot, J., Adamowicz, S. and Robin, P. (1998) Modelling plant nutrition of horticultural crops: a review. *Scientiae Horticulturae* 74, 47–82.
Lemaire, G. and Denoix, A. (1987) Croissance estivale en matière sèche de peuplements de fétuque élevée (*Festuca arundinacea* Schreb.) et de dactyle (*Dactylis glomerata* L.) dans l'Ouest de la France. II – Interaction entre les niveaux d'alimentation hydrique et de nutrition azotée. *Agronomie* 7(6), 381–389.
Lemaire, G. and Gastal, F. (1997) N uptake and distribution in plant canopies. In: Lemaire, G. (ed.) *Diagnosis on the Nitrogen Status in Crops*. Springer-Verlag, Heidelberg, pp. 3–43.
MacAdam, J.W., Volenec, J.J. and Nelson, C.J. (1989) Effects of nitrogen on mesophyll cell division and epidermal cell elongation in tall fescue leaf blades. *Plant Physiology* 89, 549–556.
Martre, P., Bogeat-Triboulot, B. and Durand, J.-L. (1999) Measurement of a growth-induced water potential gradient in tall fescue leaves. *New Phytologist* 142, 435–439.
Matthew, C., Xia, J.X., Chu, A.C.P., Mackay, A.D. and Hodgson, J. (1991) Relationship between root production and tiller appearance rates in perennial ryegrass (*Lolium perenne* L.). In: Atkinson, D. (ed.) *Plant Root Growth, and Ecological Perspective*. Blackwell Scientific Publications, Oxford, pp. 281–290.
Monteith, J.L. (1995) A reinterpretation of stomatal response to humidity. *Plant, Cell and Environment* 18, 357–364.
Muller, B. and Touraine, B. (1992) Inhibition of NO_3^- uptake by various phloem-translocated amino acids in soybean seedlings. *Journal of Experimental Botany* 43, 617–623.
Onillon, B., Durand, J.-L., Gastal, F. and Tournebize, R. (1995) Drought effects on growth

and carbon partitioning in a tall fescue sward grown at different nitrogen rates. *European Journal of Agronomy* 4(1), 91–100.

Palmer, S.J., Berridge, D.M., McDonald, A.J.S. and Davies, W.J. (1996) Control of leaf expansion in sunflower (*Helianthus annuus* L.) by nitrogen nutrition. *Journal of Experimental Botany* 296, 350–368.

Parsons, R., Stanforth, A., Raven, J.A. and Sprent, J.I. (1993) Nodule growth and activity may be regulated by a feedback mechanisms involving phloem nitrogen. *Plant, Cell and Environment* 16, 125–136.

Passioura, J.B. (1988) Root signals control leaf expansion in wheat seedlings growing in drying soils. *Australian Journal of Plant Physiology* 15, 687–693.

Robson, M.J. and Parsons, A.J. (1978) Nitrogen deficiency in small closed communities of S24 ryegrass. I. Photosynthesis, respiration, dry matter production and partition. *Annals of Botany* 42, 1185–1197.

Sage, R.F., Pearcy, R.W. and Seeman, J.R. (1987) The nitrogen use efficiency of C_3 and C_4 plants. III. Leaf nitrogen effects on the activity of carboxylating enzymes in *Chenopodium album* (L.) and *Amaranthus retroflexus* (L.). *Plant Physiology* 85, 355–359.

Schnyder, H. and Nelson, C.J. (1988) Diurnal growth of tall fescue leaf blades. I. Spatial distribution of growth, deposition of water, and assimilate import in the elongation zone. *Plant Physiology* 86, 1070–1076.

Serraj, R., Sinclair, T.R. and Purcell, L.C. (1999) Symbiotic N_2 fixation response to drought. *Journal of Experimental Botany* 50, 143–155.

Sheehy, J.E., Minchin, F.R. and Witty, J.F. (1985) Control of nitrogen fixation in a legume nodule: an analysis of the role of oxygen diffusion in relation to nodule structure. *Annals at Botany* 55, 549–562.

Sheehy, J.E., Gastal, F., Mitchell, P.L., Durand, J.-L., Lemaire, G. and Woodward, F.I. (1996) A nitrogen-led model of grass growth. *Annals of Botany* 77, 165–177.

Spollen, B. and Nelson, C.J. (1994) Response of fructans to water deficits in growing leaves of tall fescue. *Plant Physiology* 106, 329–336.

Tabourel-Tayot, F. and Gastal, F. (1998a) MecaNiCAL, a supply–demand model of carbon and nitrogen partitioning applied to defoliated grass. 1. Model description and analysis. *European Journal of Agronomy* 9, 223–241.

Tabourel-Tayot, F. and Gastal, F. (1998b) MecaNiCAL, a supply–demand model of carbon and nitrogen partitioning applied to defoliated grass. 2. Parameter estimation and model evaluation. *European Journal of Agronomy* 9, 243–258.

Tardieu, F., Zang, J., Katerji, N., Bethenod, O., Palmer, S. and Davies, W. J. (1992) Xylem ABA controls the stomatal conductance of field-grown maize subject to compaction or soil drying. *Plant, Cell and Environment* 15, 193–197.

Thornley, J.H.M. (1972) A balanced quantitative model for root : shoot ratios in vegetative plants. *Annals of Botany* 36, 431–441.

Thornley, J.H.M. and Johnson, I.R. (1990) *Plant Crop Modelling: a Mathematical Approach to Plant and Crop Physiology*. Clarendon Press, Oxford.

Touraine, B., Clarckson, D.T. and Muller, B. (1994) Regulation of nitrate uptake at the whole plant level. In: Roy, J. and Garnier, E. (eds) *A Whole Plant Perspective on Carbon–Nitrogen Interactions*. SPB Academic Publishing, The Hague, pp. 11–30.

Tournebize, R. (1990) *Contribution à l'étude de l'effet de la sécheresse sur la croissance de la fétuque élevée à deux niveaux d'azote*. Mémoire de DEA, Orsay, 37 pp.

Troughton, A. (1957) *The Underground Organs of Herbage Grasses*. Bulletin no. 44, Commonwealth Bureau of Pastures and Field Crops, Hurley, UK.

Troughton, A. (1981) Length of life of grass roots. *Grass and Forage Science* 36, 117–120.

Van den Honert, T.H. (1948) Water transport in plants as a catenary process. *Discussions of the Faraday Society* 3, 146–153.

Viratelle, L. (1992) *Caractérisation de l'enroulement foliaire induit par la sécheresse chez deux graminées cultivées*. Mémoire de maîtrise, Université d'Angers, 41 pp.

Volenec, J.J. and Nelson, C.J. (1984) Carbohydrate metabolism in leaf meristems of tall fescue. II. Relationship to leaf elongation rates modified by nitrogen fertilization. *Plant Physiology* 74, 595–600.

3

An Integrated View of C and N Uses in Leaf Growth Zones of Defoliated Grasses

Hans Schnyder,[1] Rudolf Schäufele,[1] Ries de Visser[2] and C. Jerry Nelson[3]

[1]*Department of Grassland Science, Technische Universität München, Germany;* [2]*Research Institute of Agrobiology and Soil Fertility (AB-DLO), Wageningen, The Netherlands;* [3]*Department of Agronomy, University of Missouri, Columbia, USA*

Introduction

Grass leaves have an important role in the ecology and economy of grazed grassland systems: they produce the assimilate needed for growth and maintenance of the plant, as well as the food for heterotrophic organisms that thrive in the grazed ecosystem, including cattle. For the largest part of the growing season, most of the assimilate produced by the grass plant is used in leaf production. Conversion of this assimilate into leaf biomass mainly occurs in the leaf growth and differentiation zone (Fig. 3.1), where the tissue is heterotrophic, as it is completely enclosed by sheaths of older leaves.

Grassland utilization essentially consists in defoliation of grass, i.e. in removal of leaf laminae. Hence, the ability to maintain leaf production in the face of periodic defoliation is essential for both sustained grassland production and survival of the grazed plants. Therefore, it is necessary that we understand the fundamental processes that are directly involved in the growth of a grass leaf and how leaf growth is controlled by resource availability within and to the plant and by genetic, management and climatic factors. The last two decades have seen great progress in the understanding of leaf growth in the *Gramineae* and much of this work was performed with forage grasses.

This chapter first gives a brief description of the growth process in grass leaves and related cellular dynamics, and then C and N metabolism in the functionally distinct zones of expanding and differentiating leaf tissue is investigated. Lastly, the effects of and responses to defoliation at the level of the leaf growth zone are

 Grassland Ecophysiology and Grazing Ecology (eds G. Lemaire, J. Hodgson, A. de Moraes, C. Nabinger and P.C. de F. Carvalho)

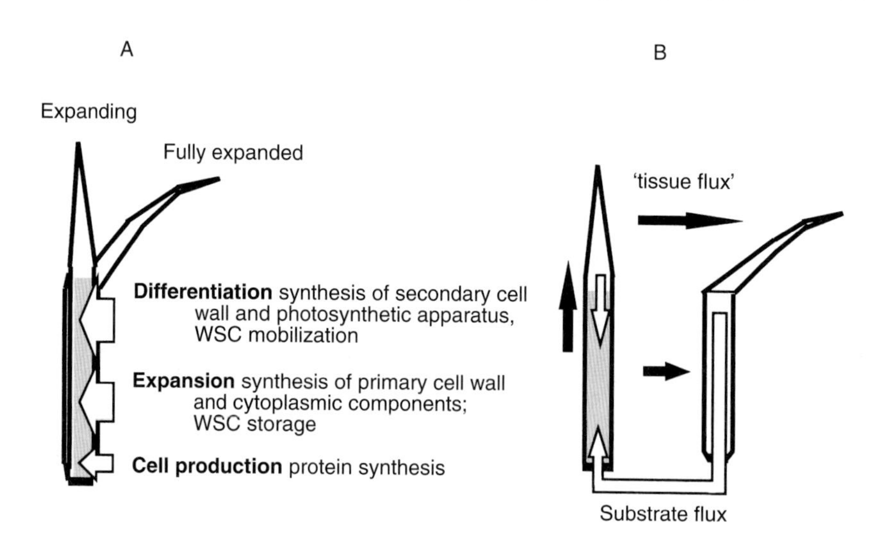

Fig. 3.1. In grasses, leaf growth and differentiation take place in the enclosed basal region of the leaf (A). Growth at the base generates a flux of tissue to the exposed (grazed/mown) zone (B). A vegetative grass tiller can be divided into two functionally distinct leaf categories: expanding leaves and fully expanded leaves. Following cessation of expansion, the leaf is 'displaced' to the fully expanded leaf category (B). C substrate fluxes are opposite to the direction of tissue flux: C supply to the leaf growth and differentiation zone (shaded area) comes from the exposed zone of expanding leaves, as well as from the fully expanded leaves. WSC, water-soluble carbohydrate.

discussed. The chapter closes with an identification of critical areas of ignorance about defoliation responses of leaf growth zones, suggesting new approaches to exploring these areas.

Components of Leaf Growth

Identification of the factors controlling the expansion of a leaf and the elaboration of its physiological and anatomical properties require investigations at the level of the growing and differentiating tissue. In contrast to the situation in dicotyledonous species (Granier and Tardieu, 1998), leaf growth in grasses is confined to the basal region of the leaf, which is completely enclosed within the sheaths of older leaves (Davidson and Milthorpe, 1966; Kemp, 1980). Cells are produced by an intercalary meristem which is located near the point of attachment of the leaf to the shoot axis (Volenec and Nelson, 1981; Figs 3.1 and 3.2A). Meristematic cells of the sheath are present relatively early in the development of the leaf (Skinner and Nelson, 1995). However, initial leaf expansion is confined to the lamina part, and the sheath only starts to expand actively when

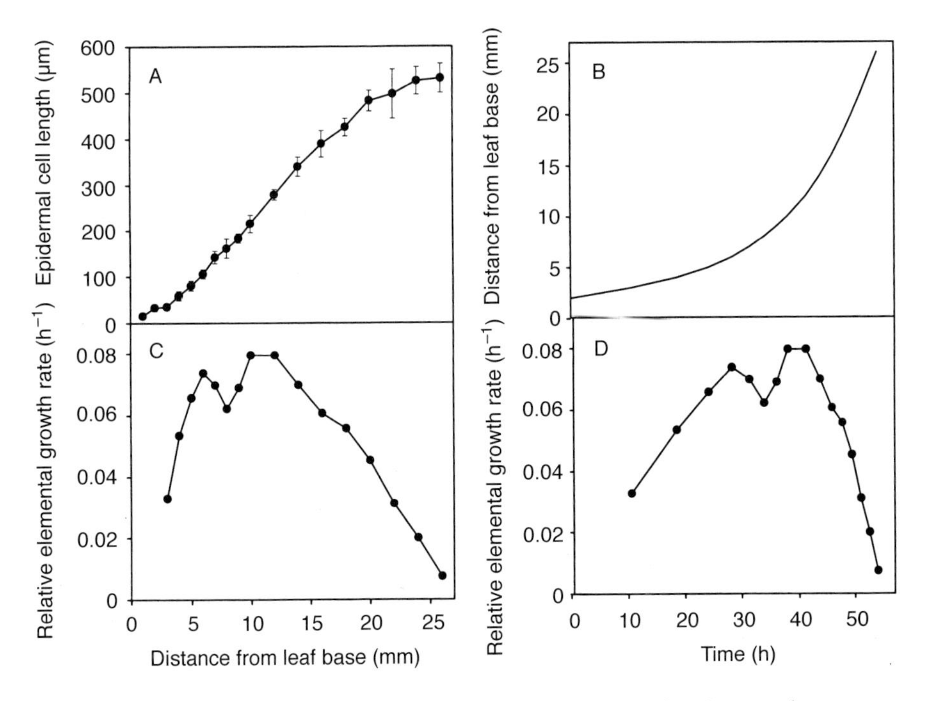

Fig. 3.2. Epidermal cell lengths along the basal part of expanding leaves of perennial ryegrass (A). Cell division takes place near the leaf base. Expansion occurs within 26 mm of the base. As cells expand they are displaced away from the base, due to cell production and expansion at more proximal positions (time–position relationship shown in B). Displacement of a cell is slow near the base, but increases with distance up to the distal limit of the leaf growth zone (26 mm), where the rate of displacement equals the leaf elongation rate (LER) (B). The cell lengths data (A) can be used to derive the spatial distribution of relative elemental growth rates (REGR) (C) and the temporal pattern of REGR experienced by an expanding element (D). REGR increases for a considerable period of time (D), but deceleration of REGR and cessation of growth occur rapidly. (From data presented by Schnyder *et al.*, 1990.)

lamina expansion slows down (Schnyder *et al.*, 1990; Skinner and Nelson, 1995). This transition from lamina to sheath expansion is easily recognized by the displacement of the ligule through the growth zone and away from the leaf base. Transition has no marked effect on the leaf elongation rate (LER, mm h^{-1}) (Kemp, 1980; Schnyder *et al.*, 1990).

Grass leaves expand predominantly in the longitudinal direction (Schnyder and Nelson, 1988; Maurice *et al.*, 1997). The meristematic region of the leaf epidermis is generally very short, in the order of a tenth or less of the length of the leaf growth zone (MacAdam *et al.*, 1989; Ben-Haj-Salah and Tardieu, 1995; Durand *et al.*, 1999). However, mesophyll cells continue to divide within the proximal one-third to one-half of the length of the leaf growth zone (MacAdam *et al.*, 1989). Any cell produced by the meristem is displaced away from the leaf base as a result of continued cell production and expansion of cells at more basal positions in the leaf (Fig. 3.2B). It is evident that the distance between a cell and the meristem from which it originated is a function of the age and developmental stage of a cell. However, the relationship between position and time is not linear in the growth zone: displacement is slow near the leaf base and increases as the number of expanding elements located between the origin (meristem, leaf base) and the cell increases with time (Fig. 3.2B). After a cell has reached the distal limit of the leaf growth zone the rate of its displacement equals the rate of leaf elongation (Durand *et al.*, 1995).

During displacement, the cells expand and differentiate. These processes result in the formation of a typical spatial gradient of cell length with distance from the leaf base (Fig. 3.2A). Where growth is steady (i.e. LER and the spatial distribution of cell lengths in the growth zone are constant with time), the spatial gradient of cell lengths in the 'elongation-only' region of the epidermis can be used to derive the spatial distribution of relative (elemental) growth rates (REGR, mm (mm leaf length)$^{-1}$ h^{-1}; Erickson, 1976) in the growth zone (Schnyder *et al.*, 1990; MacAdam *et al.*, 1992a; Schünmann *et al.*, 1997; Fig. 3.2C). Such analysis usually shows an approximately sinusoidal spatial pattern of REGR in the leaf growth zone.

An alternative method for the assessment of the growth distribution consists in placing equidistant marks along the growth zone and measuring the displacement of the marks after a short growth interval (Davidson and Milthorpe, 1966; Schnyder *et al.*, 1987). While this technique is destructive (as it requires pushing a needle through the base of a tiller (Schnyder and Nelson, 1987; Peters and Bernstein, 1997) or removing the surrounding sheaths for placement of ink marks (Volenec and Nelson, 1981)), it allows an assessment of the growth distribution even including periods of non-steady growth (Schnyder and Nelson, 1988; Spollen and Nelson, 1994; Durand *et al.*, 1995).

Basically, there are two different ways of analysing growth and associated C and N metabolism in leaf growth zones of grasses. Growth can be treated as a property of a specified material element (e.g. a cell), which is followed over time (e.g. Fig. 3.2D): from its formation, through the periods of expansion and differentiation, until attainment of the fully mature state (Silk, 1984). Alternatively, growth can be analysed in spatial terms (Fig. 3.2C). Growth is then treated as a function of the position in the growth zone. Where growth is steady, the temporal (i.e. material) growth patterns can be derived directly from the spatial patterns (and vice versa), as was done for the data shown in Fig. 3.2. Although both approaches have been used in the past, most analyses of leaf growth and associated metabolism have used the spatial description (e.g. Schnyder and Nelson, 1987).

When a growing cell is followed over time, it can be seen that the REGR (of that cell) increases continuously over an extended period of time (MacAdam *et al.*, 1992a; Fig. 3.2D). This indicates a sustained increase in the relative rate of axial cell wall expansion. In contrast, deceleration of REGR, i.e. cessation of cell wall expansion, is a very rapid process.

The above type of analysis can be extended to assess further functional parameters of leaf expansion, including: (i) the size (length) of the leaf growth zone; (ii) the spatial distribution of cell expansion rates (the product of cell size and REGR at any location in the growth zone); (iii) the duration of cell expansion (the time needed for displacement between the distal limit of the cell division zone and the distal end of the leaf growth zone); (iv) the flux of cells into the growth zone (rate of cell production); and (v) the flux of cells out of the growth zone (rate of formation of fully expanded cells). In a steady-state cell influx equals cell efflux, and cell flux per cell file is calculated as LER divided by final cell length.

LER is the product of two components: growth zone size (i.e. length) and average REGR of the tissue in the growth zone. Conversely, LER can be defined as the product of the rate of cell production and the rate and duration of cell expansion. If genotype or environment has an effect on LER, this must result from effects on one or more of these growth components. For example, temperature mainly affects REGR, while drought reduces both the size and the REGR of the growth zone (see below).

Apart from the monodimensional description of leaf growth in terms of elongation rate, which is often adequate when dealing with monocotyledonous species, leaf growth can be described as: (i) leaf area expansion (such as in dicots (e.g. Granier and Tardieu, 1998)); (ii) leaf volume expansion, including changes in leaf thickness (Maurice *et al.*, 1997); and (iii) leaf fresh or dry matter accumulation (Schnyder and Nelson, 1988). It is important to note here that the response of leaf growth to defoliation varies greatly depending on the type of growth description applied.

C and N Metabolism Associated with Leaf Growth

As it is completely concealed by the sheaths of older leaves (see Fig. 3.1A), the growth zone tissue of the grass leaf is heterotrophic. Hence, the C and N substrate for growth must be imported from other parts of the plant. During early stages of leaf growth or when the plant is defoliated severely, the C and N assimilate must enter the leaf growth zone at the leaf base, where cell division is active (see Fig. 3.1A). Conversely, when the leaf tip has emerged above the whorl of enclosing leaf sheaths, assimilate produced by the exposed, photosynthetically active zone of the expanding leaf can contribute substantially to the substrate needs of its own growth zone (Allard and Nelson, 1991). In fact, C may be imported into the leaf growth zone from both directions simultaneously (Brégard and Allard, 1999).

In grasses, carbohydrate transport in the phloem occurs mainly in the form of sucrose (e.g. Fisher and Gifford, 1987), while most of the N is imported in a reduced form, probably as amino acids or amides (Gastal and Nelson, 1994). Therefore, both sucrose and amino acids contribute to C import into the leaf growth zone. This is a factor that requires attention when interpreting the results of C labelling experiments (de Visser *et al.*, 1997) and doing carbon budgeting experiments.

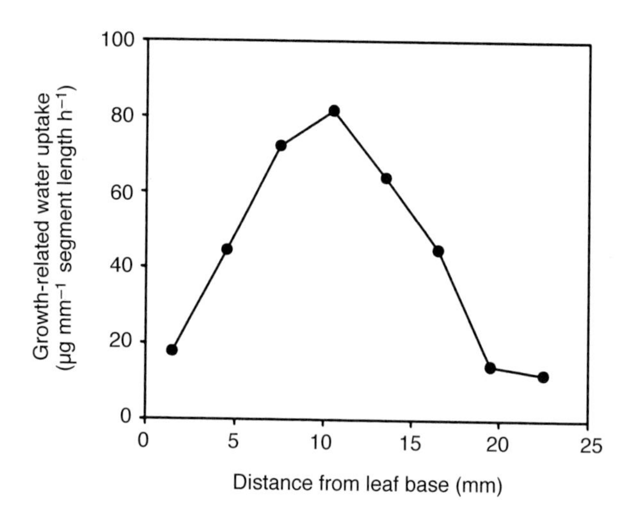

Fig. 3.3. Spatial distribution of growth-related water uptake in expanding tall fescue leaves. Water uptake is maximum near the centre of the growth zone, where REGR is maximum. (Redrawn from data presented by Schnyder and Nelson, 1989.)

Furthermore, investigation of C and N metabolism in the leaf growth zone necessitates consideration of the fact that the expanding tissue elements are continuously displaced relative to the leaf base. During displacement, biomass is 'diluted' as water enters the expanding tissue. This effect is counteracted by concomitant import of assimilate and biosynthetic processes. These may occur at very high relative rates, as is illustrated by the fact that leaf growth zones may 'reproduce' their own size within a day or less under favourable conditions (e.g. Volenec and Nelson, 1984a, b; Schnyder and Nelson, 1989).

Local rates of substance synthesis or deposition (e.g. water deposition in expanding tissue (Fig. 3.3)) can be estimated using the continuity equation and knowledge of the spatial and temporal changes in growth rate and substance content in the growth zone (Silk, 1984). Such investigations have shown that the spatial patterns of C and N concentration (Fig. 3.4A) and deposition rates in leaf growth zones diverge to a considerable extent.

The N deposition rate (μg (mm leaf length)$^{-1}$ h^{-1}) is highest in the basal half of the leaf growth zone, where cell division is active (Gastal and Nelson, 1994), and the rate declines sharply in the distal half of the leaf growth zone. Also, the deposition rate of N is relatively small distal to the position where cells reach their final size. N deposited in the basal region is used rapidly in the synthesis of protein and nucleic acids (Gastal and Nelson, 1994). Also, even when plants were fertilized with ample N, the concentration of nitrate in the leaf growth zone accounted for only a small fraction of total N. Synthesis of Rubisco starts near the position where cells reach their final length and where N deposition is low, indicating that N may be recycled among functionally different proteins during cell/tissue growth and differentiation (Gastal and Nelson, 1994).

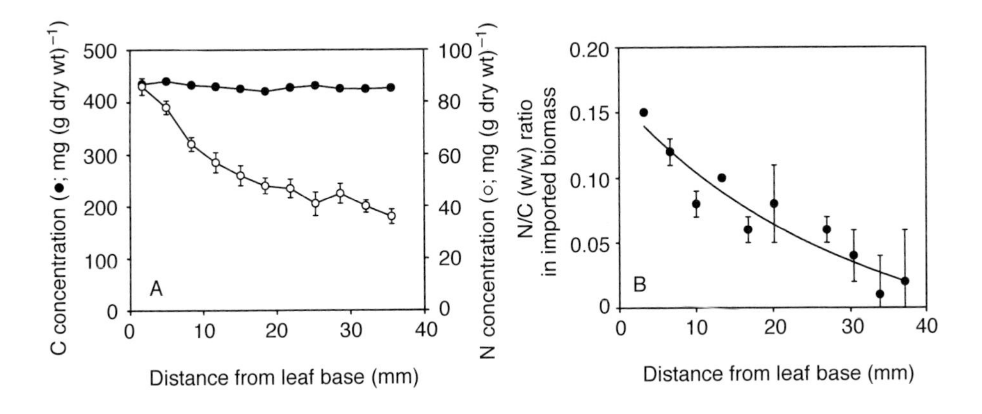

Fig. 3.4. N (○) and and C (●) concentration in mg (g dry mass)$^{-1}$ along the leaf growth zone of perennial ryegrass (A). N concentration decreases with distance from the leaf base, due to a spatial gradient in the N/C ratio of substrate imported in the growth zone (B). C use in respiration was ignored when calculating C import. (R. Schäufele and H. Schnyder, unpublished data.)

These relationships were qualitatively similar in plants grown with contrasting N fertilizer supply, although N deposition in the leaf growth zone was decreased strongly where N supply was low. Interestingly, N concentration in the basal 3-mm-long segment of the leaf growth zone was closely related to LER as modified by N fertilizer supply. In these experiments, final cell size was affected little by N supply, indicating that the main effect of N was on the rate of cell production (MacAdam *et al.*, 1989; Gastal and Nelson, 1994).

Whereas N deposition is at a maximum in the basal part of the leaf growth zone (Gastal and Nelson, 1994), the rate of carbohydrate deposition is generally highest near the centre of the leaf growth zone, where REGR is maximal (Schnyder and Nelson, 1989). As a result, the N is effectively 'diluted' in leaf tissue during expansion and differentiation, involving vacuolation and cell-wall synthesis. This phenomenon is expressed in a strong spatial gradient of N concentration and of the N/C (w/w) ratio in imported assimilate (Fig. 3.4). Although respiratory losses were not accounted for in the above analysis, the magnitude of these changes in deposition strongly indicates that the ratio of amino acid to carbohydrate import must change dramatically with time for a given expanding and developing tissue element. Moreover, the data suggest that the relative rates of import of sucrose and amino acids differ very strongly over a very short distance in the leaf growth zone.

Carbohydrates imported into the leaf growth zone can be used for: (i) carbon skeletons and respiration in biosynthetic processes; (ii) respiration associated with protein turnover and maintenance of cellular structures, ion gradients and metabolite gradients; (iii) maintaining carbohydrate concentrations that are being diluted when water is deposited in expanding cells; and (iv) storage of reserve carbohydrates (mainly fructan in temperate grasses) (Schnyder and Nelson, 1987). Carbohydrate use in maintenance respiration apparently accounts for, at most, a few per cent of the total rate of carbohydrate import in the leaf growth zone of tall fescue, whereas respiration associated with biosynthetic processes may be in the order of 10% (Ryle *et al.*, 1973; Penning de Vries, 1975; Moser *et al.*, 1982; Volenec and Nelson, 1984b; Schnyder and Nelson, 1987). However, respiration is high in the cell division zone, where N import and protein synthesis are very active (Pearen and C.J. Nelson, unpublished).

Synthesis of fructan, a vacuolar storage carbohydrate, often accounts for a large fraction of the total assimilate (including nitrogenous compounds) imported into the growth zone of grass leaves (up to 40% of total biomass (Schnyder and Nelson, 1987, 1989; Schnyder *et al.*, 1988)). This is interesting in view of the high relative rates of assimilate consumption in the biosynthesis of structural biomass in this tissue. Moreover, the degree of polymerization (DP) of the fructan synthesized in growth zones of tall fescue is low (Schnyder and Nelson, 1987, 1989), contributing up to 0.2 MPa to the osmotic potential. In contrast, the fructan in mature leaf blade tissue of tall fescue is predominantly in the form of large polymers (Spollen and Nelson, 1988). However, a low DP of fructan in leaf growth zones is not a universal feature in grasses (Spollen and Nelson, 1988). When subject to drought, the fructan in the growth zone may be hydrolysed to yield hexoses, thus facilitating osmotic adjustment (Spollen and Nelson, 1994).

Sucrose is the immediate substrate for fructan synthesis (Pollock and Cairns, 1991) and glucose is released from sucrose during fructosyl transfer in fructan synthesis. The latter may serve as substrate for respiration and synthesis of structural biomass in the growth zone (Lüscher and Nelson, 1995). In fact, under conditions of high fructan synthesis, the associated release of glucose may be the most important source of substrate for respiration and synthesis of structural biomass.

In a number of studies, the spatial distribution of fructan synthesis in the growth zone was closely related to the spatial distribution of REGR (Schnyder *et al.*, 1988; Schnyder and Nelson, 1989). Fructan synthesis may thus contribute to maintaining a low sucrose concentration in the cytosol of expanding cells and hence sustain sucrose import. Such a mechanism could also facilitate the influx of other solutes (e.g. amino acids) and water from the phloem, by sequestration of sucrose that is not used in respiration or synthesis of structural biomass. In fact, efflux of water from the phloem could account for much (if not all) of the water uptake by expanding cells (Schmalstig and Cosgrove, 1990; Farrar *et al.*, 1995; Patrick and Offler, 1996). However, to our knowledge, these putative mechanisms have not been studied in growth zones of grass leaves.

Fructan synthesis in the leaf growth zone of tall fescue was significant (20% of dry mass), even when plants were grown at low irradiance (Schnyder and Nelson, 1989). Yet the cumulative rate of fructan synthesis in leaf growth zones does not exhibit a correlation with the total (volumetric) rate of expansion in the growth zone, when results obtained in different experiments with tall fescue are compared (H. Schnyder and C.J. Nelson, unpublished data). The same was true for the comparison of expansion and total water-soluble carbohydrate (WSC)

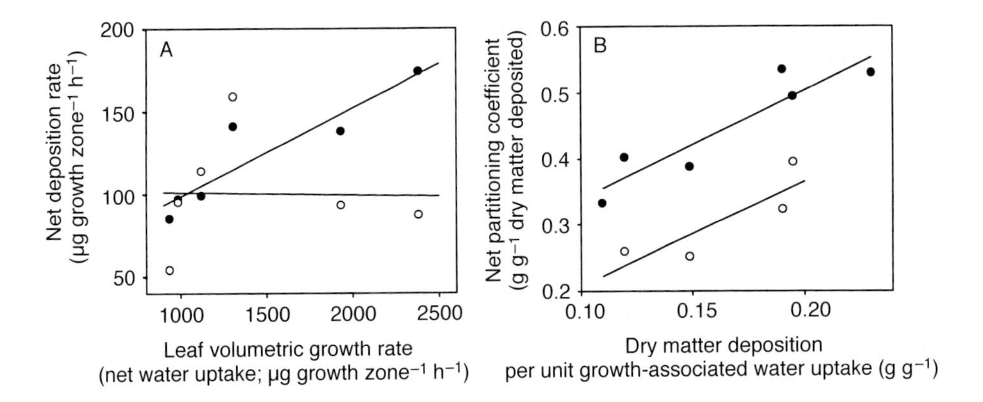

Fig. 3.5. Relationship between structural dry matter deposition rate (SDM, ●), water-soluble carbohydrate deposition rate (WSC, ○) and the total rate of growth-related water uptake in leaf growth zones of tall fescue (A). SDM was defined as total dry matter less WSC. Growth rate was correlated with SDM deposition rate, but not with WSC deposition rate. Partitioning of dry matter to WSC deposition (● in B) was enhanced when total dry matter deposition (i.e. substrate import) in the growth zone was large relative to expansion (i.e. water uptake) (B). Increased WSC deposition was essentially due to fructan synthesis (○). (Derived from data presented by Schnyder and Nelson, 1988, 1989; Schnyder *et al.*, 1988.)

deposition (Fig. 3.5A). Notably, the rate of total dry matter deposition in growth zones was also poorly related to expansion, i.e. growth-related water deposition in the growth zone ($r^2 = 0.30$).

In contrast, there was a highly significant correlation ($r^2 = 0.85$) between total expansion and the cumulative rate of synthesis of structural biomass (defined as dry matter deposition less deposition of water-soluble carbohydrates, including fructan (Fig. 3.5A)). Structural biomass, as defined here, included cell-wall material and protein. The close relationship between synthesis of structural biomass and expansion may not be surprising, since cytoplasmic components and cell-wall material must be synthesized to prevent excessive dilution and to allow for continued growth.

It does seem from these studies that processes involved in the synthesis of structural biomass have a higher priority for the imported assimilate than storage of non-structural carbohydrates. WSC (including fructan) deposition was stimulated in situations where dry matter deposition was large relative to expansion (at high irradiance or during the light period of a diurnal cycle) (Fig. 3.5B). The relative contributions of N (amino acids) and carbohydrates to assimilate import was not investigated in the above studies. Thus, one may wonder whether the increased partitioning of imported assimilate to synthesis of fructan was related to a limitation in N, i.e. a relative abundance of carbohydrates in phloem sap. If this was the case, use of the surplus sucrose in fructan synthesis may have contributed to maintaining the influx of the growth-limiting N.

When investigating the spatial gradient of structural dry matter concentration (g g^{-1} tissue water or g per unit leaf area), a dilution of structural biomass during expansion of tissue is generally evident (MacAdam and Nelson, 1987). This may be (at least partially) due to the fact that expansion of the vacuole contributes relatively more to expansion than the cytoplasmic compartment, where the density of biomass may be higher. Also, when results from different experiments were compared, the total rate of synthesis of structural biomass in the growth zone was not directly proportional to expansion. Where the cumulative rate of expansion was large, less structural biomass was deposited per unit volumetric growth (Fig. 3.5A). Other factors, such as regulation of the activity of enzymes mediating the expansion of cell walls (expansins, hydrolases, xyloglucanendotransglycosylase, peroxidases), perhaps acting via hormonal control, may be involved (e.g. MacAdam *et al.*, 1992a, b; Schünmann and Ougham, 1996; Schünmann *et al.*, 1997; Cosgrove, 1998).

Clearly, the definition of structural biomass as used here requires refinement, since it includes structurally and biochemically very diverse components. In this context, it is interesting to note the similarity that exists in the composition of monomeric sugars released *in vitro* by enzymatic hydrolysis of cell walls from the base of expanding *Holcus lanatus* (high LER) and *Deschampsia flexuosa* (low LER) leaves (Groeneveld and Bergkotte, 1996). Thus, the species difference in LER was not due to a difference in the chemical composition of cell walls. However, the rate of cell-wall synthesis was substantially larger in the leaf bases of the fast-growing *H. lanatus*. Differences in the composition of cell walls were evident in mature tissue, but these arose from cell-wall deposition after termination of expansion (Groeneveld *et al.*, 1998).

Synthesis of structural material, including cell walls, continues at a relatively high rate after cell expansion has ceased (MacAdam and Nelson, 1987; Groeneveld *et al.*, 1998). This process may continue even after the tissue has emerged from the surrounding sheaths (Groeneveld *et al.*, 1998). However, most of the differentiation of the photosynthetic apparatus takes place before the tissue emerges, so that the tissue is photosynthetically competent when it is first exposed to light (Wilhelm and Nelson, 1978; Boffey *et al.*, 1980; Dean and Leech, 1982; Gastal and Nelson, 1994). The fructan stored in tissue while it was expanding is mobilized near the time when expansion stops and may contribute substantially to the substrate needs of these local processes (MacAdam and Nelson, 1987; Allard and Nelson, 1991).

Effects of Defoliation on Leaf Growth Zones

The relationships described above are characteristic of plants that were not disturbed during growth or which have recovered from defoliation. Although the short-term response of grasses to defoliation has been subject to much investigation, very few studies have focused on the effects and responses at the level of the leaf growth zone.

Defoliation may cause a rapid and drastic (and at least transient) reduction of LER (Davidson and Milthorpe, 1966). In a recent study with perennial ryegrass, this fast reduction of LER was associated with both a rapid decrease in the length of the growth zone (due to a 'premature' cessation of cell expansion) and a reduction in cell production. The biochemical and physiological bases of these effects have not been analysed. Interestingly, however, defoliation had no effect on the expansion of cells in the proximal part of the growth zone (Schäufele and Schnyder, 2000).

Longer-term effects of defoliation on cellular dynamics in leaf growth zones were analysed by Volenec and Nelson (1983). In this study, with two contrasting genotypes of tall fescue, frequent defoliation led to reduced LER. This was also related to decreases in the rate of cell production and expansion. Both mechanisms contributed to a shortening of the leaf growth zone with frequent defoliation. These effects were apparent after plants had recovered from the immediate effect of defoliation.

Longer-term effects of defoliation on LER are probably related to the overall morphogenetic response of plants to defoliation and the associated reduction of tiller size (Chapman and Lemaire, 1993). Tiller size, LER and growth zone size are closely related in perennial ryegrass (Schnyder *et al.*, 1990). Strong decreases in LER of continuously grazed versus infrequently mown tall fescue were observed by Mazzanti *et al.* (1994).

Refoliation Mechanisms

In grasses, during vegetative growth, the active shoot meristems and leaf growth zones are not damaged when a plant is grazed or mown, but photosynthesis and nutrient uptake may be strongly decreased (Richards, 1993). In such a situation, it is generally held that refoliation is controlled by the availability of substrate (source control), although the involvement of other factors has not been ruled out.

A number of mechanisms may facilitate initial refoliation (de Visser *et al.*, 1997; Schnyder and de Visser, 1999). These include: (i) increasing the efficiency of substrate use for new leaf area expansion and exposure ('dilution mechanisms'); (ii) use of stored carbohydrates already present in the leaf growth zone at defoliation; (iii) preferential partitioning of current assimilate to the leaf growth zone; and (iv) transient import of mobilized reserves.

An increased efficiency of resource use for leaf area expansion and exposure was observed in several experiments with perennial ryegrass. In the studies of Sheehy *et al.* (1980), Grant *et al.* (1981) and van Loo (1993), the specific leaf area (SLA, m^2 g^{-1}) was increased strongly after defoliation, thus indicating that leaf area produced per unit substrate consumed was maximized after defoliation. Also, in the studies of van Loo (1993), SLA was increased when perennial ryegrass was subject to frequent relative to infrequent defoliation. This effect was particularly strong when defoliation occurred at a low stubble height. Such increases in the efficiency of substrate use for leaf area expansion may be due to decreased substrate use per unit expansion in the leaf growth zone or to decreased substrate investments for differentiation and maturation of leaf tissue. In one experiment with perennial ryegrass, defoliation was followed by a rapid and significant decrease in the concentration of structural biomass (defined as g WSC-free C g^{-1} tissue water) in the immature leaf zone, which included the growth and differentiation zone of the leaf (Schnyder and de Visser, 1999). Thus, structural biomass was effectively diluted, due to a promotion of expansion relative to synthesis of structural biomass.

'Spatial dilution' (de Visser *et al.*, 1997) may also contribute to initial refoliation: fresh mass per unit leaf length was transiently (2 days) decreased in both the leaf growth zone and the differentiation zone of perennial ryegrass shortly after defoliation. These results suggest that axial relative to cross-sectional (width and/or thickness) expansion was transiently increased immediately after defoliation (but see Schäufele and Schnyder, 2000). This mechanism would also facilitate the exposure of light-intercepting leaf area by reducing 'costs' in terms of both total (volumetric) expansion and associated substrate consumption. This process of enhancing longitudinal growth relative to width and thickness also occurs at low radiation (Schnyder and Nelson, 1989).

The potential importance of reserve carbohydrates that are already present in the leaf growth zone at defoliation was first recognized by Davidson and Milthorpe (1966). In their pioneering study, Davidson and Milthorpe observed a rapid and significant decrease in the concentration of soluble carbohydrates (mainly fructan) in the leaf growth zone of *Dactylis glomerata* when plants were defoliated. Strong and rapid decreases in the concentration of fructan in the immature (enclosed) zone of expanding leaves following defoliation were also observed in tall fescue (Volenec, 1986) and perennial ryegrass (Morvan *et al.*, 1997). In the experiment of Morvan and coworkers, the decrease in the fructan concentration was related to a strong stimulation in the activity of fructan exohydrolase, the enzyme responsible for fructan mobilization. Furthermore, the activity of sucrose–sucrose fructosyltransferase (a key enzyme of fructan synthesis) was decreased simultaneously. It seems likely that the decrease in fructan concentration in the leaf base is due to hydrolysis and consequent use of the liberated fructose in processes associated with cell expansion (e.g. cell-wall synthesis). However, as expansion continues, the fructan-mobilizing cells are displaced away from the leaf growth zone to the zone of secondary cell-wall synthesis. Thus, mobilization perhaps needs to be followed by export and basipetal transport of the mobilized carbohydrates to feed the tissue in the leaf growth zone.

In a steady state (such as may be the case during undisturbed growth), the tissue-bound efflux of C from the immature zone (cf. 'tissue flux' to the exposed zone in Fig. 3.1B) is balanced by the import of C substrate. However, if leaf expansion is maintained after defoliation while C import is reduced, the C balance of the growth and differentiation zone of leaves becomes transiently negative. This phenomenon was observed in perennial ryegrass and was associated with both 'spatial' and 'chemical' dilution of C in the leaf growth and differentiation zone (de Visser *et al.*, 1997; Schnyder and de Visser, 1999).

Preferential allocation of current assimilate to the growth zone of expanding leaves may also contribute to rapid refoliation (Richards, 1993). Such preferential allocation is at least partially due to a 'privileged' position of leaf growth zones relative to the path of current photosynthate exported from the newly exposed leaf area. In the investigations of Davidson and Milthorpe (1966), 94% of the leaf area exposed during the first 4 days after defoliation of *D. glomerata* was produced by leaves that were (partially) defoliated at the cut. In another study (Schnyder and de Visser, 1999), 82% of the new exposed foliage at 2 days after defoliation was contributed by cut leaves, which were still actively expanding at that time. For export from these leaves, the carbohydrates and reduced N produced would first need to pass their own growth zone (see Fig. 3.1A). However, the leaf growth zone is a strong sink for the assimilate produced by its own photosynthetically active tissue and it would have first access to this source of substrate (Allard and Nelson, 1991). Experiments with tall fescue during undisturbed growth have shown that leaves do not export C before their own lamina has expanded to 80% of the final length (Brégard and Allard, 1999), a stage near when cell production has stopped (Skinner and Nelson, 1994).

The photosynthetic activity of the leaf area exposed in 2 days may cover a large fraction (if not all) of the C needs of the leaf growth zone. Thus, Volenec and Nelson (1984b) estimated that it takes about 1.3 days of photosynthesis near light saturation to produce the carbohydrate equivalents for production (cell division and expansion) of that unit of leaf area. Additional costs are associated with differentiation of tissue to fully develop its function. Hence, about 3 days of photosynthesis may be required to offset the total costs for growth and differentiation of the same leaf tissue (C.J. Nelson, unpublished).

A steady-state $^{13}C/^{12}C$-, $^{15}N/^{14}N$-labelling experiment demonstrated a rapid transition from reserve-dependent to current assimilation-driven leaf growth in severely defoliated perennial ryegrass. Current photosynthesis contributed about 63% of the total C incorporation and about 87% of the carbohydrate incorporation in regrowing tillers during the second day after defoliation (Schnyder and de Visser, 1999). Also, the C in the leaf growth zone was rapidly replaced by currently fixed C following defoliation (de Visser *et al.*, 1997; Fig. 3.6). In another study with perennial ryegrass, the sucrose pool in the leaf growth zone was almost completely exchanged by currently fixed C within 2 days after defoliation (Morvan-Betrand *et al.*, 1999), indicating that reserve carbohydrates contributed little to leaf growth 1 day after defoliation.

These results are in apparent conflict with conclusions from many studies suggesting substantial and sustained contributions of mobilized carbohydrate reserves to regrowth. These studies did not investigate the actual influx of reserve-derived C into the (re-)growing shoot sink, and most did not take account of tissue heterogeneity in the stubble. Often, a contribution to regrowth was inferred from the loss of WSC from total stubble tissue, which would include the carbohydrates present within the expanding, basal region of growing leaves. Also, even when using steady-state labelling to quantify the fluxes of reserve- and current

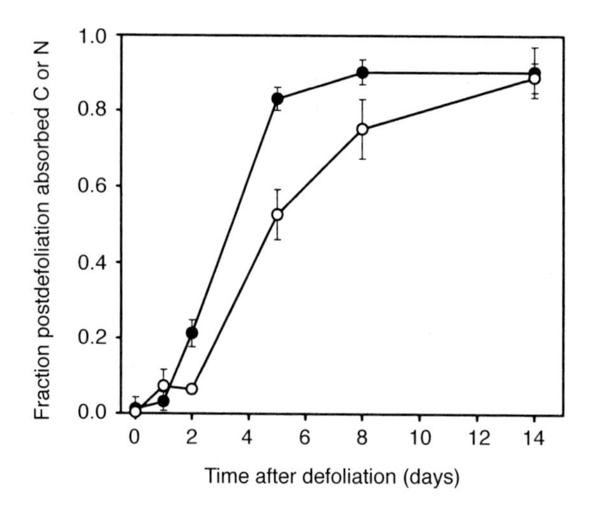

Fig. 3.6. Incorporation of currently assimilated C (●) and N (○) in the leaf growth zone of severely defoliated perennial ryegrass. Current photosynthate was the main C source already on day 2 after defoliation. (From data presented by de Visser *et al.*, 1997.)

assimilation-derived C to regrowing tissue, it must be realized that much of the reserve-derived C imported into the growth zone is in the form of amino acids. Thus, (at least) labelling of C must be combined with labelling of N to allow an estimation of the fluxes of reserve-derived amino-C and carbohydrate-C into regrowing shoot tissue. Such double-labelling studies indicated that much of the reserve-derived C influx was in the form of amino acids (Avice *et al.*, 1996; de Visser *et al.*, 1997; Schnyder and de Visser, 1999).

Clearly, the functional and morphological heterogeneity of the stubble left behind after defoliation (see Fig. 3.1) must be considered in experiments investigating reserve-derived C and N fluxes into (re-)growing leaves (de Visser *et al.*, 1997). The first leaf tissue exposed above the defoliation level is near-fully differentiated at defoliation (when it was still located below the defoliation level). Its exposure results from tissue flux, i.e. expansion of tissue near the base of growing leaves. Basically, the exposure to light of the near-mature, enclosed tissue only requires an influx of water into expanding cells at the leaf base. Thus, for a short period (perhaps depending on the substrate level in the growth zone at defoliation), expansion could occur without a concurrent import of organic substrate into the growing leaf base (see discussion on 'dilution' above).

In labelling experiments, it must be realized that the C and N in the structural biomass of the first-exposed tissue has the same isotopic signature as the reserve-derived C and N that is imported into the regrowing tiller. Therefore, neglect of tissue flux can lead to a large overestimation of the contribution of reserves to (re-)growth of the shoot. For the same reason, it may also overestimate the duration of reserve-dependent growth (Schnyder and de Visser, 1999).

When compared with carbohydrates the reserve dependence of leaf (re-) growth seems to be more important for N and for a substantially longer duration (Avice *et al.*, 1996; Volenec *et al.*, 1996; Schnyder and de Visser, 1999). This may be related to the extended period of time required for recovery of N uptake.

Conclusions

In grasses, the production, expansion and much of the differentiation (including synthesis of structural biomass) of leaf tissue takes place in the basal zone of the leaf, which is enclosed by the sheaths of older leaves. During vegetative growth, this zone forms part of the stubble left behind after defoliation. This factor is probably one of the key elements explaining the relative defoliation tolerance of grasses and their abundance in grasslands. Identification of the factors controlling leaf expansion and related C and N metabolism requires investigations at the level of the production, expansion and differentiation of leaf tissue. Knowledge about these fundamental relationships has greatly increased during the last two decades.

More recently, work in this area has expanded dramatically, and significant advances are now being made in the understanding of the factors that potentially control leaf expansion in grasses. These include studies of: (i) enzymes mediating cell-wall loosening (expansins (for reviews, see McQueen-Mason, 1995; Cosgrove, 1998)) and hardening (see above); (ii) the chemical composition of cell walls in the successive zones of cell development (e.g. Groeneveld and Bergkotte, 1996); and (iii) osmotica and water deposition in expanding tissue (e.g. Schnyder and Nelson, 1987; Spollen and Nelson, 1994; Hu and Schmidhalter, 1998a, b). Also, knowledge of the effects of temperature (Ben-Haj-Salah and Tardieu, 1995; Thomas and Stoddart, 1995; Durand *et al.*, 1999), nutritional factors (e.g. nitrogen (Gastal and Nelson, 1994; Fricke *et al.*, 1997)) and stress (e.g. salinity (Bernstein *et al.*, 1993; Marcum and Murdoch, 1994; Hu and Schmidhalter, 1998a, b)); drought (Spollen and Nelson, 1994; Durand *et al.*, 1995; Thompson *et al.*, 1997; Thomas *et al.*, 1999); and soil compaction (Beemster *et al.*, 1996) on leaf growth zones has increased strongly.

Progress in other areas of leaf expansion has been more limited, notably those of gene expression (see, however, Schünmann and Ougham, 1996; Blomstedt *et al.*, 1998), hormonal control, effects of mechanical stresses, such as trampling and wounding, and the range of species studied. The latter is particularly true for forage and rangeland grasses.

Very few studies have investigated the immediate effects of defoliation on leaf growth zones. Knowledge of the short-term effects of defoliation on the different components of leaf growth (cell production, expansion and differentiation) is very restricted. Furthermore, C and N metabolism in the different parts of the leaf growth zone has not been studied. Most studies have evaluated bulk segments of tissue (e.g. the immature leaf base). In addition, much more needs to be learnt on the short- and long-term effects of the frequency and intensity of defoliation and of predefoliation and postdefoliation growth conditions on cellular dynamics and the C and N economy in leaf growth zones. These C and N studies should include roots, as they are very sensitive to defoliation and could be important sources of hormones, such as cytokinins and abscisic acid (ABA), that affect leaf growth.

New techniques have now become available to open up these areas, including stable isotope labelling (Schnyder and de Visser, 1999), acoustic emission (Shimotashiro *et al.*, 1998) and laser scanning confocal microscopy (Lemon and Posluszny, 1998). Single-cell sampling and analysis techniques are a powerful tool for testing cellular metabolism in different cell types, e.g. epidermis, mesophyll and bundle sheath (Koroleva *et al.*, 1998).

The few studies that have been conducted at the level of the leaf growth zone suggest that refoliation – more precisely, continued leaf elongation – is highly flexible in terms of substrate use. 'Dilution' mechanisms, use of substrate already present in the growth zone at defoliation and preferential substrate partitioning to leaf growth as opposed to root growth may all be important. Now, comprehensive comparative studies among species with contrasting growth strategies and differing in defoliation tolerance must be performed to assess the role of the above-suggested mechanisms for conferring or limiting tolerance to defoliation. Also, these mechanisms should be assessed systematically across a range of ecological situations.

References

Allard, G. and Nelson, C.J. (1991) Photosynthate partitioning in basal zones of tall fescue leaf blades. *Plant Physiology* 95, 663–668.

Avice, J.C., Ourry, A., Lemaire, G. and Boucaud, J. (1996) Nitrogen and carbon flows estimated by ^{15}N and ^{13}C pulse-chase labelling during regrowth of alfalfa. *Plant Physiology* 112, 282–290.

Beemster, G.T.S., Masle, J., Williamson, R.E. and Farquhar, G.D. (1996) Effects of soil resistance to root penetration on leaf expansion in wheat (*Triticum aestivum* L.): kinematic analysis of leaf elongation. *Journal of Experimental Botany* 47, 1663–1678.

Ben-Haj-Salah, H. and Tardieu, F. (1995) Temperature affects expansion rate of maize leaves without change in spatial distribution of cell length. *Plant Physiology* 109, 861–870.

Bernstein, N., Silk, W.K. and Läuchli, A. (1993) Growth and development of sorghum leaves under conditions of NaCl stress. *Planta* 191, 433–439.

Blomstedt, C.K, Gianello, R.D, Gaff, D.F, Hamill, J.D. and Neale, A.D. (1998) Differential gene expression in desiccation-tolerant and desiccation-sensitive tissue of the resurrection grass, *Sporobolus stapfianus*. *Australian Journal of Plant Physiology* 25, 937–946.

Boffey, S.A., Selldén, G. and Leech, R.M. (1980) Influence of cell age on chlorophyll formation in light-grown and etiolated wheat seedlings. *Plant Physiology* 65, 680–684.

Brégard, A. and Allard, G. (1999) Sink to source transition in developing leaf blades of tall fescue. *New Phytologist* 141, 45–50.

Chapman, D.F. and Lemaire, G. (1993) Morphogenetic and structural determinants of plant regrowth after defoliation. In: Baker, M.J. (ed.) *Grasslands for Our World*. SIR Publishing, Wellington, New Zealand, pp. 55–64.

Cosgrove, D.J. (1998) Cell wall loosening by expansins. *Plant Physiology* 118, 333–339.
Davidson, J.L. and Milthorpe, F.L. (1966) Leaf growth of *Dactylis glomerata* L. following defoliation. *Annals of Botany* 30, 173–184.
Dean, C. and Leech, R. (1982) Genome expression during normal leaf development. I. Cellular and chloroplast numbers and DANN, RNA, and protein levels in tissues of different ages within a seven-day-old wheat leaf. *Plant Physiology* 69, 904–910.
de Visser, R., Vianden, H. and Schnyder, H. (1997) Kinetics and relative significance of remobilized and current C and N incorporation in leaf and root growth zones of *Lolium perenne* after defoliation: assessment by ^{13}C and ^{15}N steady-state labelling. *Plant, Cell and Environment* 20, 37–46.
Durand, J.-L., Onillon, B., Schnyder, H. and Rademacher, I. (1995) Drought effects on cellular and spatial parameters of leaf growth in tall fescue. *Journal of Experimental Botany* 46, 1147–1155.
Durand, J.-L., Schäufele, R. and Gastal, F. (1999) Grass leaf elongation rate as a function of developmental stage and temperature: morphological analysis and modelling. *Annals of Botany* 83, 577–588.
Erickson, R.O. (1976) Modeling of plant growth. *Annual Review of Plant Physiology* 27, 407–434.
Farrar, J.F., Minchin, P.E.H. and Thorpe, M.R. (1995) Carbon import into barley roots: effects of sugars and relation to cell expansion. *Journal of Experimental Botany* 46, 1859–1865.
Fisher, D.B. and Gifford, R.M. (1987) Accumulation and conversion of sugars by developing wheat grains. VII. Effects of changes in sieve tube and endosperm cavity sap concentrations on the grain filling rate. *Plant Physiology* 84, 341–347.
Fricke, W., McDonald, A.J.S. and Mattson-Djos, L. (1997) Why do leaf cells of N-limited barley elongate at reduced rates? *Planta* 202, 522–530.
Gastal, F. and Nelson, C.J. (1994) Nitrogen use within the growing leaf blade of tall fescue. *Plant Physiology* 105, 191–197.
Granier, C. and Tardieu, F. (1998) Spatial and temporal analyses of expansion and cell cycle in sunflower leaves. *Plant Physiology* 116, 991–1001.
Grant, S.A., Bartram, G.T. and Torvell, L. (1981) Components of regrowth in grazed and cut *Lolium perenne* swards. *Grass and Forage Science* 36, 155–168.
Groeneveld, H.W. and Bergkotte, M. (1996) Cell wall composition of leaves of an inherently fast- and an inherently slow-growing grass species. *Plant, Cell and Environment* 19, 1389–1398.
Groeneveld, H.W., Bergkotte, M. and Lambers, H. (1998) Leaf growth in the fast-growing *Holcus lanatus* and the slow-growing *Deschampsia flexuosa*: tissue maturation. *Journal of Experimental Botany* 49, 1509–1517.
Hu, Y. and Schmidhalter, U. (1998a) Spatial distributions of inorganic ions and sugars contributing to osmotic adjustment in the elongating wheat leaf under saline soil conditions. *Australian Journal of Plant Physiology* 25, 591–597.
Hu, Y. and Schmidhalter, U. (1998b) Spatial distribution of mineral elements and their net deposition rates in the elongating wheat leaf under saline soil conditions. *Planta* 204, 212–219.
Kemp, D.R. (1980) The location and size of the extension zone of emerging wheat leaves. *New Phytologist* 84, 729–737.
Koroleva, O.A., Farrar, J.F., Tomos, A.D. and Pollock, C.J. (1998) Carbohydrates in individual cells of epidermis, mesophyll, and bundle sheath in barley leaves with changed export or photosynthetic rate. *Plant Physiology* 118, 1525–1532.

Lemon, G.D. and Posluszny, U. (1998) A new approach to the study of apical meristem development using laser scanning confocal microscopy. *Canadian Journal of Botany* 76, 899–904.

Lüscher, M. and Nelson, C.J. (1995) Fructosyltransferase activities in the leaf growth zone of tall fescue. *Plant Physiology* 107, 1419–1425.

MacAdam, J.W. and Nelson, C.J. (1987) Specific leaf weight in zones of cell division, elongation and maturation in tall fescue leaf blades. *Annals of Botany* 59, 369–376.

MacAdam, J.W., Volenec, J.J. and Nelson, C.J. (1989) Effects of nitrogen on mesophyll cell division and epidermal cell elongation in tall fescue leaf blades. *Plant Physiology* 89, 549–556.

MacAdam, J.W., Nelson, C.J. and Sharp, R.E. (1992a) Peroxidase activity in the leaf elongation zone of tall fescue. I. Spatial distribution of ionically bound peroxidase activity in genotypes differing in length of the elongation zone. *Plant Physiology* 99, 872–878.

MacAdam, J.W., Sharp, R.E. and Nelson, C.J. (1992b) Peroxidase activity in the leaf elongation zone of tall fescue. II. Spatial distribution of apoplastic peroxidase activity in genotypes differing in length of the elongation zone. *Plant Physiology* 99, 879–885.

McQueen-Mason, S.J. (1995) Expansins and cell wall expansion. *Journal of Experimental Botany* 46, 1639–1650.

Marcum, K.B. and Murdoch, C.L. (1994) Salinity tolerance mechanisms of six C-4 turfgrasses. *Journal of the American Society for Horticultural Science* 119, 779–784.

Maurice, I., Gastal, F. and Durand, J.-L. (1997) Generation of form and associated mass deposition during leaf development in grasses: a kinematic approach for non-steady growth. *Annals of Botany* 80, 673–683.

Mazzanti, A., Lemaire, G. and Gastal, F. (1994) The effect of nitrogen fertilization upon herbage production of tall fescue swards continuously grazed with sheep. 1. Herbage growth dynamics. *Grass and Forage Science* 49, 111–120.

Morvan, A., Challe, G., Prud'homme, M.-P., Le Saos, J. and Boucaud, J. (1997) Rise of fructan exohydrolase activity in stubble of *Lolium perenne* after defoliation is decreased by uniconazole, an inhibitor of the biosynthesis of gibberellins. *New Phytologist* 136, 81–88.

Morvan-Bertrand, A., Pavis, N., Boucaud, J. and Prud'homme, M.-P. (1999) Partitioning of reserve and newly assimilated carbon in roots and leaf tissues of *Lolium perenne* during regrowth after defoliation: assessment by ^{13}C steady-state labelling and carbohydrate analysis. *Plant, Cell and Environment* 22, 1097–1108.

Moser, L.E., Volenec, J.J. and Nelson, C.J. (1982) Respiration, carbohydrate content and leaf growth of tall fescue. *Crop Science* 22, 781–786.

Patrick, J.W. and Offler, C.E. (1996) Post-sieve element transport of photoassimilates in sink regions. *Journal of Experimental Botany* 47, 1165–1177.

Penning de Vries, F.W.T. (1975) The cost of maintenance processes in plant cells. *Annals of Botany* 39, 77–92.

Peters, W.S. and Bernstein, N. (1997) The determination of relative elemental growth rate profiles from segmental growth rates: a methodological evaluation. *Plant Physiology* 113, 1395–1404.

Pollock, C.J and Cairns, A.J. (1991) Fructan metabolism in grasses and cereals. *Annual Review of Plant Physiology and Plant Molecular Biology* 42, 77–101.

Richards, J.H. (1993) Physiology of plants recovering from defoliation. In: Baker, M.J. (ed.) *Grasslands for Our World.* SIR Publishing, Wellington, New Zealand, pp. 46–54.

Ryle, G.J.A., Brockington, N.R., Powell, C.E. and Cross, B. (1973) The measurement and prediction of organ growth in a uniculm barley. *Annals of Botany* 37, 233–246.

Schäufele, R. and Schnyder, H. (2000) Cell growth analysis during steady and non-steady growth in leaves of perennial ryegrass (*Lolium perenne* L.) subject to defoliation. *Plant, Cell and Environment* (in press).

Schmalstig, J.G. and Cosgrove, D.J. (1990) Coupling of solute transport and cell expansion in pea stems. *Plant Physiology* 94, 1625–1634.

Schnyder, H. and de Visser, R. (1999) Fluxes of reserve-derived and currently assimilated carbon and nitrogen in perennial ryegrass recovering from defoliation: the regrowing tiller and its component functionally distinct zones. *Plant Physiology* 119, 1423–1435.

Schnyder, H. and Nelson, C.J. (1987) Growth rates and carbohydrate fluxes within the elongation zone of tall fescue leaf blades. *Plant Physiology* 85, 548–553.

Schnyder, H. and Nelson, C.J. (1988) Diurnal growth of tall fescue leaf blades. I. Spatial distribution of growth, deposition of water, and assimilate import in the elongation zone. *Plant Physiology* 86, 1070–1076.

Schnyder, H. and Nelson, C.J. (1989) Growth rates and assimilate partitioning in the elongation zone of tall fescue leaf blades at high and low irradiance. *Plant Physiology* 90, 1201–1206.

Schnyder, H., Nelson, C.J. and Coutts, J.H. (1987) Assessment of spatial distribution of growth in the elongation zone of grass leaf blades. *Plant Physiology* 85, 290–293.

Schnyder, H., Nelson, C.J. and Spollen, W.G. (1988) Diurnal growth of tall fescue leaf blades. II. Dry matter partitioning and carbohydrate metabolism in the elongation zone and adjacent expanded tissue. *Plant Physiology* 86, 1077–1083.

Schnyder, H., Seo, S., Rademacher, I.F. and Kühbauch, W. (1990) Spatial distribution of growth rates and of epidermal cell lengths in the elongation zone during leaf development in *Lolium perenne* L. *Planta* 181, 423–431.

Schünmann, P.H.D. and Ougham, H.J (1996) Identification of three cDNA clones expressed in the leaf extension zone and with altered patterns of expression in the slender mutant of barley: a tonoplast intrinsic protein, a putative structural protein and protochlorophyllide oxidoreductase. *Plant Molecular Biology* 31, 529–537.

Schünmann, P.H.D., Smith, R.C., Lang, V., Matthews, P.R. and Chandler, P.M. (1997) Expression of XET-related genes and its relation to elongation in leaves of barley (*Hordeum vulgare* L.). *Plant, Cell and Environment* 20, 1439–1450.

Sheehy, J.E., Cobby, J.M. and Ryle, G.J.A. (1980) The use of a model to investigate the influence of some environmental factors on the growth of perennial ryegrass. *Annals of Botany* 46, 343–365.

Shimotashiro, T., Inanaga, S., Sugimoto, Y., Matsuura, A. and Ashimori, M. (1998) Non-destructive method for root elongation measurement in soil using acoustic emission sensors. II. Spatial measurement of single root elongation. *Plant Production Science* 1, 248–253.

Silk, W.K. (1984) Quantitative descriptions of development. *Annual Review of Plant Physiology* 35, 479–518.

Skinner, R.H. and Nelson, C.J. (1994) Epidermal cell division and the coordination of leaf and tiller development. *Annals of Botany* 74, 9–15.

Skinner, R.H. and Nelson, C.J. (1995) Elongation of the grass leaf and its relationship to the phyllochron. *Crop Science* 35, 4–10.

Spollen, W.G. and Nelson, C.J. (1988) Characterization of fructan from mature leaf blades and elongation zones of developing leaf blades of wheat, tall fescue, and timothy. *Plant Physiology* 88, 1349–1353.

Spollen, W.G. and Nelson, C.J. (1994) Response of fructan to water deficit in growing leaves of tall fescue. *Plant Physiology* 106, 329–336.

Thomas, H., and Stoddart, J.L. (1995) Temperature sensitivities of *Festuca arundinacea* Schreb. and *Dactylis glomerata* L. ecotypes. *New Phytologist* 130, 125–134.

Thomas, H., James, A.R. and Humphreys, M.W. (1999) Effects of water stress of leaf growth in tall fescue, Italian ryegrass and their hybrid: rheological properties of expansion zones of leaves, measured on growing and killed tissue. *Journal of Experimental Botany* 50, 221–231.

Thompson, D.S., Wilkinson, S., Bacon, M.A. and Davies, W.J. (1997) Multiple signals and mechanisms that regulate leaf growth and stomatal behaviour during water deficit. *Physiologia Plantarum* 100, 303–313.

van Loo, E.N. (1993) On the relation between tillering, leaf area dynamics and growth of perennial ryegrass (*Lolium perenne* L.). Doctoral thesis, Wageningen Agricultural University, The Netherlands.

Volenec, J.J. (1986) Nonstructural carbohydrates in stem base components of tall fescue during regrowth. *Crop Science* 26, 122–127.

Volenec, J.J. and Nelson, C.J. (1981) Cell dynamics in leaf meristems of contrasting tall fescue genotypes. *Crop Science* 21, 381–385.

Volenec, J.J. and Nelson, C.J. (1983) Responses of tall fescue leaf meristems to N fertilization and harvest frequency. *Crop Science* 23, 720–724.

Volenec, J.J. and Nelson, C.J. (1984a) Carbohydrate metabolism in leaf meristems of tall fescue. I. Relationship to genetically altered leaf elongation rates. *Plant Physiology* 74, 590–594.

Volenec, J.J. and Nelson, C.J. (1984b) Carbohydrate metabolism in leaf meristems of tall fescue. II. Relationships to leaf elongation rates modified by nitrogen fertilization. *Plant Physiology* 74, 595–600.

Volenec, J.J., Ourry, A. and Joern, B.C. (1996) A role for nitrogen reserves in forage regrowth and stress tolerance. *Physiologia Plantarum* 97, 185–193.

Wilhelm, W.W. and Nelson, C.J. (1978) Leaf growth, leaf aging and photosynthetic rate of tall fescue genotypes. *Crop Science* 18, 769–772.

4

Effects of Grazing on the Roots and Rhizosphere of Grasses

L.A. Dawson, S.J. Grayston and E. Paterson

Plant Science Group, Macaulay Land Use Research Institute, Craigiebuckler, Aberdeen, UK

Introduction

Fertilizer inputs are currently being reduced in many areas (Commission, European Communities, 1992) and the resultant drop in pasture fertility will reduce the stock carrying capacity. As a consequence, individual plants will be defoliated less frequently (Curll and Wilkins, 1982) and less nitrogen will be deposited as urine (Thomas *et al.*, 1988), thus influencing plant competition and composition. Research to date has concentrated on the effects of grazing on above-ground aspects, and it has recently been stated that 'the effects of herbivory on the timing, mass and quality of below-ground inputs remains one of the greatest unresolved issues of the dynamics of nutrient cycling' (Ruess and Seagle, 1994). The soil microbiota in grasslands consists of populations of microorganisms, including bacteria, fungi, protozoa, nematodes, and micro- and macroarthropod groups (Ingham and Detling, 1986). These all rely for their growth, at least in part, on carbon or nitrogen substrates via litter, root production, sloughage and exudation. Figure 4.1 illustrates the main links in the detrital trophic food web and shows the primary role of plant roots, the connectivity and some of the many trophic interactions. Although nearly all soil organisms belong to the detrital food web, significant numbers of root herbivores exist in grassland soil (Curry, 1994), also relying on plant roots for their survival. Any alteration in plant-derived carbon, such as through defoliation, will have consequences at many levels in the food web (Fig. 4.1). Since microbial activity, supported in part by root-derived carbon, drives soil nutrient cycling, the production and use of carbon from root systems is also a key issue in the functioning of soil ecosystems (van Veen *et al.*, 1989).

(eds G. Lemaire, J. Hodgson, A. de Moraes, C. Nabinger and P.C. de F. Carvalho)

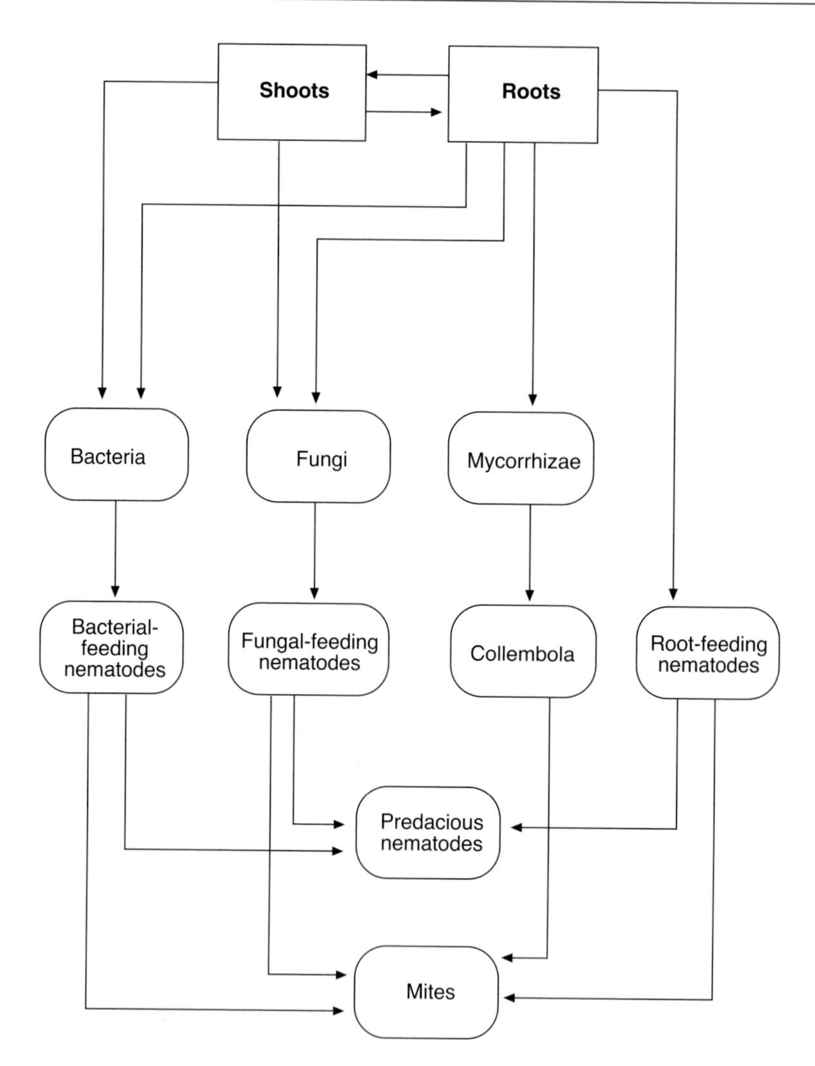

Fig. 4.1. Conceptual diagram of detrital food web for short-grass prairie (modified from Ingham *et al.*, 1986 and Hunt *et al.*, 1987), showing flows and connectivity between trophic levels. Omitted are flows from every organism through death and defaecation.

The coupling of rhizodeposition processes to microbial activity and nutrient cycling is affected both by the quantity and quality of plant inputs and by soil nutrient availability. Rhizodeposition, in particular the exudate components, stimulates the growth of distinct microbial communities. Shifts in quantity and quality of rhizodeposition, as occur following defoliation, affect microbial community structure and, therefore, also microbially driven processes in soil. Soil nutrient availability is a strong determinant of both the quality and quantity of rhizodeposition. Low availability of nutrients generally results in an increase in rhizodeposition. This can be related to release of specific compounds that increase the availability of phosphorus, zinc, iron and aluminium (Pellet *et al.*, 1995; Rengel, 1997). In the case of N limitation, increased rhizodeposition may non-specifically increase N cycling processes, ultimately increasing availability to the plant. However, it is unclear to what degree effects of N supply on plant partitioning and root architecture are contributory to these shifts in rhizodeposition, which highlights the need for an integrated approach to developing an understanding of productivity in grazed systems.

As a consequence of the many complex interactions between the plant and the soil, results obtained from field observations examining the effects of alteration in grazing habit on species composition are hard to extrapolate. In general, it has been predicted that, as pastures become more extensive and nutrient limitations increase, species that can maximize assimilation of nutrients through a high biomass allocation to structures that enhance nutrient absorption (Tilman, 1988) or minimize loss of nutrients (Chapin, 1980) will increase. However, competition for nutrients could be modified by species differences in resistance or tolerance to grazing. Under grazing, it has been suggested that minimizing loss of nutrients could be of greater importance than a high biomass allocation to roots (Berendse *et al.*, 1992). It was shown that nitrogen losses through cutting were greater in species from nutrient-rich habitats (e.g. *Arrhenatherum elatius*) than in species from nutrient-poor habitats (e.g. *Festuca rubra*) (Berendse *et al.*, 1992). According to Elberse and Berendse (1993), plant morphology and architecture seemed more relevant adaptations to habitats with contrasting nutrient supplies than plant allocation. As a consequence, it is important to understand the physiological response of the whole plant to grazing at the individual species level. Grazing effects will also be separated into the two main factors, defoliation and urine deposition, to understand the individual processes involved.

This review identifies the main effects of shoot herbivory on individual species and long-term effects on the plant root system, such as root biomass distribution, morphology and architecture, which have direct implications for competition between individual plants in a grazed grassland. This chapter also investigates the short-term changes in root exudation and effects on the soil microbiota. The effects of grazing on below-ground plant properties and soil microbiota are important in understanding how sustainable a system is or how likely it is to change with altered grazing pressure.

Plant Root System Responses to Defoliation

Root biomass as affected by defoliation

It has long been recognized that regular defoliation can alter root biomass and distribution. In general, root mass has been shown to reduce with defoliation (Ennik and Hofman, 1983; Holland and Detling, 1990; Matthew *et al.*, 1991). A reduction in root biomass with increasing defoliation intensity has been found in grasses from several environments, from temperate grasslands (Mawdsley and Bardgett, 1997) to semi-arid environments (McNaughton *et al.*, 1983; Ruess, 1988; Seagle *et al.*, 1992). As well as root biomass being affected by defoliation, root length and elongation rate have been shown to be reduced by defoliation (Evans, 1971; Brouwer, 1983; Ennik and Hofman, 1983; Richards, 1984; Jarvis and Macduff, 1989; Matthew *et al.*, 1991).

Root mass reduction can be related to the intensity and frequency of defoliation (Ennik and Hofman, 1983; Danckwerts and Gordon, 1987; Karl and Doescher, 1991; Wilsey, 1996). In prairie grassland, Holland and Detling (1990) studied a grazing chronosequence and found root biomass decreased in relation to increased grazing impact; they attributed this to a reduced allocation of carbon to the root system, although they did not assess the role of rhizodeposition or root turnover. However, contrasting responses have been found. For example, Milchunas and Lauenroth (1993), from data analysed and modelled in their study of the effects of grazing, found that, while differences in above-ground net primary productivity with grazing were related to differences in species composition, differences in root mass between grazed and ungrazed sites were not related to differences in above-ground net primary productivity. Negative impacts of grazing on above-ground net primary productivity were accompanied by both increases and decreases in root mass. McNaughton *et al.* (1998) found no evidence that grazing inhibited root biomass or productivity in Serengeti grasslands, where water is the major limitation for growth. In the field, the response of plants to grazing can be affected by other factors, such as light, nutrients, temperatures and water (McNaughton, 1979). As a consequence of these many interrelated factors operating in the field, studies on the physiology of individual grass species under controlled conditions need to be conducted to allow an understanding of these interacting responses.

Species, herbivore and fertility interactions

Within a grazed sward, there can be many individual species, heterogeneously distributed, which differ in size, age, growth rate and reproductive capacity (Bullock, 1996). They also present contrasting preferences to the grazing animal, as shown by Fraser and Gordon (1997).

Different species and varieties (Jones, 1983) and genotypes (Harris, 1973) can show contrasting above-ground morphological adaptations to variations in cutting height and frequency. Plants grazed frequently can develop a short, prostrate canopy, which can be more resistant to grazing if less biomass is available to herbivores, and a greater amount of photosynthetic and meristematic tissue remains available for regrowth following defoliation. Under increasing grazing pressure, pasture mass decreases and the grass structure changes to a high density of smaller tillers per unit area of pasture (Grant *et al.*, 1983). As well as this grazing avoidance behaviour, plants can also exhibit grazing tolerance, e.g. storage of reserves, compensatory photosynthesis and increased nutrient uptake (Rosenthal and Kotanen, 1994). It has been suggested that grasslands with a long history of grazing tend to be dominated by grazing-tolerant species (Milchunas *et al.*, 1988). In grazing-intolerant plants with relatively weak shoot meristematic activity after defoliation, it has been suggested that the proportion of carbon resources allocated to roots is higher than in tolerant species (Richards, 1993). Root growth reductions might be a mechanism to reduce below-ground carbon demand in defoliated plants, allowing greater allocation of carbon to the shoot (Richards, 1984). Also related to this is an increased nutrient absorption capacity in defoliated plants (Rosenthal and Kotanen, 1994). However, this may not be sufficient to compensate for losses when soil nutrient availability is low (Chapin and McNaughton, 1989).

Individual species responses

Root biomass

In order to try to understand the contrasting responses to grazing found in mixed pastures, an experiment was conducted in a greenhouse on monocultures of five grass species found on pastures ranging in intensity of management from unimproved to improved by the addition of lime and fertilizer. The species *Festuca ovina*, *F. rubra*, *Agrostis capillaris*, *Poa trivialis* and *Lolium perenne* were subjected to three clipping treatments for a total of 1 year; unclipped (UC), clipped every week to 4 cm from the stem base (WC) and clipped every 8 weeks to 4 cm from the stem base (IC). For all species, clipping each week removed a significant amount of leaf material, thereby having a direct impact on the plant (Fig. 4.2). In general, clipping reduced root biomass, particularly for the slower-growing species *F. ovina*, *F. rubra* and *A. capillaris* (Fig. 4.2). In contrast, however, defoliation had no significant effect on the root biomass of *P. trivialis*, a species found in productive pastures. For *L. perenne*, a species found on improved pastures, weekly clipping (WC) significantly reduced root biomass relative to the less intensive defoliation (IC), but not relative to the uncut treatment (UC). Deinum (1985) found that the root biomass of *L. perenne* was greater where only parts of the foliage were removed periodically, in contrast to where defoliation was frequent and complete. These results illustrate the importance of considering intensity and frequency of defoliation when investigating physiological responses to defoliation. Another aspect is the cutting treatment used to simulate herbivory in the field. It is very difficult to make comparisons between studies where plants have been defoliated in different ways. For example, a constant grazing height was used in this study of British pasture grasses, but the contrasting vertical distribution of plant parts and the species grazing preference that has been observed (Bakker *et al.*, 1998) throw some doubt on a constant sward height being the most appropriate simulation of defoliation under field conditions. This again illustrates the need for an integrated approach in the understanding of grazing effects.

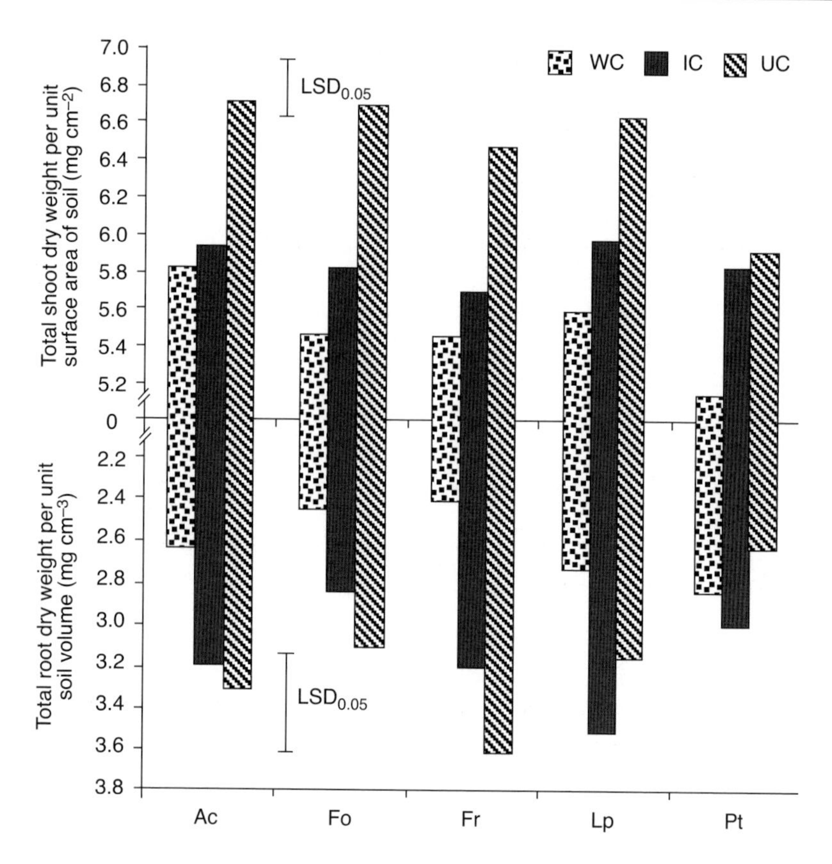

Fig. 4.2. Shoot biomass per surface area of ground (mg cm^{-2}) and root biomass per volume of soil (mg cm^{-3}) of five grass species, Ac (*Agrostis capillaris*), Fo (*Festuca ovina*), Fr (*Festuca rubra*), Lp (*Lolium perenne*) and Pt (*Poa trivialis*). Plants had been grown for 1 year from seed and given one of three defoliation treatments: unclipped (UC), weekly clipped (WC) and clipped every 8 weeks (IC), both to 4 cm from the stem base. LSD, least significant difference.

Genotypic responses

Responses to defoliation can also vary within a species. For example, Fig. 4.3 shows some results for two genotypes from a range of genotypes as an illustration of the contrasting responses to defoliation. Vegetative tillers of genotypes AC1 and AK5 of *A. capillaris* both showed a similar reduction in shoot biomass after five twice-weekly clippings to 4 cm above the stem base (Fig. 4.3a). Only genotype AK5 showed a significant reduction in root biomass with clipping (Fig. 4.3b), and only AC1 showed an increase in root-tip number and root length with clipping (Fig. 4.3c and d). Having established that significant differences exist in root responses between genotypes to defoliation, future studies will examine the importance and consequences of these responses for competition in a mixed population.

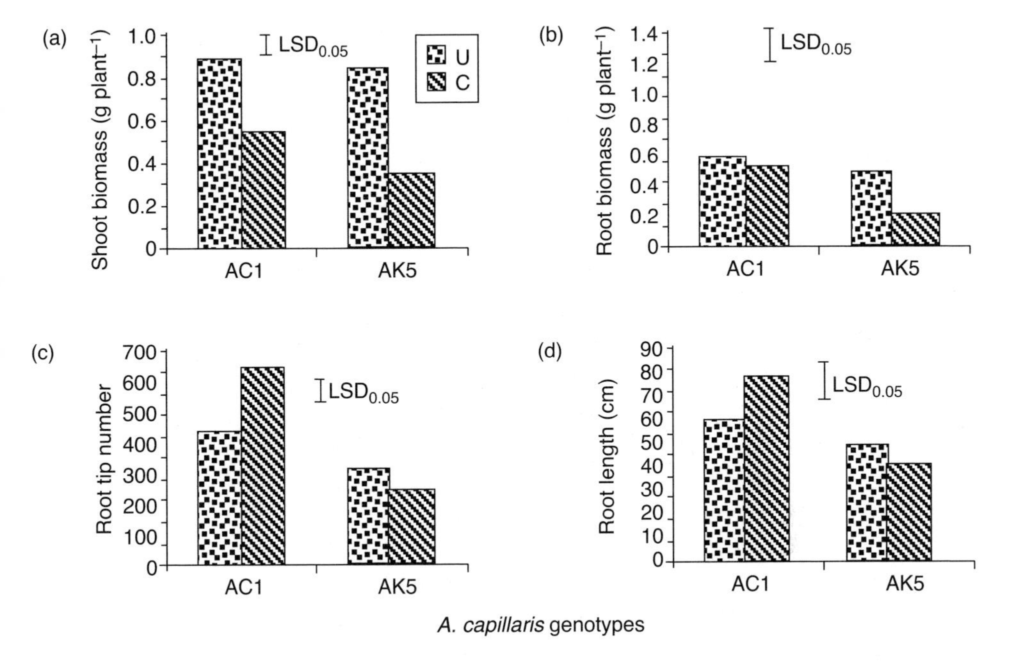

Fig. 4.3. Genotypic differences for shoot biomass (a), root biomass (b), root tip number (c) and root length (d) for uncut (U) and twice weekly cut (C) genotypes of *Agrostis capillaris,* ACI and AK5. LSD, least significant difference.

The number of root tips and branching characteristics could increase the ability, in the short term, to compete for nutrients, but, in the longer term, an ability to reallocate growth to the shoots would be a more beneficial characteristic. Richards (1984) found that, after clipping, root length extension was reduced by half in grazing-tolerant genotypes but not in grazing-intolerant species, as had been observed in the field.

Root distribution

The majority of grass roots can be found in the top 15 cm of soil in the field (Macklon *et al.*, 1994). In a study of three perennial grasses, *Cenchrus ciliaris*, *Digitaria cummutata* and *Stipa lagascae*, Chaieb *et al.* (1996) found that repeated clipping resulted in a reduced biomass and a more superficially distributed root system. This even greater than normal surface location of roots could expose the plant to water stress during summer drought. The same authors found that, with continual overgrazing (repeated cutting of all shoots to ground level), more than 65% of the roots of all three species were found in the top 15 cm of soil. However, when medium grazing was simulated (two to three cuttings), the root system was increasingly superficial for only one species.

Table 4.1. Percentage of total root biomass in the 0–6 cm soil zone. Mean values from five replicate pots and SE in parentheses. Five pasture species grown in miniswards for a period of 1 year. Defoliation treatments were weekly clipped (WC), clipped every 8 weeks (IC) and undefoliated (UC). LSD, least significant difference = 4.

	Grass species				
Clipping treatment	*Agrostis capillaris*	*Festuca ovina*	*Festuca rubra*	*Lolium perenne*	*Poa trivialis*
WC	40 (7)	36 (2)	48 (4)	55 (8)	77 (5)
IC	60 (7)	47 (6)	65 (4)	64 (5)	68 (6)
UC	53 (6)	52 (5)	66 (4)	52 (5)	67 (3)
Maximum depth of rooting (cm)	38	36	36	37	29

In our study of British pasture species (see Fig. 4.2), only one species, *P. trivialis*, showed an increase in the percentage of root biomass found in the surface soil with regular clipping (Table 4.1). *Poa trivialis* can be drought-sensitive (Grime *et al.*, 1988) and has a low persistence under close cutting, probably due to this increased surface root distribution. For the three slower-growing species, *F. rubra*, *F. ovina* and *A. capillaris*, the percentage of roots in the surface zone decreased with regular clipping.

Root diameter

In general, root diameter has been shown to be reduced by repeated defoliation (Table 4.2; Evans, 1971; Chapin and Slack, 1979; Chapin, 1980), but Arredondo and Johnson (1998) found that, with defoliation, diameter decreased in one grass species but increased in another. This response also depends upon N supply (Table 4.2).

Root architecture

In a study using a slant-board system, the effects of repeated defoliation on the root length, morphology and architecture of *L. perenne* and *F. ovina* were examined. In *L. perenne*, defoliation had no significant effect on the total root axis length, but the length of the primary axis increased significantly with defoliation under low N supply (Table 4.2a). In contrast, for *F. ovina*, the slower-growing species, root length was inhibited by defoliation (Table 4.2b), mainly manifested by a reduction in the length of first-order laterals at both N levels. The proportion of total root length as first-order laterals and the total number of links were significantly reduced by defoliation. The low-N undefoliated *F. ovina* root system, where the carbon-to-nitrogen ratio was highest, represents a lower topological index, which would increase total resource transport efficiency (Fitter, 1985,

Table 4.2. Topological analysis of root axes of *Lolium perenne* (a) at day 14 and *Festuca ovina* (b) at day 28, either defoliated twice weekly (D) or undefoliated (U) and either grown under high-N (HN) (2.0 mM NH_4NO_3) or low-N (LN) 0.02 mM NH_4NO_3) nutrition. Least significant difference (LSD); $P < 0.05$.

	Treatments (mean)					
	UHN	DHN	ULN	DLN	LSD_1	LSD_2
(a) *Lolium perenne*						
Total root length (mm)	2720	1736	2053	3067	2163	2761
Primary root axis length (mm)	266	372	343	512	118	133
First-order lateral length (mm)	2241	1325	1661	2421	1930	2462
Root fresh weight (mg)	137	76	111	119	62	51
Log altitude/log magnitude	0.95	0.99	0.99	0.98	0.05	0.06
Root diameter (mm)	306	225	259	217	41	37
(b) *Festuca ovina*						
Total root length (mm)	2677	903	3816	1529	1549	1760
Primary root axis length (mm)	303	203	252	232	78	67
First-order lateral length (mm)	2235	670	3091	1178	1308	1473
Root fresh weight (mg)	101	33	89	48	39	44
Log altitude/log magnitude	0.97	0.98	0.92	0.98	0.05	0.06
Root diameter (mm)	217	151	127	144	34	24

LSD_1 when comparing effect of N at same level of defoliation.
LSD_2 when comparing effect of defoliation at same level of N.

1986, 1987). This increased branching has also been observed in relation to elevated CO_2 (Berntson and Woodward, 1992). Under low-N conditions with defoliation, the root system of *F. rubra* was less branched, reflecting the reduced propensity to branch when carbon was limited due to shoot defoliation (Fig. 4.4). In a study of the effects of clipping on seedlings of grasses, Arredondo and Johnson (1998), found that the root branching of 'Hycrest', a cultivar of hybrid crested couch grass (*Agropyron desertorum* (Fisch. ex Link) Schult. × *Agropyron cristatum* (L.) Gaert.), which is grazing-tolerant, was unaffected. In contrast, defoliation significantly increased the number of second-order laterals in 'Whitmar', a cultivar of bluebunch couch grass (*Pseudoroegneria spicata* (Pursh) A. Löve), which is grazing-sensitive. In a later study, they found that the largest plasticity in root architecture was observed for 'Whitmar' for defoliated and undefoliated plants, and suggested this plasticity as the main mechanism allowing it to forage in heterogeneous soil conditions (Arredondo and Johnson, 1999).

Specific architectural responses to defoliation have also been found among species for arid-zone perennials (Hodgkinson and Baas Becking, 1977). For *Danthonia caespitosa*, root branching was reduced with defoliation, and this caused it to be susceptible to drought. The root system of *Medicago sativa*, on the other hand, was little affected by defoliation, which the authors attribute to the reserve

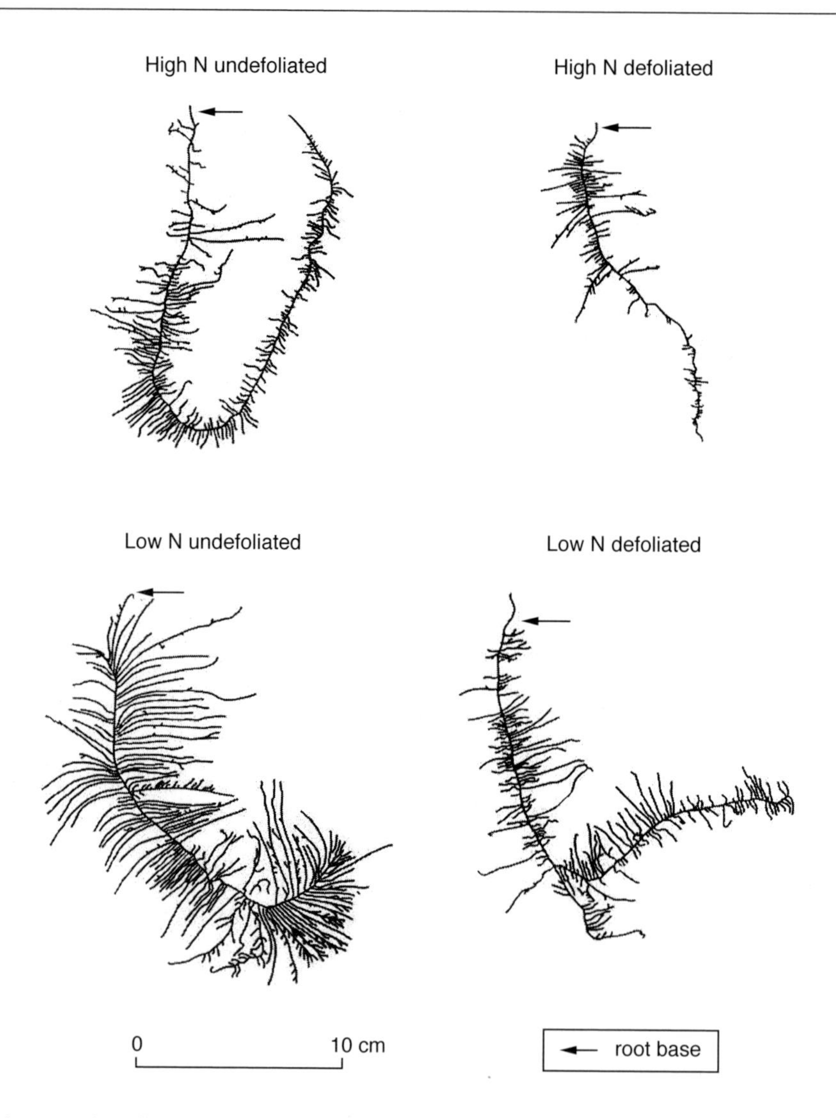

Fig. 4.4. The effect of repeated defoliation for 28 days under contrasting levels of nitrogen supply (High N – 2.0 mM NH_4NO_3, and low N – 0.02 mM NH_4NO_3) on the branching characteristics of root axes of *Festuca ovina*.

organic compounds in the tap root (Hodgkinson, 1969) and early resumption of current photosynthate supply to the lateral roots, thus allowing *Medicago* to tolerate drought conditions. When assessing contrasting species responses to defoliation, the contribution of the tap root and other structural compartments needs also to be considered.

Some workers have reported that there is, following grazing, a short-term storage of carbon in roots, allowing a rapid mobilization of carbon (Dyer *et al.*,

1991; Holland *et al.*, 1996), and of other nutrients (Polly and Detling, 1989) for plant regrowth. However, other workers, e.g. Oesterheld and McNaughton (1988), have suggested that root reserves do not contribute significantly to plant regrowth, although this reliance is very much species-dependent. For a more detailed appraisal of this, see Thornton *et al.* (Chapter 5, this volume). The arguments that, in grazing-tolerant species, reduction in root growth represents an adaptation for rapid regrowth of shoots (Richards, 1984) may not always be valid. As can be seen, a rapid reduction in root growth is not always advantageous, and all aspects of root morphology, distribution and architecture need to be considered. These results highlight the difficulty in making generalizations at the mixed-sward level about the effects of defoliation on root characteristics. Species and genotype composition in a sward and its interaction with defoliation intensity and frequency and soil fertility need to be considered more fully.

Mycorrhizae

Grass roots form arbuscular mycorrhizae in most natural and agricultural ecosystems (Newman and Reddell, 1987), and consequently the effect of defoliation on the mycorrhizal symbiosis is of concern. Much of the research has been in the field with moderate defoliation treatments and has tended to show no effect (Wallace, 1987; Allen *et al.*, 1989). Other studies have shown a decrease in arbuscular mycorrhizal colonization under heavy grazing (Bethlenfalvay and Dakessian, 1984). In a study of three perennial grasses, of similar morphology, Allsopp (1998) found that in *Digitaria eriantha* and *L. perenne* mycorrhizal infection was reduced by defoliation, but hyphal densities were not. In contrast, in *Themeda triandra*, which is a species intolerant of heavy grazing, defoliation severely affected plant regrowth, P accumulation, arbuscular colonization and external hyphal production. Allsopp (1998) suggested that in some grass species, such as *Digitaria* and *Lolium*, an external hyphal network may help compensate for the loss of root mass due to defoliation.

Root dynamics

The effects of grazing on fine root growth and dynamics have been identified as a crucial link in understanding herbivore interactions within carbon and nutrient cycling processes (Holland and Detling, 1990; Pregitzer *et al.*, 1993; Ruess *et al.*, 1998). Immediately after defoliation, rapid and extensive root death can occur (Evans, 1971; Hodgkinson and Baas Becking, 1977; Allsopp, 1998). However, as it is difficult to extract and distinguish roots that are actively growing from those that are inactive or dead, many studies have ignored this distinction. Also, root biomass can be a poor reflection of below-ground growth, as root production and root mortality occur simultaneously. To complete the picture on the effects of grazing on roots, the dynamics of root growth also has to be considered. In a sward box study by the authors on *A. capillaris*, regular defoliation reduced root biomass density and also increased the percentage of necromass (Table 4.3), agreeing with

Table 4.3. Effect of regular defoliation twice weekly (RC) and uncut (UC) for 3 months on root weight density and percentage necromass in a sward box experiment of *Agrostis capillaris*. Analysis of variance on untransformed data, with $n = 3$.

	Treatment mean		
	RC	UC	$LSD_{0.05}$
Dry weight density ($mg\ cm^{-3}$) in 0–15 cm zone	0.37	1.47	0.58
% necromass in 0–5 cm zone	17.6	8.3	5.1

LSD, least significant difference.

results in the field, where heavy grazing increased the dead root mass on mixed-grass steppes (van der Maarel and Titlyanova, 1989). This increase in proportion of dead root material could act as a resource for decomposer organisms, potentially change microorganism community structure and act as a subsequent source of nutrients for plant growth.

Carbon Flow in the Rhizosphere

Roots release a broad range of organic compounds to the soil, which are utilized as substrates by rhizosphere microorganisms. These organic inputs to soil comprise exudates (primarily sugars, amino acids and organic acids), secretions (including enzymes and metallophores), lysates and sloughed-off cells, collectively termed rhizodeposition. The amount and relative proportions of rhizodeposition components vary dependent on a host of factors, including plant species, developmental stage, climatic conditions and availability of nutrients and water in the soil matrix (Kraffczyk *et al.*, 1984; Meharg and Killham, 1988; Grayston *et al.*, 1996; Paterson *et al.*, 1997). In particular, factors affecting the balance of plant partitioning of resources, such as defoliation and soil mineral nutrient availability, strongly affect the process of rhizodeposition. Therefore defoliation is a factor which affects C input to soil and, consequently, the coupling of plant productivity to microbial activity and soil nutrient cycling. To understand the effects of grazing on microbial communities and consequently on nutrient cycling, it is necessary to determine the quantitative and qualitative impacts of grazing on rhizodeposition.

Partial shoot defoliation has been found to result in increased release of organic compounds from roots, in particular, the low-molecular-weight soluble exudates (Hamlen *et al.*, 1972; Bokhari and Singh, 1974; Dyer and Bokhari, 1976; Holland *et al.*, 1996). Increased rhizodeposition of sugars and amino acids would be expected to stimulate microbial activity in the rhizosphere, and this mecha-

Table 4.4. Effect of defoliation (weekly to 4 cm) and N supply (2 mM (High N) or 0.01 mM (Low N) NH_4NO_3) on root growth and rhizodeposition of *Lolium perenne* and *Festuca rubra* grown for 36 days in percolated axenic sand culture systems. Results are means of six replicates ± standard errors.

		Total root length (m)	Specific root length (m g^{-1})	Cumulative rhizodeposition (% plant net C assimilate)*
High N	*L. perenne* non-defoliated	17.7 ± 3.6	69.3 ± 4.9	0.81 ± 0.15
	L. perenne defoliated	13.7 ± 1.8	77.1 ± 11.4	1.83 ± 0.60
	F. rubra non-defoliated	12.7 ±1.8	108.1 ± 10.8	1.7 ± 0.18
	F. rubra defoliated	9.7 ± 1.3	123.6 ± 8.2	4.1 ± 0.52
Low N	*L. perenne* non-defoliated	11.0 ± 1.2	229.8 ± 11.5	40.1 ± 1.4
	L. perenne defoliated	11.3 ± 1.2	222.4 ± 15.0	44.1 ± 2.5
	F. rubra non-defoliated	8.1 ± 0.7	198.2 ± 22.1	12.2 ± 0.91
	F. rubra defoliated	6.2 ± 0.3	247.5 ± 27.3	18.3 ± 1.27

*Net plant assimilate was calculated as the amount of C in the plant biomass at harvest plus the amount of C collected as rhizodeposition throughout the growth period.

nism has been invoked to explain increased microbial activity in grazed systems (Bardgett *et al.*, 1998). Increased rhizodeposition following defoliation has been suggested to be a consequence of increased partitioning of assimilate below ground, as a means of grazing tolerance (Bardgett *et al.*, 1998). The correlation of increased assimilate allocation to roots with increased rhizodeposition following defoliation has been demonstrated for several species (Holland *et al.*, 1996; Hamlen *et al.*, 1972). However, as stated earlier, it is more common for grasses to have a net export of carbon to support shoot regrowth (Miller and Rose, 1992; Thornton *et al.*, Chapter 5, this volume). Despite this, rhizodeposition from grasses has also been found to increase following defoliation (Bokhari and Singh, 1974). Paterson and Sim (1999) found transient (1–2-day) increases in release of organic compounds from roots of *L. perenne* and *F. rubra* following defoliations at weekly intervals, at both high and low N supply (Table 4.4).

The physiological bases for increased rhizodeposition from grasses following defoliation have yet to be elucidated. Several possible mechanisms merit further investigation: (i) degradation of C and N storage compounds in roots during remobilization may transiently increase the concentrations of sugars and amino acids, promoting diffusive release of these exudates; (ii) defoliation reduces the energy status of roots (Ofosubudu *et al.*, 1995), and this may perturb the reuptake of root-released organic compounds (Mühling *et al.*, 1993), increasing net exudation; and (iii) the physical damage caused by defoliation may initiate an electrical action or slow wave potential propagated through the plant, which is capable of depolarizing root transmembrane potentials (Stahlberg and Cosgrove, 1996; Stankovic *et al.*, 1998). Such depolarization would remove chemipotential gradients (primarily of H^+) required to energize reuptake of rhizodeposits (Mühling *et al.*, 1993).

As discussed in previous sections, mineral nitrogen supply strongly affects assimilate allocation in grasses and also the quantity of rhizodeposition, with release of organic compounds generally increasing with reduced N supply (Bowen, 1969; Hodge *et al.*, 1996). For *L. perenne* and *F. rubra*, low N supply significantly increased the proportion of plant-assimilated C released through rhizodeposition (Table 4.4). The effect of N supply on rhizodeposition was greater for *L. perenne*, with cumulative release of 5.8 ± 0.6 and 16.6 ± 1.2 mg C during the first 36 days of growth of non-defoliated plants at high and low N, respectively. As biomass accumulation was greatly reduced at low N, rhizodeposition as a proportion of the plant C budget increased from 0.81% at high N to 40.1% at low N, for non-defoliated plants (Table 4.4). Low N supply also increased the proportion of plant C released by roots of *F. rubra*, from 1.7 to 12.2% of plant-assimilated C, for high and low N, respectively. However, this effect was less than for *L. perenne*, and for *F. rubra* absolute amounts released from the smaller plants at low N were not significantly greater than those from plants grown at high N. For both species, an increased release of plant C as rhizodeposition at low N supply was concurrent with an increased specific root length, associated with optimization of biomass investment for nutrient capture. Increased root length relative to root mass has been found previously to increase the proportion of plant C that is released through rhizodeposition (Xu and Juma, 1994). Increased rhizodeposition from the finer roots produced under nutrient limitation may provide a means by which N cycling and returns to the plant are increased via the microbial loop (Clarholm, 1985).

Grazing Effects on Soil Microorganisms

As grazing systems become more extensive, plant growth and competition become more dependent on the use of available soil resources. Interactions between roots and associated microorganisms, mediated by rhizodeposition, will be central to our understanding of fluxes to and from plant-available carbon pools. As carbon is the main driving force for microbial activity in the soil, variations in the quantity and quality of root and shoot material and plant root exudates may result in the selection of microorganisms specific to the rhizosphere of individual plant species (Grayston *et al.*, 1996). Impacts on microbial community size, activity and diversity would affect nutrient availability and thus plant species competition.

In upland grassland, the size of the soil microbial community is commonly found to be greater under low-fertility than under high-fertility conditions (Bardgett *et al.*, 1993, 1996, 1997; Grayston *et al.*, 2000), closely related to trends in root biomass observed in the field. Using the National Vegetation Classification (NVC), gradients of grassland types were chosen in ten different biogeographical areas of the UK. These ranged from improved, intensively managed *Lolium–Cynosurus* grassland (NVC – MG6), through semi-improved *Festuca–Agrostis–Galium* grasslands, *Holcus–Trifolium* subcommunity (NVC – U4b), to unimproved, extensive *Festuca–Agrostis–Galium* grassland (NVC – U4a)

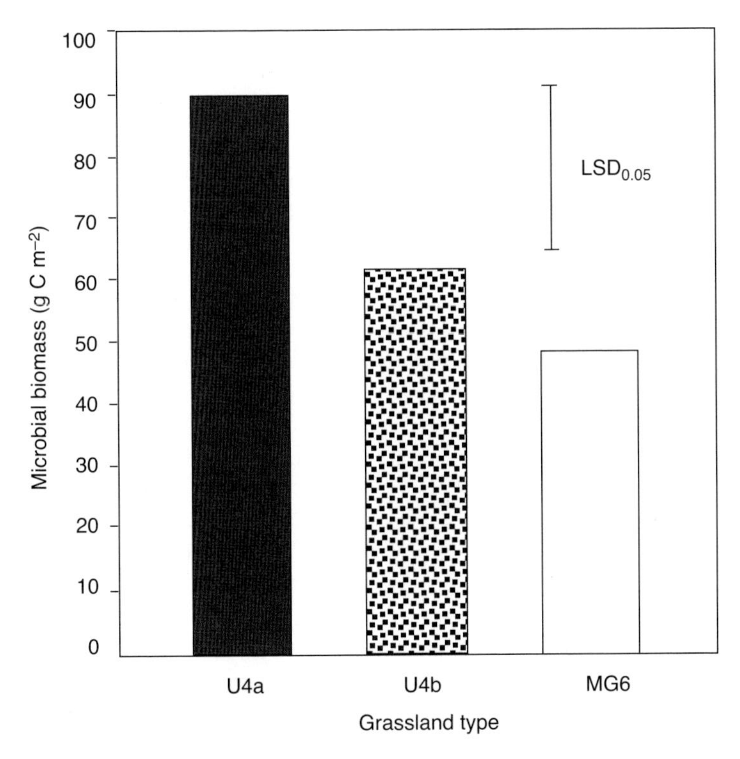

Fig. 4.5. Microbial biomass measured using fumigation technique in unimproved (U4a), semi-improved (U4b) and improved (MG6) grasslands, across ten sites in the UK. LSD, least significant difference.

(Grayston *et al.*, 2000). Soil microbial biomass increased as the soil fertility decreased, moving from improved to unimproved grasslands (Fig. 4.5). This was accompanied by a shift in microbial community structure, with an increase in the proportion of fungi to bacteria in the unimproved grasslands (Grayston *et al.*, 2000). These changes in microbial community structure under varying fertility levels have been recorded in other grassland studies (Bardgett *et al.*, 1993, 1996, 1998). Quantitative and qualitative differences in rhizodeposition, due to variation in the plant species, growth rate, root biomass, root architecture and variation in litter and root turnover, are all likely to have had an impact on the microbial communities in these grasslands (Grayston *et al.*, 1998a). For example, clover, which is found in improved grasslands, has a greater exudation rate than grasses (Martin, 1971). The carbon released as exudates in improved grasslands may be readily decomposable substrates, part of the 'fast pool' of carbon, resulting in preferential stimulation of bacterial growth. In contrast, unimproved grasslands contain more recalcitrant compounds, which are the preferred substrates of fungi (Grayston *et al.*, 2000). The use of community-level physiological profiles (CLPP) to characterize microbial communities based on their metabolic profiles (Grayston

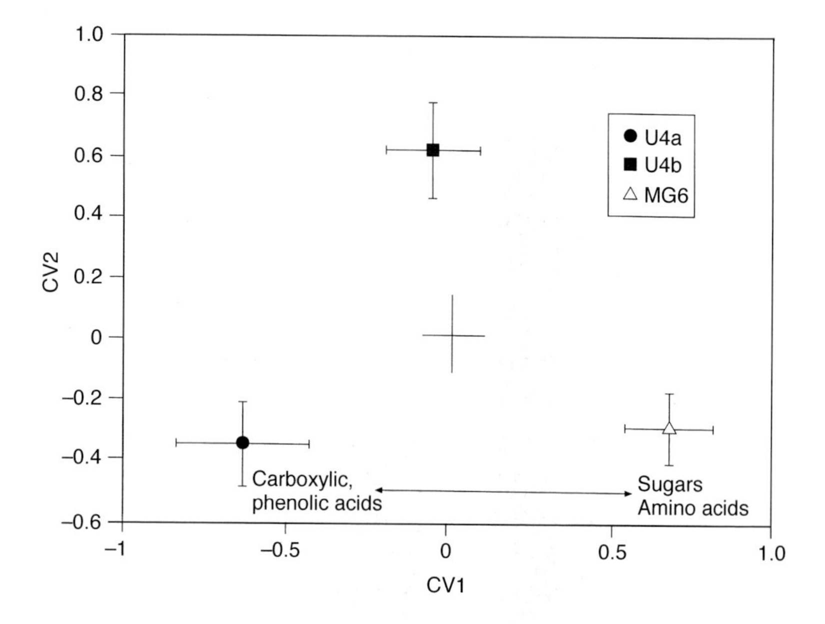

Fig. 4.6. Canonical variate analysis of metabolic profiles of microbial communities from unimproved (U4a), semi-improved (U4b) and improved (MG6) grasslands (bars are SE).

and Campbell, 1996; Grayston *et al.*, 1998a) and as a possible indicator of carbon source availability in the rhizosphere (Grayston *et al.*, 1998b) supports the above hypotheses. Microbial communities from improved grasslands have greater utilization of sugars and amino acids, whereas those from unimproved grasslands have greater utilization of phenolic and carboxylic acids (Fig. 4.6). This may reflect the greater exudation and growth rate of plant species in improved grasslands and the higher organic matter (phenolic acid) content of unimproved grassland (Grayston *et al.*, 2000).

The root turnover rates of contrasting species found in pastures of different fertility and productivity could also help us to understand carbon and nutrient cycling in grassland ecosystems (Caldwell, 1979). In a study by Schläpfer and Ryser (1996), the root turnover of the faster-growing species (*A. elatius*) was faster than that of the slow-growing species characteristic of nutrient-poor grasslands (*Bromus erectus*). Associated with a faster turnover rate would be a greater input of decomposing roots into the soil, thus providing more substrate for bacterial populations to flourish in. This has been suggested as being part of an efficient nutrient conservation strategy in slow-growing species, which helps to explain the long-term success of slow-growing species at low nutrient supply (Grime, 1977; Berendse and Aerts, 1984). It has also been suggested that fast-growing species in general contain less lignin and hemicellulose and more organic N compounds than slow-growing species (Poorter and Bergkotte, 1992; van Arendonk and Poorter, 1994), thereby providing a faster litter decomposition rate in the faster-

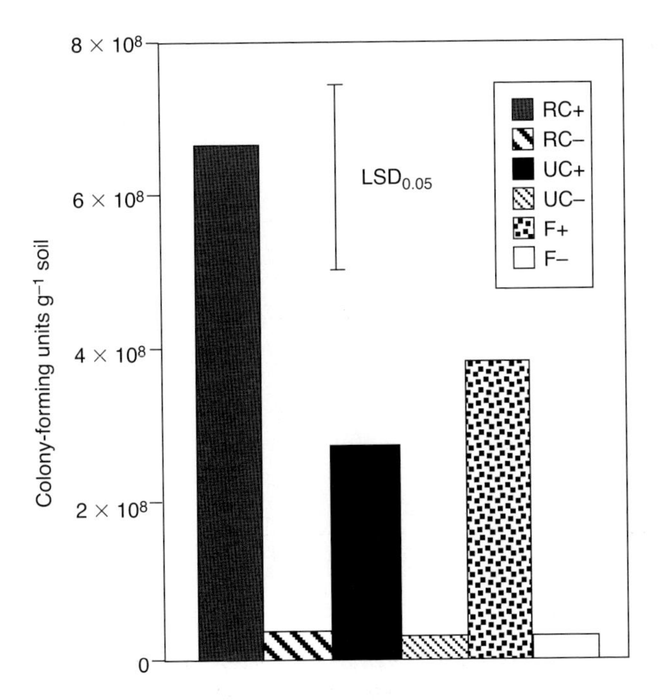

Fig. 4.7. Populations of bacteria (colony-forming units) in the rhizosphere of *Agrostis capillaris* subject to twice-weekly defoliation (RC), or left uncut (UC) and in fallow (F) soil, and in the presence (+) or absence (−) of simulated urine addition, 7 days after urine application. LSD, least significant difference.

growing species (Pastor *et al.*, 1987; Enriquez *et al.*, 1993). The quality of the plant litter, including leaf material, and its relation to the soil microbiota are another important consideration when trying to understand the many interactions observed in the field, as has been discussed in Bardgett *et al.* (1998).

Changes in microbial community structure can also be related to input directly from the grazing animal, which may also account for the differences in communities between grassland types. As improved grasslands typically support more grazing animals than unimproved grasslands, grazing will increase soil nutrient availability, due to increased direct inputs from animal excreta, which will favour bacterial growth. The equivalent of 510 kg N ha^{-1} can be deposited within a single urine patch by sheep (Thomas *et al.*, 1988). This can increase soil heterogeneity and create patches with higher levels of soil and plant N than surrounding areas (Ledgard *et al.*, 1982), increase rates of nutrient cycling (Floate, 1981) and change species composition (Marriott *et al.*, 1987). Scorching is common in urine patches (Floate, 1981) and has been attributed to root death from exposure to NH_3 (Richards and Wolton, 1975). In the experiment described in Table 4.3, a single application of artificial urine accelerated root death in the undefoliated treatment, with 50% of roots disappearing within 8 days of urine deposition, compared with 13 days without ($P < 0.05$). In the 0–15 cm zone of the soil, 15.5% of the roots were dead where

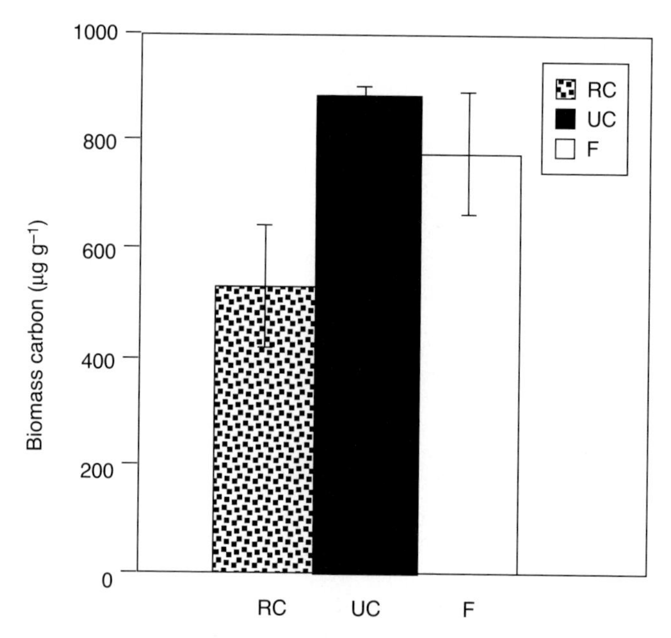

Fig. 4.8. Microbial biomass measured using fumigation technique in the rhizosphere of *Agrostis capillaris* subject to either regular defoliation twice weekly (RC), or left undefoliated (UC) and in fallow (F) soil 2 months after the start of defoliation (bars are SE).

urine had been applied, in contrast to 6.8% in the no-urine treatment ($P < 0.05$). The application of urine stimulated growth of culturable soil bacterial communities (Fig. 4.7), which could be related to this accelerated root death or to a rapid rise in soil pH after urine application, which favours bacterial growth (B.L. Williams, 1999, unpublished results). The application of artificial urine significantly decreased total fungal communities ($P < 0.05$) but stimulated soil bacterial communities. In this same study, regular defoliation decreased the total soil microbial biomass (Fig. 4.8), which was positively correlated with root biomass (see Table 4.3), while defoliation significantly increased the bacterial population only in the presence of urine (see Fig. 4.7). These results illustrate both the direct and the indirect effects of urine and defoliation as individual factors in shifting the community structure of the microbiota. The size of the microbial biomass appears to be related to resource availability, i.e. root biomass, whereas the microbial structure appears to be related to the quality of the resource, i.e. exudate quality and inputs from the grazing animal.

Conclusions

Pasture systems have diverse and dynamic plant communities, consisting of many species, which can respond in contrasting ways to defoliation and urine deposition. To be able to understand and predict change in these communities through time, species selectivity by the grazing animal and species-specific responses, both above and below ground, need to be considered together in multidisciplinary projects. Grazing animals not only shape the above-ground biomass that they feed on but also both directly and indirectly affect the roots and soil microbial communities below ground through factors such as defoliation and urine deposition. Information on the many interactions between plants and their associated microbiota will be essential to our understanding of the grazed ecosystem and how it will respond to change in management inputs. This chapter stresses the importance of considering responses at the whole-plant level in terms of the nutrient and carbon dynamics of the root system, of rhizodeposition and of their relationships with the associated soil microbiota. It also highlights the need for a multidisciplinary approach to the understanding of such a multifaceted system. It will be only through such an approach that a true understanding of the effects of manipulating animal densities on the hills and uplands can be made.

Acknowledgements

Scottish Executive Rural Affairs Department is acknowledged for funding this research, as are members of the Plant Science Group at the Macaulay Land Use Research Institute (MLURI), Aberdeen for carrying out some of the research reported here.

References

Allen, M.F., Richards, J.H. and Busso, C.A. (1989) Influence of clipping and soil water status on vesicular-arbuscular mycorrhizae of two semi-arid tussock grasses. *Biology and Fertility of Soils* 8, 285–289.

Allsopp, N. (1998) Effect of defoliation on the arbuscular mycorrhizas of three perennial pasture and rangeland grasses. *Plant and Soil* 202, 117–124.

Arredondo, J.T. and Johnson, D.A. (1998) Clipping effects on root architecture and morphology of 3 range grasses. *Journal of Range Management* 51, 207–213.

Arredondo, J.T. and Johnson, D.A. (1999) Root architecture and biomass allocation of three range grasses in response to non-uniform supply of nutrients and shoot defoliation. *New Phytologist*, 143, 373–385.

Bakker, M.L., Gordon, I.J. and Milne, J.A. (1998) Effects of sward structure on the diet selected by guanacos (*Lama guanicoe*) and sheep (*Ovis aries*) grazing a perennial ryegrass-dominated sward. *Grass and Forage Science* 53, 19–30.

Bardgett, R.D., Frankland, J.C. and Whittaker, J.B. (1993) The effects of agricultural practices on the soil biota of some upland grasslands. *Agriculture, Ecosystems and Environment* 45, 25–45.

Bardgett, R.D., Hobbs, P.J. and Frostegard, A. (1996) Changes in fungal : bacterial biomass ratios following reductions in the intensity of management on an upland grassland. *Biology and Fertility of Soils* 22, 261–264.

Bardgett, R.D., Leemans, D.K., Cook, R. and Hobbs, P.J. (1997) Seasonality of the soil biota of grazed and ungrazed hill grasslands. *Soil Biology and Biochemistry* 29, 1285–1294.

Bardgett, R.D., Wardle, D.A. and Yeates, G.W. (1998) Linking above-ground and below-ground interactions: how plant responses to foliar herbivory influence soil organisms. *Soil Biology and Biochemistry* 30, 1867–1878.
Berendse, F. and Aerts, R. (1984) Competition between *Erica tetralix* L. and *Molinia caerulea* (L.) Moench as affected by the availability of nutrients. *Acta Oecologia* 5, 3–14.
Berendse, F., Elberse, W.T. and Geerts, R.H.M.E. (1992) Competition and nitrogen loss from plants in grassland ecosystems. *Ecology* 73, 46–53.
Berntson, G.M. and Woodward, F.I. (1992) The root system architecture and development of *Senecio vulgaris* in elevated CO_2 and drought. *Functional Ecology* 6, 324–333.
Bethlenfalvay, G.J. and Dakessian, S. (1984) Grazing effects on mycorrhizal colonisation and floristic composition of vegetation on a semi-arid range in northern Nevada. *Journal of Range Management* 37, 312–316.
Bokhari, U.G. and Singh, J.S. (1974) Effects of temperature and clipping on growth, carbohydrate reserves and root exudation of western couch grass in hydroponic culture. *Crop Science* 14, 790–794.
Bowen, G.D. (1969) Nutrient status effects on loss of amides and amino acids from pine roots. *Plant and Soil* 30, 139–141.
Brouwer, R. (1983) Functional equilibrium: sense or nonsense? *Netherlands Journal of Agricultural Science* 10, 361–376.
Bullock, J.M. (1996) Plant competition and population dynamics. In: Hodgson, J. and Illius, A.W. (eds) *The Ecology and Management of Grazing Systems*. CAB International, Wallingford, UK, pp. 37–67.
Caldwell, M.M. (1979) Root structure: the considerable cost of below-ground function. In: Solbrig, O.T., Jain, S., Johnson, G.B. and Raven, P.H. (eds) *Topics in Plant Population Biology*, Columbia University Press, New York, pp. 408–427.
Chaieb, M., Hendri, B. and Boukhris, M. (1996) Impact of clipping on root systems of 3 grass species in Tunisia. *Journal of Range Management* 49, 336–339.
Chapin, F.S. (1980) The mineral nutrition of wild plants. *Annual Review Ecology and Systematics* 11, 233–260.
Chapin, F.S. and McNaughton, S.J. (1989) Lack of compensatory growth under phosphorus deficiency in grazing-adapted grasses from the Serengeti Plains. *Oecologia* 79, 551–557.
Chapin, F.S. and Slack, M. (1979) Effect of defoliation upon root growth, phosphate absorption and respiration in nutrient-limited tundra graminoids. *Oecologia* 42, 67–79.
Clarholm, M. (1985) Interactions of bacteria, protozoa, and plants leading to mineralisation of soil nitrogen. *Soil Biology and Biochemistry* 17, 181–187.
Commission, European Communities (1992) *Europe in Figures*, 3rd edn. Office for the Official Publication of the European Communities, Luxemburg.
Curll, M.L. and Wilkins, R.J. (1982) Frequency and severity of defoliation of grass and clover by sheep at different stocking rates. *Grass and Forage Science* 37, 291–297.
Curry, J.P. (1994) *Grassland Invertebrates, Ecology, Influence on Soil Fertility and Effects on Plant Growth*. Chapman and Hall, London, 437 pp.
Danckwerts. J.E. and Gordon. A.J. (1987) Long-term partitioning, storage and remobilization of ^{14}C assimilated by *Lolium perenne* (cv. Melle). *Annals of Botany* 59, 55–66.
Deinum, B. (1985) Root mass of grass swards in different grazing systems. *Netherlands Journal of Agricultural Science* 33, 377–384.
Dyer, M.I. and Bokhari, U.G. (1976) Plant–animal interactions: studies of the effects of grasshopper grazing on blue gamma grass. *Ecology* 57, 762–772.

Dyer, M.I., Acra, M.A., Wang, G.M., Coleman, D.C., Freckman, D.W., McNaughton, S.J. and Strain B.R. (1991) Source–sink carbon relations in two *Panicum coloratum* ecotypes in response to herbivory. *Ecology* 72, 1472–1483.

Elberse, W.T. and Berendse, F. (1993) A comparative study of the growth and morphology of eight grass species from habitats with different nutrient availabilities. *Functional Ecology* 7, 223–229.

Ennik, G.C. and Hofman, T.B. (1983) Variation in the root mass of ryegrass types and its ecological consequences. *Netherlands Journal of Agricultural Science* 31, 325–334.

Enriquez, S., Duarte, C.M. and Sand-Jensen, K. (1993) Patterns in decomposition rates among photosynthetic organisms: the importance of detritus C : N : P content. *Oecologia* 94, 457–471.

Evans, P.S. (1971) Root growth of *Lolium perenne* L. II. Effects of defoliation and shading. *New Zealand Journal of Agricultural Research* 14, 552–562.

Fitter, A.H. (1985) Functional significance of root morphology and root system architecture. In: Fitter, A.H., Read, D.J., Atkinson, D. and Usher, M.B. (eds) *Ecological Interactions in Soil.* Blackwell Scientific Publications, Oxford, pp. 87–106.

Fitter, A.H. (1986) The topology and geometry of plant root systems: influence of watering rate on root system topology in *Trifolium pratense*. *Annals of Botany* 58, 91–101.

Fitter, A.H. (1987) An architectural approach to the comparative ecology of plant root systems. *New Phytologist* 106, 61–77.

Floate, M.J. (1981) Effects of grazing by large herbivores on nitrogen cycling in agricultural ecosystems. In: Clark, F.E. and Rosswall, T. (eds) *Terrestrial Nitrogen Cycle Processes, Ecosystem Strategies and Management Impacts*. Ecological Bulletin 33. Swedish Natural Science Research Council, Stockholm, pp. 585–601.

Fraser, M.D. and Gordon, I.J. (1997) The diet of goats, red deer (*Cervus elaphus*) and South American camelids feeding on three contrasting Scottish upland vegetation communities. *Journal of Applied Ecology* 32, 668–686.

Grant, S.A., Barthram, G.T., Torvell, L., King, J. and Smith, H.K. (1983) Sward management, lamina turnover and tiller population density in continuously stocked *Lolium perenne* dominated swards. *Grass and Forage Science* 38, 333–344.

Grayston, S.J. and Campbell, C.D. (1996) Functional biodiversity of microbial communities in the rhizospheres of hybrid larch (*Larix eurolepis*) and Sitka spruce (*Picea sitchensis*). *Tree Physiology* 16, 1031–1038.

Grayston, S.J., Vaughan, D. and Jones, D. (1996) Rhizosphere carbon flow in trees, in comparison with annual plants: the importance of root exudation and its impact on microbial activity and nutrient availability. *Applied Soil Ecology* 5, 29–56.

Grayston, S.J., Wang, S., Campbell, C.D. and Edwards, A.C. (1998a) Selective influence of plant species and microbial diversity in the rhizosphere. *Soil Biology and Biochemistry* 30, 369–378.

Grayston, S.J., Campbell, C.D., Lutze, J.L. and Gifford, R.M. (1998b) Impact of elevated CO_2 on the metabolic diversity of microbial communities in N-limited grass swards. *Plant and Soil* 203, 289–300.

Grayston, S.J., Griffith, G.S., Mawdsley, J.L., Campbell, C.D. and Bardgett, R.D. (2000) Accounting for variability in soil microbial communities of temperate upland grassland ecosystems. *Soil Biology and Biochemistry* (in press).

Grime, J.P. (1977) Evidence for the existence of three primary strategies in plants and its relevance to ecological and evolutionary theory. *America Naturalist* 111, 1169–1194.

Grime, J.P., Hodgson, J.G. and Hunt, R. (1988) *Comparative Plant Ecology: a Functional Approach to Common British Species.* Unwin Hyman, London, 742 pp.

Hamlen, R.A., Lukezic, F.L. and Bloom, J.R. (1972) Influence of clipping height on the neutral carbohydrate levels of root exudates of lucerne plants grown under gnotobiotic conditions. *Canadian Journal of Plant Science* 52, 643–649.

Harris, W. (1973) Ryegrass genotype–environment interactions in response to density, cutting height, and competition with white clover. *New Zealand Journal of Agricultural Research* 16, 207–222.

Hodge, A., Grayston, S.J. and Ord, B.G. (1996) A novel method for characterization and quantification of plant root exudates. *Plant and Soil* 184, 97–104.

Hodgkinson, K.C. (1969) The utilisation of root organic compounds during the regeneration of lucerne. *Australian Journal of Biological Science* 22, 1113.

Hodgkinson, K.C. and Baas Becking, H.G. (1977) Effect of defoliation on root growth of some arid zone perennial plants. *Australian Journal of Agricultural Research* 29, 31–42.

Holland, E.A. and Detling, J.K. (1990) Plant response to herbivory and below-ground nitrogen cycling. *Ecology* 71, 1040–1049.

Holland, J.N., Cheng, W. and Crossley, D.A., Jr (1996) Herbivore-induced changes in plant carbon allocation: assessment of below-ground C fluxes using carbon-14. *Oecologia*, 107, 87–94.

Hunt, H.W., Cleman, D.C., Ingam, E.R., Ingham, R.E., Elliott, E.T., Moore, J.C., Rose, S.L., Reid, C.P.P. and Morley, C.R. (1987) The detrital food web in a shortgrass prairie. *Biology and Fertility of Soils* 3, 57–68.

Ingham, R.E. and Detling, J.K. (1986) Effects of defoliation and nematode consumption on growth and leaf gas exchange in *Bouteloua curtipendula*. *Oikos* 46, 23–28.

Ingham, R.E., Trofymow, J.A., Ames, R.N., Hunt, H.W., Morley, C.R., Moore, J.C. and Coleman, D.C. (1986) Trophic interactions and nitrogen cycling in a semi-arid grassland soil. *Journal of Applied Ecology* 23, 577–614.

Jarvis, S.C. and Macduff, J.H. (1989) Nitrate nutrition of grasses from steady-state supplies in flowing solution culture following nitrate deprivation and/or defoliation. *Journal of Experimental Botany* 40, 965–975.

Jones, E.L. (1983) The production and persistency of different grass species cut at different heights. *Grass and Forage Science* 38, 79–87.

Karl, M.G. and Doescher, P.S. (1991) Monitoring roots of grazed rangeland vegetation with the root periscope/minirhizotron technique. *Journal of Range Management* 44, 296–298.

Kraffczyk, I., Trollendier, G. and Beringer, H. (1984) Soluble root exudates of maize: influence of potassium supply and rhizosphere microorganisms. *Soil Biology and Biochemistry* 16, 315–322.

Ledgard, S.J., Steele, K.W. and Saunders, W.M.H. (1982) Effects of cow urine and its major constituents on pasture properties. *New Zealand Journal of Agricultural Research* 25, 61–68.

Macklon, A.E.S., Mackie-Dawson, L.A., Sim, A., Shand, C.A. and Lilly, A. (1994) Soil P resources, plant growth and rooting characteristics in nutrient poor upland grasslands. *Plant and Soil* 163, 257–266.

McNaughton, S.J. (1979) Grazing as an optimisation process: grass–ungulate relationships in the Serengeti. *American Naturalist* 113, 691–703.

McNaughton, S.J., Wallace, L.L. and Coughenour, M.B. (1983) Plant adaptation in an ecosystem context: effects of defoliation, nitrogen and water on growth of an African C_4 sedge. *Ecology* 64, 307–318.

McNaughton, S.J., Banyikwa, F.F. and McNaughton, M.M. (1998) Root biomass and productivity in a grazing ecosystem: the Serengeti. *Ecology* 79, 587–592.

Marriott, C.A., Smith, M.A. and Baird, M.A. (1987) The effect of sheep urine on clover performance in a grazed upland sward. *Journal of Agricultural Science* 109, 177–185.
Martin, J.K. (1971) ^{14}C-labelled material leached from the rhizosphere of plants supplied with $^{14}CO_2$. *Australian Journal of Biological Science* 24, 1131–1142.
Matthew, C., Xia, J.X., Chu, A.C.P., MacKay, A.D. and Hodgson, J. (1991) Relationship between root production and tiller appearance rates in perennial ryegrass (*Lolium perenne* L.). In: Atkinson, D. (ed.) *Plant Root Growth: An Ecological Perspective*. Special Publication Series of the British Ecological Society no. 10, pp. 281–290.
Mawdsley, J.L. and Bardgett, R.D. (1997) Continuous defoliation of perennial ryegrass (*Lolium perenne*) and white clover (*Trifolium repens*) and associated changes in the microbial population of an upland grassland soil. *Biology and Fertility of Soils* 24, 52–58.
Meharg, A.A. and Killham, K. (1988) Factors affecting the pattern of carbon flow in a grass soil system. *Journal of the Science of Food and Agriculture* 45, 135–136.
Milchunas, D.G. and Lauenroth, W.K. (1993) Quantitative effects of grazing on vegetation and soils over a global range of environments. *Ecological Monographs* 63, 327–366.
Milchunas, D.G., Sala, O.E. and Lauenroth, W.K. (1988) A generalised model of the effects of grazing by large herbivores and grassland community structure. *American Naturalist* 132, 87–106.
Miller, R.F. and Rose, J.A. (1992) Growth and carbon allocation of *Agropyron desertorum* following autumn defoliation. *Oecologia* 89, 482–486.
Mühling, K.H., Schubert, S. and Mengel, K. (1993) Mechanism of sugar retention by roots of intact maize and field bean plants. *Plant and Soil* 155/156, 99–102.
Newman, E.I. and Reddell, P. (1987) The distribution of mycorrhizae among families of vascular plants. *New Phytologist* 106, 745–751.
Oesterheld, M. and McNaughton, S.J. (1988) Intraspecific variation in the response of *Themeda triandra* to defoliation: the effect of time of recovery and growth rates on compensatory growth. *Oecologia* 77, 181–186.
Ofosubudu, K.G., Saneoka, H. and Fujita, K. (1995) Factors controlling the release of nitrogenous compounds from roots of soybean. *Soil Science and Plant Nutrition* 41, 625–633.
Pastor, J., Stillwell, M.A. and Tilman, D. (1987) Little bluestem litter dynamics in Minnesota old fields. *Oecologia* 72, 327–330.
Paterson, E. and Sim, A. (1999) Rhizodeposition and C-partitioning of *Lolium perenne* in axenic culture affected by nitrogen supply and defoliation. *Plant and Soil* 216, 155–164.
Paterson, E., Hall, J.M., Rattray, E.A.S., Griffiths, B.S., Ritz, K. and Killham, K. (1997) Effect of elevated CO_2 on rhizosphere carbon flow and soil microbial processes. *Global Change Biology* 3, 363–377.
Pellet, D.M., Grunes, D.L. and Kochian, L.V. (1995) Organic acid exudation as an aluminium-tolerance mechanism in maize (*Zea mays* L.). *Planta* 196, 788–795.
Polly, H.W. and Detling, J.K. (1989) Defoliation, nitrogen and competition effects on plant growth and nitrogen nutrition. *Ecology* 70, 721–727.
Poorter, H. and Bergkotte, M. (1992) Chemical composition of 24 wild species differing in relative growth rate. *Plant, Cell and Environment* 15, 221–229.
Pregitzer, K.S., Hendrick, R.L. and Fogel, R. (1993) The demography of fine roots in response to patches of water and nitrogen. *New Phytologist* 125, 575–580.
Rengel, Z. (1997) Root exudation and microflora populations in rhizosphere of crop genotypes differing in tolerance to micronutrient deficiency. *Plant and Soil* 196, 255–260.
Richards, I.R. and Wolton, K.M. (1975) A note on urine scorch caused by grazing animals. *Journal of British Grassland Society* 30, 187–188.

Richards, J.H. (1984) Root growth response to defoliation in two *Agropyron* bunchgrasses: field observations with an improved root periscope. *Oecologia* 64, 21–25.
Richards, J.H. (1993) Physiology of plants recovering from defoliation. In: *Proceedings of the XVIIth International Grassland Congress*. SIR Publishing, Wellington, New Zealand, pp. 85–94.
Rosenthal, J.P. and Kotanen, P.M. (1994) Terrestrial plant tolerance to herbivory. *Tree* 9, 145–148.
Ruess, R.W. (1988) The interaction of defoliation and nutrient uptake in *Sporobolus kentrophyllus*, a short-grass species from the Serengeti Plains. *Oecologia* 77, 550–556.
Ruess, R.W. and Seagle, S.W. (1994) Landscape patterns in soil microbial processes in the Serengeti National Park, Tanzania. *Ecology* 75, 892–904.
Ruess, R.W., Hendrick, R.D. and Bryant, J.P. (1998) Regulation of fine root dynamics by mammalian browsers in early successional Alaskan Taiga forests. *Ecology* 79, 2706–2720.
Seagle, S.W., McNaughton, S.J. and Ruess, R.W. (1992) Simulated effects of grazing on soil nitrogen and mineralisation in contrasting Serengeti grasslands. *Ecology* 73, 1105–1123.
Schläpfer, B. and Ryser, P. (1996) Leaf and root turnover of three ecologically contrasting grass species in relation to their performance along a productivity gradient. *Oikos* 75, 398–406.
Stahlberg, R. and Cosgrove, D.J. (1996) Induction and ionic basis of slow wave potentials in seedlings of *Pisum sativum* L. *Planta* 200, 416–425.
Stankovic, B., Witters, D.L., Zawadzki, T. and Davies, E. (1998) Action of potentials and variation potentials in sunflower: an analysis of their relationships and distinguishing characteristics. *Physiologia Plantarum* 103, 51–58.
Thomas, R.J., Logan, K.A.B., Ironside, A.D. and Bolton, G.R. (1988) Transformations and fate of sheep urine-N applied to an upland UK pasture at different times during the growing season. *Plant and Soil* 107, 173–181.
Tilman, D. (1988) *Plant Strategies and the Dynamics and Structure of Plant Communities*. Princeton University Press, Princeton, New Jersey, 360 pp.
van Arendonk, J.J.C.M. and Poorter, H. (1994) The chemical composition and anatomical structure of leaves and grass species differing in relative growth rate. *Plant, Cell and Environment* 17, 963–970.
van der Maarel, E. and Titlyanova, A. (1989) Above-ground and below-ground biomass relations in steppes under different grazing conditions. *Oikos* 56, 364–370.
van Veen, J.A., Merckx, R. and van de Geijn, S.C. (1989) Plant and soil related controls of the flow of carbon from roots through the soil microbial biomass. *Plant and Soil* 115, 179–188.
Wallace, L.L. (1987) Effects of clipping and soil compaction on growth, morphology and mycorrhizal colonisation of *Schizachyrium scoparium*, a C4 bunchgrass. *Oecologia* 72, 423–428.
Wilsey, B.J. (1996) Urea additions and defoliation affect plant responses to elevated CO_2 in a C_3 grass from Yellowstone National Park. *Oecologia* 108, 321–327.
Xu, J.G. and Juma, N.G. (1994) Relations of shoot C, root C and root length with root-released C of two barley cultivars and the decomposition of root-released C in soil. *Canadian Journal of Soil Science* 74, 17–22.

5

Reserve Formation and Recycling of Carbon and Nitrogen during Regrowth of Defoliated Plants

B. Thornton, P. Millard and U. Bausenwein

Plant Science Group, Macaulay Land Use Research Institute, Craigiebuckler, Aberdeen, UK

Introduction

Most herbaceous plants store carbon and nitrogen. There are several possible ecological advantages that the ability to store resources confers upon a species, which can be particularly important when considering plant competition and vegetation dynamics. These advantages include: (i) allowing growth to occur when either the availability of the external nitrogen is low (e.g. in the spring for growth of *Molinia caerulea* (Thornton and Millard, 1993)) or a short growing season means that uptake of soil nitrogen alone is not sufficient for growth (Jaeger and Monson, 1992); (ii) supporting reproduction by recycling resources from vegetative to reproductive growth – for example, in species exhibiting monocarpic senescence (Millard, 1988) and particularly in biennials (Heilmeier *et al.*, 1986); and (iii) enabling more rapid recovery from catastrophic events, such as defoliation (e.g. Thornton *et al.*, 1993a). It is this last role for storage that is considered in this review. Before we consider the quantitative significance of storage of both carbon and nitrogen in grasses in relation to defoliation, we shall first examine briefly the physiological strategies for storage used by herbaceous species. Experiments to quantify the storage and remobilization of both carbon and nitrogen will then be discussed and their limitations in determining the ecological significance of these processes in relation to vegetation dynamics highlighted. In order to understand competitive interactions between defoliated plants there is a need for field experiments that quantify storage and remobilization. Recent research is discussed which has developed a range of techniques that might allow such field experimentation in the future.

 Grassland Ecophysiology and Grazing Ecology (eds G. Lemaire, J. Hodgson, A. de Moraes, C. Nabinger and P.C. de F. Carvalho)

Carbon and Nitrogen Storage

Several definitions of storage occur in the literature (e.g. Millard, 1996), and a generally accepted definition is that storage can occur if nitrogen or carbon can be 'remobilized from one tissue and subsequently used for the growth or maintenance of another'. Two kinds of storage can be distinguished (Lemaire and Millard, 1999), reserve formation and recycling. Reserve formation involves nitrogen or carbon deposition in discrete storage organelles, such as the vacuole (Raven, 1987) or amyloplast (Zamski, 1996). Alternatively, reserve formation of nitrogen can utilize discrete protein bodies, as often found in the seeds of many species (e.g. Higgins, 1984). Production of vegetative storage proteins (VSPs) is another example of reserve formation. Cyr and Bewley (1990) defined VSPs as proteins which: (i) are synthesized preferentially during development of storage organs; (ii) are depleted during plant growth; and (iii) are more abundant than other protein in the storage tissues. VSPs are found in many herbaceous species (Volenec *et al.*, 1996) and have been reported in the tap roots of forage species, such as *Medicago sativa* (Avice *et al.*, 1996). Starch deposition in tap roots is also an example of carbon storage by reserve formation (Avice *et al.*, 1997). Allocation of resources to reserve formation often requires active transport across a membrane and will only occur when the availability of carbon and nitrogen in the plant exceeds the requirements for growth and maintenance (Lemaire and Millard, 1999). Reserve formation tends only to occur when the growth rate of the plant is low, either, for example, in the autumn or when growth is constrained by the availability of another resource (e.g. water or other nutrients). Under these conditions reserve formation can sequester substrates that might otherwise cause down-regulation of acquisition and assimilation. Storage by reserve formation tends to be very seasonally dependent and is often used to augment supply of further resources in addition to uptake capacity during periods of rapid growth (e.g. in the spring) or, in the case of some forage species, following defoliation (Avice *et al.*, 1996).

The other form of storage, recycling, is often more dynamic than reserve formation. Recycling of carbon and particularly nitrogen usually involves pools that are metabolically active. An example of nitrogen storage by recycling is the role of Rubisco as a storage protein (Millard, 1988). Rubisco activity is under complex regulation, with active and inactive pools (e.g. Parry *et al.*, 1997). However, the role of Rubisco as a storage protein is probably a consequence of its abundance and turnover characteristics (Millard, 1988). During leaf senescence, there is no evidence for a preferential loss of Rubisco protein, with autolysis of chloroplasts causing a parallel decline in the amounts of several other Calvin cycle enzymes (Crafts-Brandner *et al.*, 1990). Rubisco clearly, therefore, does not meet the criteria specified by Cyr and Bewley (1990) for VSPs. Recycling of carbon and nitrogen usually occurs as a consequence of tissue development or turnover and provides a dynamic form of storage, which can make resources available in the short term. As such, recycling is often the main source of nitrogen used by grasses for regrowth after defoliation. Remobilization of nitrogen following defoliation often involves tissue senescence, particularly root turnover (e.g. Ourry *et al.*, 1988). Despite several attempts to identify the turnover of VSPs following defoliation of grasses, there is only one report of putative VSPs in *Lolium perenne* (Louahlia *et al.*, 1999), which even then only contributed a small proportion of the total N remobilized. It is likely, therefore, that nitrogen stored by grasses and remobilized following defoliation comes from recycling, due to protein turnover associated with tissue senescence.

Quantification of Carbon and Nitrogen Remobilization by Defoliated Grasses

Carbon remobilization

There have been many studies that have correlated carbohydrate status with the amount of shoot growth achieved following defoliation. For example, Davies (1965) exposed *L. perenne* to various treatments designed to result in plants containing different amounts of carbohydrates with minimum differences in plant morphology at the time of defoliation, and concluded that the yield of *L. perenne* leaf material after cutting was, at least in part, associated with the level of carbohydrate present in the stubble at the time of cutting. Fulkerson and Slack (1994) also observed a positive correlation between the water-soluble carbohydrate content of the stubble of *L. perenne* and the amount of leaf growth achieved in 6 days following defoliation, regrowth being more closely related to the content of water-soluble carbohydrate in stubble, rather than carbohydrate concentration. Comparing the growth of *L. perenne* after defoliation, achieved either with 6 days in the light or 7 days in darkness, Donaghy and Fulkerson (1997) estimated that remobilization of carbohydrate stores contributed 33% and current assimilation 66% to the regrowth.

The severity of defoliation will affect the relative use of current assimilation and mobilization of stores to supply regrowing leaves with carbon. Ryle and Powell (1975) commented upon the fact that, if in the extreme case all leaf (photosynthetic) material is removed by defoliation, it is inevitable that plants must initially use mobilized stores for regrowth. Relationships between stored carbohydrate and amount of growth achieved by *L. perenne* following defoliation have been observed at a variety of defoliation heights, ranging from 2.5 cm (Davies, 1965) to 5 cm (Davies, 1965; Donaghy and Fulkerson, 1997).

Studies of defoliated grasses, looking only at changes in the carbohydrate content of structures remaining after defoliation, cannot demonstrate remobilization of carbon to growing leaves. Davidson and Milthorpe (1966) measured carbon dioxide exchange of the root and shoot of *Dactylis glomerata* following defoliation and concluded that the root remained a net sink for carbon, irrespective of defoliation severity. When defoliation was not severe, changes in carbohydrates could account for respiratory losses and some carbon remobilization to new tissue may have occurred. However, with severe defoliation, reductions in carbohydrates were insufficient to account for respiration (Davidson and Milthorpe, 1966). Therefore, even in moderately defoliated grasses, reductions in the carbohydrate content of roots and stubble may be due to their respiration.

In order to quantify carbon remobilization, it is necessary to use isotopes. Ryle and Powell (1975) used short-term pulse labelling of ^{14}C in barley to quantify remobilization of carbon to growing leaves following a moderate defoliation. Their defoliation removed approximately half of the leaf area of the plant, leaving the oldest two leaves intact. These authors concluded that, during the first 2 days after defoliation, current assimilate fixed by the two remaining leaves supplied sufficient carbon to account for growth of new leaf, stem and root material. Beyond this time, new leaves were capable of supplying themselves with required assimilate and therefore no indication that remobilization supplied carbon to growing leaves was observed . Also using short-term pulse labelling of ^{14}C, though followed by a 2-day period prior to defoliation to allow 'allocation of carbon to proceed to near equilibrium', Atkinson (1986) investigated carbon partitioning in severely defoliated *Festuca ovina*. Depletion of ^{14}C in both roots and stubble over a 50-day period following defoliation was found to occur, suggesting use in either regrowth or respiration. Danckwerts and Gordon (1987) also used pulse labelling with ^{14}C prior to defoliation. They showed that depletion of ^{14}C occurred in the stubble but not in the roots of severely defoliated *L. perenne*. Some of the depleted ^{14}C was subsequently incorporated in new growth, proving that remobilization of carbon had occurred. However, as this did not fully account for the observed depletion, it implied that some ^{14}C was also respired (Danckwerts and Gordon, 1987).

De Visser *et al.* (1997) advocated the use of steady-state labelling in preference to pulse labelling of carbon for determination of the relative use of stores and current assimilation in supplying growing leaves following defoliation. An accurate evaluation of the use of the two possible sources of carbon for leaf growth requires a homogeneous labelling of one of the sources, the specific activity of which should be known, and they considered that these criteria could not be met using pulse-labelling techniques (De Visser *et al.*, 1997). Using such steady-state labelling, Johansson (1993) estimated that 21% of the carbon remaining in plants of *Festuca pratensis* after cutting was remobilized to new shoot material, with both roots and stubble contributing to the supply. In this study, new shoot material was taken to be that appearing above the height of clipping, and remobilization might have been overestimated, due to the physical movement of shoot material from below the height of cut, at the time of defoliation, to above the height of cut by subsequent growth. Steady-state ^{13}C labelling was used to investigate the use of remobilized and currently assimilated carbon into leaf and root growth zones of *L. perenne* after a single defoliation (De Visser *et al.*, 1997). These authors estimated that, in growth zones 3 days after defoliation, 50% of their carbon content was derived from remobilization, falling to 10% after 5 days. Moreover root and leaf growth zones were shown to exhibit similar kinetics of incorporation of currently assimilated carbon.

The growth of grass is often limited to a greater extent than photosynthesis under conditions of low nutrient status, low temperatures or moisture stress.

Under these conditions, often encountered in the field, carbohydrate in excess of that required for current growth can accumulate in the grass tissues (Brown and Blaser, 1965) as a form of reserve formation to avoid down-regulation of photosynthesis (Lemaire and Millard, 1999). If carbohydrate has accumulated in excess, any relationship between plant carbohydrate status and post-defoliation regrowth may break down. It should also be borne in mind, that where nutrient limitation has led to accumulation of carbohydrate, the excess carbohydrate may be utilized to maintain nutrient uptake following defoliation (Richards, 1993, and references therein).

Millard *et al.* (1990) considered that, whilst remobilization of nitrogen containing compounds such as amino acids is a mechanism primarily supplying nitrogen to leaves when root uptake is reduced, it could also act as a carbon source at times of decreased supply of current photosynthate. A similar view was expressed by Volenec *et al.* (1996), who considered that, with plants in which the carbon is labelled prior to defoliation, the appearance of label in new shoot material after defoliation is difficult to interpret. The labelling of almost all the organic chemical entities within the plant make it impossible to determine which chemical forms of carbon have been remobilized. The appearance of labelled carbon in the new shoot material, therefore, may in part be due to remobilization of compounds whose primary function is to supply nitrogen (for example, proteins and amino acids) but which also contain carbon. Indeed, Schnyder and De Visser (1999) have estimated that 60% of the net influx of carbon into regrowing *L. perenne* tillers during the first 2 days after defoliation was derived from amino-C. With this in mind, it is interesting to note that remobilization of carbon (Gonzalez *et al.*, 1989) and nitrogen (Ourry *et al.*, 1988) exhibit broadly similar dynamics in *L. perenne* following defoliation under similar conditions – namely, an initial phase in the first 4–6 days when regrowth is heavily dependent on remobilization, and then a later phase in which growth becomes progressively more dependent on current assimilation, concomitant with replenishment of stores. When the remobilization of both carbon and nitrogen to the growth zones of *L. perenne* following shoot defoliation was measured by De Visser *et al.* (1997), both leaf and root growth zones exhibited a stronger dependence on current assimilation of carbon than on that of nitrogen. One interpretation of these data is that remobilization of nitrogen after defoliation of grasses is more important than that of carbon.

Nitrogen remobilization

Turning now to a consideration of the remobilization of nitrogen, pulse chase labelling (5-day labelling period) with ^{15}N was used by Ourry *et al.* (1988) to provide evidence that, in *L. perenne* supplied with ammonium nitrate, nitrogen was remobilized from both roots and stubble to growing leaves following a single defoliation, with the majority of nitrogen being mobilized from stubble. Remobilization formed the major source of nitrogen for growing leaves in the first 4 days following defoliation (Ourry *et al.*, 1988). Similar techniques were used by the same authors to demonstrate that the total amount of nitrogen remobilized in *L. perenne* following a single defoliation was independent of the current nitrogen supply after defoliation (Ourry *et al.*, 1990). In this respect, defoliated grasses behave in an analogous manner to the seasonal remobilization of nitrogen observed in trees, where steady-state ^{15}N labelling showed that the amount remobilized in spring is dependent on the previous season's nitrogen supply and is independent of the current supply (Millard, 1996). However, for the reasons described above for carbon, the concerns of De Visser *et al.* (1997) regarding the use of pulse labelling to accurately determine post-defoliation mobilization apply equally to studies of nitrogen mobilization.

Notwithstanding differences in techniques, steady-state labelling of ^{15}N also showed that, in plants of *L. perenne*, remobilization was the major source of nitrogen immediately following a single defoliation (Ourry *et al.*, 1989b). Steady-state labelling has also been used to demonstrate that, in regularly defoliated *L. perenne*, the amount of nitrogen mobilized following defoliation was dependent on the form of nitrogen supplied and temperature. A greater amount of nitrogen remobilization was observed in plants supplied with ammonium, compared with those receiving an equivalent supply of nitrate, and in plants grown at 20°C, compared with those at 12°C (Thornton *et al.*, 1993b). These authors also showed that the proportion of the total nitrogen supplied to the growing leaves derived from remobilization was greater at the lower temperature but unaffected by the form of nitrogen supplied .

Remobilization of nitrogen in response to defoliation has been observed in grass species other than *L. perenne*. Thornton *et al.* (1993a) showed that the species order *L. perenne* < *Poa trivialis* < *Agrostis castellana* < *Festuca rubra* had an increasing proportion of the total nitrogen in regrowing leaves derived from remobilization. Whilst the reliance on remobilization increased in all species when the nitrogen supply was reduced, the relationship between species in relative use of remobilization was unaltered (Thornton *et al.*, 1994). Bakken *et al.* (1998) showed that both *Phleum pratense* and *F. pratensis* remobilized more nitrogen when grown under long- compared with short-day conditions. Species differences were also observed, with *F. pratensis* remobilizing a greater amount of nitrogen to growing leaves than *P. pratense*. They also considered that post-defoliation remobilization of nitrogen could act, in addition to supplying nitrogen to growing leaves, by reinforcing the down-regulation of mineral nitrogen uptake. This would avoid energy costs associated with the uptake and assimilation of mineral nitrogen at a time when photosynthate may be limited (Bakken *et al.*, 1998).

As with studies on carbon remobilization, many of the earlier investigations into remobilization of nitrogen had separations of plant material following defoliation, which were not ideal in terms of source–sink relationships. Differing approaches have been used to address this problem. In one approach (Thornton and Millard, 1997), the entire two youngest, i.e. growing, leaves from all tillers of plants of *L. perenne* and *F. rubra* were dissected out. In a second approach (De Visser *et al.*, 1997; Schnyder and de Visser, 1999), leaf growth zones have been dissected out of *L. perenne* tillers. Steady-state labelling with a series of destructive harvests was used in both studies to estimate the source of nitrogen (remobilization or current uptake) present in either the growth zone or the two youngest leaves at various times following defoliation. Whilst these methods represent improvements on earlier separations, both have their own particular disadvantages.

In the study of Thornton and Millard (1997), leaf appearance rate was ignored. Sampling the two youngest leaves after a new leaf has been initiated is clearly not sampling the same tissue as the two youngest leaves prior to initiation of the new leaf. Whilst this problem may not be significant immediately following defoliation, it will be exacerbated through time. Schnyder and De Visser (1999) successfully accounted for post-defoliation leaf appearance, describing it as 'tissue flux'. When sampling the basal 0–5 cm of a leaf (De Visser *et al.*, 1997), equivalent tissue will only be sampled if the residence time of cells within the particular leaf zone is unaffected by defoliation. Whilst leaf elongation rate was constant following defoliation in the study of De Visser *et al.* (1997), this is not always the case; Volenec (1986) observed large changes in leaf elongation rate of *Festuca arundinacea* in the days immediately following defoliation.

The Need for Field Experiments

The laboratory experiments described above have demonstrated that the storage and remobilization of nitrogen are quantitatively important for the regrowth of defoliated grasses. The situation for carbon is less clear, but what is certain is that stored carbon is important in providing respiratory substrates. To understand how plant responses to defoliation regulate competitive interactions between different individuals or species, it is important that measurements of remobilization can ultimately be made in the field. This poses several difficulties for experiments. Given the complex dynamics of the availability of soil nitrogen pools for uptake by plants (Williams *et al.*, 1999b), it is not possible to use ^{15}N-enriched tracers, because the added nitrogen will not necessarily equilibrate with the native soil nitrogen pools. Therefore, remobilization may be underestimated, due to uptake of soil nitrogen at natural abundance simultaneously with the fertilizer nitrogen (Millard, 1996). In addition, the spatial distribution of soil nitrogen is often very heterogeneous (Marriott *et al.*, 1997), particularly in grazed pastures, where dung and urine returns will have a localized impact on soil nutrient availability (Williams *et al.*, 1999a). Notwithstanding such problems, ^{15}N-enriched tracers have been used to estimate remobilization of plants growing in soil (Phillips *et al.*, 1983).

An alternative to using enriched tracers for quantifying remobilization is possibly the use of natural abundance variations in $\delta^{15}N$ and $\delta^{13}C$. Plants often have a $\delta^{15}N$ signal that differs from that of the source nitrogen taken up by the roots. There are a variety of reasons for these discrepancies, which have been reviewed by Robinson *et al.* (1998). An important aspect of this discrimination is that essentially all the enzymes, except nitrogenase, which transform nitrogen compounds show discrimination against ^{15}N (Handley and Raven, 1992). The $\delta^{15}N$ signature of a grass leaf will therefore depend upon that of the source pool, on $^{15}N/^{14}N$ fractionations during nitrogen assimilation and on nitrogen transport within and loss from the plant, and can be modelled in relation to physiological processes (Robinson *et al.*, 1998). It might be possible to quantify nitrogen remobilization using $\delta^{15}N$ signatures if that of the soil nitrogen pool is significantly different from that of the stored nitrogen in the roots and stubble. Conceptually, this approach is attractive, but there are several practical limitations at present, which will require further research to resolve.

First, grasses can take up nitrate, ammonium and organic nitrogen (such as amino acids (Cliquet *et al.*, 1997)). Each of these sources of nitrogen in soil will have their own $\delta^{15}N$ signature and, while a reliable method for measuring soil nitrate $\delta^{15}N$ is now available (Johnston *et al.*, 1999), measurement of the ammonium and organic nitrogen signatures is still problematic. Secondly, grasses can exhibit a wide range of genotypic plasticity in their $\delta^{15}N$ signatures. Table 5.1 gives data for *Agrostis capillaris*. Five genotypes were isolated from an upland pasture and genetically characterized with random amplified polymorphic DNA (RAPD) primers to ensure that they were distinct genotypes. Plants were grown in sand culture, with NO_3^-N of constant $\delta^{15}N$ signature as the only source of nitrogen. The data in Table 5.1 show the range and variation of $\delta^{15}N$ values measured in the shoots after 8 weeks of growth and allow comparison with variation in $\delta^{15}N$ reported in the literature for plants growing in a range of ecosystems. The variation in $\delta^{15}N$ in the five *A. capillaris* genotypes is greater than that reported for a range of wild barley genotypes grown in hydroponics and six species of savannah grasses, but is close to that found between 14 species of rain-forest trees. While the data in Table 5.1 do not represent an extensive review of the literature, they stress the fact that genotypic variability in $\delta^{15}N$ signature alone can be as great as variability measured in plants grown in the field, where genetic factors and an array of environmental factors act together. In part, variability of $\delta^{15}N$ may reflect genotypic variability in nitrogen remobilization, though the contribution of remobilization to the total nitrogen supplied to leaves was shown to be similar in four varieties of *L. perenne* (Wilkins *et al.*, 1997). Such variability might make interpretation of signatures in mixed plant communities in the field difficult.

Other possible approaches to measuring N remobilization in the field would be to take advantage of physiological changes occurring in plants subject to defoliation, which may correlate with the amount of nitrogen mobilized. The activity of enzymes involved in the degradation of storage compounds offers one such opportunity. During recycling of nitrogen from vegetative to reproductive tissues in wheat, Dalling *et al.* (1976) found a linear relationship between actual nitrogen loss from different tissues and losses estimated from the tissue protease activity. Remobilization of nitrogen to supply spring growth in *M. caerulea* is related to protease activity (Fig. 5.1). Clones of *M. caerulea* that remobilized most nitrogen were also the clones with greatest protease activity, as indicated by their ability to degrade azocasein (B. Thornton and U. Bausenwein, unpublished). The specific

Table 5.1. The variation and range of $\delta^{15}N$ signature of *Agrostis capillaris* genotypes compared with those of a range of other plant species.

Plant type	Environment	$\delta^{15}N$ range in foliage	Variation	Reference
Agrostis capillaris (five genotypes)	Controlled experiment	+1.4 to +3.9	2.5	U. Bausenwein (unpublished data)
Hordeum spontaneum (28 genotypes) and *Hordeum vulgare* x *H. spontaneum* crosses	Controlled experiment			
	Control plants		1.9	Handley *et al.* (1997)
	Salt-stressed plants		2.2	
Savannah grasses (six species)	Savannah in Côte d'Ivoire	−2.0 to −0.3	1.7	Abbadie *et al.* *(1992)*
Picea abies	German spruce forest	−4.1 to−2.5	1.6	Gebauer and Schulze (1991)
Pinus sylvestris	Swedish pine forest	−1.5 to +2.9	4.4	Högberg and Johannisson (1993)
Rain-forest trees (14 species)	Amazonian rain forest	+0.8 to +3.5	2.7	Guehl *et al.* (1998)
Taiga trees and shrubs (six species)	Alaskan taiga forest (three forest types)	−8.5 to +0.5	9.0	Kielland *et al.* (1998)
Mulga vegetation (~100 species of different life forms)	Eastern and western Australian mulga woodlands	+7.5 to +15.5	8.0	Pate *et al.* (1998)

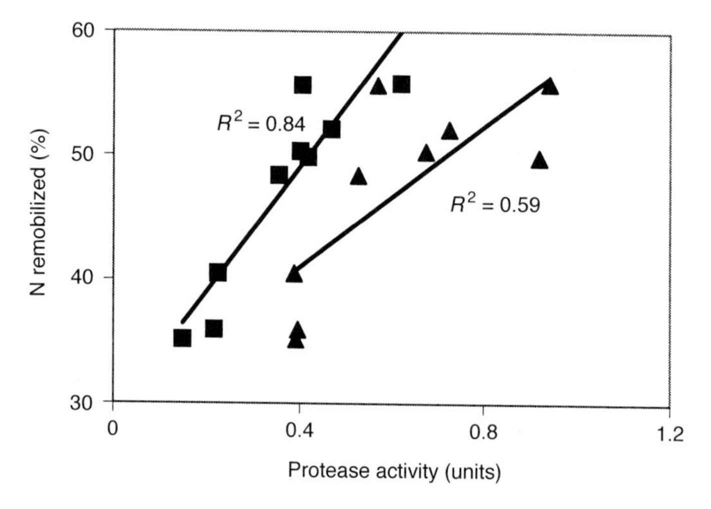

Fig. 5.1. The percentage of nitrogen in *Molinia caerulea* roots at the start of the season subsequently remobilized to shoot, against root protease activity (ability to degrade azocasein), expressed per g fresh weight root (▲) or per mg root protein (■). A unit of protease activity is as defined by Weckenmann and Martin (1984). Symbols are individual clones on Julian day 132 of the season.

activity of several protease enzymes was shown by Ourry *et al.* (1989a) to increase in the roots of *L. perenne* following defoliation. The dynamics of endopeptidase activity, in particular, is closely related to the dynamics of observed nitrogen remobilization (Ourry *et al.*, 1989a). Such changes in enzyme activity may provide a means, albeit indirect, of estimating nitrogen remobilization in field-grown plants. An analogous approach could also be adopted for estimation of carbon remobilization in the field, using, for example, the activity of fructan exohydrolase (FEH). This enzyme is involved in remobilization of fructan, and its activity increases following defoliation of grasses (Prud'homme *et al.*, 1992; Simpson and Bonnett, 1993). Increased concentration of sucrose in roots, resulting from increased activity of FEH, has been suggested by Paterson and Sim (1999) to be the cause of increased exudation of carbon from grass roots following defoliation.

The integrated physiological changes that occur in defoliated grasses, including those in enzyme activity highlighted above (Richards, 1993), suggest that some control at the molecular level is involved. Ourry *et al.* (1996) showed that changes in the translation products of mRNA could be found within 1 h after defoliation of *Lolium temulentum*. Major changes in the abundance of transcripts occur between 12 h and 72 h following defoliation. Ourry *et al.* (1996) accepted that the functional significance of the up-regulated mRNA could not be established; however, a tentative suggestion implicating FEH was made, based on correlated changes in enzyme activity. Ultimately, the isolation of genes that switch on in specific response to defoliation may offer a means of estimating remobilization of carbon and nitrogen in field-grown plants.

Table 5.2. The concentration of Asn and Gln and the Asn : Gln ratio in the xylem exudate of *Lolium perenne* following defoliation.

Time after defoliation (h)	Xylem [Asn] (μg Asn g^{-1})	Xylem [Gln] (μg Gln g^{-1})	Asn:Gln ratio
0	33.86	740.0	0.046
7.5	0.22	0.62	0.36
15.5	2.32	2.50	0.93

Following defoliation the shift in the source of nitrogen used to supply growing leaves from current uptake and assimilation towards use of remobilized stores may result in changes in the amino acid pools within plants. In turn, such changes may offer a means of indirectly estimating nitrogen remobilization. Bigot *et al.* (1991) observed an increase in the Asn : Gln ratio in the xylem exudate of *L. perenne* for 3 to 7 days following defoliation. They considered that the change was associated with a switch in the origin of nitrogen in the xylem to that derived from hydrolysis of protein, though possible interference with the root–shoot–root cycling of nitrogen described by Cooper and Clarkson (1989) was also suggested. However, due to data variability, Bigot *et al.* (1991) only presented relative amounts of amino acids in the xylem, the absolute changes in amino acid contents behind the shift in the Asn : Gln ratio being unreported. Using shorter xylem sampling periods, B. Thornton and J.H. Macduff (unpublished) were able to obtain reliable data on the concentration of amino acids in the xylem exudate of *L. perenne* following defoliation. They confirmed that an increase in the Asn : Gln ratio did occur in the xylem exudate of *L. perenne* following defoliation (Table 5.2), and were able to prove that in the first 7.5 h following defoliation both [Asn] and [Gln] fell, an observed 7.8 times increase in the Asn : Gln ratio being the result of a greater reduction in [Gln]. Over the following 8 h, both [Asn] and [Gln] showed slight recovery, but they did not approach their predefoliated values (Table 5.2). Therefore, at least in the first 15.5 h following defoliation, the increased Asn : Gln ratio is not due to increased [Asn] in the xylem resulting from hydrolyses of storage protein. This result does not preclude the possibility that, over longer time periods following defoliation, hydrolysis of protein may contribute in part to the increased xylem Asn : Gln ratio; however, it does suggest that the Asn : Gln ratio is unlikely to be a reliable surrogate for the amount of nitrogen remobilized.

Conclusion

The storage and subsequent remobilization of carbon and nitrogen can offer plants several ecological advantages, including enhanced recovery from catastrophic events, such as defoliation. The use of carbon and nitrogen tracers under laboratory conditions has provided much information, suggesting an important role of storage and subsequent remobilization in the growth of defoliated grasses.

However, to understand how plant responses to defoliation regulate competitive interactions between different individuals or species, measurement of remobilization in the field is required. To date, this step has not been satisfactorily taken, though exciting possibilities exist for ultimately achieving this goal. Only then will the true role of carbon and nitrogen remobilization in the vegetation dynamics of grasslands be known.

Acknowledgement

This work was funded by the Scottish Office Agriculture, Environment and Fisheries Department.

References

Abbadie, L., Mariotti, A. and Menaut, J.C. (1992) Independence of savanna grasses from soil organic matter for their nitrogen supply. *Ecology* 73, 608–613.

Atkinson, C.J. (1986) The effect of clipping on net photosynthesis and dark respiration rates of plants from an upland grassland, with reference to carbon partitioning in *Festuca ovina*. *Annals of Botany* 58, 61–72.

Avice, J.C., Ourry, A., Volenec, J.J., Lemaire, G. and Boucaud, J. (1996) Defoliation-induced changes in abundance and immuno-localization of vegetative storage proteins in tap roots of *Medicago sativa*. *Plant Physiology and Biochemistry* 34, 561–570.

Avice, J.C., Lemaire, G., Ourry, A. and Boucaud, J. (1997) Effects of the previous shoot removal frequency on subsequent shoot regrowth in two *Medicago sativa* L. cultivars. *Plant and Soil* 188, 189–198.

Bakken, A.K., Macduff, J.H. and Collison, M. (1998) Dynamics of nitrogen remobilization in defoliated *Phleum pratense* and *Festuca pratense* under short and long photoperiods. *Physiologia Plantarum* 103, 426–436.

Bigot, J., Lefevre, J. and Boucaud, J. (1991) Changes in the amide and amino acid composition of xylem exudate from perennial ryegrass (*Lolium perenne* L.) during regrowth after defoliation. *Plant and Soil* 136, 59–64.

Brown, R.H. and Blaser, R.E. (1965) Relationships between reserve carbohydrate accumulation and growth rate in orchard grass and tall fescue. *Crop Science* 5, 577–582.

Cliquet, J.B., Murray, P.J. and Boucaud, J. (1997) Effect of the arbuscular mycorrhizal fungus *Glomus fasciculatum* on the uptake of amino nitrogen by *Lolium perenne*. *New Phytologist* 137, 345–349.

Cooper, H.D. and Clarkson, D.T. (1989) Cycling of amino-nitrogen and other nutrients between shoots and roots in cereals – a possible mechanism integrating the shoot and root in the regulation of nutrient uptake. *Journal of Experimental Botany* 40, 753–762.

Crafts-Brandner, S.J., Salvucci, M.E. and Egli, D.B. (1990) Changes in ribulosebiphosphate carboxylase/oxygenase and ribulose 5-phosphate kinase abundances and photosynthetic capacity during leaf senescence. *Photosynthesis Research* 23, 223–230.

Cyr, D.R. and Bewley, J.D. (1990) Proteins in the roots of the perennial weed chicory (*Cichorium intybus* L.) and dandelion (*Taraxcum officinale weber*) are associated with overwintering. *Planta* 182, 370–374.

Dalling, M.J., Boland, G. and Wilson, J.H. (1976) Relationship between acid proteinase activity and redistribution of nitrogen during grain development in wheat. *Australian Journal of Plant Physiology* 3, 721–730.

Danckwerts J.E. and Gordon A.J. (1987) Long-term partitioning, storage and remobilization of ^{14}C assimilated by *Lolium perenne* (cv. Melle). *Annals of Botany* 59, 55–66.

Davidson, J.L. and Milthorpe, F.L. (1966) The effect of defoliation on the carbon balance in *Dactylis glomerata*. *Annals of Botany* 30, 185–198.

Davies, A. (1965) Carbohydrate levels and regrowth in perennial rye-grass. *Journal of Agricultural Science* 65, 213–221.

De Visser, R., Vianden, H. and Schnyder, H. (1997) Kinetics and relative significance of remobilized and current C and N incorporation in leaf and root growth zones of *Lolium perenne* after defoliation: assessment by ^{13}C and ^{15}N steady-state labelling. *Plant, Cell and Environment* 20, 37–46.

Donaghy, D.J. and Fulkerson, W.J. (1997) The importance of water-soluble carbohydrate reserves on regrowth and root growth of *Lolium perenne* (L.). *Grass and Forage Science* 52, 401–407.

Fulkerson, W.J. and Slack, K. (1994) Leaf number as a criterion for determining defoliation time for *Lolium perenne*. 1. Effect of water-soluble carbohydrates and senescence. *Grass and Forage Science* 49, 373–377.

Gebauer, G. and Schulze, E.D. (1991) Carbon and nitrogen isotope ratios in different compartments of a healthy and a declining *Picea abies* forest in the Fichtelgebirge, NE Bavaria. *Oecologia* 87, 198–207.

Gonzalez, B., Boucaud, J., Salette, J., Langlois, J. and Duyme, M. (1989) Changes in stubble carbohydrate content during regrowth of defoliated perennial ryegrass (*Lolium perenne* L.) on two nitrogen levels. *Grass and Forage Science* 44, 411–415.

Guehl, A.M., Domenach, M., Bereau, T.S., Barigah, H., Casabianca, A., Ferhi, A. and Garbaye, J. (1998) Functional diversity in an Amazonian rainforest of French Guyana: a dual isotope approach ($\delta^{15}N$ $\delta^{13}C$). *Oecologia* 116, 316–330.

Handley, L.L. and Raven, J.A. (1992) The use of natural abundance of nitrogen isotopes in plant physiology and ecology. *Plant, Cell and Environment* 15, 965–985.

Handley, L.L., Robinson, D., Forster, B.P., Ellis, R.P., Scrimgeour, C.M., Gordon, D.C., Nevo, E. and Raven, J.A. (1997) Shoot $\delta^{15}N$ correlates with genotype and salt stress in barley. *Planta* 201, 100–102.

Heilmeier, H., Schulze, E.D. and Wale, D.M. (1986) Carbon and nitrogen partitioning in the biennial monocarp *Arctium tomentosum* Mill. *Oecologia* 70, 466–467.

Higgins, T.V.J. (1984) Synthesis and regulation of major proteins in seeds. *Annual Review of Plant Physiology* 35, 191–221.

Högberg, P. and Johannisson, C. (1993) N-15 abundance of forests is correlated with losses of nitrogen. *Plant and Soil* 157, 147–150.

Jaeger, C.H. and Monson, R.K. (1992) Adaptive significance of nitrogen storage in *Bistorta bistortoides*, an alpine herb. *Oecologia* 92, 578–585.

Johansson, G. (1993) Carbon distribution in grass (*Festuca pratensis* L.) during regrowth after cutting – utilization of stored and newly assimilated carbon. *Plant and Soil* 151, 11–20.

Johnston, A.M., Scrimgeour, C.M., Henry, M.O. and Handley, L.L. (1999) Isolation of NO_3-N as 1-phenylazo-2-naphthol (Sudan-1) for measurement of $\delta^{15}N$. *Rapid Communications in Mass Spectrometry* 13, 1531–1534.

Kielland, K., Barnett, B. and Schell, D. (1998) Intraseasonal variation in the delta N-15 signature of taiga trees and shrubs. *Canadian Journal of Forest Research – Revue Canadienne de Recherche Forestière* 28, 485–488.

Lemaire, G. and Millard, P. (1999) An ecophysiological approach to modelling resource fluxes in competing plants. *Journal of Experimental Botany* 50, 15–28.

Louahlia, S., Macduff, J.H., Ourry, A., Humphreys, M. and Boucaud, J. (1999) Nitrogen reserve status affects the dynamics of nitrogen remobilization and mineral nitrogen uptake during recovery from defoliation by contrasting cultivars of *Lolium perenne*. *New Phytologist* 142, 451–462.

Marriott, C.A., Hudson, G., Hamilton, D., Neilson, R., Boag, B., Handley, L.L., Wishart, J., Scrimgeour, C.M. and Robinson D. (1997) Spatial variability of total C and N and their stable isotopes in an upland Scottish grassland. *Plant and Soil* 196, 151–162.

Millard, P. (1988) The accumulation and storage of nitrogen by herbaceous plants. *Plant Cell and Environment* 11, 1–8.

Millard, P. (1996) Ecophysiology of the internal cycling of nitrogen for tree growth. *Journal Plant Nutrition and Soil Science* 159, 1–10.

Millard, P., Thomas, R.J. and Buckland, S.T. (1990) Nitrogen supply affects the remobilization of nitrogen for the regrowth of defoliated *Lolium perenne* L. *Journal of Experimental Botany* 41, 941–947.

Ourry, A., Boucaud, J. and Salette, J. (1988) Nitrogen mobilization from stubble and roots during re-growth of defoliated perennial ryegrass. *Journal of Experimental Botany* 39, 803–809.

Ourry, A., Bigot, J. and Boucaud, J. (1989a) Protein mobilization from stubble and roots, and proteolytic activities during post-clipping re-growth of perennial ryegrass. *Journal of Plant Physiology* 134, 298–303.

Ourry, A., Gonzalez, B., Bigot, J., Boucaud, J. and Salette, J. (1989b) Nitrogen and carbohydrate mobilization during regrowth of defoliated *Lolium perenne* L. In: *Proceedings of the XVIth International Grassland Congress, Nice*, INRA. AFPF, Versailles, France, pp. 513–514.

Ourry, A., Boucaud, J. and Salette, J. (1990) Partitioning and remobilization of nitrogen during regrowth in nitrogen-deficient ryegrass. *Crop Science* 30, 1251–1254.

Ourry, A., Macduff, J.H. and Ougham, H.J. (1996) The relationship between mobilization of N reserves and changes in translatable messages following defoliation in *Lolium temulentum* L. and *Lolium perenne* L. *Journal of Experimental Botany* 47, 739–747.

Parry, M.A.J., Andralojc, P.J., Parmar, S., Keys, A.J., Habash, D., Paul, M.J., Alred, R., Quick, W.P. and Servaites, J.C. (1997) Regulation of Rubisco by inhibitors in the light. *Plant, Cell and Environment* 20, 528–534.

Pate, J.S., Unkovich, M.J., Erskine, P.D. and Stewart, G.R. (1998) Australian mulga ecosystems – C-13 and N-15 natural abundances of biota components and their ecophysiological significance. *Plant, Cell and Environment* 221, 1231–1242.

Paterson, E. and Sim, A. (1999) Rhizodeposition and C-partitioning of *Lolium perenne* in axenic culture affected by nitrogen supply and defoliation. *Plant and Soil* 216, 155–164.

Phillips, D.A., Center, D.M. and Jones, M.B. (1983) Nitrogen turnover and assimilation during regrowth in *Trifolium subterraneum* L. and *Bromus mollis* L. *Plant Physiology* 71, 472–476.

Prud'homme, M.-P., Gonzalez, B., Billard, J.-P. and Boucaud, J. (1992) Carbohydrate content, fructan and sucrose enzyme activities in roots, stubble and leaves of ryegrass (*Lolium perenne* L.) as affected by source/sink modification after cutting. *Journal of Plant Physiology* 140, 282–291.

Raven, J.A. (1987) The role of vacuoles. *New Phytologist* 106, 357–422.
Richards, J.H. (1993) Physiology of plants recovering from defoliation. In: *Proceedings of the XVIIth International Grassland Congress*. SIR Publishing, Wellington, New Zealand, pp. 85–94.
Robinson, D., Handley, L.L. and Scrimgeour, C.M. (1998) A theory for $^{15}N/^{14}N$ fractionation in nitrate-grown vascular plants. *Planta* 205, 397–406.
Ryle, G.J.A. and Powell, C.E. (1975) Defoliation and regrowth in the graminaceous plant: the role of current assimilate. *Annals of Botany* 39, 297–310.
Schnyder, H. and De Visser, R. (1999) Fluxes of reserve-derived and currently assimilated carbon and nitrogen in perennial ryegrass recovering from defoliation: the regrowing tiller and its component functionally distinct zones. *Plant Physiology* 119, 1423–1435.
Simpson, R.J. and Bonnett, G.D. (1993) Fructan exohydrolase from grasses. *New Phytologist* 123, 453–469.
Thornton, B. and Millard, P. (1993) The effects of nitrogen supply and defoliation on the seasonal internal cycling of nitrogen in *Molinia caerulea*. *Journal of Experimental Botany* 44, 531–536.
Thornton, B. and Millard, P. (1997) Increased defoliation frequency depletes remobilization of nitrogen for leaf growth in grasses. *Annals of Botany* 80, 89–95.
Thornton, B., Millard, P., Duff, E.I. and Buckland, S.T. (1993a) The relative contribution of remobilization and root uptake in supplying nitrogen after defoliation for regrowth of laminae in four grass species. *New Phytologist* 124, 689–694.
Thornton, B., Millard, P. and Galloway, S. (1993b) The effects of temperature and form of nitrogen supply on the relative contribution of root uptake and remobilization in supplying nitrogen for laminae regrowth of *Lolium perenne* L. *Journal of Experimental Botany* 44, 1601–1606.
Thornton, B., Millard, P. and Duff, E.I. (1994) Effects of nitrogen supply on the source of nitrogen used for regrowth of laminae after defoliation of four grass species. *New Phytologist* 128, 615–620.
Volenec, J.J. (1986) Nonstructural carbohydrates in stem base components of tall fescue during regrowth. *Crop Science* 26, 122–127.
Volenec, J.J., Ourry, A. and Joern, B.C. (1996) A role for nitrogen reserves in forage regrowth and stress tolerance. *Physiologia Plantarum* 97, 185–193.
Weckenmann, D. and Martin, P. (1984) Endopeptidase activity and nitrogen mobilization in senescing leaves of *Nicotiana rustica* in light and dark. *Physiologia Plantarum* 60, 333–340.
Wilkins, P.W., Macduff, J.H., Raistrick, N. and Collison, M. (1997) Varietal differences in perennial ryegrass for nitrogen use efficiency in leaf growth following defoliation: performance in flowing solution culture and its relationship to yield under simulated grazing in the field. *Euphytica* 98, 109–119.
Williams, B., Shand, C., Sellers, S. and Young, M. (1999a) Impact of synthetic sheep's urine on N and P in two pastures in the Scottish uplands. *Plant and Soil* 214, 93–103.
Williams, B., Silcock, D. and Young, D. (1999b) Seasonal dynamics of N in two *Sphagnum* moss species and the underlying peat treated with $^{15}NH_4$ $^{15}NO_3$. *Biogeochemistry* 45, 285–302.
Zamski, E. (1996) Anatomical and physiological characteristics of sink cells. In: Zamski, E. and Schaffer, A.A. (eds) *Photoassimilate Distribution in Crop Plants: Source–Sink Relationships*. Marcel Dekker, New York, pp. 283–310.

6

Shoot Morphological Plasticity of Grasses: Leaf Growth vs. Tillering

C.J. Nelson

Department of Agronomy, University of Missouri, Columbia, MO 65211, USA

Introduction

Members of the grass family, i.e. the *Poaceae*, whether they are annuals or perennials, are widespread in their adaptation, due largely to the vast diversity they possess in physiology and growth form. Knowledge about the growth characteristics of grasses, especially perennial pasture grasses, is essential for effective management and genetic improvement. Thus, basic processes need to be understood to evaluate the growth potentials of plants within a given environment, how they adapt morphologically to grow and persist as environments change and how they display herbage to optimize grazing.

Some recent reviews have covered parts of this topic, but at a different level or from a different viewpoint. For example, Chapman (1996) covered many growth aspects of grasses from a biologist's perspective. In a novel way, he related development of agriculture over several centuries with the human-assisted evolution of grasses towards increased utility. Söderstrom *et al.* (1987) edited the proceedings from an international symposium on grass systematics and evolution and Cheplick (1998) edited the proceedings from a symposium on the population biology of grasses. Both volumes address the anatomy, growth and development of grasses in some detail. Mooney *et al.* (1991) considered stress responses and Tilman (1988) covered growth and structure in plant communities.

While being drawn primarily from specific disciplines, these volumes bring to the agriculturalist the findings and analyses of the anatomist, systematist and ecologist, who also contribute basic principles to our understanding of grassland adaptation and management. The recent emphasis on sustainable and environmentally friendly approaches to agriculture has brought the biologist and

(eds G. Lemaire, J. Hodgson, A. de Moraes, C. Nabinger and P.C. de F. Carvalho)

agriculturalist closer together in scientific goals and methodology. A recent monograph on cool-season grasses (Moser *et al.*, 1996) focused on their use as forages and included a chapter on physiology and developmental morphology (Nelson, 1996). Other reviews related to grassland management emphasized developmental morphology (Briske, 1991; Briske and Derner, 1998) and plant growth following defoliation (Chapman and Lemaire, 1993; Richards, 1993).

The goal of this chapter is to examine the growth processes of perennial grasses at the fundamental and whole-plant levels, to evaluate the anatomical and developmental interrelationships of leaf growth and tillering and to relate these responses to genotypic and phenotypic plasticity. The emphasis is on plasticity of growth zones, their roles and interactions during growth and canopy development and assessments of potentials for pasture management and genetic improvement. Other authors address the influence of carbon and nitrogen (N) supplies (see Schnyder *et al.*, Chapter 3, this volume) and management strategies (see Matthew *et al.*, Chapter 7, this volume) on these processes.

Plasticity in Perspective

Fundamental aspects, such as climatic adaptation, depend on plasticity of plant growth habit and how the environment affects its expression. The plastic response may also affect presentation of herbage to the grazing animal. Huber *et al.* (1999) presented a perspective on plasticity that begins with the structural blueprint, a species-specific set of traits that determine the basic growth form and structural organization. These traits generally form the basis for classical taxonomy and show genetic variation within species. For example, certain species or genotypes within a species tiller more vigorously or produce leaves faster, which gives better adaptation to frequent and close grazing or resistance to animal traffic.

Above-ground plasticity

Within the structural blueprint, Huber *et al.* (1999) separated the phenotype response due to ontogeny from other forms of phenotypic plasticity. Plasticity that occurs due to differences from normal ontogeny plays a major role in deviating from the structural blueprint in response to the environment, especially for grasses that alter resource acquisition strategies by changes in rate of development among organs due to stress. These ontogeny events are also very evident during the management of grasses that are progressing from reproductive to vegetative growth stages (see Matthew *et al.*, Chapter 7, this volume).

The phyllochron of grasses is an ontogeny event that is strongly affected by temperature (Frank and Bauer, 1995; Wilhelm and McMaster, 1995) and it directly alters the rate of tiller site production. If site usage – the proportion of axillary buds that develop into tillers – is similar among environments, it appears that the tillering rate is altered if it is compared on a time basis. Similarly, genotypes of the same species at the same chronological age may be different ontogenetically due to environment (Skinner and Nelson, 1994b). Since tillering often changes with ontogenetic stage, the number of tillers produced by plants at the same age, but at different stages of ontogeny, may differ. In these examples, however, the effect of the environment on the tillering response is mainly a shift in the rate of normal ontogenetic development, which also contributes to plasticity among meristems or in resource allocation.

Phenotypic plasticity is defined by Huber *et al.* (1999) as environmental effects on morphology and architecture that are not due to ontogeny, and includes changes in size, structure and spatial positioning of organs. These include timing of meristem release and the fate of meristems, which lead to changes in plant architecture and, in addition to ontogeny, provide adaptation to different habitats, which further enhances resource acquisition. While being separate conceptually, a change in ontogeny can influence the degree of subsequent plasticity expressed. Diggle (1994) described ontogenetic contingency, a condition in which meristems produced at different stages of ontogeny may respond differently to the same environmental stimulus. An example would be the growth of tillers of different ages after defoliation that respond differently to the partial removal of the canopy and the change in the red/far red (R/FR) spectral properties of light.

Below-ground plasticity

Casper and Jackson (1997) considered plasticity in plant competition underground and divided it conceptually into morphological and physiological plasticity. In both cases, plasticity was viewed as altering the ability of the root system to capture resources, either by exploring new areas morphologically or altering uptake per area physiologically. Roots of grasses in nutrient- or water-stressed conditions are generally smaller in diameter, but have near-equal longitudinal growth rates to those with adequate supplies, a strategy that favours exploration, whereas in nutrient-rich areas the roots branch more profusely to favour exploitation. Root lifespan may also be longer in nutrient-rich and adequate water environments.

For roots, Casper and Jackson (1997) suggest that physiological plasticity is related to enzyme activity and includes such processes as osmoregulation to lower cell water potential and maintain water uptake and growth in drying soils. Similarly, plants can adjust by exuding organic acids, which may serve as effective chelators for mineral nutrient uptake when in short supply. For example, roots in soils with low phosphorus (P) supply can exude citric acid, which increases the P in solution and allows for more P uptake. Shoots can also respond physiologically for adaptation without a change in morphology, such as during osmoconditioning or hardening for winter, but plasticity generally indicates a morphological adaptation.

In general, plasticity in root systems needs to be considered differently from plasticity of shoot systems. The root apex is responsible for root elongation, but there are no nodes, internodes or axillary buds, so root growth can be considered to be continuous. The location and timing of branch formation on roots are less predictable than on shoots. Even so, the rate of root initiation on grass tillers appears to be an important factor in the growth and survival of the tiller, and may be regulated by the tiller (de Ropp, 1945), an area that needs to be researched. Once initiated, roots grow into soil, after which they have a rather steady growth rate and branching behaviour, depending on the soil environment and nutrient supply. In contrast, shoot growth is more stepwise and start/stop, following a well-defined sequence of events, beginning with leaf initiation and followed by tiller site initiation and internode elongation. Each shoot tissue has an initiation mechanism and a finite growth period, so shoot growth is the summation of a connected series of partially independent organs and is based on several meristems. The root needs to be considered, however, as the degree of independence among meristems or growth zones of the root and shoot helps govern the amount of plasticity a plant can express.

Plasticity in production systems

In modern agriculture, the structural blueprint of Huber *et al.* (1999) is rarely expressed, as it is generally modified, due to plant spacing and, in the case of forages, repeated defoliation. For example, Moulia *et al.* (1999) evaluated the structural blueprint of modern maize when grown in the field as widely spaced plants. Basal tillers appeared, similar to other grasses, after a delay of seven phyllochrons. They concluded that the basal tillering of maize was regulated similarly to that of other grasses, but with much less effect on the development of ear shoots and tassel branches. This differential plasticity allows maize to be grown for high grain production at densities that inhibit basal tillering.

Skinner and Nelson (1992) measured tiller appearance as tall fescue seedlings at low density grew and developed more leaf area. As the canopy developed, the first strategy was for a reduction in the rate of ontogeny as rates of leaf appearance and leaf elongation slowed. This reduced the rate of production of tillering sites. In addition, the number of developed leaves above the site of tiller appearance increased from about 1.7 to 2.5, largely because the slower elongation of the emerging tiller delayed its emergence. Site usage, a true indicator of plasticity (Huber *et al.*, 1999), decreased from near 90% during the early stages to less than 20% at 70 days after planting. If site production and site usage during the first 35 days had continued in the same log-linear manner until day 70, each plant would have had over 2000 tillers.

As with maize, tiller production in forage grasses is heavily down-regulated in a canopy. Similar responses were found in rice (Nemoto *et al.*, 1995) and perennial ryegrass (Davies and Thomas, 1983). But forage grass cultivars are highly heterogeneous, with genetic variation for morphological characters in the population. Thus, the architectural blueprint of individual plants is masked in the population and, due to close spacing, can rarely be expressed. In production systems, forage grasses are usually operating in an environmentally sensitive manner and are simultaneously expressing both ontogenetic and phenotypic plasticity. My

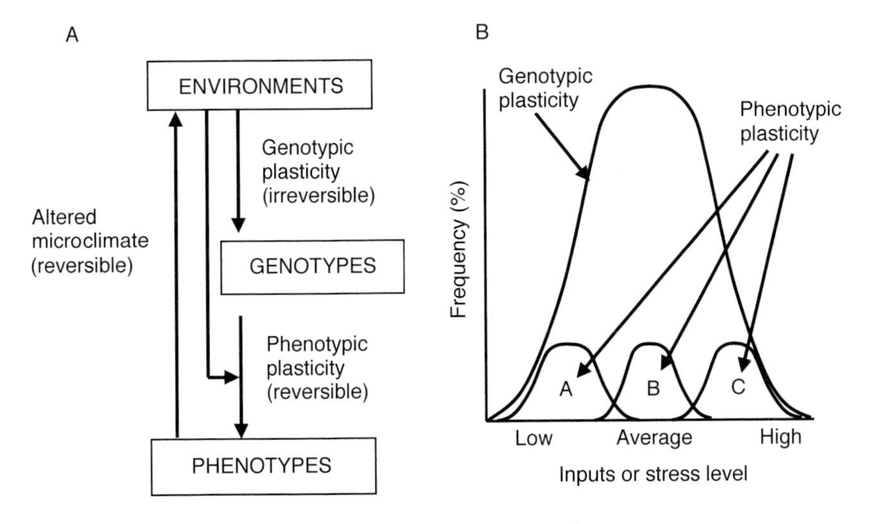

Fig. 6.1. Genotypic and phenotypic plasticity influence adaptation of pasture grasses. (A) Environments affect the survival of genotypes in a population and interact with each genotype to alter the morphology and adaptation of the phenotype. (B) Genotypic plasticity depends on the survival of the genotypes that make up the population. Phenotypic plasticity is expressed as morphological and physiological changes of a genotype that improve its survival and broaden its adaptation.

approach with outcrossing forage grasses is to divide plasticity into genotypic and phenotypic components (Fig. 6.1), i.e. how the genotypes and environment interact to give the resulting form (Nelson, 1998).

Genotypes, Phenotypes and Plasticity

Most cool-season grasses are cross-pollinated. Therefore, seed lots are highly heterogeneous and represent the range of morphological and physiological characters the breeder selected. The population of plants includes some variability for most characters, including leaf growth rates, leaf size and tillering habits, which are all controlled by additive gene action (Sleper *et al.*, 1977), i.e. several genes are involved in regulating the expression. The long-term success of the sward will depend on the degree of adaptation that can be accomplished by the mixture of genotypes and on the relative competitiveness, survival and capacity for reproduction of individuals. Breeders, recognizing the value of adaptation for outcrossing forage grasses, generally combine several parents in crossing blocks to maintain genetic heterozygosity in the selected population and cultivar. The heterozygosity or genotypic plasticity is represented by a range of plants, each having its unique structural blueprint. Changes in growth form due to ontogeny or adaptation are considered to be phenotypic plasticity.

Genotypic plasticity

Over time, genotypes in the heterogeneous population or cultivar that are best suited for the particular environment and the management condition will survive and constitute the predominant portion of the population (Hazard and Ghesquière, 1995). The ability and degree to which the population can shift is termed genotypic plasticity (Fig. 6.1). It is generally not reversible, because the gene complements of plants that are not adapted are lost from the population as plants die (Nelson, 1998). Pastures that are managed intensively have minimal seed production and seedling recruitment, which, if allowed to occur with outcrossing species, would help maintain or restore genetic diversity and recover genotypic plasticity. Genotypic plasticity can ebb and flow in more extensively managed rangelands and other conditions where seed production and seedling recruitment can occur.

Maintenance of genotypic plasticity or genetic diversity in pastures depends largely on variation in physiological attributes, such as inherent stress resistance and the competitive ability of each genotype, which may involve a change in morphology. Attributes that contribute to survival include winter-hardiness, disease and insect resistance, photosynthesis rate and the manner of leaf display. Thus, genotypic diversity is maintained only if the original plants can adapt physiologically or through phenotypic plasticity to survive the range of environments and management practices employed.

The occurrence of polyploids can further expand heterozygosity and genotypic plasticity of some species (Johnson, 1972). The geographical adaptation of many cool-season forage grasses, such as *Bromus* (Ainouche *et al.*, 1995), *Dactylis* (Bretagnolle and Thompson, 1996) and *Festuca* (Sleper and West, 1996), is improved at higher ploidy levels. In contrast, many warm-season, C_4 grasses occupying native grasslands in the USA are cross-pollinated and also polyploids, but ploidy level apparently has little effect on their adaptation (Keeler, 1998).

In contrast to most perennial forage grasses, populations of self-pollinated crop plants, such as most cereals, including maize, due to its manner of seed propagation, are selected for homogeneity, which reduces genotypic plasticity. In these cases, the architectural blueprint (Huber *et al.*, 1999) is consistent among plants in the population and adaptation is heavily dependent on phenotypic plasticity.

Phenotypic plasticity

Each individual plant (genotype) within a heterogeneous or homogeneous population has the ability to adapt its growth processes to stress or to a management practice to gain resources by changing its resultant morphology, a process termed phenotypic plasticity. But, in contrast with genotypic plasticity, the change is reversible (Fig. 6.1). Examples of phenotypic plasticity include changes in root–shoot ratio, leaf orientation, final leaf size or rate of tillering in response to drought stress, altered N fertilization or change in defoliation height or frequency.

After the stress is removed or altered, the basic phenotype can be partially or fully restored.

The environment is the major factor in regulating phenotypic plasticity of a genotype (Fig. 6.1) and there is a feedback loop, because phenotypic changes, such as enhanced root growth or altered leaf angle, can offset or minimize some management or physiological stresses to improve plant growth and survival, which contributes further to genotypic plasticity. For example, Glimskär and Ericsson (1999) found phenotypic plasticity in several plants for response of specific leaf area and leaf area ratio to applications of fertilizer N. Gibson *et al.* (1992) found that leaf blades of annual ryegrass were displayed more vertically at high plant density and more horizontally at low plant density, presumably aiding radiation capture and survival. In contrast, Dallisgrass leaves were displayed horizontally at all plant densities, reflecting little phenotypic plasticity for this character.

Separating genotypic and phenotypic plasticity

In an early study, we found that the tillering capacity among tall fescue genotypes was negatively correlated with both leaf elongation rate and weight per tiller (Nelson *et al.*, 1977). Therefore, we transplanted vegetative tillers of six tall fescue genotypes into separate microplots in the field, and then fertilized them with N at 0, 90, 180 and 270 kg ha^{-1} (Fig. 6.2). All the genotypes lived, so differences were due to phenotypic plasticity. Despite the range in genotypes used there was no genotype–N interaction for either tiller density or weight per tiller. Each genotype had a yield response to N that was nearly linear, but at low tiller densities N favoured tillering for each genotype, and at high densities the expression shifted to favour weight per tiller (Nelson and Zarrough, 1981). This suggests that the mechanism and phenotypic plasticity of the N response for these two characters were similar for all six genotypes.

The degree to which phenotypic plasticity can be expressed can add to or detract from the utility of a cultivar or species. One question is whether most plants of a species in a natural heterogeneous population are similar in their phenotypic plasticity. Further, is performance of an individual genotype within a population affected by selection for greater or less phenotypic plasticity? To test this concept further, we began with a highly heterogeneous population of tall fescue and selected for high and low rates of leaf area expansion during vegetative growth stages (Reeder *et al.*, 1984). Response to selection for leaf area expansion rate was near-linear in both directions for five generations, increasing about 10.7% per cycle and decreasing about 8.8% per cycle.

Seed produced on the fourth-cycle selections was planted in field plots at two locations (Nelson *et al.*, 1985). The yield of reproductive growth showed a small positive response to selection for low leaf growth rate after two cycles, but was reduced in subsequent cycles. The density of reproductive tillers was reduced after the third and fourth cycles of selection for high leaf growth rate, while the

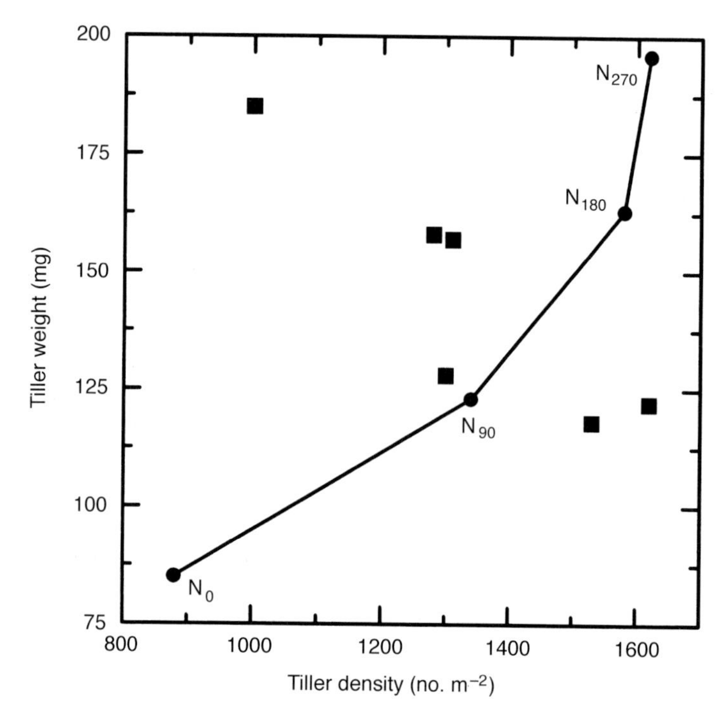

Fig. 6.2. Mean response (●) of six tall fescue genotypes to annual N fertilization rates of 0 to 270 kg ha^{-1}. Solid squares (■) are the mean response for each genotype that had been selected for a range in tiller weight. (Adapted from Nelson and Zarrough, 1981.)

density of vegetative tillers remained the same as that of the base population. Simons *et al.* (1973) also found fewer reproductive tillers in genotypes of perennial ryegrass selected for long leaves. The increased productivity due to faster leaf growth was expressed mainly during vegetative growth, because weight per tiller was highly correlated with leaf elongation rate. As in Fig. 6.2 with genotypes, weight per tiller was inversely related to tiller density (Fig. 6.3). Thus, selection for leaf growth rate shifted the genotypic frequency in the population along a response line of altered weight per tiller and altered tiller density. Further analysis of the fourth-generation material showed the 'high' (H) population had lower tiller density than the 'low' (L) population, because the 25% higher leaf elongation rate and increased weight per tiller were associated with a 12% slower plastochron and 11% lower site usage (Zarrough *et al.*, 1984).

Hazard and Ghesquière (1995) induced a genetic shift in the population of long-leafed and short-leafed perennial ryegrass plants, which depended on grazing management. In that case, some individual genotypes did not have enough phenotypic plasticity to adjust and were lost from the population. Thus, phenotypic plasticity appears to be more sensitive and usually can be expressed in shorter time spans than can genotypic plasticity. In theory, the range of phenotypic plasticity for a given genotype should be less than the range of genotypic

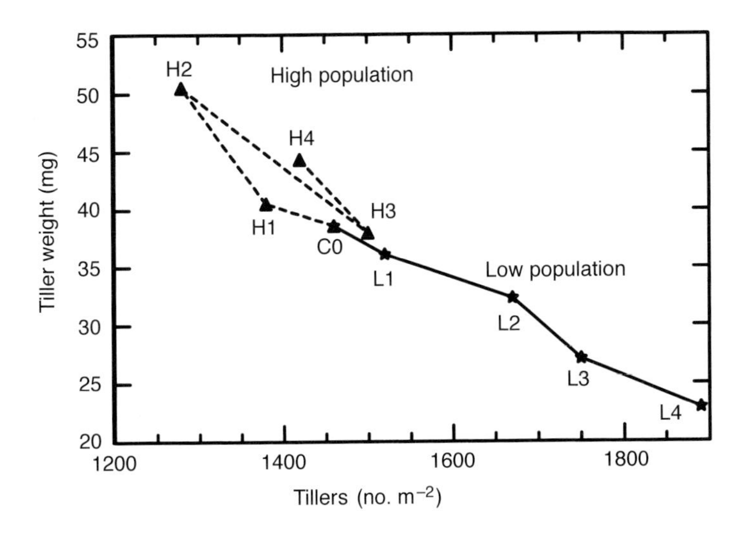

Fig. 6.3. Effect of four cycles of recurrent selection for leaf area expansion rate in tall fescue. Data are for vegetative regrowth in a field of the base population C0 and four cycles of selection in the low (L) and high (H) direction. (Adapted from Reeder *et al.*, 1984; Nelson *et al.*, 1985.)

plasticity expressed by a population (Fig. 6.1B). Unfortunately, there have been few studies to evaluate the relative ranges of these responses.

Phytomers and Growth Zones

To better understand the relationships of leaf growth and tillering, it is critical that the basic processes of initiation, development and survival be evaluated. The morphological structure of a grass plant can be divided into meristems or growing areas, which sequentially develop a series of phytomers. How these processes are interrelated structurally and physiologically affects the individual responses and the potentials for genetic improvement and response to management.

Phytomers

Shoots of grasses are organized as a series of phytomers (Fig. 6.4), which develop in a sequential manner and are connected to give morphological structure (Sharman, 1945). The phytomer consists of the leaf blade, leaf sheath, node, internode and axillary bud (Clark and Fisher, 1987). The shoot apex gives rise to new phytomers. Later, axillary buds formed as part of a phytomer may develop into a new shoot apex for a lateral branch or tiller, which can subsequently develop its own series of phytomers. The shoot apex of the tiller is very similar to

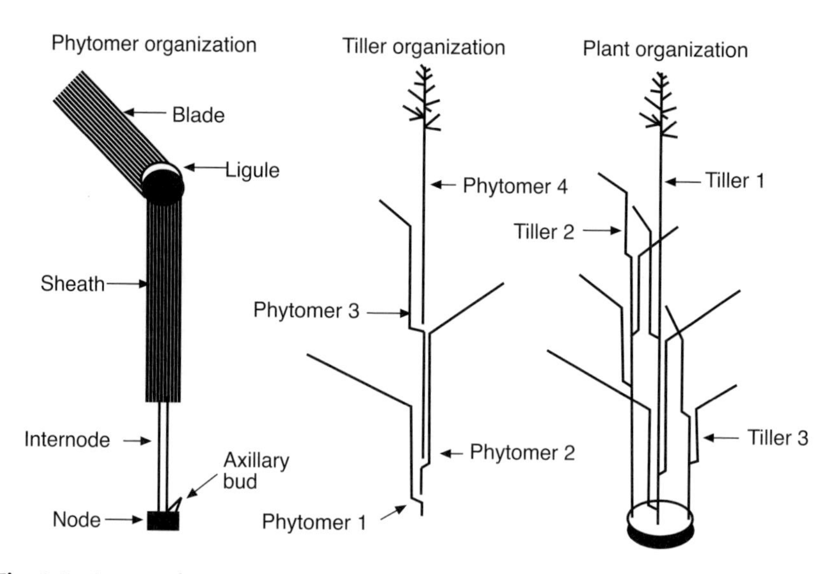

Fig. 6.4. A grass plant consists of a series of interconnected tillers, each being made up of sequential phytomers (adapted from Briske, 1991).

the one for the main shoot and apparently goes through the same embryogenesis on the shoot apex as the main shoot apex did on the developing embryo (Evans and Barton, 1997). Since the meristems are similar, the shape or morphological development of a plant can be considered in terms of interactive events within or among the interconnected phytomers (Fig. 6.4).

Some inconsistency exists about phytomers, as they can be defined in at least two and possibly three different configurations (Clark and Fisher, 1987), which are logical based on the specific purpose. Sharman (1945) defined the phytomer according to the early ontogeny of the shoot, beginning with the leaf blade, leaf sheath, the internode below the disc of leaf insertion, the node and finally the axillary bud at the base of the internode (Fig. 6.4). The axillary bud was observed to be formed last and was positioned one node below the node of the phytomer leaf. More recent molecular biology evidence supports this arrangement, as the axillary bud in maize originates from the stem side of the axil and the cell lineage is clonally related to the leaf above (McDaniel and Poethig, 1988). The Sharman configuration (Fig. 6.4), however, does not have a clear connecting node between the leaf sheath and internode.

In the Japanese literature on rice (e.g. Nemoto *et al.*, 1995), scientists want to emphasize the interconnecting vascular system of the main shoot and define the node of the phytomer as the one connecting the sheath to the stem. But this leaves the axillary bud of the phytomer at the base of the subtending internode without a nodal connection. With forage grasses, Briske (1991), Moore and Moser (1995) and others have used the Sharman model, which gives emphasis to the vascular connections between the main shoot and the axillary bud and subsequent tiller (Fig. 6.4).

To rationalize the anomaly of the node location, Clark and Fisher (1987) point out that the disc of leaf insertion on the node, formed when the leaf primordium first encircles the shoot apex, could actually divide the node. If so, the lower half of the phytomer node and disc of insertion would be at the top of the internode where the leaf is attached, and the subtending internode of the phytomer would include the upper half of the node at the top of the previous phytomer.

Regardless of whether it is anatomically or physiologically correct to split nodes or how the phytomer is defined, Chapman (1996) questions the need for the phytomer concept at all. He argues that growth or development on a phytomer cannot act independently from associated phytomers and meristems (e.g. Skinner and Nelson, 1992, 1995), and thus all should be considered collectively as the shoot. This approach may be useful in terms of the whole plant, but, for understanding and referencing developmental processes, the sequences of phytomer development and interrelationships among phytomers, once the phytomer configuration is described, offer a convenient way to delineate and measure growth events (Moore and Moser, 1995).

The shoot apex

The phytomer on the main stem of grasses begins as a leaf primordium from the shoot apex, the general meristematic area at the end of the stem. The shoot apex, a highly structured and coordinated group of cells (Evans and Barton, 1997), is defined as the meristematic region, which includes the apical dome or promeristem (Esau, 1977), the preferred term for the portion above the youngest leaf primordium or youngest node (Fig. 6.5). The apex also includes the leaf primordia, all non-expanding leaves and associated stem tissues (Esau, 1977). The apex functions to maintain a group of cells to perpetuate the apex and to produce derivative cells, which develop and differentiate to form the plant body.

The shoot apex can be divided into functional layers, or groups of cell types that perform a given function (Nelson, 1996). Grasses nearly always have one or two tunica layers. For example, it appears that tall fescue has one cell layer (Fig. 6.5a), whereas couch grass and wheat have two (Williams, 1960). Regardless of the number of layers, the lateral tunica initiates leaf primordia by periclinal divisions (cell plate parallel to the apex surface) and remains meristematic, adding a few more cells, which lengthen the internodes. The role of the axial tunica is less clear; some anatomists believe it functions mainly to elongate the shoot apex above the last primordium and to develop the inflorescence. Most anatomists, however, believe that cell division occurs at a slow and regular pace in the axial tunica, giving rise to the cells of the lateral tunica (Esau, 1977; Fahn, 1990). Lateral tunica cells have an increasing gradient of mitotic activity as they are displaced from the axial tunica to the region of leaf primordia initiation.

A few corpus initial cells, located just below the axial tunica, provide derivative cells to develop the flank corpus and the pith-like cells of the rib corpus (Fig. 6.5). Corpus initials tend to divide randomly and are multisided; the rib corpus appears more structural and gives rise to the pith in the centre of the stem. The

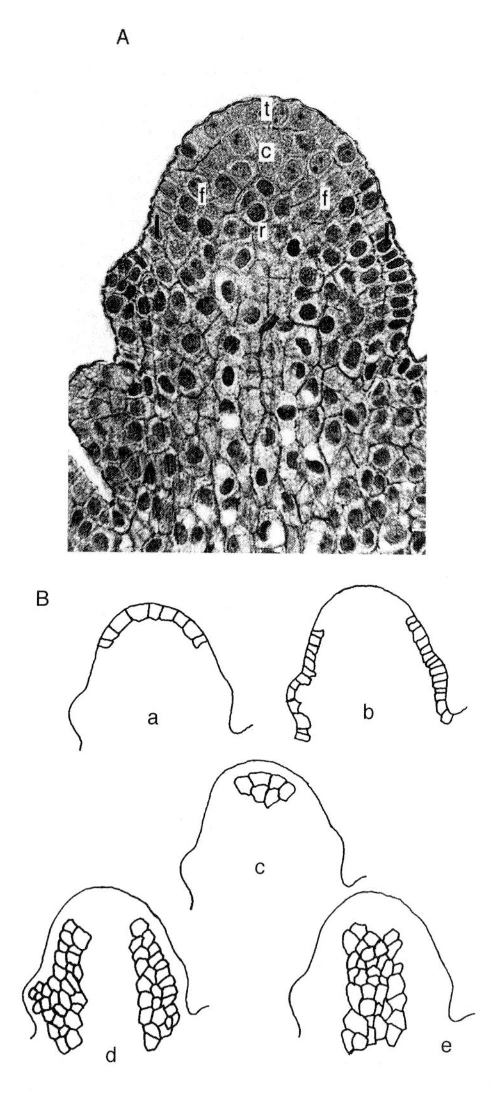

Fig. 6.5. (A) Longitudinal section through the vegetative shoot apex on tall fescue. (B) Line drawings circumscribe the axial tunica (a), lateral tunica (b), corpus initials (c), flank corpus (d) and rib corpus (e). (From Vassey, 1986.)

flank corpus is made up of irregular-shaped cells, which are generally larger than those of the adjacent lateral tunica, increase in size as they are displaced from the corpus initials and have large vacuoles. They also have slower cell division than do the cells of the flank corpus, which contribute the inner cells of the leaf primordium, fibre bundles and the outer regions of stem pith.

Clearly, the shoot apex plays a major role in leaf initiation, early leaf development and tillering regulation. Murphy and Briske (1992) examined the role of apical dominance through growth regulators, such as auxin. Branching of many dicots is regulated by auxin from the shoot apex, and auxin has been implicated in grasses (Jewiss, 1972; Yeh *et al.*, 1976), but newer evidence suggests that the regulation is more complex. Physiological and molecular evidence indicates that cytokinin is involved (Evans and Barton, 1997). The current concept is that auxin in the shoot apex inhibits the synthesis or activity of cytokinin in the nearby axillary bud (Murphy and Briske, 1992). Knowing the relative location of the synthesis of auxin and cytokinin on or near the shoot apex will be helpful for developmental studies to understand plasticity.

Plasticity for leaf initiation

Leaves of grasses are initiated by cell division in the lateral tunica to form a primordium at a position 180° from the previous primordium. Once initiated, a lateral wave of cell divisions encircles the shoot apex to form the young primordium. The cells in the epidermal files of tall fescue leaves divide repeatedly until the primordium is about 0.3 mm long, while maintaining a maximum cell length of about 20 μm (Fig. 6.6). The shoot apex of cool-season grasses usually

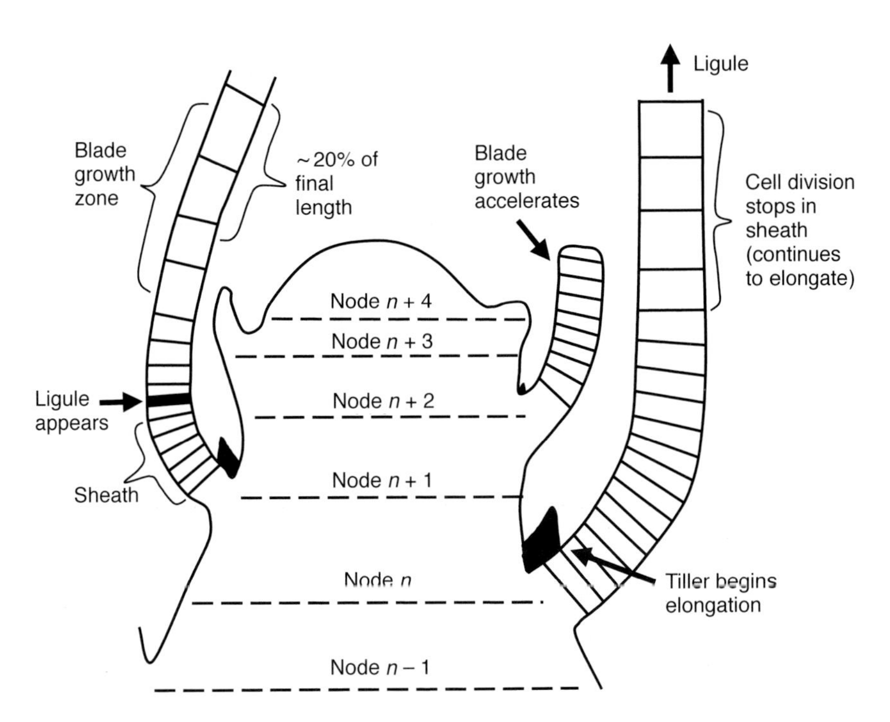

Fig. 6.6. Diagram of longitudinal section through a shoot apex showing synchrony among events occurring at different positions (adapted from Skinner and Nelson, 1994c, 1995).

has three or more leaf primordia, which are shorter than 1 mm and consist of dividing cells. About the time the fourth primordium is initiated on tall fescue seedlings, in a sequential manner the oldest primordium begins or accelerates cell elongation to form the typical leaf growth zone. Cell elongation at the leaf base pushes the leaf tip away from the shoot apex. The ligule between the blade and sheath forms when the leaf is about 1.0 mm long (Skinner and Nelson, 1994c, 1995).

Leaf initiation on the shoot apex is rarely, if ever, skipped (Evans and Barton, 1997), which limits plasticity, although the rate may change as the plant develops (Skinner and Nelson, 1994b). The plastochron – the time interval between successive leaf appearances – cannot be monitored easily on intact plants, so most investigators measure the time between appearances of leaf tips or ligules of elongating leaves above the whorl. The assumption is that rates of primordia initiation and the appearance of leaf structures are correlated (e.g. Nemoto *et al.*, 1995; McMaster *et al.*, 1999), i.e. the number of primordia associated with the shoot apex remains constant, which may not be the case with perennial cool-season grasses (Rogan and Smith, 1975). In early vegetative growth and under-water-stress conditions, when leaf elongation rate is lower, the shoot apex may continue to develop primordia and build up a supply. The abundance of primordia on the apex may be an asset when the stress is relieved (Horst and Nelson, 1979) and so shorten the phyllochron during early regrowth after cutting (Volaire *et al.*, 1998).

Temperature is the major factor regulating the phyllochron (Wilhelm and McMaster, 1995), as temperatures above or below the optimum lengthen the phyllochron. Nutrient availability at non-extreme levels has little influence, except that low N (Longnecker *et al.*, 1993) and low P (Rodríguez *et al.*, 1998) may slow the phyllochron of wheat. In tall fescue, high N tends to increase leaf elongation rate and leaf length, factors known to slow the phyllochron (Zarrough *et al.*, 1984), but in field studies different N rates and water stress did not affect the phyllochron of crested and intermediate wheatgrass (Frank and Bauer, 1995). Water stress tended to lengthen the phyllochron of perennial ryegrass (Volaire *et al.*, 1998), whereas exposure to high atmospheric CO_2 generally shortens the phyllochron (McMaster *et al.*, 1999). The phyllochron of tall fescue lengthens as the canopy develops, probably due to shading factors and tiller development (Skinner and Nelson, 1994a). Overall, genetic differences occur within populations (Frank and Bauer, 1995), but, except for temperature, rates of leaf appearance are generally less plastic in response to the environment than are rates of leaf growth and tillering.

Plasticity for leaf elongation

Leaf elongation rate and leaf length are genetically controlled in tall fescue (Nelson *et al.*, 1977), perennial ryegrass (Rhodes and Mee, 1980) and most cool-season grasses that have been evaluated. Except for water stress, the response is due largely to regulation of cell production, rather than epidermal cell length (Volenec and Nelson, 1983). In addition, there is considerable phenotypic plasticity due to effects of shade (Allard *et al.*, 1991), temperature (McWilliam, 1978), N supply (Volenec and Nelson, 1983; Gastal *et al.*, 1992) and water stress (Volaire *et al.*, 1998).

Genetic increases in leaf photosynthetic rate are generally unrelated or somewhat negatively associated with changes in leaf elongation rate (Wilhelm and Nelson, 1978; Jones *et al.*, 1979), but are positively associated with specific leaf area. Instead of enhancing leaf growth, the increased photosynthate is used to support additional root growth and tillering (Nelson, 1988; McMaster *et al.*, 1999). In the same manner, increasing substrate available for growth by selection for slow dark respiration had little influence on leaf elongation, but enhanced tillering and root growth (Robson, 1982). These data suggest that, in normal conditions, the leaf growth zone has a high priority for the use and accumulation of carbohydrate (Volenec and Nelson, 1984). The leaf growth zone is rarely carbohydrate-deficient, as carbohydrate is accumulated even under low light conditions (Schnyder and Nelson, 1989), which markedly reduce tillering and root growth. In contrast, leaf growth rate is very responsive to N supply, as N strongly affects cell production (MacAdam *et al.*, 1989).

Responses of leaf growth to temperature change occur in a few minutes, due to the enzymatic control of cell division and cell elongation (Durand *et al.*, 1999). The response is slightly slower for decreases in external N supply, as N is redistributable among tissues, and the time for cell cycles ranges from 12–13 h for mesophyll cells (MacAdam *et al.*, 1989) to 28 h for epidermal cells (Skinner and Nelson, 1994c). It may take 3–6 days to respond to water stress, depending on how rapidly the soil water is depleted (Volaire *et al.*, 1998). Recovery can be rapid for all three stresses, however, and in some cases the recovery growth of stressed plants can be more rapid than the growth of non-stressed plants (Horst and Nelson, 1979; Volaire *et al.*, 1998).

During vegetative growth stages, cool-season grasses, such as tall fescue and perennial ryegrass, generally have two leaves elongating at the same time. As each leaf reaches its full length, the tip of the next leaf is emerging above the previous sheaths and is about 50% of its final length, and the primordium at the next node is beginning rapid elongation. Both leaves elongate at a similar rate. Since the plastochron and phyllochron do not change rapidly, i.e. tend not to be plastic, the leaf growth rate of cool-season grasses needs to be regulated in a sensitive manner, probably to aid in survival and recovery after stress. Usually N is the most limiting nutrient in natural and extensively managed grasslands, so the close association between N and cell production establishes a moderately sensitive resource control for leaf growth.

In contrast to tall fescue and perennial ryegrass, several C_4 grasses in good environments are able to up-regulate so as to have three or more leaves growing at one time, perhaps in response to greater photosynthate, water or nutrient supplies. Increasing or decreasing the numbers of growing leaves would contribute to phenotypic plasticity, but probably as a coarse or long-term control compared with short-term responses like leaf elongation rate. The relative response of C_4 grasses needs to be evaluated and compared with that of cool-season grasses.

Plasticity for tillering

In contrast to leaf growth, which is overlapping and appears to be continuous, tillering is regulated independently in an on/off manner at each site. Three major regulatory steps are involved. First, the axillary bud needs to be developed, then the bud must be activated to elongate and develop the visible tiller and finally the environment must be adequate to ensure survival and the continued growth of the tiller.

Initiation of axillary buds

Sharman (1945) described the initiation process of axillary buds in couch grass. Since the axillary bud is the last part of the phytomer to be formed, its initiation can be delayed by about one plastochron (Fig. 6.6). By then, the leaf primordium of the phytomer is developing in size and the leaf primordium for the next phytomer, which occurs on the opposite side of the shoot apex, is initiated. This delay and the positioning of the axillary bud one node below the disc of insertion of the primordium make it unclear as to whether the anticlinal divisions to initiate the axillary bud occur in lateral tunica tissue that is still dividing to expand the internode, or whether it originates from partially differentiated tissues along the edge of the internode (Evans and Barton, 1997). In either case, the initiation event is specific to a region, positioning the axillary bud just above the disc of insertion, near the initiation point of the previous primordium (Fig. 6.6). Similarly to the development of the leaf primordium, the corpus fills in the axillary bud to form a characteristic shell zone of undifferentiated tissue.

It is generally accepted that every phytomer forms the axillary initials and the shell zone, and thus has the potential to form a tiller (e.g. Simons *et al.*, 1973; Zarrough *et al.*, 1984; Skinner and Nelson, 1992). After the initial cell divisions, the tissue in the shell zone must gradually organize an apical meristem and a prophyll, the modified leaf that encloses the shoot apex of the tiller. The regulation of this important first step is not understood, and yet it is vitally important and probably involves cytokinins (Evans and Barton, 1997; Briske and Derner, 1998). When it occurs and how long it takes are also considerations. Our experience with young seedlings of tall fescue (Skinner and Nelson, 1994c) suggests that each axillary bud normally develops through this stage until it has a characteristic shape.

Release of the tiller

The second stage is the release of the axillary bud and its associated tissue to develop and elongate the tiller, an event we (Skinner and Nelson, 1992, 1994c) have indicated is coordinated with several nearly simultaneous events (Fig. 6.6). The release of the axillary bud at node n appears to be coordinated with the stage when cell division stops in the sheath at the same node (axillary bud formed by phytomer $n + 1$). In agreement with Porter (1985), our observations (not enough data for an unequivocal conclusion) are that, if an axillary bud misses this window, the probability of its release at each new window is markedly reduced. Rarely does a tiller emerge during canopy development after it has passed two or three windows, but it may be released later, e.g. after defoliation. Apparently, the uppermost developed tiller has priority in the developing canopy.

Most cool-season grasses studied have three developed leaves on a vegetative shoot or tiller – in this case, at nodes $n-1$, $n-2$, and $n-3$ (Fig. 6.6). As the fourth leaf develops at node n, the oldest one at $n-3$ is senescing. Thus, the first window for tiller development occurs when there are about two complete live leaves originating below its axillary node (Skinner and Nelson, 1992) and, in most cases, the leaf at the node of any axillary buds that missed the window would be senescing. The flux of materials leaving the senescing leaf may be available to support growth or it may contain inhibitors, but the senescing leaf at node $n-3$ is on the opposite side of the shoot apex from the axillary bud that is developing. Evans and Barton (1997) point out the significance of cell lineages from the apex to the tiller axillary bud at node n. In canopies that are defoliated, however, there is often a greater degree of tiller release, perhaps because leaf removal alters the light environment at the base of the canopy (Gautier *et al.*, 1999). It is unknown if tiller releases after defoliation are as well coordinated according to the leaf growth processes as are those during canopy development.

Recognizing the close linkage between leaf growth and tillering, we evaluated the interrelationship further in tall fescue at the whole plant level. Genetic selection for high leaf elongation rate reduced the tillering rate (Zarrough *et al.*, 1984), mostly due to slowed ontogeny and the development of fewer sites for tillering, but also due to lower site usage. Apparently, faster leaf elongation, due to more active cell division, reduced site use. In another experiment, we trimmed the tillers away from developing plants (Skinner and Nelson, 1994b) before they could contribute significantly to leaf area. Presence or absence of the tiller had little influence on leaf elongation rate or sheath length on the main stem, again suggesting that leaf growth rate was a primary factor in the relationship.

We also noted that the presence of tillers caused a slowing in the rate of leaf initiation and, therefore, through altered ontogeny, reduced the number of sites for tiller production. The mechanistic relationship between leaf growth rate and the rate of primordia initiation is less clear. Collectively, these data and observations indicate that the leaf elongation rate has more influence on the tillering rate than the tillering rate has on leaf elongation. This is consistent with many of the other plastic relationships between leaf elongation and tillering, such as the response to shade, N deficiency and water stress.

Tiller survival

Once initiated, the tiller grows leaves and develops phytomers, similar to the main shoot, but again it may not survive to contribute to dry-matter accumulation (Langer, 1963; Zarrough *et al.*, 1983). Tiller death is an important contribution to the dynamic process of tiller regulation in canopies and is probably regulated by another series of factors.

Many cool-season grasses tiller almost continuously, with tiller density

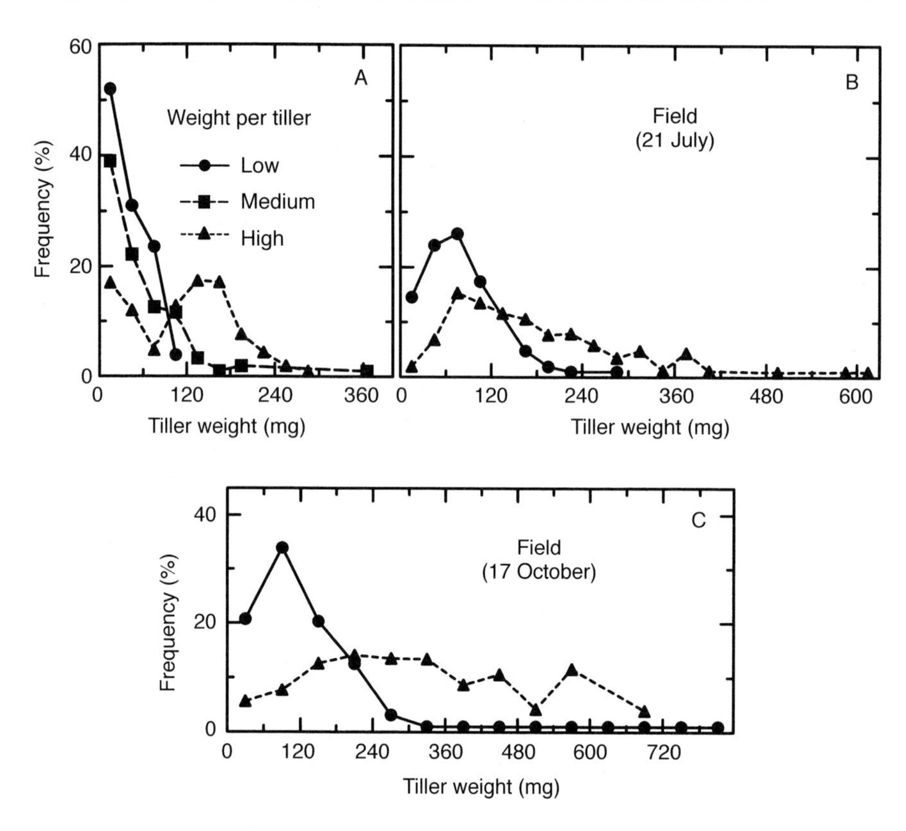

Fig. 6.7. Frequency distribution of tiller weights of vegetative tall fescue in a growth chamber (A) and in the field on two dates (B, C). The growth chamber experiment included a genotype expressing intermediate weight per tiller. (Adapted from Jones *et al.*, 1979.)

reaching a stable level, which depends on the species and management (Hart *et al.*, 1971; Zarrough *et al.*, 1984). At equilibrium, the rate of tiller death is equal to the rate of tiller initiation (Langer, 1963) and a turnover of small tillers occurs (Ong, 1978). Within canopies of tall fescue genotypes selected for differences in weight per tiller, there was a high proportion of tillers that were small, and frequency at the smallest weight was related directly to tillering rate (Fig. 6.7). The average tiller size increased as the season progressed, especially for the genotypes selected for high weight per tiller, but the smaller tillers in the canopy were dying and being replaced (Zarrough *et al.*, 1983). At some critical tiller size or at a minimum light intensity, perhaps also a function of tiller density and leaf area index, the young tiller either survived or died.

Regulating tillering by altering turnover tends to be inefficient in terms of carbon investment, but offers the advantage of having young tillers available in the canopy for recovery after defoliation. It also offers a secondary mechanism for environmental regulation of tiller density beyond that of site usage, which occurs several days or even weeks earlier. The relative importance of the effects of light quality (i.e. R/FR ratio (Casal *et al.*, 1985; Gautier *et al.*, 1999)) and light quantity on tiller survival is not fully known.

Ong (1978) suggests that the shoot can provide resources via vascular connections to the developing tillers, but the capacity is finite. Thus, most investigators suggest that adequate light for carbon acquisition by the new tiller is critical for survival. In contrast, de Ropp (1945) found that root initiation from the shoot stimulated apex activity, those apices that developed roots having an enhanced development of leaves. Nemoto *et al.* (1995) suggested that root initiation and leaf development in rice are highly synchronized at each phytomer. Subsequent root growth is less synchronized. Interestingly, root initiation and root growth are often directly related to sucrose supplies (Williams and Farrar, 1990). Thus, there is probably a role for light intensity in carbohydrate production, as well as light quality for regulation. Roots on tillers would also be potential sources of cytokinin.

Unfortunately, there is little information regarding the timing and regulation of root initiation on grass shoots, especially from developing tillers. Yet initiated roots do not emerge and grow into dry soils and, if root growth is impaired, the acquisition of water and minerals is reduced and supplies or balances of growth hormones from growing roots, such as cytokinins and abscisic acid, are altered. Subsequently, without being able to develop an independent root system, the newly emerged tiller may not survive. Several investigators have concluded that tiller death is an important factor involved in regulating tiller density in swards (Zarrough *et al.*, 1983). Detailed studies on tiller birth and death (see Matthew *et al.*, Chapter 7, this volume) should help us to understand the problem and the potentials for exploiting plasticity for tiller survival.

Dynamics of plasticity for tillering

Based on the above, it is clear that grasses overproduce axillary buds and tillers and then down-regulate to match the environment and management regime. In this sense, tillering would appear to be an exploitative or opportunity response and not an aggressive response regarding plant growth and competitiveness, especially if light intensity and light quality are considered to be the dominant environmental factors.

Recently, Durand *et al.* (1999) published a developmental model for leaf growth, from leaf initiation to final size, for tall fescue seedlings at 21°C. Figure 6.8 shows the time relationship for leaf length (Durand *et al.*, 1999) and the developmental sequence of tiller development at the same node (Skinner and Nelson, 1994c, 1995). Using a phyllochron of 12 days (Durand *et al.*, 1999) and beginning on day zero with primordia development of phytomer 4 on the fourth node, leaf 4 reached its final length on about day 42. Both Skinner and Nelson (1994c) and Durand *et al.* (1999) determined that cellular development for the leaf primordium was slow until the primordium reached a length of 1.0 to 2.5 mm, primarily

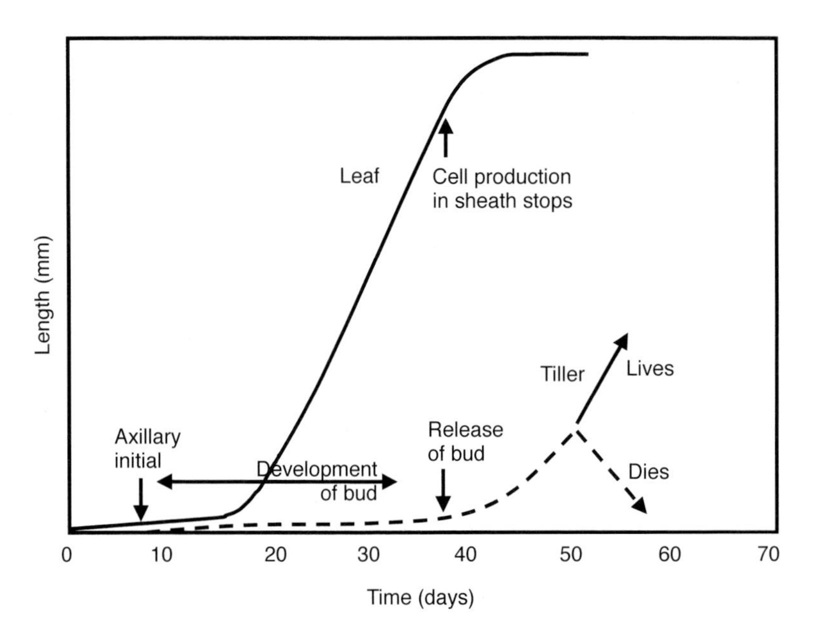

Fig. 6.8. Tiller development on node *n* related to leaf growth on node *n*. Leaf growth is adapted from Durand *et al.* (1999), tiller growth from Skinner and Nelson (1994c) and tiller death from Zarrough *et al.* (1983).

because growth was due to continued division to add more cells. There was a lag period during active cell division, followed by slow (Durand *et al.*, 1999) or arrested growth (Skinner and Nelson, 1994c). Active cell elongation and rapid leaf elongation began on day 15, as the leaf growth zone was forming and cells were elongating as well as dividing (see Fig. 6.6). Tall fescue very typically has two leaves elongating at one time, so active elongation of leaf 3 would begin about day 13 (Fig. 6.8). Active elongation of leaf 5 would begin two phyllochrons later, about day 27, and that of leaf 6 about day 39. These latter times would correspond closely with the times for active tiller elongation at nodes 3 and 4, respectively (see Fig. 6.6).

The initiation of the axillary bud associated with node 4 would arise as the last step in the formation of the next phytomer (Sharman, 1945), but the exact timing is still unknown. It is known that the tiller initials have several cell divisions and a lag phase, similar to the leaf primordia, and that release of the tiller bud and rapid tiller elongation begin at a time similar to the activation of leaf elongation two nodes higher (Fig. 6.6). Therefore, active elongation of the tiller at node 4 would also correspond with the initiation of rapid elongation of leaf 6, which would occur two phyllochrons later, i.e. on about day 39. Near this same date, the cell production and elongation of leaf blade 4 would be complete, cell production of the leaf sheath would be ceasing and final growth of leaf 4 would be due to elongation of sheath cells to collapse the growth zone (Fig. 6.6).

During the events described above on day 39, the ligule for the leaf at node 5 would be forming, just after the associated blade had begun forming a growth zone (Fig. 6.6). Ligule development effectively divides the sheath and blade meristems (Schnyder *et al.*, 1990), an event that is not fully understood. It is known, however, that little, if any, photosynthate from the developing leaf is exported until shortly after the beginning of sheath elongation (Bregard and Allard, 1999). This timing of sink-to-source transition occurs near that of active tiller elongation at the associated node.

Early in plant development, the elongation rate of the first tiller leaf is generally near 90% of that for the main stem leaf, and then, as the canopy develops, the elongation rate of the tiller leaf decreases proportionally. This gradually slows the phyllochron, which is based upon appearance of the leaf above the surrounding sheaths. This response may be governed by light quality or by the supply of resources from the main stem (Ong, 1978). Subsequent survival probably depends on the ability of the young tiller to become self-sufficient in carbon supply and to develop roots.

Within a given plant, the above events appear to be synchronized between sequential phytomers, and the synchrony extends to other phytomers, especially among the main stem tillers (Skinner and Nelson, 1992). Site usage was higher when high synchrony among tillers was maintained, suggesting that there may be common signals from the main stem that are regulatory. Tillers that are of higher order and developmentally further from the main stem tend to lose synchrony with the main stem and may not be at the appropriate developmental stage to initiate rapid growth when the window for growth of synchronized tillers is available.

Conclusion

Leaf growth and tillering appear to be interconnected in most of the cool-season grasses studied in detail, and both exhibit phenotypic plasticity. Selecting for high rates of leaf growth reduces tillering through altering ontogeny by lengthening the phyllochron, perhaps due to increased shade from the larger leaves and higher leaf area index. It also reduces site usage. Providing there is adequate radiation intensity to meet carbohydrate needs, site usage is probably determined as an early warning system based on changes in light quality, especially the R/FR balance (Casal *et al.*, 1985; Gautier *et al.*, 1999). The mechanism for site usage is unknown, but it is associated with the ability for rapid elongation of the tiller bud to develop a visible tiller. The subsequent fate of the tiller in terms of living or dying probably depends on its obtaining sufficient light energy to give the tiller a reasonable chance to be independent. Rooting of the tiller may also be critical.

Hopefully, this insight will lead to more research on the sensitive areas in the system that may be regulatory. Of particular interest is the role of cytokinin (Murphy and Briske, 1992), as it and other growth regulators have been implicated, and molecular biologists can offer new tools (Evans and Barton, 1997). We attempted to evaluate the levels of cytokinins in root tips (source) and leaf growth zones in tall fescue as affected by N fertilization (Nojima *et al.*, 1998); but, unfortunately, our method of immunoaffinity purification and high-performance liquid chromatography (HPLC) analysis for cytokinin was not sensitive enough with the small samples for definitive conclusions.

In general, the cool-season grasses studied, mainly perennial ryegrass and tall fescue, tend to be very conservative in regulation of leaf growth rates and tillering interrelationships. One of the goals is to relate technology to other species, especially warm-season grasses. Unfortunately, except for some range grasses (e.g. Briske and Derner, 1998), there is less comparable information on C_4 grasses, and there is very little information on C_4 pasture grasses adapted to humid areas. For understanding the ability to transfer technology from the C_3 grasses, several significant questions need to be answered:

1. How many leaves on the C_4 grass are growing at one time? What is the plasticity for this character?
2. Are the plastochron and phyllochron related in C_4 species and can they change rapidly?
3. What proportion of potential sites in C_4 species is used for tiller formation?
4. What is the magnitude of tiller turnover in C_4 canopies by continued death and birth?
5. Do shoot apices of C_4 grasses function like those of C_3 grasses?
6. Is tillering plasticity mostly related to changes in ontogeny or to site usage?
7. What are the relative effects of radiation density and radiation quality on tillering in C_4 species?
8. What is the regulatory process of stolon and rhizome initiation and growth?
9. How are endogenous growth hormones involved in regulating leaf growth and tillering?

References

Ainouche, M., Misset, M.-T. and Houn, A. (1995) Genetic diversity in Mediterranean diploid and tetraploid *Bromus* L. (section *Bromus* Sm.) populations. *Genome* 38, 879–888.

Allard, G., Nelson, C.J. and Pallardy, S.G. (1991) Shade effects on growth of tall fescue. I. Leaf anatomy and dry matter partitioning. *Crop Science* 31, 163–167.

Bregard, A. and Allard, G. (1999) Sink to source transition in developing leaf blades of tall fescue. *New Phytologist* 141, 45–50.

Bretagnolle, F. and Thompson, J.D. (1996) An experimental study of ecological differences in winter growth between sympatric diploid and autotetraploid *Dactylis glomerata*. *Journal of Ecology* 84, 343–351.

Briske, D.D. (1991) Developmental morphology and physiology of grasses. In: Heitschmidt, R.K. and Stuth, J.W. (eds) *Grazing Management: An Ecological Perspective*. Timber Press, Portland, Oregon, pp. 85–108.

Briske, D.D. and Derner, J.D. (1998) Clonal biology of caespitose grasses. In: Cheplick, G.P. (ed.) *Population Biology of Grasses*. Cambridge University Press, Cambridge, UK, pp. 106–135.

Casal, J.J., Deregibus, V.A. and Sanchez, R.A. (1985) Variations in tiller dynamics and

morphology in *Lolium multiflorum* Lam. vegetative and reproductive plants as affected by differences in red/far-red irradiation. *Annals of Botany* 56, 553–559.
Casper, B.B. and Jackson, R.B. (1997) Plant competition underground. *Annual Review of Ecology and Systematics* 28, 545–570.
Chapman, C.R. (1996) *The Biology of Grasses*. CAB International, Wallingford, UK, 273 pp.
Chapman, D.F. and Lemaire, G. (1993) Morphogenic and structural determinants of plant regrowth after defoliation. In: Baker, M.J. (ed.) *Grasslands for Our World*. SIR Publishing, Wellington, New Zealand, pp. 55–64.
Cheplick, G.P. (ed.) (1998) *Population Biology of Grasses*. Cambridge University Press, Cambridge, UK, 399 pp.
Clark, L.G. and Fisher, J.B. (1987) Vegetative morphology of grasses: shoots and roots. In: Söderstrom, T.R., Hilu, K.W., Campbell, C.S. and Barkworth, M.E. (eds) *Grass Systematics and Evolution*. Smithsonian Institute Press, Washington, DC, pp. 37–45.
Davies, A. and Thomas, H. (1983) Rates of leaf and tiller production in young spaced perennial ryegrass plants in relation to soil temperature and solar radiation. *Annals of Botany* 57, 591–597.
de Ropp, R.S. (1945) Studies in the physiology of leaf growth. I. The effect of various accessory growth factors on the growth of the first leaf of isolated stem tips of rye. *Annals of Botany* 9, 370–381.
Diggle, P.K. (1994) The expression of andromonoecy in *Solanum hirtum* (Solanaceae): phenotypic plasticity and ontogenetic contingency. *American Journal of Botany* 81, 1354–1365.
Durand, J.-L., Schäufele, R. and Gastal, F. (1999) Grass leaf elongation rate as a function of developmental stage and temperature: morphological analysis and modeling. *Annals of Botany* 83, 577–588.
Esau, K. (1977) *Anatomy of Seed Plants*, 2nd edn. Wiley, New York, 550 pp.
Evans, M.S. and Barton, M.K. (1997) Genetics of angiosperm shoot apical meristem development. *Annual Review of Plant Physiology and Plant Molecular Biology* 48, 673–701.
Fahn, A. (1990) *Plant Anatomy*, 4th edn. Pergamon Press, New York, 588 pp.
Frank, A.B. and Bauer, A. (1995) Phyllochron differences in wheat, barley, and forage grasses. *Crop Science* 35, 19–23.
Gastal, F., Belanger, G. and Lemaire, G. (1992) A model of the leaf extension rate of tall fescue in response to nitrogen and temperature. *Annals of Botany* 70, 437–442.
Gautier, H., Varlet-Grancher, C. and Hazard, L. (1999) Tillering responses to the light environment and to defoliation in populations of perennial ryegrass (*Lolium perenne* L.) selected for contrasting leaf length. *Annals of Botany* 83, 423–429.
Gibson, D., Casal, J.J. and Deregibus, V.A. (1992) The effect of plant density on shoot and leaf lamina angles in *Lolium multiflorum* and *Paspalum dilatatum*. *Annals of Botany* 70, 69–73.
Glimskär, A. and Ericsson, T. (1999) Relative nitrogen limitation at steady-state nutrition as a determinant of plasticity in five grassland plant species. *Annals of Botany* 84, 413–420.
Hart, R.H., Carlson, G.E. and McCloud, D.E. (1971) Cumulative effect of cutting management on forage yield and tiller densities of tall fescue and orchardgrass. *Agronomy Journal* 63, 895–898.
Hazard, L. and Ghesquière, M. (1995) Evidence from the use of isozyme markers of

competition in swards between short-leaved and long-leaved perennial ryegrass. *Grass and Forage Science* 50, 241–248.

Horst, G.L. and Nelson, C.J. (1979) Compensatory growth of tall fescue following drought. *Agronomy Journal* 71, 559–563.

Huber, H., Lukács, S. and Watson, M.S. (1999) Spatial structure of stoloniferous herbs: an interplay between structural blueprint, ontogeny and phenotypic plasticity. *Plant Ecology* 141, 107–115.

Jewiss, O.R. (1972) Tillering in grasses – its significance and control. *Journal of the British Grassland Society* 27, 65–82.

Johnson, L.B. (1972) Polyploidy as a factor in the evolution of grasses. In: Younger, V.B. and McKell, C.M. (eds) *The Biology and Utilization of Grasses*. Academic Press, New York, pp. 18–35.

Jones, R.J., Nelson, C.J. and Sleper, D.A. (1979) Seedling selection for morphological characters associated with yield of tall fescue. *Crop Science* 19, 631–634.

Keeler, K.H. (1998) Population biology of intraspecific polyploidy in grasses. In: Cheplick, G.P. (ed.) *Population Biology of Grasses*. Cambridge University Press, Cambridge, UK, pp. 183–206.

Langer, R.H.M. (1963) Tillering in herbage grasses. *Herbage Abstracts* 33, 141–148.

Longnecker, N., Kirby, E.J.M. and Robson, A. (1993) Leaf emergence, tiller growth, and apical development of nitrogen-deficient spring wheat. *Crop Science* 33, 154–160.

MacAdam, J.W., Volenec, J.J. and Nelson, C.J. (1989) Effects of nitrogen on mesophyll cell division and epidermal cell elongation in tall fescue leaf blades. *Plant Physiology* 89, 549–556.

McDaniel, C.N. and Poethig, R.S. (1988) Cell-lineage patterns in the shoot apical meristem of the germinating maize embryo. *Planta* 175, 13–22.

McMaster, G.S., LeCain, D.R., Morgan, J.A., Aiguo, L. and Hendrix, D.L. (1999) Elevated CO_2 increases CER, leaf and tiller development, and shoot and root growth. *Journal of Agronomy and Crop Science* 183, 119–128.

McWilliam, J.W. (1978) Response of pasture plants to temperature. In: Wilson, J.R. (ed.) *Plant Relations in Pastures*. CSIRO, East Melbourne, Australia, pp. 17–34.

Mooney, H.A., Winner, W.E. and Pell, E.J. (1991) *Response of Plants to Multiple Stresses*. Academic Press, San Diego, California, 427 pp.

Moore, K.J. and Moser, L.E. (1995) Quantifying developmental morphology of perennial grasses. *Crop Science* 35, 37–43.

Moser, L.E., Buxton, D.R. and Casler, M.D. (1996) *Cool-season Forage Grasses*. Agronomy Monograph 34, American Society of Agronomy, Madison, Wisconsin, 841 pp.

Moulia, B., Loup, C., Chartier, M., Allirand, J.M. and Edelin, C. (1999) Dynamics of architectural development of isolated plants of maize (*Zea mays* L.), in a non-limiting environment: the branching potential of modern maize. *Annals of Botany* 84, 645–656.

Murphy, J.S. and Briske, D.D. (1992) Regulation of tillering by apical dominance: chronology, interpretive value, and current perspectives. *Journal of Range Management* 45, 419–429.

Nelson, C.J. (1988) Genetic association between photosynthetic characteristics and yield: review of the evidence. *Plant Physiology and Biochemistry* 26, 543–554.

Nelson, C.J. (1996) Physiology and developmental morphology. In: Moser, L.E., Buxton, D.R. and Casler, M.D. (eds) *Cool-season Forage Grasses*. Agronomy Monograph 34 American Society of Agronomy, Madison, Wisconsin, pp. 87–125.

Nelson, C.J. (1998) Methods of analysis of quantitative data in crop research: an overview.

In: Chopra, V.L., Singh, R.B. and Varma, A. (eds) *Crop Productivity and Sustainability – Shaping the Future. Proceedings of 2nd International Crop Science Congress*. Oxford and IBH Publishing Company, New Delhi, pp. 753–758.

Nelson, C.J. and Zarrough, K.M. (1981) Tiller density and tiller weight as yield determinants in vegetative swards. In: Wright, C.E. (ed.) *Plant Physiology and Herbage Production*. British Grassland Society Occasional Symposium 13, British Grassland Society, Hurley, UK, pp. 25–29.

Nelson, C.J., Asay, K.H. and Sleper, D.A. (1977) Mechanisms of canopy development of tall fescue genotypes. *Crop Science* 17, 449–452.

Nelson, C.J., Sleper, D.A. and Coutts, J.H. (1985) Field performance of tall fescue selected for leaf-area expansion rate. In: *Proceedings of the XV International Grassland Congress*. The Science Council of Japan, Nishi Nasuno, Japan, pp. 320–322.

Nemoto, K., Morita, S. and Baba, T. (1995) Shoot and root development in rice related to the phyllochron. *Crop Science* 35, 24–29.

Nojima, H., Nelson, C.J. and Coutts, J.H. (1998) Cytokinins in root tips of tall fescue under different N conditions. In: *American Society of Agronomy Abstracts*. American Society of Agronomy. Madison, Wisconsin, p. 90.

Ong, C.K. (1978) The physiology of tiller death in grasses. I. The influence of tiller age, size and position. *Journal of the British Grassland Society* 33, 197–203.

Porter, J.R. (1985) Approaches to modeling canopy development in wheat. In: Day, W. and Atkin, R.K. (eds) *Wheat Growth Modeling*. Plenum Press, New York, pp. 69–81.

Reeder, L.R., Sleper, D.A. and Nelson, C.J. (1984) Response to selection for leaf area expansion rate of tall fescue. *Crop Science* 24, 97–100.

Rhodes, I. and Mee, S.S. (1980) Changes in dry matter yield associated with selection for canopy characters in ryegrass. *Grass and Forage Science* 35, 35–39.

Richards, J.H. (1993) Physiology of plants recovering after defoliation. In: Baker, M.J. (ed.) *Grasslands for Our World*. SIR Publishing, Wellington, New Zealand, pp. 46–54.

Robson, M.J. (1982) The growth and carbon economy of selection lines of *Lolium perenne* cv. 323 with differing rates of dark respiration. I. Grown as simulated swards during a regrowth period. *Annals of Botany* 49, 321–329.

Rodríguez, D., Pomar, M.C. and Goudriaan, J. (1998) Leaf primordium initiation, leaf emergence and tillering in wheat (*Triticum aestivum* L.) grown under low-phosphorus conditions. *Plant and Soil* 202, 149–157.

Rogan, P.G. and Smith, D.L. (1975) Rates of leaf initiation and leaf growth in *Agropyron repens* (L.) Beauv. *Journal of Experimental Botany* 26, 70–78.

Schnyder, H. and Nelson, C.J. (1989) Growth rates and assimilate partitioning in the elongation zone of tall fescue leaf blades at high and low irradiance. *Plant Physiology* 90, 1201–1206.

Schnyder, H., Seo, S., Rademacher, I.F. and Kuhbauch, W. (1990) Spatial distribution of growth rates and epidermal cell lengths in the elongation zone during leaf development in *Lolium perenne* L. *Planta* 181, 423–431.

Sharman, B.C. (1945) Leaf and bud initiation in the Gramineae. *Botanical Gazette* 106, 269–289.

Simons, R.G., Davies, A. and Troughton, A. (1973) Effect of spacing on the growth of genotypes of perennial ryegrass. *Journal of Agricultural Science (Cambridge)* 80, 495–502.

Skinner, R.H. and Nelson, C.J. (1992) Estimation of potential tiller production and site usage during tall fescue canopy development. *Annals of Botany* 70, 493–499.

Skinner, R.H. and Nelson, C.J. (1994a) Role of leaf appearance rate and the coleoptile tiller in regulating tiller production. *Crop Science* 34, 71–75.

Skinner, R.H. and Nelson, C.J. (1994b) Effect of tiller trimming on phyllochron and tillering regulation during tall fescue development. *Crop Science* 34, 1267–1273.

Skinner, R.H. and Nelson, C.J. (1994c) Epidermal cell division and the coordination of leaf and tiller development. *Annals of Botany* 74, 9–15.

Skinner, R.H. and Nelson, C.J. (1995) Elongation of the grass leaf and its relationship to the phyllochron. *Crop Science* 35, 4–10.

Sleper, D.A. and West, C.P. (1996) Tall fescue. In: Moser, L.E., Buxton, D.R. and Casler, M.D. (eds) *Cool-season Forage Grasses*. Agronomy Monograph 34, American Society of Agronomy, Madison, Wisconsin, pp. 471–502.

Sleper, D.A., Nelson, C.J. and Asay, K.H. (1977) Diallel and path coefficient analysis of tall fescue (*Festuca arundinacea*) regrowth under controlled conditions. *Canadian Journal of Genetics and Cytology* 19, 557–564.

Söderstrom, T.R., Hilu, K.W., Campbell, C.S. and Barkworth, M.E. (eds) (1987) *Grass Systematics and Evolution*. Smithsonian Institute Press, Washington, DC, 473 pp.

Tilman, D. (1988) *Plant Strategies and the Dynamics and Structure of Plant Communities*. Monographs in Population Biology No. 26, Princeton University Press, Princeton, New Jersey, 351 pp.

Vassey, T.L. (1986) Morphological, anatomical, and cytohistological evaluation of terminal and axillary meristems of tall fescue. PhD dissertation, University of Missouri, Columbia, Missouri (Dissertation Abstract no. 87–16730), 96 pp.

Volaire, F., Thomas, H. and Lelievre, F. (1998) Survival and recovery of perennial forage grasses under prolonged Mediterranean drought. I. Growth, death, water relations and solute content in herbage and stubble. *New Phytologist* 140, 439–449.

Volenec, J.J. and Nelson, C.J. (1983) Responses of tall fescue leaf meristems to nitrogen fertilization and harvest frequency. *Crop Science* 23, 720–724.

Volenec, J.J. and Nelson, C.J. (1984) Carbohydrate metabolism in leaf meristems of tall fescue. II. Relationship to leaf elongation modified by nitrogen fertilization. *Plant Physiology* 74, 595–600.

Wilhelm, W.W. and McMaster, G.S. (1995) Importance of the phyllochron in studying development and growth in grasses. *Crop Science* 35, 1–3.

Wilhelm, W.W. and Nelson, C.J. (1978) Leaf growth, leaf aging, and photosynthetic rates of tall fescue genotypes. *Crop Science* 18, 769–772.

Williams, J.H.H. and Farrar, J.F. (1990) Control of barley root respiration. *Physiologia Plantarum* 79, 259–266.

Williams, R.F. (1960) The physiology of growth in the wheat plant. I. Seedling growth and the pattern of growth at the shoot apex. *Australian Journal of Biological Science* 13, 401–428.

Yeh, R.Y., Matches, A.G. and Larson, R.L. (1976) Endogenous growth regulators and summer tillering of tall fescue. *Crop Science* 16, 409–413.

Zarrough, K.M., Nelson, C.J. and Coutts, J.H. (1983) Relationship between tillering and forage yield of tall fescue. II. Pattern of tillering. *Crop Science* 23, 358–342.

Zarrough, K.M., Nelson, C.J. and Sleper, D.A. (1984) Interrelationships between rates of leaf appearance and tillering in selected tall fescue populations. *Crop Science* 24, 565–569.

Tiller Dynamics of Grazed Swards

7

C. Matthew,[1] S.G. Assuero,[2] C.K. Black[3] and N.R. Sackville Hamilton[4]

[1] *Institute of Natural Resources, Massey University, Private Bag 11–222, Palmerston North, New Zealand;* [2] *Universidad Nacional de Mar del Plata, CC 276, 7620 Balcarce, Argentina;* [3] *Williams and Kettle Ltd, PO Box 501, Feilding, New Zealand;* [4] *Institute of Grassland and Environmental Research, Plas Gogerddan, Aberystwyth SY23 3EB, UK*

Introduction

Historically, the study of grass sward behaviour has attracted significant research effort in countries spanning almost every geographical region of the world. In the UK literature, there are a number of reviews dealing wholly or partly with the tillering behaviour of grass swards, including Langer (1963), Dorrington Williams (1970), Davies (1977) and Davies (1988), the last of these having 182 references to research in some seven countries. This English literature tends to be dominated by studies on *Lolium perenne* (perennial ryegrass), is usually directed towards meeting the needs of intensive production systems and often uses artificial swards or swards subject to management that varies widely from farming practice. Literature from other geographical regions tends to cover a more diverse range of species and is often directed at more extensive production systems.

This chapter refers to information that has been presented in previous reviews, and draws on the rich diversity in the wider international literature, but seeks primarily to identify themes and principles that lead to a more general understanding of sward responses and the current thinking about reasons for these responses. To provide a framework for such an understanding, sward responses can conveniently be considered from three perspectives: tiller morphology, regulation of the canopy leaf area index (LAI) and tiller demography.

The tiller morphology perspective views leaf, tiller and root formation as complementary processes occurring at different points on the tiller axis and examines the similarities, differences and interactions between the three processes. The canopy LAI perspective recognizes that increase or decrease in tiller population

 Grassland Ecophysiology and Grazing Ecology (eds G. Lemaire, J. Hodgson, A. de Moraes, C. Nabinger and P.C. de F. Carvalho)

density is an important mechanism for adjustment and optimization of canopy leaf area. The tiller demography perspective recognizes that persistence in grass swards is directly determined by the combined effect of seasonal patterns of tiller natality and mortality.

Tiller Morphology

Interrelationship between leaf, root and tiller formation

Tiller morphology and the relationships and feedback between leaf and tiller formation have been described by Nelson (Chapter 6, this volume), so coverage here will be limited to expansion of selected aspects for completeness or as a basis for the discussion that follows.

Site filling

One point made by Nelson (Chapter 6, this volume) and germane to a discussion of tiller dynamics is that the number of leaves formed determines the potential tiller appearance rate, there being one tiller bud in each leaf axil. The ratio of tiller appearance to leaf appearance, termed site filling, was the first widely used measure of the proportion of buds formed that later grow into tillers (Davies 1974), and has a theoretical maximum of 0.693 (Neuteboom and Lantinga, 1989), assuming strictly sequential tiller appearance and a delay of one phyllochron between leaf and tiller appearance at a given phytomer. More recently, alternative measures with a maximum of 1.0 (or 100%) have also been proposed: site usage (Skinner and Nelson, 1992); nodal probability (Matthew *et al.*, 1998); and specific site usage (Bos and Neuteboom, 1998).

Site filling or site usage values approaching the biological maximum are frequently recorded in *L. perenne* plants in the establishment phase (Neuteboom *et al.*, 1988; Hume, 1991; Van Loo, 1992; Bahmani *et al.*, 2000) or with high nutrient levels – for example, in urine patches (Matthew *et al.*, 1998). In contrast, in established swards, each tiller needs to form only one replacement tiller in its lifetime to maintain the population (Parsons and Chapman, 1999). If the average perennial ryegrass tiller life were 100 days (half life 36–143 days, Table 5 of Korte, 1986) and the average leaf appearance interval 13 days (Fig. 3 of Chapman *et al.*, 1983), this would imply 7.7 leaves produced during the life of an average tiller and a site usage of 13% to produce one tiller in that time. It is interesting, then, that values of site filling reported in field swards tend to be higher than this. For example, Chapman *et al.* (1983) observed site filling values of 0.14–0.43 in hill swards, and Simon and Lemaire (1987) observed values ranging from 0.1 to a little over 0.5 in lowland swards. This suggests measurement bias, perhaps through disturbance effects associated with measurement (Matthew, 1992) or arising where population density counts include smaller classes of tiller but corresponding demographic studies do not. Langer (1956) has indicated that younger tillers contribute little to tiller appearance.

Increased tiller appearance in response to a range of stimuli, together with the capacity for decrease in the tiller population density, are important considerations in regulation of canopy leaf area – for example, in response to change in canopy leaf area (see below). The capacity for rapid tillering in *L. perenne* seems to arise in part from development of tillers at prophyll buds (Mitchell, 1953; Neuteboom and Lantinga, 1989). Not all species show this rapid tillering capacity. For example, *Bromus willdenowii* has a lower site filling rate (around 0.3) in the establishment phase because tillers do not develop from prophyll buds (Hume, 1991).

Equivalent ratios can be calculated for the proportion of root sites used. Klepper *et al.* (1984) described four root sites per phytomer for wheat, and the same appears true for perennial ryegrass (Matthew *et al.*, 1998), though the number of root sites per phytomer can be as high as eight in *Panicum maximum* (D.D. Carvalho, Brazil, 1999, unpublished data) and ten in maize (Demotes-Mainard and Pellerin, 1992). The limited data available on numbers of roots formed per node (Hunt and Thomas, 1985; Matthew and Kemball, 1997; Matthew *et al.*, 1998; Yang *et al.*, 1998) suggests a typical site usage of 40–60% and rather less responsiveness to environmental conditions than tiller formation, although it is known that the number of roots formed is greatly reduced in dry conditions (Troughton, 1980).

It is relevant to the discussion below on canopy leaf area regulation to note that, of the three processes (leaf, tiller and root formation), tiller formation is the most plastic.

True stem

The grass tiller has a jointed, branched true stem, continuous with and extending below the pseudostem (see Fig. 6.6 of Nelson, Chapter 6, this volume). The jointing reflects the individual phytomers laid down below the apical meristem as successive leaves are formed (Silsbury, 1970; Hitch and Sharman, 1971). The branches arise from daughter tiller formation and subsequent true stem formation by the daughter tillers. The true stem varies greatly in morphology between grass species, leading to differences in growth habit. Three major categories recognized in most textbooks are: (i) bunch grasses, which have short true stem segments orientated almost vertically; (ii) creeping grasses, in which there is above-ground plagiotropic internode elongation in the vegetative phase, forming prostrate stems of varying length, usually rooting at the nodes; and (iii) those grasses spreading by underground stems. The fact that tillers of bunch grasses possess a basal segment of true stem is often overlooked.

In current usage, where the true stem of bunch grasses and creeping grasses is referred to, it is commonly termed a stolon, even when buried or comprising short lengths orientated near-vertically (Harris *et al.*, 1979; Matthew *et al.*, 1989a; Brock and Fletcher, 1993; Brock *et al.*, 1997), and an underground horizontal true stem is termed a rhizome (Etter, 1951; Vignolio *et al.*, 1994), However, naming of basal stems in this way creates anomalies. For example, under this usage, the 'rhizomes' of *Paspalum dilatatum* and *Poa pratensis* have very different morphologies, with the former having characteristics closer to those of the 'stolon' of *L. perenne*.

The original definition of the term 'stolon' is 'a prostrate or reclined branch which strikes root at the tip and then develops an ascending growth which becomes an independent plant' (*Oxford English Dictionary*, 1989; attributed to A. Gray, 1880). Clearly, true stems of grasses do not end in orthotropic shoot axes as described in the *Oxford English Dictionary* definition of a stolon, and it has been noted, in respect of equivalent structures in white clover (J.L. Harper, UK, 1999, personal communication), that the distinction is especially important if one wishes to make realistic models of vegetative dispersal and spread. Thus, it would be helpful to develop a more universally applicable terminology.

The true stem provides the vascular connection between roots and leaves (Yang *et al.*, 1998) and reciprocal transport between parent and daughter tillers (Clifford *et al.*, 1973). At times, these structures also appear to provide a connection between young tillers without roots and surviving roots of old tillers that have lost their apical meristem and their leaves (Matthew, 1992).

Internode length is very variable and in ryegrass is increased in response to tiller burial or heavy shade. In one experiment, ryegrass tillers enclosed in artificial sheaths made from opaque plastic tubes produced 150 mm internode elongation in approximately 6 weeks (C. Matthew, unpublished data). Even in a tufted grass like *L. perenne*, formation of true stem tissue provides a mechanism for the spread of dominant genotypes over time (Harris *et al.*, 1979), occurs on a large number of tillers (Korte and Harris, 1987) and accounts for 1.0 t dry matter (DM) ha^{-1} or more (Matthew *et al.*, 1989a) that is not readily harvestable by grazing animals. True stem structures have been described in detail on a per plant basis for ryegrass, tall fescue and cocksfoot (Brock and Fletcher, 1993; Brock *et al.*, 1996; Hume and Brock, 1997).

In *P. dilatatum*, detailed comparison of true stem in grazed and ungrazed swards has shown that grazing increased the number of true stem segments formed, the area occupied per plant, branching order and branching angle, but decreased internode length, diameter and volume and rhizome weight. Grazing also increased the number of internodes per rhizome, possibly through suppression of flowering. It was concluded that these responses could play an important role in persistence (Vignolio *et al.*, 1994, 1998). In the coastal species *Ammophila arenaria*, internode elongation of true stem tissue binds wind-blown sand trapped at stem bases, forming large mounds over time. Observation suggests that more complex patterns also exist. For example, *Cynodon dactylon* appears to have a true-stem structure comprising regular units of three phytomers.

Bud site examination

Depending on the grass species and soil conditions, older or dead lengths of true stem may endure for months or years before eventually decomposing. If carefully separated from surrounding soil, observation of the true stem allows hierarchical branching relationships between neighbouring tillers to be traced back, often for several generations (Etter, 1951; Vignolio *et al.*, 1994). Historical tiller appearance for a limited preceding period can be evaluated (Ryle, 1964; Ito and Nakamura, 1974; Hendrickson and Briske, 1997; Matthew *et al.*, 1998), and in some species (e.g. *P. dilatatum*) individual leaf scars also remain visible, so the same principle could be applied to leaf appearance.

Also relevant is the question of whether or not older, dormant buds have the capacity to form tillers. Retrospective characterization of tiller appearance in *L. perenne* following urine deposition or creation of sward gaps (Matthew *et al.*, 1998) could only be possible where there is a comparatively narrow time window for bud initiation, and findings of Hendrickson and Briske (1997) suggest that the two rangeland species they studied also had a narrow time window for bud initiation. Brock *et al.* (1997) indicate that, in *Festuca arundinacea*, old buds are more likely to form rhizomes.

Canopy Leaf Area Optimization

Continuously grazed swards

A component analysis of sward LAI (equation 1) indicates three variables: tiller population density, leaf number per tiller and leaf length or leaf area.

$$\text{LAI} = \text{Tillers m}^{-2} \times \text{Leaves per tiller} \times \text{Area per leaf (m}^2\text{)} \qquad (1)$$

Leaf area in grasses is largely determined by leaf length, which in turn is controlled by defoliation height. Leaf number per tiller is remarkably constant for a given species, usually averaging between 2.5 and 3.5 leaves per tiller for *L. perenne* (Davies, 1977; Chapman *et al.*, 1983; Yang *et al.*, 1998). (Species with a somewhat higher leaf number per tiller include *Cynosurus cristatus*, *Pennisetum clandestinum* and some *Bromus* spp.) This leaves tiller population density as the component of sward LAI where changes in LAI can readily be expressed. Therefore, at low grazing heights a higher population density of smaller tillers optimizes sward LAI and, conversely, at higher grazing heights a lower population density of larger tillers optimizes sward LAI. This principle of size–density compensation (SDC) in grass swards has been recognized with increasing clarity in successive studies (Langer, 1963; Bircham and Hodgson, 1983; Davies, 1988; Chapman and Lemaire, 1993; Matthew *et al.*, 1995; Hernández Garay *et al.*, 1999).

Initially it was thought that SDC in continuously grazed swards would follow the $-3/2$ self-thinning rule of Yoda *et al.* (1963), but self-thinning trajectories define SDC for an approximately constant canopy leaf area (Mohler *et al.*, 1978), whereas swards grazed at different heights vary systematically in LAI (Bircham and Hodgson, 1983; Matthew *et al.*, 1995), and the SDC slope is normally steeper than $-3/2$ (Matthew *et al.*, 1995). The latter authors calculated an SDC slope correction for change in sward LAI with defoliation height (C_a) and a second correction for packing density implications of change in tiller morphology or shape (C_r) with defoliation height. To explain C_a, consider that, as tiller size increases with increased defoliation height, canopy LAI simultaneously increases. If the $-3/2$ self-thinning line is conceptualized as defining a constant canopy LAI at which self-thinning occurs (Matthew *et al.*, 1995), the change in SDC slope from the theoretical value of $-3/2$ is defined by the movement in the intercept of the self-thinning line represented by the difference in LAI between a long and a short sward (Matthew *et al.*, 1995).

A measure of tiller shape, *R*, is defined by Sackville Hamilton *et al.* (1995) as the ratio (leaf area per tiller)$^{3/2}$: (volume per tiller). Under this definition, *R* has units m^3 m^{-3} and is therefore a dimensionless constant, unaffected by tiller size, denoting the tiller area : volume ratio associated with a particular tiller shape. A ryegrass tiller has an *R* value around 50 (see below). When considering implications for tiller dynamics of tiller shape, a size-independent measure is highly desirable to eliminate anomalies. For example, in ryegrass swards, smaller tillers tend to have a higher leaf DM : pseudostem DM ratio than larger tillers, but when a size correction is introduced, as above, the larger tiller can often have the greater value of *R* (Hernández Garay *et al.*, 1999).

For *L. perenne* data sets examined by Matthew *et al.* (1995), C_a was around 2.0 and C_r around -0.2, indicating that the leaf area correction was the larger of the two and that *R*, or 'leafiness', increased somewhat in longer swards, partially reducing the slope increase predicted by C_a. These two corrections almost exactly explained the discrepancy between the theoretical $-3/2$ slope and the empirical slope. Empirical SDC slopes are observed to be close to $-5/2$ (Matthew *et al.*, 1995), and earlier data, such as in Fig. 3 of Davies (1988), also support this conclusion. In another study, *R* varied from 39.0 at 20 mm defoliation height to 54.8 at 160 mm defoliation height (Hernández Garay *et al.*, 1999). That is to say, leaf length increases proportionately more than pseudostem length as sward height increases. However, not all grass species follow this pattern. For example, in a recent study of *C. dactylon* (Sbrissia *et al.*, 1999), the SDC mechanism appeared to be different. In this case C_r was around +1.0, indicating that stem volume increased proportionately more than the leaf length with increasing sward height, although the method of tiller volume estimation may have contributed to this result.

This theoretical approach to SDC may also have other applications. The distance of plotted tiller size/density coordinates from an arbitrarily placed theoretical SDC line of slope $-3/2$, or the distance of biomass density coordinates from a line of slope $-1/2$, on a double-logarithmic plot, appears to correlate closely with sward LAI, and to have value as a productivity index when comparing otherwise similar swards subjected to different treatments (Bahmani *et al.*, 1998; Hernández Garay *et al.*, 1999). When a certain 'environmental ceiling' sward LAI (ECLA) is reached, SDC appears to switch quite abruptly and not gradually from a steeper slope to $-3/2$ self-thinning (Matthew *et al.*, 1995). The point at which this change occurs appears to have biological meaning, in that it defines an upper limit to plasticity of tiller size for a given cultivar, similar but not identical to that illustrated conceptually by Chapman and Lemaire (1993). This point occurred at densities of 19,950 tillers m^{-2} and 5010 tillers m^{-2} in British and New Zealand data sets, respectively, indicating a tendency for New Zealand plant material to have larger tillers and lower tiller density than British material (Matthew *et al.*, 1995), an observation that agronomists would concur with. The increase in *R* value in long swards in ryegrass, mentioned above, implies a flatter SDC slope during this phase of self-thinning, meaning that the rate of tiller loss from the sward is expected to be accelerated when the sward moves into this phase of growth (Hernández Garay *et al.*, 1999).

It will be interesting in future to explore the ecological implication of inter-species differences in *R* value. *Dactylis glomerata* is able to develop a much higher *R* value than *L. perenne* (C. Matthew, New Zealand, 1998, unpublished data), perhaps explaining the observed dominance of *D. glomerata* over *L. perenne* when a mixed sward containing the two species is subject to mild defoliation. *C. dactylon*, on the other hand, appears to have a low *R* value compared with *L. perenne* (Sbrissia *et al.*, 1999). Additionally, *R* on a per plant basis changes with the level of clonal integration. For example, an interconnected unit of three tillers, each with an individual *R* value of 50, has a combined *R* value of 86.7, and it is not known if this has ecological significance for the performance of clonally integrated plants.

Finally, it would appear that, under very severe defoliation, the plant lacks sufficient energy resources for tillering, and sward tiller population density may fall (Bircham and Hodgson, 1983; Van Loo, 1992; Matthew *et al.*, 1995).

Rotationally grazed swards

Tiller population size/density values, when plotted as coordinates to show movement over time, normally follow a curved trajectory, as herbage accumulates following defoliation. Population density usually increases during early regrowth, but then decreases (Fig. 7.1; see also Kays and Harper, 1974). The rate of tiller appearance at any point in time (i.e. site filling or site usage) is controlled by various internal and external signals and stimuli. Increased tiller appearance in response to an increased red : far red ratio of incident light has been demonstrated for a number of grasses, including *Lolium multiflorum* (Casal *et al.*, 1985) and *L. perenne* (Gautier *et al.*, 1999). In many grass species, tillering tends to be reduced at higher temperatures, suggesting that depletion of carbohydrate reserves through rapid growth and/or higher dark respiration rate acts to reduce tiller bud initiation.

From Fig. 7.1 it is evident that tiller production ceases and tiller death occurs before ECLA is attained. Therefore, in contrast to some other population equilibria reported in the ecological literature, oscillation around ECLA is not expected in tiller population dynamics. It also follows that there will be, at any particular time, a constraint to leaf area determined by rate of growth following defoliation. This has been called elsewhere (G. Lemaire, personal communication) the morphological ceiling leaf area (MCLA), representing the LAI limit

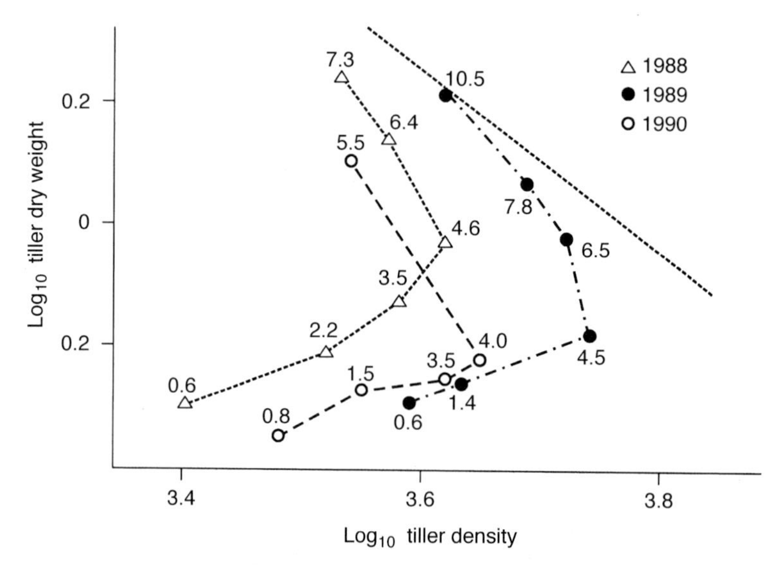

Fig. 7.1. Regrowth trajectory for size/density coordinates of *F. arundinacea*. Numbers adjacent to symbols are LAI values for successive destructive harvests. Note that distances from the arbitrarily placed −3/2 boundary line (dashed line) correlate closely with sward LAI ($r = 0.94$, $P > 0.001$). Unpublished data from experiment of Onillon *et al.* (1995).

determined by the maximum leaf area per tiller, together with the maximum number of tillers that can be attained following one defoliation event and before the next occurs. In rotationally grazed swards, tiller death will not occur if the defoliation height or defoliation interval is such that MCLA is much less than ECLA. Thus successive regrowth events can be expected to result in movement over time towards an equilibrium tiller density, such that MCLA and ECLA will approximately coincide. This is illustrated by data of Fan *et al.* (1995) for ryegrasss swards defoliated at 14, 28 or 42 days to 25, 50, 75 or 100 mm height (Fig. 7.2). The published paper has data on sward tiller population density measured monthly through the growing season and herbage accumulation. To compile Fig. 7.2, monthly tiller population density data for each treatment were averaged and approximate pregrazing herbage mass calculated by multiplying the herbage accumulation rate by the defoliation interval and adding 15 kg mm^{-1} stubble below the defoliation height. Herbage mass estimated in this way and seasonal mean tiller density were then plotted on a log scale. The arbitrary reference line above the data has a slope of −3/2. The expected value of the SDC slope, when expressed on a biomass/density basis in this way, is −1/2, not −3/2 (Weller, 1987), but we assumed a value of 1.0 for the LAI-related slope correction (C_a + C_r), thus increasing the theoretical SDC slope to −3/2. This data set of Fan *et al.* (1995) shows a reduction in tiller density for swards defoliated to 14 mm every 25 days, compared with milder defoliation at the same frequency. Thus, the tiller

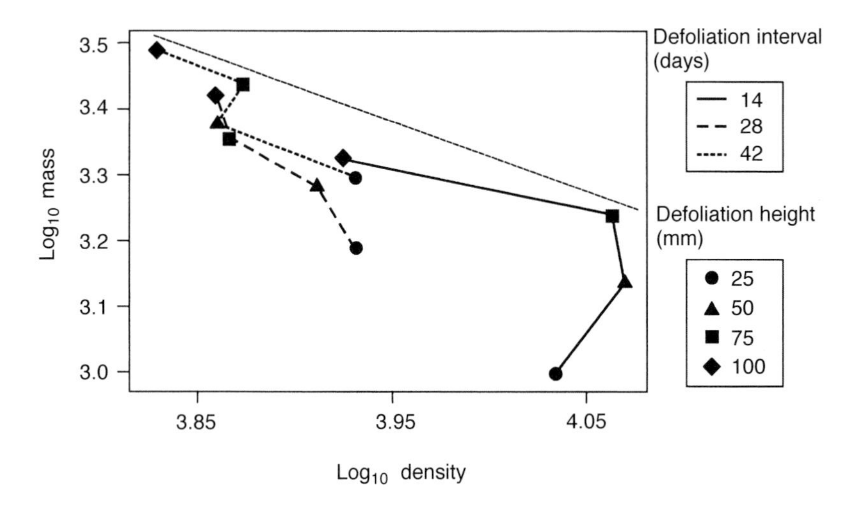

Fig. 7.2. Effect of defoliation height and defoliation interval on sward biomass/density coordinates. Tiller density data of Fan *et al.* (1995). In order to prepare this figure from the published data (Fan *et al.*, 1995), approximate pregrazing herbage mass was calculated as (stubble height (mm) × 15) + (defoliation interval × treatment mean accumulation rate) (kg DM ha^{-1}). Arbitrarily placed reference line has slope −3/2 representing −1/2 theoretical biomass density SDC relationship and assuming $(C_a + C_r) = 1.0$

population density declines under more severe defoliation (Fig. 7.2), as reported by Bircham and Hodgson (1983). In this sense, SDC in rotationally grazed swards is remarkably similar to phases 1 and 2 of SDC as described for a series of continuously grazed swards (Matthew *et al.*, 1995).

Distance from the SDC line

Matthew *et al.* (1995) suggested that distance between plotted tiller size/density coordinates for a sward and an arbitrarily positioned SDC line of slope −3/2 might have value as a sward productivity index. This follows logically from the conceptualization of the SDC line as a boundary line defining a constant leaf area or ECLA. This has since been confirmed both for experimental miniature swards (Hernández Garay *et al.*, 1999) and for field swards (Hernández Garay, 1995; Bahmani *et al.*, 1998), where data are for otherwise similar swards within the same experiment, but receiving different treatments. Indeed, visual inspection of Fig. 7.1 shows the link between canopy leaf area and distance from the SDC line very clearly. On statistical analysis, this relationship had a correlation of 0.94 ($F_{1,15} = 113.4$, $P < 0.001$). For extension purposes, tiller size/density coordinates are probably easier to interpret when plotted in a biomass/density format, as in Fig. 7.2 (see also McKenzie *et al.*, 1999), rather than a size or tiller dry weight/density format, as in Fig. 7.1.

Seasonal effects

Most size/density studies are conducted within a season with assumed constant environmental conditions (e.g. Kays and Harper, 1974) or on perennial species (Mohler *et al.*, 1978), in which case the population density is assumed not to fluctuate on a seasonal basis. In most, but not all, grass species the average tiller life is much less than a year (see below, Fig. 7.4), so that population dynamics are subject to seasonal perturbations, the most obvious of these being seasonal fluctuation in incoming light energy.

Bishop-Hurley (1999) estimated the effect of seasonal change in insolation on ECLA, by rearranging equations 1.1 and 1.2 of Lemaire and Chapman (1996) to give:

$$L_c = \frac{\ln\left\{\left(\frac{R_i k_1 - R_g}{R_g k_1}\right) - 1\right\}}{k_2} \text{ or equivalently } L_c = \frac{\ln\left\{\left(\frac{R_i - R_g}{R_g}\right) - \frac{1}{k_1}\right\}}{k_2} \quad (2)$$

where L_c is the leaf area index of the sward, k_1 represents the albedo or reflectance of the canopy, k_2 is a canopy light extinction coefficient, R_i is the incident light and R_g is the light not absorbed by the canopy and reaching the ground. It can then be assumed that L_c = ECLA would occur when a leaf at the bottom of the canopy received just enough light to be at the photosynthetic compensation point. If this compensation point and the incident radiation are known, equation 2 can be solved by using these values for R_g and R_i respectively, and then an estimation of ECLA can be done. Seasonal fluctuation in ECLA was then incorporated into a model based on equations 1 and 5 of Sackville Hamilton *et al.* (1995) and solar radiation relationships reported by Bonhomme (1994).

This model has limitations. In particular, it is not clear from published studies that senescence of leaves is linked to their carbon balance in this way, or exactly what the value for the photosynthetic compensation point is, and the predicted ECLA is sensitive to the value assumed for the compensation point. Also, the model does not take account of reproductive tillers, and the equation of Matthew *et al.* (1995) has three interrelated variables, tiller weight, tiller density and herbage mass, as well as the shape parameter *R*, with the result that complex interactions between the variables are possible.

Even so, when constant herbage mass is assumed and tiller population density is predicted from the intersection of the herbage mass and −1/2 biomass/density SDC lines, the model predicts a pattern of seasonal change in tiller population density, as shown by the solid line in Fig. 7.3. The predicted seasonal change in tiller density is very similar to that in Fig. 3.6 of Davies (1988), obtained by averaging results from several experiments. Under this application of the model, herbage mass is fixed and tiller weight is the quotient of biomass and tiller population density and also changes on a seasonal basis, being lower in summer and higher in winter (Fig. 7.3).

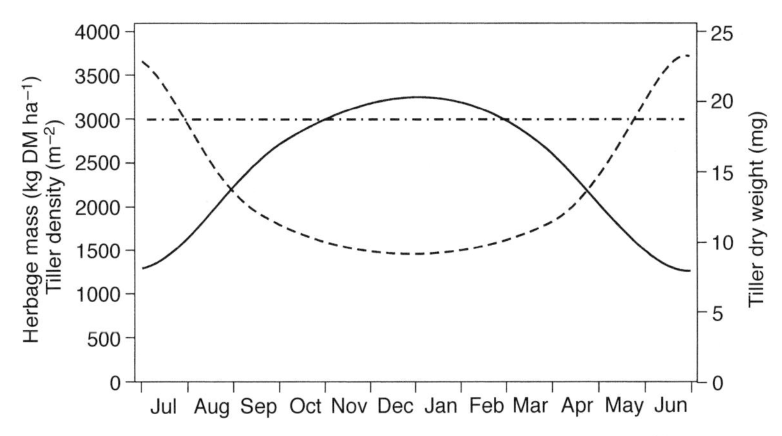

Fig. 7.3. Predicted response of sward tiller density (solid line) and tiller size (dashed line) to seasonal change in light energy intercepted for Palmerston North, New Zealand (Bishop Hurley, 1999). Model is based on equations 1.1 and 1.2 of Lemaire and Chapman (1996), and equation 5 of Sackville Hamilton *et al.* (1995).

An alternative approach would be to input observed seasonal values of tiller weight and sward herbage mass in order to predict tiller density. We have only briefly explored this possibility. It is clear from the preliminary appraisal, however, that the model also predicts that tiller population density will be highly sensitive to a change in sward biomass. For example, a fall in herbage mass from 3000 to 2100 kg DM ha^{-1} is predicted to double tiller population density. Thus, where herbage mass is used as an animal feed buffer and declines in winter, an increase rather than a decrease in tiller population density in winter is predicted by the model. Peak seasonal tiller population density in winter was reported by Brock and Hay (1993) for rotationally grazed swards.

Tiller Population Demography

Persistence strategies

Various methodologies have been employed to monitor tiller population dynamics of swards. In the simplest case, tiller populations may be monitored by counting tiller number per plant or tiller population density at regular time intervals, and for some experimental objectives this is all that is required. Secondly, further insight can be gained by monitoring marked tillers in order to resolve change in the tiller population density into components of tiller appearance and tiller death. This is analogous to resolution of net herbage accumulation into components of leaf elongation and leaf senescence by tissue turnover techniques (Bircham and Hodgson, 1983). Thirdly, a still greater level of insight is achieved when the survival within successive age cohorts of marked tillers in fixed quadrats is monitored over a period of time. Finally, for a complete demographic analysis, it is necessary

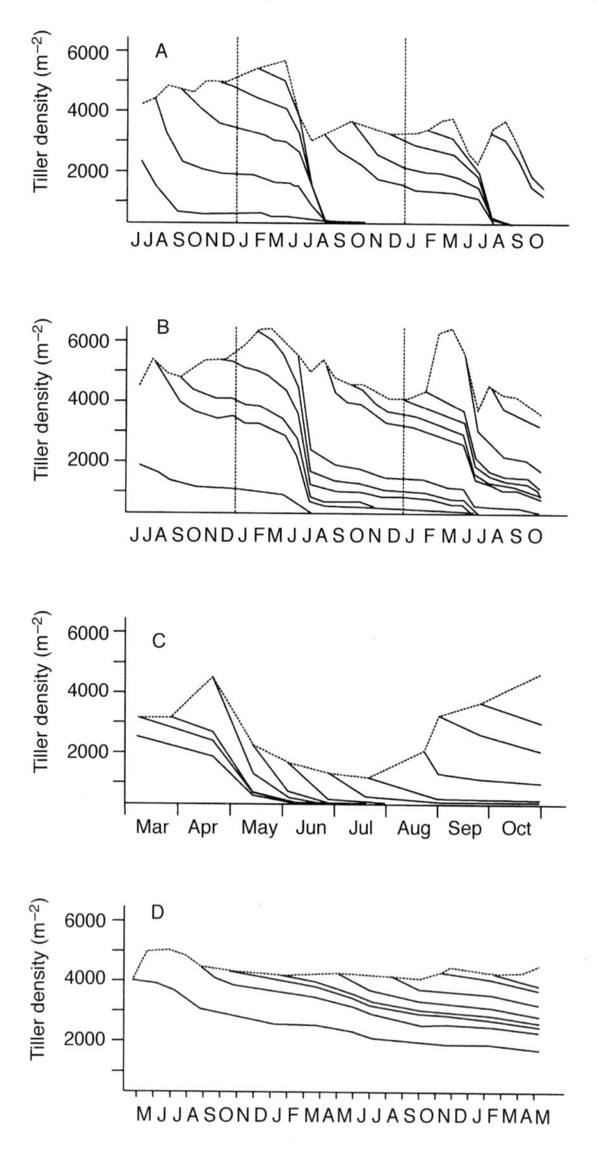

Fig. 7.4. Survival diagrams for marked tiller age cohorts illustrating different perennation strategies. A. *Phleum pratense*, sward renewal occurs mainly in association with flowering. Data from *L. perenne* cultivars with a similar pattern of tiller appearance indicate majority of daughter tillers are formed from bases of decapitated flowering tillers. (Data of Jewiss, 1966.) B. *Festuca pratensis*, spring tillers formed preflowering show a propensity to persist until flowering the following year (data of Jewiss, 1966.) C. *L. multiflorum*, high mortality of young tillers in summer (unpublished data of A. Davies and R.G. Simon). D. *Paspalum notatum*, new tillers are formed at any time of the year, presumably from non-flowering tillers (Pakiding and Hirata, 1998). Note also the longer life of tillers for this subtropical species, compared with the temperate species in A, B and C.

not only to monitor survival in each age cohort, but also to categorize new tillers according to the age cohort of the tillers that produced them. This usually requires destructive harvesting of fixed quadrats, in which case the number of fixed quadrats that must be set up at the start of an experiment is several times greater than for simple non-destructive recording of tiller survival.

Elsewhere, we summarize 13 experiments known to the authors which have generated tiller demographic information at the third level of complexity, measurement of survivorship of age cohorts in fixed quadrats (C. Matthew and N.R. Sackville Hamilton, unpublished manuscript). There may well be other data sets not included. A common way of presenting the data is in the form of age-cohort survival diagrams (Fig. 7.4). It is clear from these diagrams that, in many cases, there are distinct seasonal increases or decreases in rates of tiller birth and death, which may be thought of as persistence strategies or perennation pathways, with different grass species having different persistence strategies.

Phleum pratense (Fig. 7.4A, from Jewiss, 1966) shows a seasonal pattern of tiller birth and death consistent with that described by Matthew *et al.* (1993) as a 'reproductive' pathway. In this persistence strategy, sward renewal occurs mainly in association with flowering and the majority of the replacement tillers are formed from bases of decapitated flowering tillers. Meadow fescue (Fig. 7.4B, from Jewiss, 1966) has a persistence strategy based on a high spring tiller appearance rate, with the spring-formed tillers tending to persist until flowering the following year. *L. multiflorum* cv. 'S22' (Italian ryegrass) does not show an absence of tiller appearance in summer as has been popularly supposed, but does show high mortality of young tillers through the summer period, which can be attributed to flowering of unvernalized tillers within a matter of weeks of their appearance (Fig. 7.4C, unpublished data of A. Davies and R.G. Simons, Aberystwyth, UK). *Paspalum notatum* (Fig. 7.4D, from Pakiding and Hirata, 1998) shows a persistence strategy whereby new tillers are formed at an approximately steady rate throughout the year, presumably the majority of these being from non-flowering tillers. A similar strategy was observed for *L. perenne* cv. 'Grasslands Ruanui' by Matthew (1992), and was referred to by Matthew *et al.* (1993) as a 'vegetative' persistence pathway, to distinguish it from the contrasting persistence strategy described above (Fig. 7.4A). However, despite the similar pattern of turnover, tillers of *P. notatum* had a longer half-life (321–902 days) than tillers of 'Grasslands Ruanui' (approximately 90–150 days). As would be theoretically expected, seasonal variation in tiller population density appears to be greatly reduced when individual tillers are perennial (compare Fig. 7.4D with Fig. 7.4A, B and C).

Comparing the wider body of tiller demography data (C. Matthew and N.R. Sackville Hamilton, unpublished manuscript), lighter grazing in spring consistently increases turnover of the tiller population in summer, presumably through an increase in spring tiller size leading to an increase in percentage of tillers flowering. An increased magnitude of the turnover event in year 2, compared with the establishment year, is seen clearly in two of the data sets and possibly in a third.

These data sets are by nature very large, and the commonly used analysis of variance (ANOVA) techniques can only achieve a partial analysis – for example, rates of tiller birth or tiller death for specified observation periods. Perhaps partly as a result of the analytical difficulty, a number of the data sets remain unpublished (C. Matthew and N.R. Sackville Hamilton, unpublished manuscript). Some authors have fitted exponential decay curves to tiller numbers in individual cohorts (Korte, 1986; Pakiding and Hirata, 1998). This certainly provides helpful insight, but does not allow examination of seasonal events affecting groups of cohorts for a short part of their life, which is characteristic of particular persistence strategies. More detailed statistical insight into tiller demography patterns has been obtained by multivariate analysis techniques, including canonical correlation of 'neighbourhood' and 'tiller behaviour' data (Laterra *et al.*, 1997b).

Two other approaches to the analysis of such data, aimed at quantifying persistence strategies, are explored by C. Matthew and N.R. Sackville Hamilton (unpublished manuscript) and it is hoped to apply this methodology to other data sets in the near future. The unpublished analysis mentioned above includes a stability diagram, developed from a matrix algebra approach, and indicates that species such as *P. pratense*, which have a high tiller mortality associated with flowering, require a level of site filling near the biological maximum to maintain the tiller population through the flowering period.

Manipulation of seasonal population demography

Modern New Zealand perennial ryegrass cultivars usually have a high summer tiller turnover, similar to that shown in Fig. 7.4A for *P. pratense* (Korte, 1986; Matthew *et al.*, 1989b), although this pattern was not seen in year 1 of Korte's (1986) data, or when 'Ellett' perennial ryegrass was grown in subtropical Africa at latitude 30°S, with less seasonal variation in day length than in New Zealand (McKenzie, 1997). Older swards at Hamilton, New Zealand, renovated with 'Ellett' and 'Grasslands Nui' ryegrass also showed a high tiller turnover in summer (L'Huillier, 1987), and in one study less than 20% of the vegetative tiller population in late October remained 2 months later (Thom, 1991). For *L. perenne* cv. 'Ellett' at Palmerston North, which had a high tiller turnover associated with flowering, 52% of the tiller population in mid-December comprised new tillers formed from stubs of reproductive tillers, 19% were new tillers formed from vegetative tillers, 15% were older tillers that had not flowered and 14% were of uncertain classification. These proportions were not significantly affected by severity of defoliation.

Recognizing this particular perennation strategy of modern New Zealand *L. perenne* cultivars, a series of experiments in New Zealand have focused on the potential for enhancement of the tiller replacement process through appropriate grazing management. At the single-plant level, later decapitation of reproductive tillers in spring increased daughter tiller production from stubs of flowering tillers, compared with earlier decapitation or no decapitation, and higher defoliation increased daughter tiller production, compared with lower defoliation. These responses were linked with increased transfer of radiocarbon from parent to daughter tillers (Matthew *et al.*, 1991; Matthew, 1992).

At the sward level, temporary relaxation of grazing pressure in spring, intended to allow accumulation of substrate in the flowering tillers before defoliation, consistently enhanced late summer tiller density and sward productivity by an average of 0.9 t DM ha^{-1} in five separate experiments. Responses in animal production were also detected in one of these experiments (Da Silva *et al.*, 1993, 1994; Hernández Garay *et al.*, 1997a, b), but these were difficult to implement at the systems level (Bishop-Hurley *et al.*, 1998). Other authors (Brock and Hay, 1993) have argued that it is more beneficial to maintain a hard defoliation regime in spring in order to establish a high tiller population density for summer.

Another pasture species, *B. willdenowii*, has been found to have high productivity but poor persistence in most New Zealand conditions. Initially, it was thought that the poor persistence might be due to elevation and grazing of apical meristems of vegetative tillers, but tiller demography data proved this not to be the case. In fact the species utilizes the 'reproductive' perennation pathway, with a high percentage of tillers flowering and persistence being dependent on the formation of replacement tillers in early summer (Fig. 7.5). Hard grazing in early October, just prior to stem elongation, resulted in very poor tiller replacement (Fig. 7.5A), whereas delay in defoliation of a month at this time, followed by hard grazing, led to vigorous tillering from decapitated flowering stubs and a healthy sward in the following season (Fig. 7.5C). A light defoliation in early October produced an intermediate result (Fig. 7.5B). Herbage production of the sown grass for a 13-month period from October until the following November was 8.7, 16.5 and 12.8 t DM ha^{-1} for the 'hard', 'hard-delayed' and 'light' defoliation treatments, respectively (Black and Chu, 1989). In this experiment, tiller appearance was recorded by age category of parent tiller (Fig. 7.6). The data show that tillers of age classes 3 and 4, formed immediately post-flowering (shaded tiller cohorts in Fig. 7.5), contributed disproportionately more than their share to new tiller production throughout the following season. For example, from August to November these tillers comprised only about 10% of the total population (Fig. 7.5), but contributed more than 50% of new tiller appearance (Fig. 7.6).

To what extent responses of a similar nature to that illustrated here for *B. willdenowii* might be seen in other species is unclear, because few studies of sufficient detail have been carried out. However, the logic for the 'hard-delayed' grazing treatment came from anecdotal information on optimal management of *P. pratense* hay meadows in Canada, and an elegant study of Hendon and Briske (1997) has shown that a herbivory-sensitive rangeland species, *Eriochloa sericia*, is sensitive to defoliation during culm elongation also. However, in the latter case, extrinsic factors, including 'drought–herbivory interactions', were postulated as factors to be considered in understanding the plant response.

Tiller population demography in mixed-species swards

Up to this point, our discussion has regarded tiller dynamics almost as a set of rate processes affecting all tillers of the same age class with the same probability. In fact,

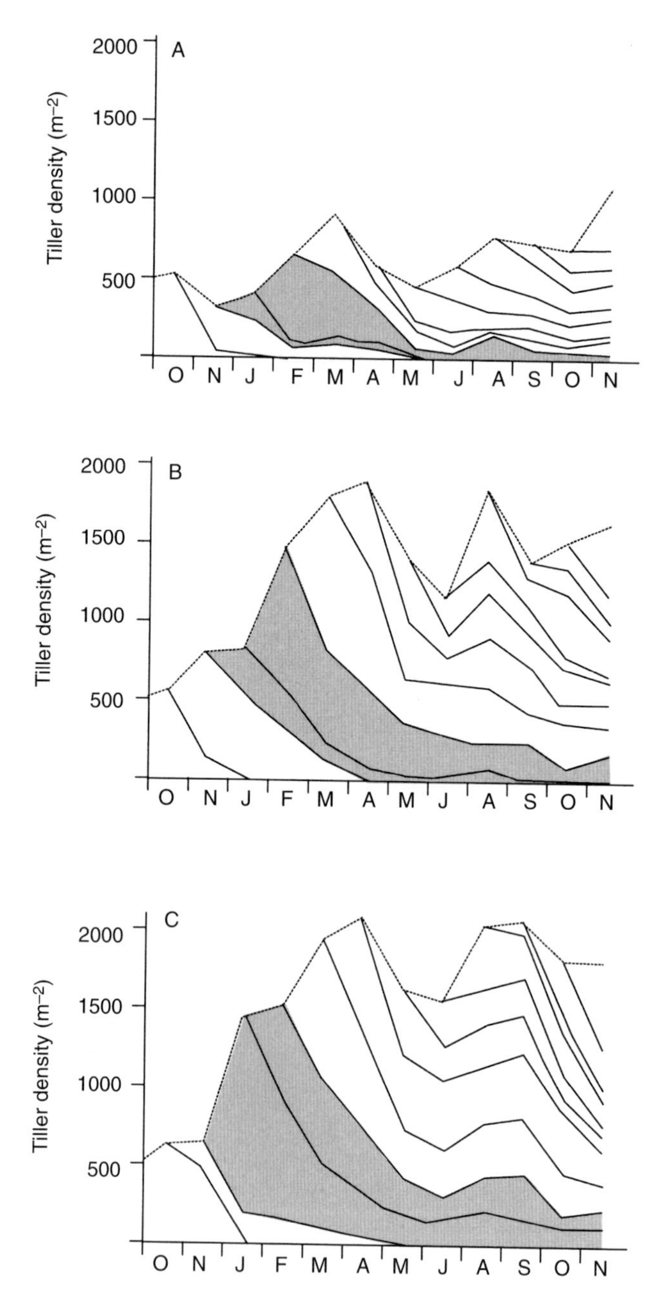

Fig. 7.5. Survival diagrams for marked tiller age cohorts of *Bromus willdenowii* grown at Palmerston North, New Zealand, under three contrasting spring defoliation regimes. A. Hard grazing prior to stem elongation. B. Light grazing prior to stem elongation. C. Hard grazing delayed until after stem elongation, intended to encourage daughter tiller formation from bases of flowering tillers.

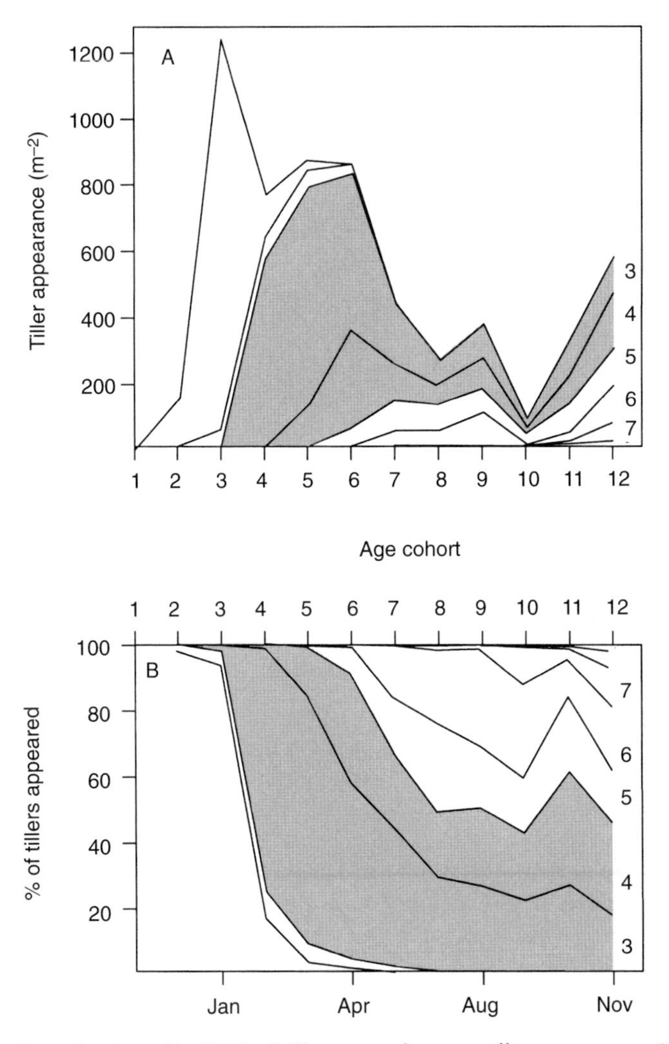

Fig. 7.6. Contribution of individual tiller age-cohorts to tiller appearance in *Bromus willdenowii*, Fig. 7.5, treatment C 'hard-delayed'. A. Number of new tillers m^{-2} at each observation. B. Number of new tillers expressed as % of total new tillers formed. Note that tillers formed immediately postflowering (shaded cohorts in Fig. 7.5) comprise approximately 10% of the population from August to November in Fig. 7.5C, but their daughter tillers make a disproportionately large contribution to tiller appearance throughout the following autumn and winter.

swards are heterogeneous, both in terms of clumping of clonally produced tillers of any one species and in terms of the interactions between species occupying neighbouring sites. More detailed studies will be important in future to understand how the more general responses discussed above might be modified in specific cases.

Briske and Butler (1989) found that recruitment was more likely at the edge of a clump than in the centre and was unaffected by severance of connections between ramets and their parent clone. Laterra *et al.* (1997a, b) have shown that variables describing tiller neighbourhood can be correlated with tiller behaviour. For example, it was ascertained by multivariate ordination that tiller numbers per plant and the tiller height of *P. dilatatum* showed greater seasonal fluctuation where there was a high neighbourhood occurrence of *L. multiflorum* and *Carex* spp. and lower seasonal fluctuation where the neighbourhood was characterized by presence of *F. arundinacea*, *Distichlis spicata* and *Stenotaphrum secundatum*. The tiller half-life was 112 and 137 days at each extreme of this neighbourhood gradient, but was 190–250 days for tillers placed in intermediate positions in the multivariate analysis. This perhaps indicates that the multivariate analysis isolated two distinct types of neighbourhood stress from the rest of the data: stress from neighbours with a seasonal period of rapid growth, and stress from neighbours with other competitive strategies that showed less seasonal variation in intensity, such as plant morphology of the competing species.

A similar analysis of tiller demography for *P. dilatatum* showed (Laterra *et al.*, 1997b) that what we have termed above the 'perennation pathway' – that is to say, the pattern of observed survival probability across age classes – is also modified by tiller size variation within an age class and by neighbourhood, so that complete definition of tiller demography patterns is extremely complex.

Other studies (e.g. Derigibus and Trlica, 1990; Bullock *et al.*, 1994) have examined the effects of defoliation on tiller demography. Again, responses observed were complex and they also differed between species. *Sporobolus indicus*, for example, was intolerant of frequent defoliation, showing greatly reduced tiller appearance under this treatment, whereas defoliation increased tiller density in *P. dilatatum* in a more favourable summer and had no effect in a less favourable summer (Derigibus and Trlica, 1990). Bullock *et al.* (1994) analysed the tiller dynamics of *L. perenne* and *Agrostis stolonifera* in response to defoliation. Tiller population density showed seasonal variation with a peak in autumn. Tiller birth and death rates were found to be density-dependent. All these responses can be explained within the multiphase SDC relationship along the environmental boundary line, as described above. However, these authors, especially Bullock *et al.* (1994), also reported additional defoliation and weather responses superimposed on the responses attributable to SDC relationships, indicating that it is highly likely that specific traits of individual species modify the SDC relationship for those particular species. An interesting area for future studies will be the characterization of these species-specific variations of the more general SDC relationship.

Conclusions

This chapter has considered tiller dynamics of grass swards from three perspectives: tiller morphology, canopy leaf area and tiller demography. Tillering behaviour cannot be considered in isolation from other plant processes, and consideration of tiller morphology provides a framework to rationalize at least some resource allocation considerations. However, the primary driving principle for tiller dynamics is the optimization of canopy leaf area in relation to defoliation intensity and available resources, such as light and water. The concept of a multiphase SDC relationship along an environmental boundary line goes a long way towards rationalizing otherwise conflicting observations on tiller density and on tiller appearance and death rate. Even so, after consideration of SDC relationships, there remain substantive, complex differences in tillering behaviour, often unique to a particular species, which are best explained from a tiller demography perspective – that is, by identifying and quantifying factors affecting birth and death rates of particular classes of tillers. SDC relationships are relevant at the ecological level to help address issues such as carrying capacity or sustainability. Tiller demography relationships provide insight for the fine-tuning of grazing management, and well-directed understanding of tiller demography has been shown to generate significant improvement in productivity in some situations.

Acknowledgements

We thank Dr Masahiko Hirata, Miyazaki University, and Professor Wang Pei, China Agricultural University, for assistance in obtaining copies of papers published in their countries. We thank Dr G. Lemaire for permission to use unpublished data in Fig. 7.1, O.R. Jewiss, A. Davies, R.G. Simons and M. Hirata for permission to reproduce survival diagrams in Fig. 7.4, Ms I. Bahmani for assistance with the literature search and Mr M.R. Alexander for the preparation of figures.

References

Bahmani, I., Thom, E.R. and Matthew, C. (1998) Effects of nitrogen and irrigation on productivity of different ryegrass ecotypes when grazed by dairy cows. *Proceedings of the New Zealand Grassland Association* 59, 117–123.

Bahmani, I., Hazard, L., Varlet-Grancher, C., Betin, M., Lemaire, G., Matthew, C. and Thom, E.R. (2000) Differences in tillering of long- and short-leaved perennial ryegrass genetic lines under full light and shade treatments. *Crop Science* (in press).

Bircham, J.S. and Hodgson, J. (1983) The influence of sward condition on rates of herbage growth and senescence in mixed swards under continuous stocking management. *Grass and Forage Science* 38, 323–331.

Bishop-Hurley, G.J. (1999) An evaluation of a dairy systems study of the effects of contrasting spring grazing managements on pasture and animal performance. Unpublished PhD thesis, Massey University, Palmerston North, New Zealand, 270 pp.

Bishop-Hurley, G.J., Matthews, P.N.P., Hodgson, J., Dake, C. and Matthew, C. (1998) Dairy systems study of the effects of contrasting spring grazing managements on pasture and animal production. *Proceedings of the New Zealand Grassland Association* 59, 209–214.

Black, C.K and Chu, A.C.P. (1989) Searching for an alternative way to manage prairie grass. *Proceedings of the New Zealand Grassland Association* 50, 219–223.

Bonhomme, R. (1994) Les rayonnements solaires et le fonctionnement du couvert végétal. In: El Hassani, T.A. and Persoons, E. (eds) *Bases Physiologiques et Agronomiques de la Production Végétale*. Hatier-Aupelf, Uref, Paris, pp. 26–48.

Bos, H.J. and Neuteboom J. (1998) Morphological analysis of leaf and tiller dynamics of wheat (*Triticum aestivum* L.): responses to temperature and light intensity. *Annals of Botany* 81, 131–139.

Briske, D.D. and Butler, J.L. (1989) Density-dependent regulation of ramet populations within the bunchgrass *Schizachyrium scoparium*: interclonal versus intraclonal interference. *Journal of Ecology* 77, 963–974.

Brock, J.L. and Fletcher, R.H. (1993) Morphology of perennial ryegrass plants in pastures under intensive sheep grazing. *Journal of Agricultural Science (Cambridge)* 120, 301–310.

Brock, J.L. and Hay, R.J.M. (1993) An ecological approach to forage management. In: *Proceedings XVII International Grassland Congress*. SIR Publishing, Wellington, New Zealand, pp. 837–842.

Brock, J.L., Hume, D.E. and Fletcher, R.H. (1996) Seasonal variation in the performance of perennial ryegrass (*Lolium perenne*) and cocksfoot (*Dactylis glomerata*) plants and populations in pastures under intensive sheep grazing. *Journal of Agricultural Science, Cambridge* 126, 37–51.

Brock, J.L., Albrecht, K.A. and Hume, D.E. (1997) Stolons and rhizomes in tall fescue under grazing. *Proceedings of the New Zealand Grassland Association* 59, 93–98.

Bullock, J.M., Clear Hill, K.B. and Silverton, J. (1994) Tiller dynamics of two grasses – responses to grazing, density and weather. *Journal of Ecology* 82, 331–340.

Casal, J.J., Deregibus, V.A. and Sánchez, R.A. (1985) Variations in tiller dynamics and morphology in *Lolium multiflorum* Lam.: vegetative and reproductive plants as affected by red/far-red radiation. *Annals of Botany* 56, 553–559.

Chapman, D.F. and Lemaire, G. (1993) Morphogenetic and structural determinants of plant regrowth after defoliation. In: *Proceedings of the XVIIth International Grassland Congress*. SIR Publishing, Wellington, New Zealand, pp. 95–104.

Chapman, D.F., Clark, D.A., Land, C.A. and Dymock, N. (1983) Leaf and tiller growth of *Lolium perenne* and *Agrostis* spp. and leaf appearance rates of *Trifolium repens* in set-stocked and rotationally grazed hill pastures. *New Zealand Journal of Agricultural Research* 26, 159–168.

Clifford, P.E., Marshall, C. and Sagar, G.R. (1973) The reciprocal transfer of radiocarbon between a developing tiller and its parent shoot in vegetative plants of *Lolium multiflorum* Lam. *Annals of Botany* 37, 777–785.

Da Silva, S.C., Matthew, C., Matthews, P.N.P. and Hodgson, J. (1993) The influence of spring grazing management on production in dairy pasture. In: *Proceedings XVIIth International Grassland Congress*. SIR Publishing, Wellington, New Zealand, pp. 859–860.

Da Silva, S.C., Hodgson, J., Matthews, P.N.P. and Matthew, C. (1994) Effects of contrasting spring grazing management on summer–autumn pasture and milk production of mixed ryegrass–clover dairy swards. *Proceedings of the New Zealand Society of Animal Production* 54, 79–82.

Davies, A. (1974) Leaf tissue remaining after cutting and regrowth in perennial ryegrass. *Journal of Agricultural Science (Cambridge)* 82, 165–172.

Davies, A. (1977) Structure of the grass sward. In: Gilsenan, B. (ed.) *Proceedings of an International Meeting on Animal Production from Temperate Grassland*. An Foras Taluntais, Dublin, pp. 36–44.

Davies, A. (1988) The regrowth of grass swards. In: Jones, M.B. and Lazenby, A. (eds) *The Grass Crop*. Chapman & Hall, London, pp. 85–127.

Demotes-Mainard, S. and Pellerin, S. (1992) Effect of mutual shading on the emergence of nodal roots and the root/shoot ratio of maize. *Plant and Soil* 147, 87–93.

Derigibus, V.A. and Trlica, M.J. (1990) Influence of defoliation upon tiller structure and demography in two warm season grasses. *Acta Oecologica* 11, 693–699.

Dorrington Williams, R. (1970) Tillering in grasses cut for conservation with special reference to perennial ryegrass. *Herbage Abstracts* 40, 383–388.

Etter, A.G. (1951) How Kentucky bluegrass grows. *Annals of the Missouri Botanical Garden* 38, 293–375.

Fan, F.C., Gao, Z.H., Han, J.G. and Wang, P. (1995) Effects of cutting on the lamina tissue turnover for perennial ryegrass. *Acta Agrestia Sinica* 3, 15–21.

Gautier, H., Varlet-Grancher, C. and Hazard, L. (1999) Tillering responses to the light environment and to defoliation in populations of perennial ryegrass (*Lolium perenne* L.) selected for contrasting leaf length. *Annals of Botany* 83, 423–429.

Harris, W., Pandey, K.K., Gray, Y.S. and Couchman, P.K. (1979) Observations on the spread of perennial ryegrass by stolons in a lawn. *New Zealand Journal of Agricultural Research* 16, 207–222.

Hendon, B.C. and Briske, D.D. (1997) Demographic analysis of a herbivory-sensitive perennial bunchgrass: does it possess an Achilles heel? *Oikos* 80, 8–17.

Hendrickson, J.R. and Briske, D.D. (1997) Axillary bud banks of two semiarid perennial grasses: occurrence, longevity, and contribution to population persistence. *Oecologica* 110, 584–591.

Hernández-Garay, A. (1995) Defoliation management, tiller density and productivity in perennial ryegrass swards. Unpublished PhD thesis, Massey University, Palmerston North, New Zealand, 228 pp.

Hernández-Garay, A., Hodgson, J. and Matthew, C. (1997a) Effect of spring grazing management on perennial ryegrass and ryegrass–white clover pastures 1. Tissue turnover and herbage accumulation. *New Zealand Journal of Agricultural Research* 40, 25–35.

Hernández-Garay, A., Matthew, C. and Hodgson, J. (1997b) Effect of spring grazing management on perennial ryegrass and ryegrass–white clover pastures 2. Tiller and growing point densities and population dynamics. *New Zealand Journal of Agricultural Research* 40, 37–50.

Hernández-Garay, A., Matthew, C. and Hodgson, J. (1999) Tiller size–density compensation in ryegrass miniature swards subject to differing defoliation heights and a proposed productivity index. *Grass and Forage Science* 54, 347–356.

Hitch, P.A. and Sharman, B.C. (1971) The vascular pattern of festucoid grass axes, with particular reference to nodal plexi. *Botanical Gazette* 132, 38–56.

Hume, D.E. (1991) Effect of cutting on production and tillering in prairie grass (*Bromus willdenowii* Kunth) compared with two ryegrass (*Lolium*) species. 1. Vegetative plants. *Annals of Botany* 67, 533–541.

Hume, D.E. and Brock, J.L. (1997) Morphology of tall fescue (*Festuca arundinacea*) and perennial ryegrass (*Lolium perenne*) plants in pastures under sheep and cattle grazing. *Journal of Agricultural Science (Cambridge)* 129, 19–31.

Hunt, W.F. and Thomas, V.J. (1985) Growth and developmental responses of perennial ryegrass grown at constant temperature II. Influence of light and temperature on leaf, tiller and root appearance. *Australian Journal of Plant Physiology* 12, 69–76.

Ito, M. and Nakamura, T. (1974) Seasonal variations of the tiller bud development in each node of orchard grass shoot. *Journal of the Japanese Society of Grassland Science* 20, 83–91.

Jewiss, O.R. (1966) Morphological and physiological aspects of the growth of grasses during the vegetative phase. In: Milthorpe, F.L. and Ivins, J.D. (eds) *The Growth of Cereals and Grasses. Proceedings of the Twelfth Easter School in Agricultural Science, University of Nottingham.* Butterworths, London, pp. 39–56.
Kays, S. and Harper, J.L. (1974) The regulation of plant and tiller density in a grass sward. *Journal of Ecology* 62, 97–105.
Klepper, E., Belford, R.K. and Rickman, R.W. (1984) Root and shoot development in winter wheat. *Agronomy Journal* 76, 117–122.
Korte, C.J. (1986) Tillering in 'Grasslands Nui' perennial ryegrass swards 2. Seasonal pattern of tillering and age of flowering tillers with two mowing frequencies. *New Zealand Journal of Agricultural Research* 29, 629–638.
Korte, C.J. and Harris, W. (1987) Stolon development in grazed 'Grasslands Nui' perennial ryegrass. *New Zealand Journal of Agricultural Research* 30, 139–148.
Langer, R.H.M. (1956) Growth and nutrition of timothy (*Phleum pratense*) I. The life history of individual tillers. *Annals of Applied Biology* 44, 166–187.
Langer, R.H.M. (1963) Tillering in herbage grasses. *Herbage Abstracts* 33, 141–148.
Laterra, P., Maceira, N.O. and Deregibus, V.A. (1997a) Neighbour influence on the tiller demography of two perennial *pampa* grasses. *Journal of Vegetation Science* 8, 361–368.
Laterra, P., Deregibus, V.A. and Maceira, N.O. (1997b) Demographic variability in populations of two perennial *pampa* grasses. *Journal of Vegetation Science* 8, 369–376.
Lemaire, G. and Chapman, D.F. (1996) Tissue flows in grazed plant communities. In: Hodgson, J. and Illius, A.W. (eds) *The Ecology and Management of Grazing Systems.* CAB International, Wallingford, pp. 3–36.
L'Huillier, P.J. (1987) Tiller appearance and death of *Lolium perenne* in mixed swards grazed by dairy cattle at two stocking rates. *New Zealand Journal of Agricultural Research* 30, 15–22.
McKenzie, B.A., Valentine, I., Matthew, C. and Harrington, K.C. (1999) Plant interactions in pastures and crops. In: White, J.G.H. and Hodgson, J. (eds) *New Zealnd Pasture and Crop Science.* Oxford University Press, Auckland, pp. 45–58.
McKenzie, F.R. (1997) Influence of grazing frequency and intensity on the density and persistence of *Lolium perenne* tillers under subtropical conditions. *Tropical Grasslands* 31, 219–226.
Matthew, C. (1992) A study of seasonal root and tiller dynamics in swards of perennial ryegrass (*Lolium perenne* L.). Unpublished PhD thesis, Massey University, Palmerston North, New Zealand.
Matthew, C. and Kemball, W.D. (1997) Allocation of carbon-14 to roots of different ages in perennial ryegrass (*Lolium perenne* L.). In: *Proceedings of the XVIIIth International Grassland Congress.* Association Management Centre, Calgary, Canada, pp. 1–2.
Matthew, C., Quilter, S.J., Korte, C.J. and Chu, A.C.P. (1989a) Stolon formation and significance for sward tiller dynamics in perennial ryegrass. *Proceedings of the New Zealand Grassland Association* 50, 255–259.
Matthew, C., Xia, J.X., Hodgson, J. and Chu, A.C.P. (1989b) Effect of late spring grazing management on tiller age profiles and summer–autumn pasture growth rates in a perennial ryegrass (*Lolium perenne* L.) sward. In: *Proceedings of the XVIth International Grassland Congress, Nice, France.* AFPF, Versailles, France, pp. 521–522.
Matthew, C., Chu, A.C.P., Hodgson, J. and Mackay, A.D. (1991) Early summer pasture control: what suits the plant? *Proceedings of the New Zealand Grassland Association* 53, 73–77.
Matthew, C., Black, C.K and Butler, B.M. (1993) Tiller dynamics of perennation in three

herbage grasses. In: *Proceedings of the XVIIth International Grasslands Congress*. SIR Publishing, Wellington, New Zealand, pp. 141–143.

Matthew, C., Lemaire, G., Sackville Hamilton, N.R. and Hernández Garay, A. (1995) A modified self-thinning equation to describe size/density relationships for defoliated swards. *Annals of Botany* 76, 579–587.

Matthew, C., Yang, J.Z. and Potter, J.F. (1998) Determination of tiller and root appearance in perennial ryegrass (*Lolium perenne*) swards by observation of the tiller axis, and potential application to mechanistic modelling. *New Zealand Journal of Agricultural Research* 41, 1–10.

Mitchell, K.J. (1953) Influence of light and temperature on the growth of ryegrass (*Lolium* spp.). 1. Pattern of vegetative development. *Physiologia Plantarum* 6, 21–46.

Mohler, C.L., Marks, P.L. and Sprugel, D.G. (1978) Stand structure and allometry of trees during self-thinning of pure stands. *Journal of Ecology* 66, 599–614.

Neuteboom, J.H. and Lantinga, E.A. (1989) Tillering potential and relationship between leaf and tiller production in perennial ryegrass. *Annals of Botany* 63, 265–270.

Neuteboom, J.H., Lantinga, E.A. and Wind, K. (1988) Tillering characteristics of diploid and tetraploid perennial ryegrass. In: *Proceedings of the 12th General Meeting of the European Grassland Federation, Dublin*. Irish Grassland Association, Belclare, pp. 498–503.

Onillon, B., Durand, J.L., Gastal, F. and Tournebize, R. (1995) Drought effects on growth and carbon partitioning in a tall fescue sward grown at different rates of nitrogen fertilization. *European Journal of Agronomy* 4, 91–99.

Oxford English Dictionary (1989) Prepared by Simpson, J.A. and Weiner, E.S.C., 2nd edn, Vol. XVI, Soot–Styx. Clarendon Press, Oxford.

Pakiding, W. and Hirata, M. (1998) Tillering in a bahiagrass (*Paspalum notatum* Flügge) pasture under cattle grazing 1. Results as meaned over quadrats. *Journal of the Japanese Society of Grassland Science* 44, 44–45.

Parsons, A.J. and Chapman, D.F. (1999) The principles of pasture growth and utilisation. In: Hopkins, A. (ed.) *Grass*, 3rd edn. Blackwell Science, Oxford.

Ryle, G.J.A. (1964) A comparison of leaf and tiller growth in seven perennial grasses as influenced by nitrogen and temperature. *Journal of the British Grassland Society* 19, 281–290.

Sackville Hamilton, N.R., Matthew, C. and Lemaire, G. (1995) In defence of the −3/2 boundary rule: a re-evaluation of self thinning concepts and status. *Annals of Botany* 76, 569–577.

Sbrissia, A.F., Da Silva, S.C., Matthew, C., Pedreira, C.G.S., Carnevalli, R.A., De Lara Fagundes, J., Pinto, L.F.De M. and Cortucci, M. (1999) Tiller size/density compensation in grazed swards of *Cynodon* spp. In: de Moraes, A., Nabinger, C., Carvalho, P.C. de F., Alves, S.J. and Lustosa, S.B.C. (eds) *Proceedings of International Symposium on Grassland Ecophysiology and Grazing Ecology*. Universidade Federal do Paraná, Curitiba, Brazil, pp. 348–352.

Silsbury, J.H. (1970) Leaf growth in pasture grasses. *Tropical Grasslands* 4, 17–36.

Simon, J.C. and Lemaire, G. (1987) Tillering and leaf area index in grasses in the vegetative phase. *Grass and Forage Science* 42, 373–380.

Skinner, R.H. and Nelson, C.J. (1992) Estimation of potential tiller production and site usage during tall fescue canopy development. *Annals of Botany* 70, 493–499.

Thom, E.R. (1991) Effect of early spring grazing frequency on the reproductive growth and development of a perennial ryegrass tiller population. *New Zealand Journal of Agricultural Research* 34, 383–389.

Troughton, A. (1980) Production of root axes and leaf elongation in perennial ryegrass in relation to dryness of the upper soil layer. *Journal of Agricultural Science (Cambridge)* 95, 533–538.

Van Loo, E.N. (1992) Tillering, leaf expansion and growth of plants of two cultivars of perennial ryegrass grown using hydroponics at two water potentials. *Annals of Botany* 70, 511–518.

Vignolio, O.R., Laterra, P. and Fernández, O.N. (1994) Caraterísticas estructurales de los rizomas de *Paspalum dilatatum* Su relación con el crecimiento aéreo y con las variaciones estacionales del clima. *Ecología Austral* 4, 177–122.

Vignolio, O.R., Laterra, P. and Fernández, O.N. (1998) Efectos del pastoreo sobre la morfología de los rizomas de *Paspalum dilatatum* (Gramineae). *Boletín de Sociedade Argentina de Botánica* 33, 129–135.

Weller, D.E. (1987) A re-evaluation of the −3/2 power rule of plant self thinning. *Ecological Monographs* 57, 23–43.

Yang, J.Z, Matthew, C. and Rowland, R.E. (1998) Tiller axis observations for perennial ryegrass (*Lolium perenne*) and tall fescue (*Festuca arundinacea*): number of active phytomers, probability of tiller appearance, and frequency of root appearance per phytomer for three cutting heights. *New Zealand Journal of Agricultural Research* 41, 11–17.

Yoda, K., Kira, T., Ogawa, H. and Hozumi, H. (1963) Self-thinning in overcrowded pure stands under cultivated and natural conditions. *Journal of Osaka City University Institute of Polytechnics* D14, 107–129.

8

Effect of Nitrogen on Some Morphogenetic Traits of Temperate and Tropical Perennial Forage Grasses

P. Cruz[1] and M. Boval[2]

[1] *INRA, Station d'Agronomie, BP 27, 31326 Castanet-Tolosan, France;* [2] *INRA, Station de Recherches Zootechniques, BP 515, 97165 Pointe-à-Pitre, France*

Introduction

The great botanical variety of forage grasses – cultivated and natural – is characterized by a wide range of morphological types. Some examples of these morphological traits, differing greatly from one species to another, are the growth habit of a plant, the level of exposure of the growth meristems and reserve organs to herbivores and the presence or absence of stolons. These different traits have huge consequences for the forage capacity of each species, but particularly as to what action to take in terms of controlling sward production through the use of nitrogen fertilization, grazing or mowing.

The morphology of a species in a given situation is the result of a process referred to as morphogenesis. Plant morphogenesis, as defined by Chapman and Lemaire (1993), is the dynamics of generation and expansion of the plant form in space. This dynamic process is the result of the rate of appearance of new organs (organogenesis) and the balance between their growth and senescence rates. In this study, we aim to analyse the effect of nitrogen on the morphogenesis of forage grasses of temperate and tropical climates. We focus on the nitrogen factor, taking into account its role in the expression of plant growth. Following a brief outline of the process of plant morphogenesis, we shall concentrate on the effect of the nitrogen nutrition level on tiller and stolon development and on the consequences this has for canopy growth dynamics. We shall also analyse the influence of the stage of development and utilization of the canopy (defoliation) on the forage value of the different morphogenetic types.

Comparing temperate and tropical grasses provides a range of morphological variability within the same botanical family, and these contrasts are enhanced by

(eds G. Lemaire, J. Hodgson, A. de Moraes, C. Nabinger and P.C. de F. Carvalho)

differences in the metabolic pathways of photosynthesis (types C_3 and C_4 in temperate and tropical species, respectively). Most perennial forage grasses of temperate climates – at least, those species which are most widely known – are C_3 tillering plants. Examples include *Lolium perenne*, *Dactylis glomerata*, *Festuca arundinacea*, *Festuca pratensis* and *Phleum pratense*. There are two main morphogenetic types of tropical perennial forage grasses: tufted plants and stoloniferous plants. The growth pattern of tropical tufted species is similar to that of temperate tillering species, but the former are different in that their stem growth occurs at almost every regrowth period during the year (Cowan and Lowe, 1998). This is due to the fact that their floral induction can occur at almost any time of the year and does not seem very dependent upon seasonal induction factors, such as temperature or day length. This is the case for *Panicum maximum*, *Andropogon gayanus*, *Pennisetum purpureum*, *Hyparrhenia rufa* (Bogdan, 1977) and *Setaria anceps* (Nada, 1985; Cruz and Sobesky, 1989), for example. Strictly stoloniferous species exhibit stem growth particular to their morphogenesis and independent of any flowering process. Examples of such species include *Axonopus compressus*, *Cynodon dactylon*, *Cynodon nlemfuensis*, *Stenotaphrum secundatum* and *Pennisetum clandestinum*. There are a number of species which can be defined as being morphological intermediates, whose propagation by stolons depends more or less on the stage of development of the plant or canopy. Examples of such species are *Dichanthium aristatum*, *Dichanthium annulatum*, *Bothriochloa pertusa* and *Digitaria decumbens* (Bogdan, 1977), even though the accuracy of referring to these species as 'intermediate' varies from one to the next. All of these species have the capacity to develop stolons to some degree when growing as isolated plants. However, in dense stands most of the stems do not reach the ground but grow laterally at the top of the canopy.

Nitrogen and Morphogenesis of Tufted Grasses

In vegetative tufted grasses, only the leaves are produced as aerial organs. The tiller produces a sequential chain of phytomers, a leaf (blade and sheath) with a node, an internode and an axillary meristem (Moore and Moser, 1995; Fick and Clark, 1998), which develop between the primordium and the maturation and senescence stages. The sequential appearance of leaves on the tiller allows the axillary meristems to develop into daughter tillers, following a well-described process for temperate grasses (Davies, 1974). During vegetative growth, the morphogenesis of a single tiller can be characterized by three variables: leaf appearance rate (LAR), leaf elongation rate (LER) and leaf lifespan. As shown by Lemaire and Chapman (1996), the combination of these variables determines sward structure by way of three other variables: leaf size, number of leaves per tiller and tiller density. These sward structural characteristics determine the leaf area index (LAI), which regulates regrowth dynamics through its effects on light interception (see Fig. 14.1 in Lemaire and Agnusdei, Chapter 14, this volume).

Leaf elongation

Nitrogen nutrition affects the expression of basic morphogenetic variables at the tiller level in a number of ways, increasing the LER and tillering rate and having a slight effect on the LAR. The LER depends largely on nitrogen nutrition, as shown for *F. arundinacea* by Gastal and Lemaire (1988) and Gastal *et al.* (1992). N application was able to increase the leaf blade elongation rate of this temperate species almost fourfold between two extreme nutrition levels. Nitrogen has also been shown to have the same positive effect on the LER of C_4 tufted species, such as *D. aristatum* (an 'intermediate' species, whose tillers and stolons develop in succession during regrowth), at the beginning of the regrowth cycle (P. Cruz, unpublished data) and *Saccharum officinalis* (Rosario, 1977).

Leaf appearance

Conversely, nitrogen only affects the LAR of tufted species slightly, LAR being a variable that plays a determinant role, as it influences each of the sward structural characteristics (Lemaire and Chapman, 1996). LAR, leaf width and leaf senescence tended to be slightly higher at high than at low nitrogen levels, but not statistically different (Gastal and Lemaire, 1988). LAR is directly influenced by temperature in temperate grasses. For this reason, and for a given species, a more or less constant leaf appearance interval can be calculated in terms of degree-days: 110°C for *L. perenne* (Davies and Thomas, 1983), 230°C for *F. arundinacea* (Lemaire, 1985) and 160°C for *D. glomerata* (Duru *et al.*, 1993). Nevertheless, Duru *et al.* (1999) pointed out that for a given nitrogen level and when climatic conditions are stable, the phyllochron, i.e. the time interval between the appearance of two successive leaves, depends on the length of the sheath of the preceding leaf (Fig. 8.1). The sheath length also depends on the height of defoliation. The different combinations of initial state after defoliation, leaf number and nitrogen availability could explain the different effects on phyllochron evolution during regrowth (Duru *et al.*, 1999).

Few data exist on the leaf emergence rate of perennial, tropical, tufted species. In *S. anceps*, a tropical tillering species, the number of leaves produced after 6 weeks of regrowth varies only from 12 to 13 as a consequence of N fertilization (Fig. 8.2). This implies that LAR is only slightly influenced by the N nutrition of the plant. The temperature factor in tropical conditions is less important because of low seasonal variability. Nevertheless, when temperatures are below the optimal temperature for C_4 species, the leaf appearance rate is reduced, as reported by Muldoon and Pearson (1979) for a hybrid, *Pennisetum americanum* × *P. purpureum*. Sato (1980) compared temperature thresholds and their effect on the LAR of temperate and tropical forage species. He showed that LAR in temperate C_3 species increased with decreasing temperature from 35°C to 15°C, and in tropical C_4 grasses LAR decreased over the same temperature range.

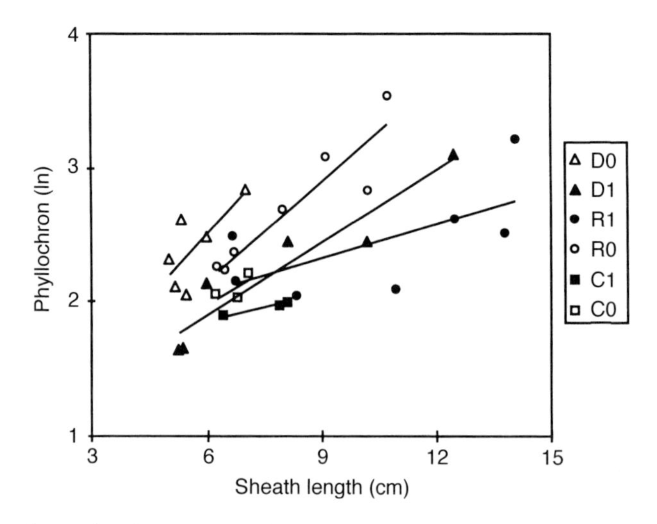

Fig. 8.1. Relationship between phyllochron and sheath size in *Dactylis glomerata*. Symbols differ with nitrogen fertilization (0 or 150 kg N ha^{-1} designated 0 and 1, respectively) and combinations of number and intensity of spring defoliation (D, one cut at 6.5 cm; R, two cuts at 6.5 cm; and C, three cuts at 7.5 cm) (from Duru *et al.*, 1999).

Leaf size

The final leaf blade size in tufted grasses is greatly increased by N nutrition. This positive N effect can be broken down into two different processes: (i) the LER; and (ii) the leaf elongation duration (LED), which can be postulated as proportional to the phyllochron (Lemaire and Chapman, 1996).

The positive effect of nitrogen on the final blade size in tropical tufted species can be demonstrated using the example of *S. anceps* (Fig. 8.2). Apart from showing the effect of nitrogen on the size of adult leaves, this figure also shows that the effect increases with the leaf number. This may be partly explained by the progressive increase in phyllochron, due to the increase in the length of successive leaf sheaths (Duru *et al.*, 1999; see also Fig. 8.1) and its consequence in increasing the leaf elongation duration. The effects of this evolution in leaf blade size on LAI, regrowth dynamics and the forage value of the canopy will be discussed later on, continuing to draw parallels between tufted and stoloniferous species.

Tillering

Numerous authors have demonstrated the positive effect of nitrogen on tillering, in temperate species (Wilman and Pearse, 1984; Simon and Lemaire, 1987; Gastal and Lemaire, 1988) as well as in tropical species (Mears and Humphreys, 1974; Corsi, 1984). Nevertheless, the opposing responses of certain morphogenetic

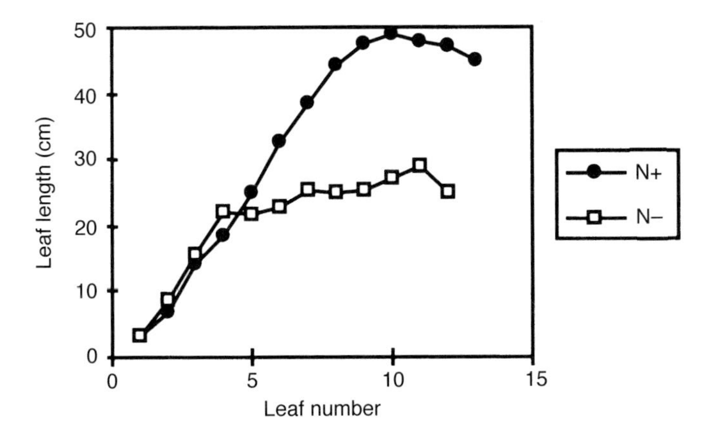

Fig. 8.2. Effect of nitrogen fertilization on lamina length of successive leaves of *Setaria anceps* (150 and 0 kg N ha^{-1} designated N+ and N−, respectively) during a period of regrowth after complete defoliation.

variables to nitrogen may appear to be contradictory. For example, the influence of N on tillering rate would be expected to be mediated through effects on LAR. However, although N use does not increase the number of leaves (i.e. the number of potential development sites in a tiller) significantly, it does increase the proportion of tillers growing in potential axillary bud sites. Not all axillary buds induce tiller development, especially in the case of grassland growing under limiting N nutrient conditions (Lemaire, 1985). Davies (1974) defined the concept of 'site filling', which makes it possible to calculate the potential tillering rate. High N availability can increase the proportion of tillers growing on potential sites, and consequently the tillering rate, without modifying the LAR.

However, and independently of any N effect, the tillering rate remains dependent on LAR and the tillering potential of each species depends on interspecific variations in LAR. For example, *L. perenne* has twice as many tillers as *F. arundinacea* at the same LAI, because the LAR in *Lolium* is twice as high (Lemaire and Chapman, 1996).

The relation between LAR and the tillering of tufted tropical species has been studied on a smaller scale. McIvor (1984), in a study carried out on two *Urochloa* species, showed that differences in leaf appearance and expansion rates exist between related tropical species, suggesting differences in potential tiller density.

Nitrogen and Morphogenesis of Stoloniferous Grasses

In vegetative stoloniferous grasses, both leaves and stems (stolons) are produced as aerial organs. As in tufted grasses, a growing point also produces a sequential chain of phytomers, but, in this case, the internodes always elongate, even when

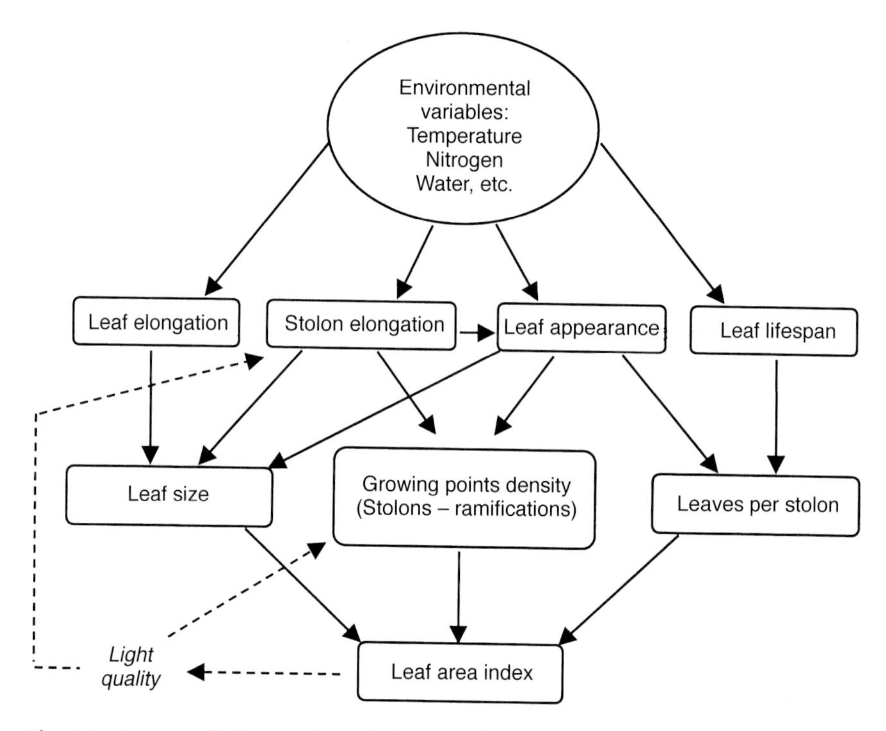

Fig. 8.3. Suggested diagram for relationships between morphogenetic variables and sward structural characteristics of a stoloniferous species (adapted from Lemaire and Chapman, 1996).

floral induction does not occur. Axillary meristems develop as horizontal stolons, extending the plant laterally (Fick and Clark, 1998). During vegetative growth, the morphogenesis of a single growing point can be characterized by the same variables described for tillers of tufted grasses (LER, LAR and leaf lifespan), as well as for the stolon (or stem) elongation rate (SER).

It is necessary to adapt the diagram of Lemaire and Chapman (1996) to describe the morphogenesis of a stoloniferous grass (Fig. 8.3). SER is the morphogenetic variable that determines the growth pattern of this type of plant, since it affects leaf variables (LAR, leaf size), whether directly or indirectly.

Stolon elongation

An increase in N availability greatly affects stolon elongation, as shown by Fig. 8.4 for *D. aristatum* and *D. decumbens*. We can see that the positive nitrogen effect appears at the beginning of regrowth, but that, in severe nitrogen deficiency, stem elongation remains virtually nil in *D. aristatum*, whose weak canopy growth means that the plants maintain a tuft-like structure. De Kroon and Hutchings

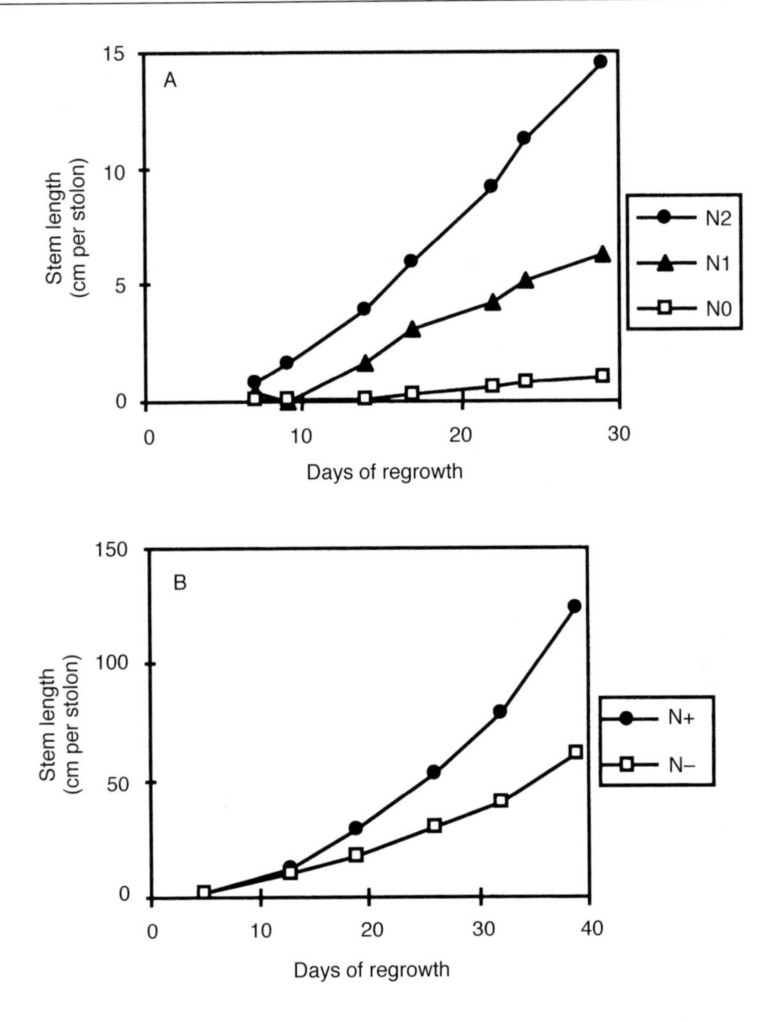

Fig. 8.4. Effect of nitrogen fertilization on stolon elongation in *Dichanthium aristatum* (A) and *Digitaria decumbens* (B) stands. Data for *D. aristatum* concern only the primary stolon. Data for *D. decumbens* include secondary stolons (ramifications). See text for characterization of N availability in each treatment (P. Cruz, unpublished data).

(1995) showed that the development of stolons and their ramifications can only occur actively when the resource supply of their environment is high.

The comparison made between the two species in Fig. 8.4 is based on data from different experiments. Therefore, the nitrogen nutrition level needs to be defined for each sward, because the plant N status depends not only on the N fertilization rate but also on the soil N availability, making comparisons difficult between experiments. The nutrient levels shown in Fig. 8.4 were calculated using the nitrogen nutrient indices (NNI) proposed by Lemaire and Gastal (1997), who refer to the

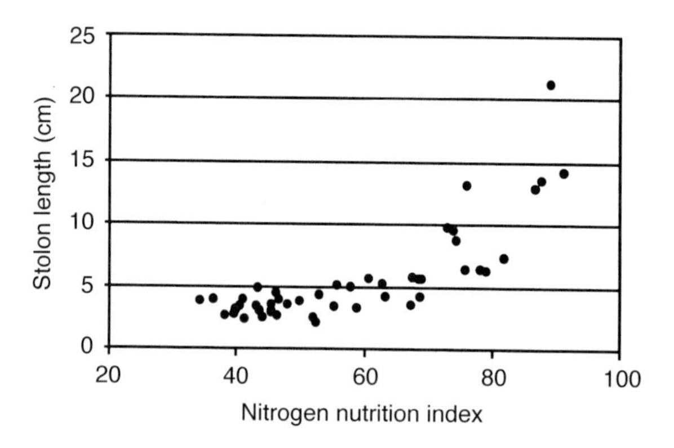

Fig. 8.5. Relationship between stolon length of a 4 weeks *Dichanthium aristatum* stand and N availability, defined in terms of nitrogen nutrition index (M. Boval, 1999, unpublished results).

concept of 'critical plant N%', using the relationship reported by Cruz and Lemaire (1996) and Duru *et al.* (1997) for C_4 species (N%crit. = 3.6 (dry matter (DM))$^{-0.34}$), which describes the decrease in the critical plant N% as the sward biomass (DM) increases. NNIs were calculated as the ratio (× 100) between the observed plant N% and the critical plant N% as calculated above for the corresponding sward biomass. A value of NNI > 100% indicates that plants are growing without N limitation and with some 'luxury' consumption, while a value of NNI < 100% indicates that plant growth is limited by N shortage, the importance of this limitation being proportional to the difference of NNI from 100%. The NNI values obtained for the different treatments reported in Fig. 8.4 are:

For *D. aristatum*:	N2 = 100	For *D. decumbens*:	N+ = 100
	N1 = 70		N− = 50
	N0 = 30		

We can see that, for the two species, the highest fertilization rate corresponds to a non-limiting N condition, whereas the lowest level is far lower in *Dichanthium*. This explains the weak elongation of the stolons in the N0 treatment in comparison with the N− treatment in *Digitaria*. A quantitative relationship between the plant N status (NNI) and stem length in *Dichanthium* is shown in Fig. 8.5 for different swards grazed by heifers after 4 weeks of regrowth and for a large range of N fertilization rates. This result suggests that a threshold NNI value of 70% exists under which stolon growth is strongly inhibited by N deficiency.

Leaf appearance and lamina size

Figure 8.6 shows that the LAR in *D. aristatum* is strongly reduced by N shortage. This effect can be the consequence of the increase in the internode elongation of stolons, which pushes each new leaf out of the sheath of the preceding leaf under

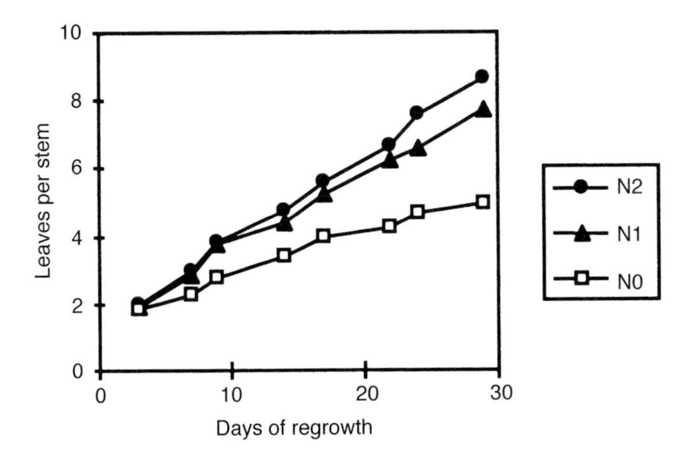

Fig. 8.6. Effect of nitrogen fertilization on cumulative leaf number per stem in *Dichanthium aristatum*. N levels similar to those in Fig. 8.4A (P. Cruz, unpublished data).

high N disposibility. So, if we consider that the leaf growth zone is restricted to the basal leaf tissue protected from light inside the sheath tube (see Schnyder *et al.*, Chapter 3, this volume), the acceleration of internode elongation may serve to reduce the duration of the elongation of leaves.

Figure 8.7 shows similar evolution of the average final size of leaf blades along a main stolon of *D. aristatum* and of *D. decumbens*. The first leaves are short (or are the leaf residue following defoliation), and then the maximum length is reached for leaf 3 or 4, corresponding to the beginning of the growth of the stolons. The size of subsequent blades is progressively reduced as a consequence of the increase in internode elongation and the resulting reduction in the duration of elongation, and then stabilizes. This evolution contrasts with that of a tufted species (see Fig. 8.2), where blade length increases progressively with leaf number.

The effect of N nutrition on leaf size is greater in *D. aristatum* than in *D. decumbens* (Fig. 8.7). In fact the effect of N nutrition on leaf size is the result of two contradictory processes: (i) an increase in LER, which tends to produce longer leaves; and (ii) a reduction in LED, which tends to produce shorter leaves. As a consequence, the overall effect of N on leaf size may be limited.

Figure 8.7 also shows that the rate of leaf appearance of *D. aristatum* is strongly affected by N nutrition (from seven to 12 leaves produced for low and high N treatments, respectively), while this effect is very small for *D. decumbens* (from 25 to 27 leaves produced in low and high N, respectively). So we can postulate that the direct consequence of the increase in stolon internode elongation on the reduction in LED could be independent of the effect on LAR. In other words, the ratio between LED and LAR, which represents the number of growing leaves per stolon, could be affected by N nutrition differently in the two species.

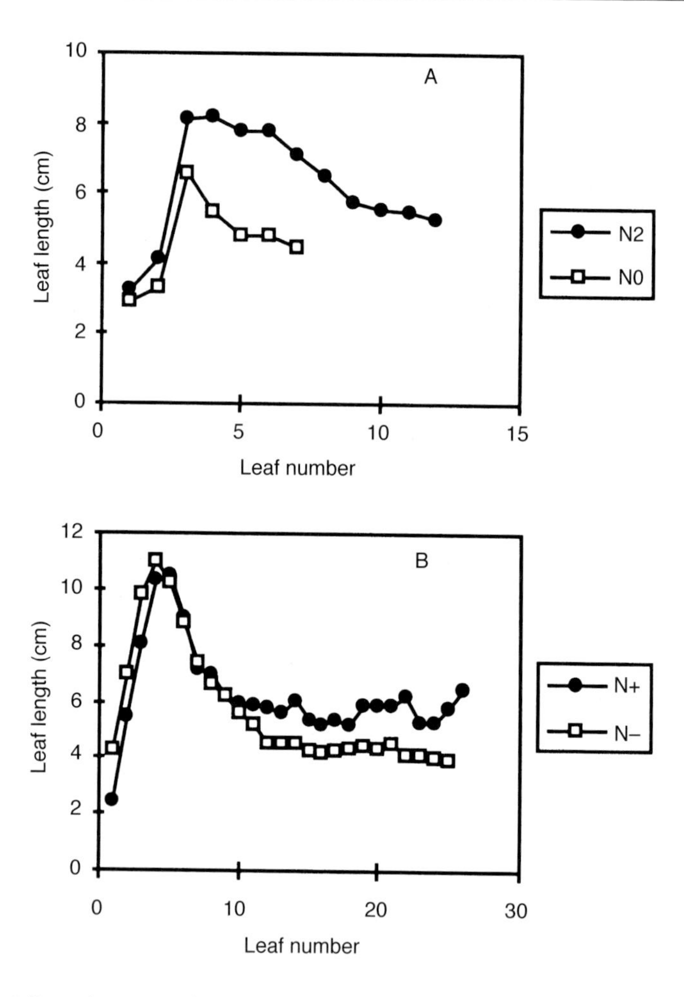

Fig. 8.7. Effect of nitrogen fertilization on lamina length of successive leaves in stoloniferous *Dichanthium aristatum* (A) and *Digitaria decumbens* (B). N levels similar to those in Fig. 8.4 (P. Cruz, unpublished data).

For *Dichanthium*, this ratio should be only slightly affected (i.e. LED should have been reduced in the same proportion as LAR, the number of growing leaves remaining almost constant), while, for *Digitaria*, LED should have been more affected than LAR. Thus the effect of N nutrition on the leaf tissue production on stoloniferous species appears to be very dependent on the response of stolon internode elongation and differs from one species to another, depending on the relative response of the three main leaf morphogenesis characteristics: LAR, LER and LED. More detailed analyses of these basic morphogenetic processes are needed for a full understanding of the overall responses of stoloniferous species to N fertilization.

Growing point density

Nitrogen increases growing point density in stoloniferous species, as it does in tufted species. Working with *D. decumbens*, P. Cruz (unpublished data) observed that increased nitrogen nutrition levels have a positive effect on the appearance of secondary ramifications on the stolon, even though the number of potential sites (number of leaves on the main stolon) is not significantly affected. By influencing canopy development, nitrogen greatly modifies the light spectrum within the canopy, with consequences for stolon differentiation (Willemoes *et al.*, 1987), but these effects are not dealt with here.

Effects on Establishment of LAI

A comparison of the two types of growth patterns, in terms of what has been said about their morphogenetic variables, will enable us to describe more accurately their respective growth dynamics. The difference between the evolution of blade size on tillers and stolons is of paramount importance. Indeed, leaf size greatly influences the kinetics of LAI establishment in the canopy. We know that the number of leaves on a growing point is the product of the balance between their appearance and senescence, and so we can deduce that, during the growth cycle for a given tufted species, up to a point the leaves emerging will be progressively larger and there will be a positive accumulation of tissues per tiller, as the senescent leaves will always be smaller than those emerging. Conversely, leaf senescence of stoloniferous species occurs very early at the expense of larger leaves, whereas the new emerging leaves will be progressively smaller. Thus, net leaf elongation on a given growing point becomes negative (Alexandre *et al.*, 1989; Cruz *et al.*, 1989; Cruz and Schemoul, 1991) and the ceiling LAI of the sward is reached very early. This ceiling LAI development stage in the canopy is reached even more rapidly when the nitrogen nutrition level is high.

Figure 8.8 shows comparisons of the kinetics of morphogenetic leaf variables during regrowth, at the level of a tiller or a stolon (gross cumulative elongation, senescence and net cumulative elongation) between two tropical grasses, *S. anceps* (tufted) and *D. decumbens* (stoloniferous), respectively. For the tufted species, leaf senescence starts 20 days after defoliation and net leaf production becomes progressively lower than gross leaf production. The ceiling leaf length per tiller is reached after 30 days of regrowth. For the stoloniferous species, leaf senescence starts 12 days after defoliation, and the maximum leaf length per stolon is reached rapidly and then declines, because senescing leaves are replaced by new leaves of smaller size (see Fig. 8.7).

After 4 weeks of growth, the loss of leaf material in a stoloniferous species cultivated under non-N-limiting conditions can account for half of the leaf tissues produced. This has been shown also in *D. aristatum* (Cruz, 1995). The morphogenesis of stoloniferous species does not enable them to reach ceiling LAI

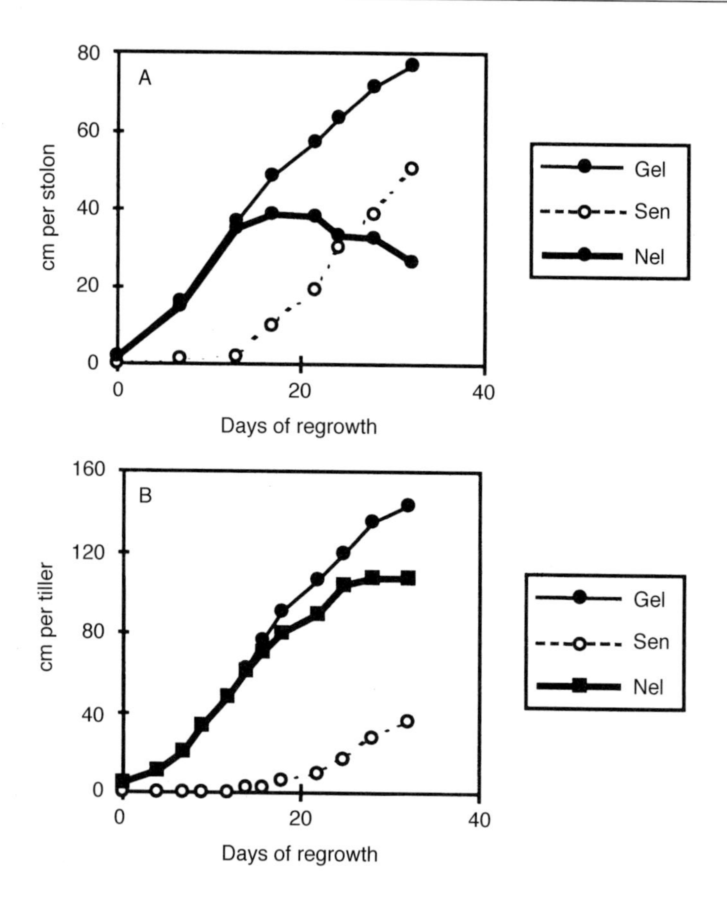

Fig. 8.8. Dynamics of gross (Gel) and net (Nel) cumulative leaf elongation and cumulative leaf senescence (Sen) in *Digitaria decumbens* (A) and *Setaria anceps* (B) growing under optimal nitrogen nutrition (P. Cruz, 1999, unpublished results).

values as high as those in tufted species. Hacker and Evans (1992) compared six tropical grasses and showed that the two stoloniferous species had lower LAIs than the tufted species. Nevertheless, the lower ceiling LAI values in stoloniferous species are enough to enable the canopy to intercept light radiation maximally, partly as a result of their planophile growth habit.

Consequences for Forage Quality and Grassland Management

The greatest impact of the contrasting patterns of leaf biomass dynamics in tufted and stoloniferous plants is not in terms of effects on light interception in stoloniferous species, but a consequence of their loss of nutritive value as forage species. This effect is exacerbated by the linear increase in stem accumulation that follows

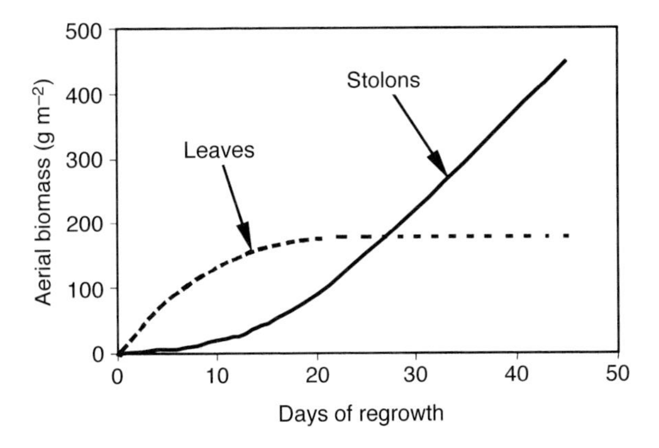

Fig. 8.9. Diagram for net cumulative biomass of leaves and stolons in a tropical stoloniferous grass.

canopy closure, stem being less digestible than leaf. The two phases of growth dynamics typical of stoloniferous species are shown in Fig. 8.9. Dry matter partitioning between the leaf and stolon compartments has been described and modelled by Overman and Wilkinson (1989) for *C. dactylon*. This simulation of leaf production seems to be somewhat unrealistic, because there is no upper limit of response in the leaf compartment to nitrogen supply. Moreover, the authors set the beginning of leaf loss by senescence at 6 weeks, without distinguishing between the different levels of N availability. Our experience shows that, in non-limiting water and mineral conditions, leaf biomass in stoloniferous species rarely exceeds 2 t DM ha^{-1} (Cruz *et al.*, 1989; Cruz and Moreno, 1992; Naves *et al.*, 1993; Cruz, 1995) and that net accumulation ends suddenly after 3 weeks of canopy growth. After this, the accumulation of stolon rapidly exceeds the accumulation of green leaf material and the leaf/stem ratio of the forage rapidly declines. This deterioration of the quality of herbage is emphasized by the accumulation of dead leaf tissues.

Because livestock production requires high-quality herbage, it is essential to avoid stolon accumulation, particularly in view of the high response of SER to nitrogen, already mentioned. At high levels of N nutrition, the only way to control stolon growth is through grazing management (Naves *et al.*, 1993) or by frequent cutting (Wilen and Holt, 1996). It has been clearly demonstrated that, unless stolon development is controlled, the utilization of biomass by sheep grazing on *D. decumbens* or *C. nlemfuensis* grasslands is limited (Boval *et al.*, 1993), as is the case for cattle on natural *D. aristatum* grasslands (M. Boval, unpublished data). Alexandre *et al.* (1989) observed that the selective nature of the intake of young goats grazing on a canopy of *D. decumbens* means it is possible to obtain high animal production performance. Stolon biomass does not limit intake for such selective animals, provided that grazing pressure is low.

The kinetics of leaf growth and stolon growth, as well as the difficulties faced

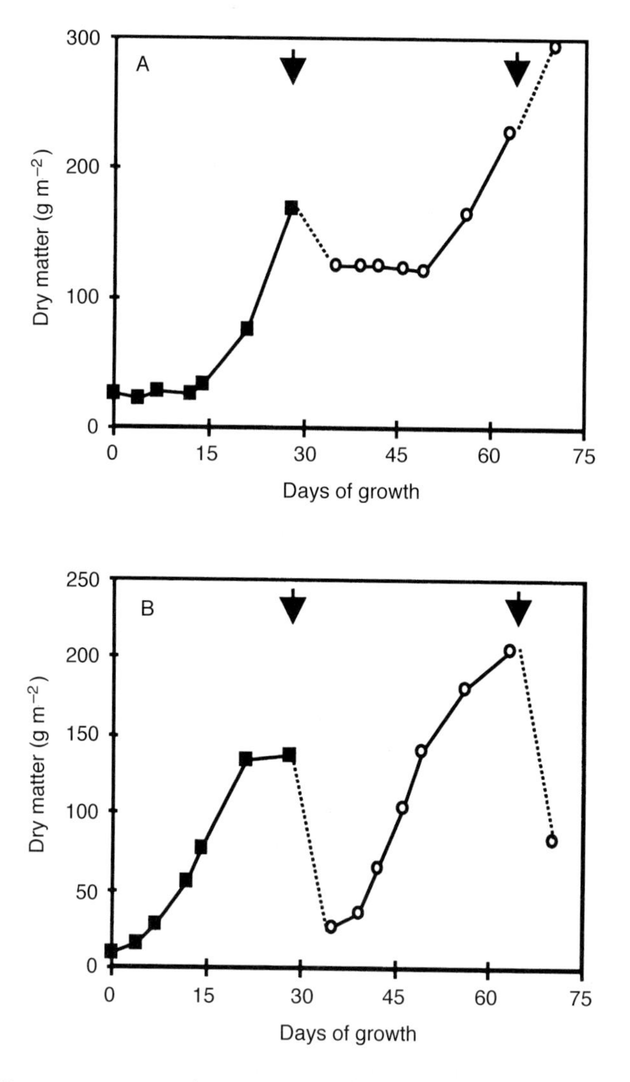

Fig. 8.10. Dry matter accumulation on (A) stolons and (B) leaves of a grazed *Digitaria decumbens* sward during two regrowth cycles. Arrows and dotted lines indicate the start and duration of the grazing periods. (From Naves *et al.*, 1993.)

when trying to control stolon accumulation through grazing, has been exemplified by the study carried out by Naves *et al.* (1993). The study shows that, at low grazing pressure, stolon biomass continues to accumulate and only leaf material is consumed. In this example, and despite a total of 14 days of grazing (two periods of 7 days each), the *Digitaria* sward accumulated a stolon biomass close to 3 t DM ha^{-1} for a 70-day period of growth (Fig. 8.10). It is difficult to maintain a stoloniferous species in the leaf growth phase by removing the stolons at the beginning of growth, due to the seasonal variations in stolon development (Salette, 1971;

Cruz *et al.*, 1989) and to the high nitrogen requirements of such practices (Cruz and Huguet, 1995). Growth variations require seasonal adjustment of stocking rate, which is awkward and difficult to apply. Combined cutting and grazing practices therefore appear to be necessary. The problem of adjusting the stocking rate to herbage growth is very different between stoloniferous and tillering species during the vegetative period. Low stocking rates of the former lead to resource degradation, because of continuous stem and dead leaf accumulation. In the case of tufted species, there is only a loss of efficiency in the use of leaf material and a decrease in the quality of the new emerging leaves (Duru *et al.*, 1999); this is less important than the degradation that occurs in stoloniferous species as a consequence of large decreases in the leaf/stem ratio.

Conclusion

Nitrogen has a positive effect on the variables defining forage plant quality, such as herbage digestibility or nitrogen concentration. Nevertheless, because of its role in the morphogenesis of stoloniferous grass, and particularly under certain grazing conditions, use of N can cancel out these positive effects, because it results in an excessively high proportion of stems in the sward. Improvement in the nutritive value of stoloniferous species on the basis of selection for morphogenetic criteria is feasible, but it is more difficult to improve the forage value of stoloniferous species. More important and rapid progress can be made if the focus is placed on controlling the morphogenesis of these grasses on the basis of: (i) grassland-adapted nitrogen fertilization; (ii) adjustment of stocking rates to herbage growth; and (iii) choosing the appropriate opportunity to defoliate by grazing or cutting during the regrowth cycle.

References

Alexandre, G., Baudot, H., Cruz, P. and Prache, S. (1989) Forward rotational grazing on *Digitaria decumbens* pasture with weaned female creole goats in Guadeloupe. In: *Proceedings of the XVIth International Grasslands Congress, Nice, France, 4–11 October 1989*. AFPF, Versailles, France, pp. 1271–1272.

Bogdan, A.V. (1977) *Tropical Pasture and Fodder Plants*. Tropical Agriculture Series, Longman, New York, 475 pp.

Boval, M., Alexandre, G., Mahieu, M., Cruz, P. and Meuret, M. (1993) Comparative use of *Digitaria decumbens* and *Cynodon nlemfuensis* by local suckling ewes in Martinique (FWI). In: *Proceedings of the XVIIth International Grassland Congress, Rockhampton, Australia*. SIR Publishing, Wellington, New Zealand, pp. 2004–2005.

Chapman, D.F. and Lemaire, G. (1993) Morphogenetic and structural determinants of plant regrowth after defoliation. In: *Proceedings of the XVIIth International Grassland Congress, Palmerston North, New Zealand*. SIR Publishing, Wellington, New Zealand, pp. 95–104.

Corsi, M. (1984) Effects of nitrogen rates and harvesting intervals on dry matter production, tillering and quality of the tropical grass *Panicum maximum* Jacq. *Dissertation Abstracts International B Sciences and Engineering* 45, 1–24.

Cowan, R.T. and Lowe, K.F. (1998) Tropical and subtropical grass management and quality. In: Cherney, J.H. and Cherney, D.J.R. (eds) *Grass for Dairy Cattle*. CAB International, Wallingford, UK, pp. 101–135.

Cruz, P. (1995) Use of RUE concept for analysing growth of pure and mixed tropical forage crops. In: Sinoquet, H. and Cruz, P. (eds) *Ecophysiology of Tropical Intercropping*. Science Update, INRA, Paris, pp. 319–330.

Cruz, P. and Huguet, J.M. (1995) Requerimientos de nitrógeno y carbono para el rebrote de una pradera de Pangola defoliada frecuentemente. In: *Proceedings of the XIV Reunión Latinoamericana de Producción Animal, Mar del Plata, Argentina*. AAPA, Balcarce, Argentina, pp. 25–28.

Cruz, P. and Lemaire, G. (1996) Diagnosis of the nitrogen status of grass stands: principles and uses of the dilution curves method. *Tropical Grasslands* 30, 166.

Cruz, P. and Moreno, J.L. (1992) Crecimiento potencial comparado de una graminea natural (*Dichanthium aristatum*) y una cultivada (*Digitaria decumbens*) sometida a variaciones de fotoperiodo. *Revista Cubana de Ciencias Agricolas* 26, 323–330.

Cruz, P. and Schemoul, E. (1991) Effet de l'azote sur l'expression du potentiel de croissance d'une prairie naturelle à base de *Dichanthium aristatum* en Guadeloupe (Antilles françaises). In: *Proceedings of the IVth International Rangeland Congress, Montpellier, France*. CIRAD, Montpellier, France, pp. 360–363.

Cruz, P. and Sobesky, O. (1989) Variations saisonnières de la croissance d'une prairie de *Setaria anceps* en Guadeloupe. In: *Proceedings of the XVIth International Grassland Congress, Nice, France*. AFPF, Versailles, France, pp. 17–18.

Cruz, P., Alexandre, G. and Baudot, H. (1989) Cinétique de la croissance foliaire et stolonifère d'un peuplement de *Digitaria decumbens* au cours de la repousse. In: *Proceedings of the XVIth International Grassland Congress, Nice, France*. AFPF, Versailles, France, pp. 499–500.

Davies, A. (1974) Leaf tissue remaining after cutting and regrowth in perennial ryegrass. *Journal of Agricultural Science (Cambridge)* 82, 165–172.

Davies, A. and Thomas, H. (1983) Rates of leaf and tiller production in young spaced perennial ryegrass plants in relation to soil temperature and solar radiation. *Annals of Botany* 57, 591–597.

de Kroon, H. and Hutchings, M.J. (1995) Morphological plasticity in clonal plants: the foraging concept reconsidered. *Journal of Ecology* 83, 143–152.

Duru, M., Justes, E., Langlet, A. and Tirilly, V. (1993) Comparaison des dynamiques d'apparition et de mortalité des organes de fétuque élevée, dactyle et luzerne (feuilles, talles et tiges). *Agronomie* 13, 237–252.

Duru, M., Lemaire, G. and Cruz, P. (1997) The nitrogen requirement of major agricultural crops: grasslands. In: Lemaire, G. (ed.) *Diagnosis of N Status in Crops*. Advanced Series in Agricultural Sciences, Springer-Verlag, Berlin, pp. 59–72.

Duru, M., Ducrocq, H. and Feuillerac, E. (1999) Effet du régime de défoliation et de l'azote sur le phyllochrone du dactyle. *Compte Rendu de l'Académie des Sciences: Sciences de la Vie* 322, 717–722.

Fick, G.W. and Clark, E.A. (1998) The future of grass for dairy cattle. In: Cherney, J.H. and Cherney, D.J.R. (eds) *Grass for Dairy Cattle*. CAB International, Wallingford, UK, pp. 1–22.

Gastal, F. and Lemaire, G. (1988) Study of a tall fescue sward grown under nitrogen deficiency conditions. In: *Proceedings of the XIIth Meeting of the European Grassland*

Federation, Dublin. Irish Grassland Association, Belclare, Ireland, pp. 323–327.

Gastal, F., Belanger, G. and Lemaire, G. (1992) A model of leaf extension rate in response to nitrogen and temperature. *Annals of Botany* 70, 437–442.

Hacker, J.B. and Evans, T.R. (1992) An evaluation of the production potential of six tropical grasses under grazing. 1. Yield and yield components, growth rates and phenology. *Australian Journal of Experimental Agriculture* 32, 19–27.

Lemaire, G. (1985) Cinétique de croissance d'un peuplement de fétuque élévée (*Festuca arundinacea* Schreb.) pendant l'hiver et le printemps. Effets des facteurs climatiques. Thèse Doctorat és Sciences Naturelles, Université de Caen, France.

Lemaire, G. and Chapman, D.F. (1996) Tissue flows in grazed plant communities. In: Hogdson, J. and Illius, A.W. (eds) *The Ecology of Management of Grazing Systems*. CAB International, Wallingford, UK, pp. 3–36.

Lemaire, G. and Gastal, F. (1997) N uptake and distribution in plant canopies. In: Lemaire, G. (ed.) *Diagnosis of N Status in Crops*. Advanced Series in Agricultural Sciences. Springer-Verlag, Berlin, pp. 3–43.

McIvor, J.G. (1984) Leaf growth and senescence in *Urochloa mosambicensis* and *U. oligotricha* in a seasonally dry tropical environment. *Australian Journal of Agricultural Research* 35, 177–187.

Mears, P.T. and Humphreys, L.R. (1974) Nitrogen response and stocking rate of *Pennisetum clandestinum* pastures. 1. Pasture nitrogen requirement and concentration, distribution of the dry matter and botanical composition. *Journal of Agricultural Science (Cambridge)* 83, 451–467.

Moore, K.J. and Moser, L.E. (1995) Quantifying developmental morphology of perennial grasses. *Crop Science* 35, 37–43.

Muldoon, D.K. and Pearson, C.J. (1979) Primary growth and re-growth of the tropical tallgrass hybrid *Pennisetum* at different temperatures. *Annals of Botany* 43, 709–717.

Nada, Y. (1985) Responses of tropical pasture grasses to photoperiod and temperature. In: *Proceedings of the XVth International Grassland Congress, Kyoto, Japan*. The Science Council of Japan, Nishi Nasuno, Japan, pp. 1256–1258.

Naves, M., Cruz, P., Malafosse, A. and Manteaux, J.P. (1993) Growth kinetics of *Digitaria decumbens* following cattle grazing at two stocking rate levels. In: *Proceedings of the XVIIth International Grassland Congress, Rockhampton, Australia*. SIR Publishing, Wellington, New Zealand, pp. 874–875.

Overman, A.R. and Wilkinson, S.R. (1989) Partitioning of dry matter between leaf and stem in coastal bermudagrass. *Agricultural Systems* 30, 35–47.

Rosario, E.L. (1977) Influence of fertility level on yield determining physiomorphological characteristics of some sugarcane varieties. *Philippine Journal of Crop Science* 2, 19–30.

Salette, J. (1971) Seasonal pattern of forage growth and related characters in humid tropical conditions. In: *Actes du Colloque sur l'Intensification de la Production Fourragère et son Utilisation par les Ruminants*. INRA, Guadeloupe, FWI, pp. 93–99.

Sato, K. (1980) Growth responses of some gramineous forage crops to daylength and temperature. *Journal of Japanese Society of Grassland Science* 25, 311–318.

Simon, J.C. and Lemaire, G. (1987) Tillering and leaf area index in grasses in the vegetative phase. *Grass and Forage Science* 42, 383–380.

Wilen, C.A. and Holt, J.S. (1996) Spatial growth of kikuyugrass (*Pennisetum clandestinum*). *Weed Science* 44, 323–330.

Willemoes, J.G., Beltrano, J. and Montaldi, E.R. (1987) Stolon differentiation in *Cynodon dactylon* (L) Pers. mediated by phytochrome. *Environmental and Experimental Botany* 27, 15–20.

Wilman, D. and Pearse, P.J. (1984) Effect of applied nitrogen on grass yield, nitrogen content, tillers and leaves in field swards. *Journal of Agricultural Science, Cambridge* 103, 201–211.

9

Modelling the Dynamics of Temperate Grasses and Legumes in Cut Mixtures

J.F. Soussana[1] and A. Oliveira Machado[2]

[1] *INRA, Unité d'Agronomie, 234 Av. du Brézet, F-63100 Clermont-Ferrand, France;* [2] *Universidade Federal do Paraná, Departamento de Fitotecnia e Fitossanitarismo, Rua dos Funcionários 1540, CEP 80.035–050, Curitiba, Parana, Brazil*

Introduction

Low-input grazing systems based on mixtures of grasses and legumes (e.g. clover) have often been proposed as a sustainable alternative to the intensive use of N fertilizer in managed grasslands. When they contain enough legumes capable of fixing atmospheric nitrogen, unfertilized pastures can sustain a high level of herbage production, despite continuous losses of nitrogen through cutting, leaching or volatilization (Ledgard and Steele, 1992). The use of legumes in pastures may also result in increased N content and digestibility and well-balanced mineral content of herbage, all of which are of importance in animal nutrition (Frame and Newbould, 1986).

A large number of factors (plant, soil and environmental) can influence the delicate species balance in a non-equilibrium pasture mixture (Haynes, 1980). However, in many cases, the proportion of pasture legumes fluctuates both from year to year and within single growth periods, which results in grass–legume mixtures often being considered as unpredictable and difficult to manage by farmers (Kessler and Nösberger, 1994). Understanding and predicting interference and competition between legumes and associated grasses is therefore of great importance for the development of sustainable pasture production in several regions of the world (see Hodgson and Da Silva, Chapter 1, this volume).

White clover (*Trifolium repens* L.) is the most abundant pasture legume in the temperate zone and almost all aspects of the growth and performance in mixtures of this species have been studied extensively (Frame and Newbould, 1986). Despite this extensive empirical knowledge, there is only a limited theoretical base to analyse and predict the changes in the relative abundance of grass and clover in pastures.

(eds G. Lemaire, J. Hodgson, A. de Moraes, C. Nabinger and P.C. de F. Carvalho)

To gain understanding, Thornley *et al.* (1995) and Schwinning and Parsons (1996a, b) have recently proposed pasture growth models that include mixed grass and legume components. These models predict that the coexistence of grasses and legumes may be based on exploitation, rather than on mutual competition. Grasses benefit from the ability of clover to fix nitrogen but simultaneously suppress legume growth through competition for light. This interaction would provide the basis for large amplitude oscillations of grass and legume densities (Schwinning and Parsons, 1996a) and the field-scale consequences of these patch-scale oscillations in legume content were also studied by these authors (Schwinning and Parsons, 1996b). These pasture models all consider a nitrogen-based competitive trade-off between grass and clover. Differences in the soil N environment (fertilizer input, leaching rate) determine whether the species can coexist, but, where coexistence is possible, the species composition is assumed to regulate soil N (Schwinning and Parsons, 1996a).

To make further progress, we develop here a detailed modelling approach to the interactions between mixed grass and clover populations at the patch scale. We do not attempt to include plant–soil interactions and herbivory in our analysis, but we try to develop a mechanistic and quantitative understanding of above-ground (light) and below-ground (inorganic nitrogen) resource partitioning among mixed grass and clover from an individual-based simulation model named Canopt (Soussana *et al.*, 1999a, b). This model was calibrated for perennial ryegrass (*Lolium perenne*) and white clover (*T. repens*) and we show here some of the model's predictions concerning the dynamics of these species in cut mixtures. From this modelling work, we argue that the coexistence of both grass and clover is favoured by morphological (plasticity) and physiological (acclimatization) adjustments to the resource level and to the neighbour densities they encounter, and we try to understand some of the major functional traits that may favour the dominance of these species in mixtures.

Modelling Approach

Model description

Canopt is an object-orientated simulation model, which is coded in C++ (Borland™ 5.02) in Windows™. It consists of five modules: (i) the management options (grazing and/or cutting mode, N fertilizer supply); (ii) the environment module, which calculates the microclimate and the inorganic N supply from the soil and fertilizer; (iii) the plant growth module, which simulates the C and N balance and the partitioning of growth among shoot structures, leaf proteins and roots; (iv) the shoot morphogenesis module, which computes the demography and the size of the leaves and axis; and (v) the competition module, which calculates photosynthetically active radiation (PAR) and inorganic N partitioning among grass and clover. This model was described in detail by Soussana *et al.* (1999a, b), and these authors have also evaluated modules iii to v, in comparison with experimental data. The main features of the model are described below.

The model simulates either pure swards or binary mixtures. Each plant population is described as a collection of axes (i.e. tillers for grasses, growing points for clover) and all axes forming one plant population are assumed to be identical. Moreover, with a binary mixture, the two plant populations are assumed to be perfectly mixed in the horizontal plane. Each plant population is described with eight state variables: the axis number (m^{-2}) and the weights per axis of the leaf proteins (W_p), shoot structures (W_s), roots (W_r), C and N substrates and C and N reserves (Fig. 9.1). The C and N substrate concentrations are assumed to be the same in the shoots and roots. The dynamics of the C and N reserves are modelled using first-order kinetics. The reserve compartments are assumed to be located below the cutting height. With clover, the weight of the stolons is assumed to be equal to a moving average of the leaf weight. Stolons are assumed to have no photosynthetic activity and only growth and maintenance respiration processes.

The carbon balance is based on the equations of Farquhar *et al.* (1980) for leaf photosynthesis, assuming a linear relationship of Vc_{max} (the maximal carboxylation activity of Rubisco) and J_{max} (the electron transport capacity) with the area-based leaf protein concentration (Field, 1983; Harley *et al.*, 1992; Nijs *et al.*, 1995). The vertical profile of leaf proteins is modelled according to Hirose *et al.* (1988). Respiration is divided into growth and maintenance components. For legumes, an additional respiration, linked with biological nitrogen fixation, is also considered, assuming a fixed carbon cost per unit N fixed.

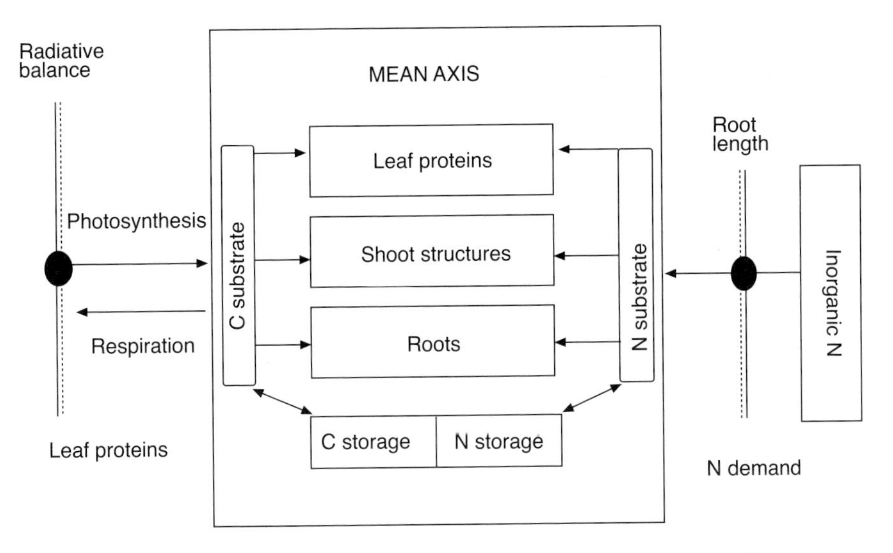

Fig. 9.1. Carbon–nitrogen submodel in Canopt. An average axis (grass tiller or clover growing point) consists of seven compartments: the leaf proteins, shoot structures, roots and the C and N substrate and reserve pools. The solid arrows depict the C and N fluxes. The dashed lines show some of the regulations affecting photosynthesis and inorganic N uptake.

The relative growth rate of the three structural compartments (W_p, W_s, W_r) is calculated through a bisubstrate (C, N) growth equation, using three partitioning variables, one for each compartment (Johnson and Thornley, 1987). The partitioning submodel specifies a target root : shoot ratio, according to the functional balance hypothesis (Brouwer, 1963; Davidson, 1969; Hilbert and Reynolds, 1991), and a target protein : structure ratio for leaves, according to the coordination theory of leaf photosynthesis (Chen *et al.*, 1993).

The potential leaf extension rate is calculated according to van Loo (1993). This rate is assumed to increase exponentially with temperature (Lemaire and Agnusdei, Chapter 14, this volume). The duration of leaf extension is proportional to the phyllochron and to the number of leaves growing simultaneously on the same axis. The potential leaf area and leaf weight are calculated from leaf length through simple allometric relationships. Whenever the supply of assimilates to the shoots does not allow this potential (see Porter, 1984), the leaf extension rate is reduced and the phyllochron is increased. Hence, the model considers a source limitation of shoot growth when the assimilate (C and N) supply to the shoots is less than their potential growth rate. In order to mimic photomorphogenetic effects on leaf growth (Thompson and Harper, 1988; Varlet-Grancher *et al.*, 1989; Varlet-Grancher and Gautier, 1995), an increased leaf extension rate and a reduced leaf width are assumed by the model when the fraction of light transmitted at the tip of the sheath (or petiole) falls below a threshold.

The leaf lifespan is calculated from the maximum number of green leaves per axis (see Lemaire and Agnusdei, Chapter 14, this volume). Roots decay according to first-order kinetics, based on thermal time. Fixed fractions of the shoot structures, the leaf proteins and the roots are recycled to the C and N substrate compartments during senescence.

Plants are assumed to consist of a collection of identical and anatomically connected axes (e.g. tillers for grasses). Maximum site-filling is assumed (Davies, 1974) to calculate the potential branching (or tillering) rate per unit thermal time. Branching is constrained by light transmission at ground level (Simon and Lemaire, 1987; Bouman *et al.*, 1996) and by the amounts per plant of substrate C and of substrate N that are available for the outgrowth of new branches (or tillers). According to the degree of physiological integration in the clonal plant, it is considered that only a fraction of the C and N substrates are made available for branching. The baseline mortality rate of axes is calculated from first-order kinetics based on thermal time. This baseline mortality is increased: (i) when assimilate supply and C and N reserves are below a threshold; and (ii) when the $-5/2$ self-thinning line is reached (see Matthew *et al.*, Chapter 7, this volume). Fixed fractions of the shoot structures, the leaf proteins and the roots are recycled to the C and N substrate compartments during axis senescence.

Individual grass leaves are assumed to have the shape of a rectangular hyperbola, the form of which varies with the canopy height and with the tiller density. The shape of clover leaves is described assuming a vertical petiole and a fixed leaf lamina angle (20°) (Faurie *et al.*, 1996; Fig. 9.2). At each cutting date, it is assumed that all the plant material above cutting height is removed from the patch.

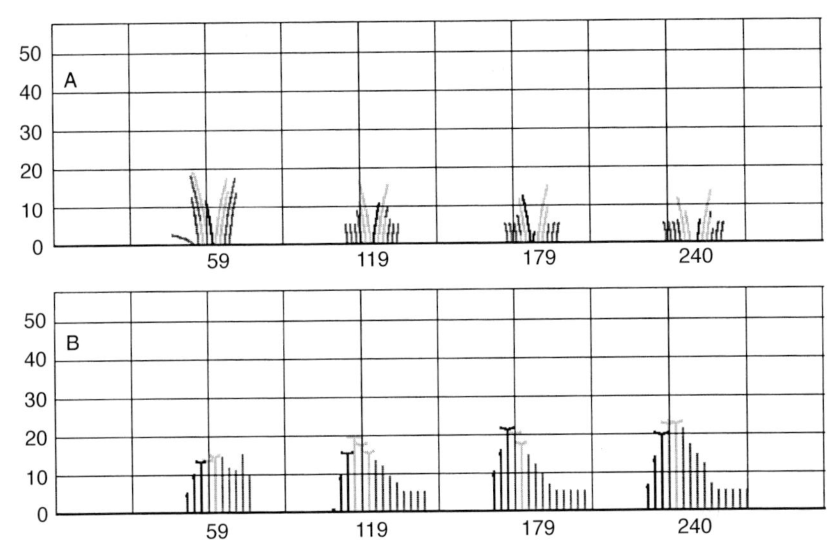

Fig. 9.2. Simulation by the Canopt model of the morphology of an average grass tiller (A) and of an average clover growing point (B) at the end of four successive regrowths (days 59, 119, 179 and 240 after the start of the simulation) of a cut mixture. The simulations were run assuming a constant environment and a daily N supply of 0.09 g N m^{-2} day^{-1}. Growing leaves are shown in black, mature leaves in light grey and senescent leaves in dark grey. The vertical scale is graduated in centimetres.

In order to compute a radiative balance, the canopy is divided into 1-cm-height horizontal layers and the light absorbed by leaves from each species in each layer is calculated assuming: (i) full diffuse radiation; and (ii) a perfect mixing of leaves in each horizontal layer. An extension of the Beer–Lambert law is used to calculate the amount of radiation transmitted below each horizontal layer from the cumulative leaf area of each plant population and from the leaf lamina angles (Varlet-Grancher and Bonhomme, 1979). Canopy photosynthesis is then calculated for each plant population by accumulating the leaf photosynthesis of each canopy layer (see Fig. 9.1).

The inorganic N uptake capacity per unit root length is down-regulated whenever the substrate N concentration exceeds a threshold (Imsande and Touraine, 1994). In the same way, for legumes the biological N_2 fixation per unit root mass is down-regulated when the substrate N concentration increases in the plant (Faurie and Soussana, 1993; Hartwig *et al.*, 1994). We assume for legumes that symbiotic N_2 fixation is relatively more sensitive to an excess substrate nitrogen than inorganic N uptake (Soussana and Faurie, 1998).

Competition for inorganic N between mixed plant populations is calculated from the total demand for N uptake (estimated from the uptake capacity and from the cumulated root length of each population). A diffusion approach (Sheehy *et al.*, 1996) is used to calculate the daily amount of N supplied by the soil to the roots. Whenever N supply per unit root length is lower than the uptake capacity, the actual uptake rate is calculated by considering that each plant population absorbs the same fraction of its potential demand per unit root length.

Calibration and use of the model for the simulation of grass–clover mixtures

The object-orientated structure of the model makes it possible to ensure that:

- processes that are common to both species (e.g. C balance, growth, partitioning, mineral N uptake) are calculated in exactly the same way by a single plant class;
- processes that are species-specific (e.g. shape of grass laminae, biological N_2 fixation of clover) are calculated by the grass and legume classes.

This model has been previously calibrated for perennial ryegrass and white clover (Soussana *et al.*, 1999a, b) and we use here the same parameter values. All parameters have the same value for grass (G) and clover (C) except for the following:

- The number of growing leaves per axis (1.3 and 3 for G and C, respectively).
- The potential leaf, or petiole, length (50 and 40 cm for G and C, respectively).
- The light extinction at the sheath, or petiole, tip inducing leaf etiolation (0.9 and 0.7 for G and C, respectively).
- The allometry between the area (A, in cm^2), weight (W, in mg) and length (L, in cm) of unetiolated leaves: $A = 0.056\ L^{1.65}$ and $W = 0.09\ L^{1.9}$ for the grass (derived from van Loo, 1993), and $A = 0.07\ L^{1.5}$ and $W = 0.12\ L^{1.8}$ (derived from Soussana *et al.*, 1995) for the clover.

The same initial values (weights of the seven compartments and axis density) were used for the state variables of the two mixed species, thereby ensuring that there was 50% clover in the mixture at the start of the simulation. This is not necessarily realistic, as, for example, the weight of an individual clover axis is, on average, greater than that of an individual tiller. On day 0, which is assumed to be just after a cut, each simulated plant population has a total of 3000 axes m^{-2}, all with the same individual structural mass (60 mg dry matter (DM)) and consisting of two-third roots and one-third shoots. The initial leaf N concentration is 1% and the initial leaf area index (LAI) of the mixture is 1.2. The initial concentrations of non-structural C and non-structural N reach 15 and 1.5%, respectively, two-thirds of which are reserves and the remaining third substrates.

The simulated responses to mineral N supply and cutting frequency reported below were obtained by running the model during one growing season (240 days) under constant and close to optimal environmental conditions (20°C, 700 μmol PAR $m^{-2}\ s^{-1}$ during 14 h, 350 p.p.m. [CO_2]). Different daily N supplies, in the range 0.03 to 0.24 g N m^{-2} day^{-1}, and three regrowth durations (20, 40 and 60 days) were compared. A cutting height of 5 cm was assumed. Unless stated otherwise, all results correspond to means calculated for the last regrowth.

Simulation Results

Productivity and clover content of cut mixtures

In order to simulate the N supply effects on the productivity and clover content of cut mixtures, the model was run close to equilibrium during 240 days, assuming constant and near-optimal environmental conditions. In agreement with the empirical evidence (Frame and Newbould, 1986), the simulation results show that the productivity of grass–clover mixtures, as estimated by the total shoot biomass at the time of the last cut, is increased by N supply. Within the range studied, the simulated increase in DM yield with nitrogen supply is greater, both in absolute and relative terms, for a long compared with a short regrowth duration (Fig. 9.3A). The productivity of the simulated mixtures (during 240 days under near-optimal light and temperature) reaches 9–22 t DM ha^{-1}, which can be compared with a range of annual production of 6–15 t DM ha^{-1} $year^{-1}$ most frequently observed in field studies (Frame and Newbould, 1986). Within the range of conditions tested, the total LAI of the simulated mixtures before the last cut varies between 1.5 and 10 and the canopy height between 9 and 30 cm (data not shown).

When grown at a low inorganic N supply, grass–legume mixtures have been reported to frequently outyield the pure stands (Haynes, 1980). The relative yield total (RYT) concept has been used to test whether the yield of a mixture exceeds that of the pure stands (Harper, 1977). The relative yield is equal to the yield of a species in mixture divided by its yield in pure stand. If the relative yield is calculated for both species in a mixture, the sum of the two values gives the RYT. Pure stands and mixtures of grass and clover were compared with the Canopt model, assuming a constant overall density at the start of the simulation. For all but the highest N supply, values of RYT above 1.0 were found (Fig. 9.3B). Therefore, in agreement with field data (de Wit *et al.*, 1966; Soussana and Lafarge, 1998), the model predicts that, when N supply is restricted, mixed grass and clover outyield pure stands and thus show some form of synergistic relationship, presumably because they make different demands on soil nitrogen.

A number of field studies have shown that high nitrogen fertilizer supplies and long regrowth durations tend to increase grass dominance within binary mixtures and cause a short-term clover decline (Laidlaw, 1980; Frame and Newbould, 1986; Soussana *et al.*, 1995). In agreement with these experimental data, the simulations indicate a reduction in the clover content of the shoots, both with increasing N supply and at a long (40 and 60 days), compared with a short (20 days) regrowth duration (Fig. 9.3C).

Therefore, within the range of conditions tested, the coexistence of grass and clover in cut mixtures is predicted by this individual-based model. Moreover, although the model describes only the competitive interactions which take place between neighbouring plants at the patch scale, it seems to account for some of the main sward management effects which have been reported to occur at the plot scale. Similar conclusions were also reached by Schwinning and Parsons (1996a) from a patch simulation model, although they did not describe explicitly the competition for resources between individual plants.

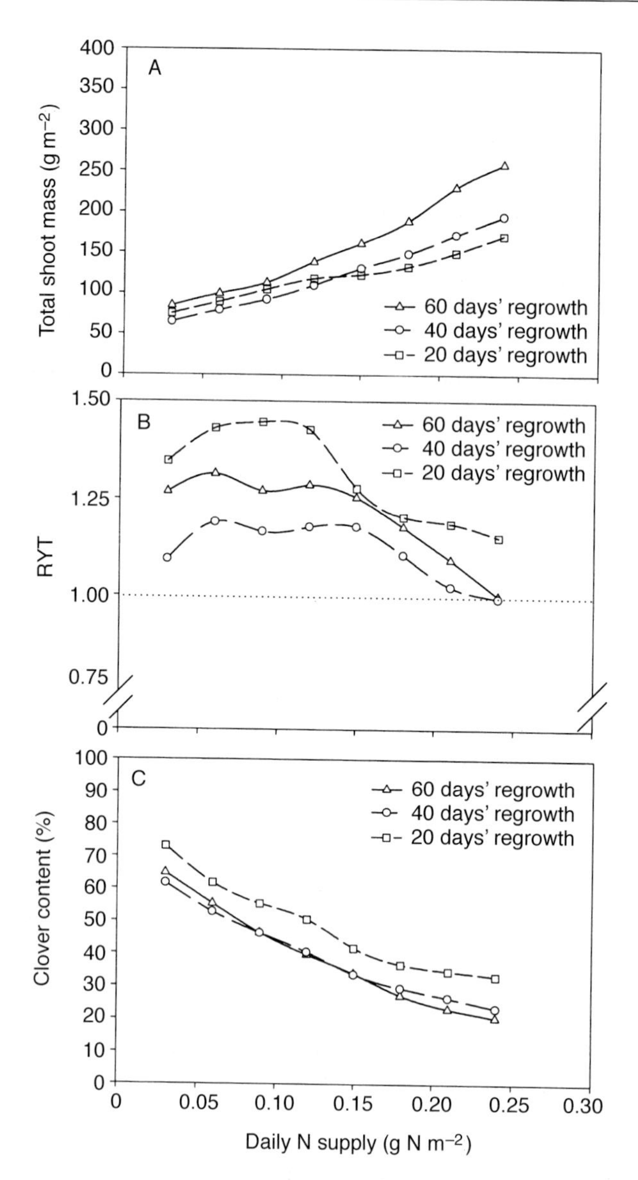

Fig. 9.3. Effects of the daily N supply (g N m^{-2}) and of the cutting frequency (20, 40 and 60 days' regrowth) on the total shoot mass (A), the relative yield total (see text) (B) and the clover content (C) of simulated grass–clover mixtures. Simulations were run for a daily N supply in the range 0.03–0.24 g N m^{-2}. The results are means calculated for the last regrowth.

Light partitioning and use in cut mixtures

Recent simulation models of grass–legume mixtures (Thornley *et al.*, 1995; Schwinning and Parsons, 1996a) derive light partitioning among mixed species from a set of fixed equations and do not attempt to describe explicitly the leaf morphology and canopy architecture of each plant population. In contrast, the Canopt model computes the amount of light absorbed by each plant species from the current number, shape and area of its individual leaves. This makes it possible to study the consequences of the phenotypic plasticity of individual plants for their ability to compete for light.

Several reports have shown that clover reacts quickly to shading by increasing its petiole length (Thompson and Harper, 1988; Varlet-Grancher *et al.*, 1989). Grass leaves also tend to increase their extension rate in response to a reduced amount of blue light (Gautier and Varlet-Grancher, 1996). Presumably, as a result of these shade avoidance strategies, the ratio of the extended length of clover petioles and grass leaves has been reported to be approximately constant, at least for mixtures between perennial ryegrass and white clover (Davies and Evans, 1990). Shade avoidance is simulated by the 'Canopt' model, which assumes that the leaf extension rate of grass and clover increases when the tip of the sheath or petiole becomes strongly shaded. With this approach, the simulated values for the ratio of the extended length of clover petioles and grass leaves are comprised between 0.5 and 1.3 in the range of conditions tested (Fig. 9.4B). Moreover, the model predicts that the vertical dominance of mixed species varies with N supply, clover being dominant in height at a low N supply and grass at a high N supply (Fig. 9.4B). Partly as a consequence of these changes in the mixed canopy structure, N supply and cutting frequency were predicted by the model to have interacting effects on the numbers of grass and clover axes. The simulated axis density increased with N supply in grass, but not in clover, and for the two species the axis density was greater at a high compared with a low cutting frequency (Fig. 9.4A).

These simulated changes in vertical dominance and axis density affect light partitioning among mixed species. According to the model, clover contributes relatively more to the total absorption of light by the mixture than to its total leaf area at a low N supply (Fig. 9.5A). In contrast, at a high N supply and a low cutting frequency, the simulated values show that light is partitioned equally among the mixed species (Fig. 9.5A). These simulation results fit well with the experimental data reported by Faurie *et al.* (1996). These authors calculated the daily average of PAR interception in a natural environment from detailed observations of the mixed canopy structure and their data indicated that white clover captures significantly more light per unit leaf area than perennial ryegrass at a low, but not at a high, nitrogen supply. This study has also shown that light partitioning among grass and clover is mostly affected by the relative area of each species in the top canopy layers, whereas differences in leaf lamina angle between the grass and the legume were reported to be of lesser importance.

In contrast to a pure stand, the fraction of leaf in the shade can reach one for a shaded species within mixture. In this case, the radiation use efficiency has been

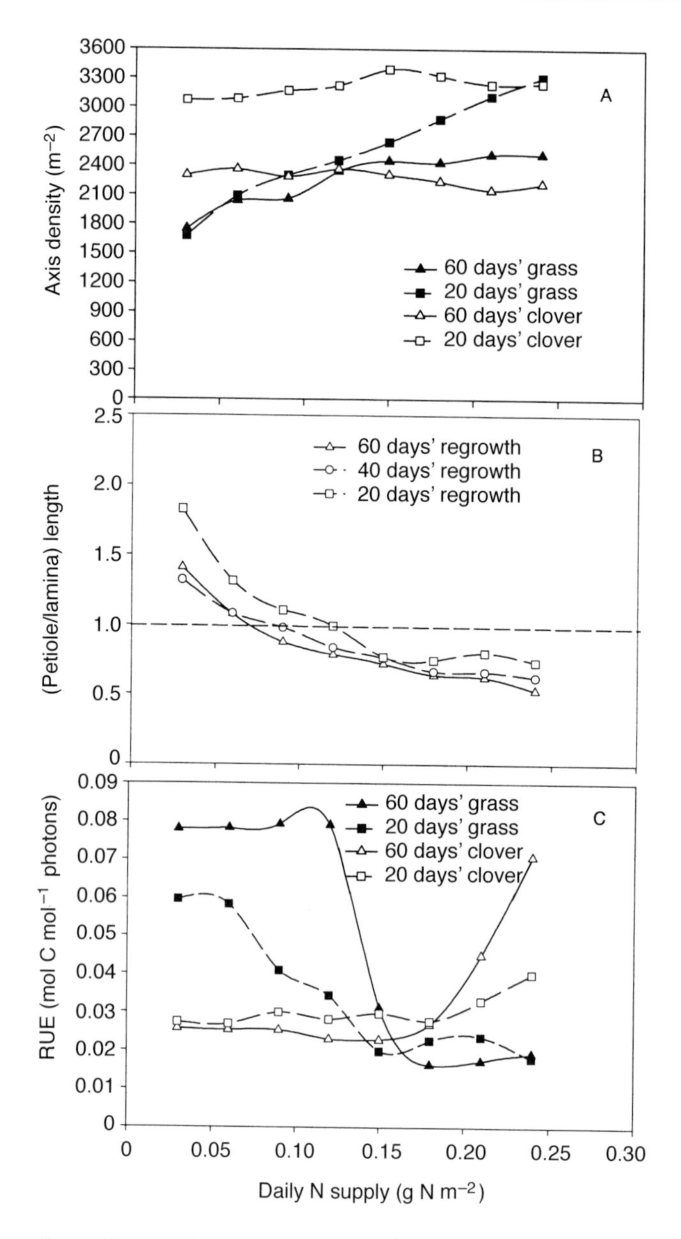

Fig. 9.4. Effects of the daily N supply (g N m^{-2}) and of the cutting frequency (20, 40 and 60 days' regrowth) on the number of axes (A), on the ratio of the extended length of clover petioles and grass leaves (B) and on the radiation use efficiency (RUE, calculated as the gross photosynthesis per unit light absorbed, mol C mol^{-1} photons) (C) of simulated grass and clover in cut mixtures. The results are means for the last regrowth (A) or are for the last day of the simulation (B, C). In A and C, for clarity, the results of the 40 days' regrowth treatment were omitted.

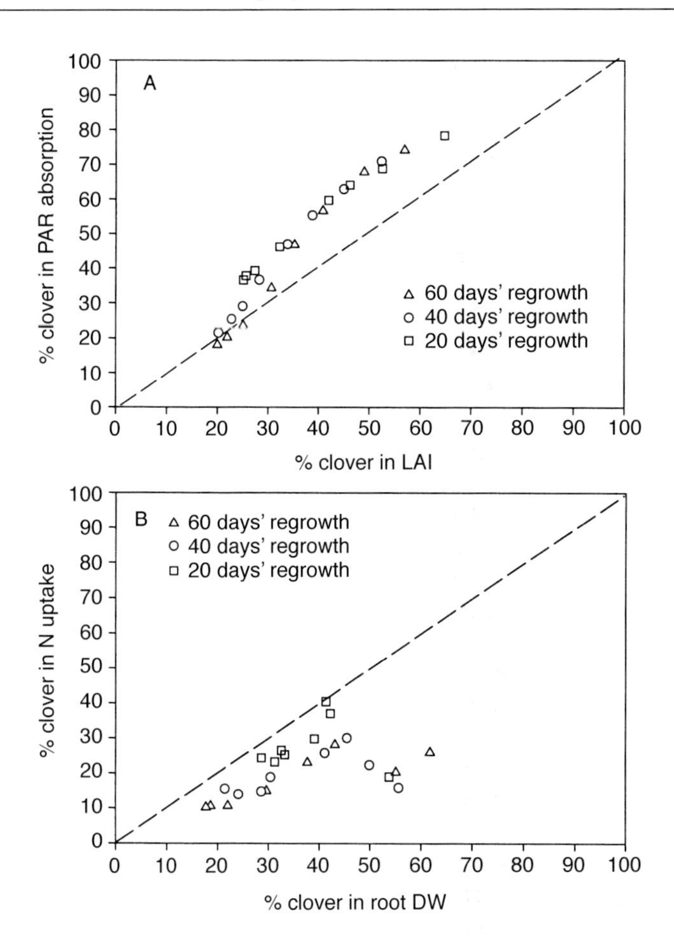

Fig. 9.5. Clover contribution (%) to light (PAR) absorption by the simulated mixture versus its contribution to the total leaf area index (LAI) (A) and percentage contribution of clover to the total N uptake versus its contribution to the total root weight of the simulated mixture (B). Simulations were run for three cutting frequencies (20, 40 and 60 days' regrowth) and for a daily N supply in the range 0.03–0.24 g N m^{-2}. The results are means calculated for the last regrowth. DW, dry weight.

reported to increase due to the avoidance of light-saturated photosynthesis (Willey, 1990; Faurie *et al.*, 1996). The Canopt model also simulates shade tolerance by adjusting the amount of leaf proteins to the irradiance experienced by leaves in the mixed canopy (Hirose and Werger, 1987; Chen *et al.*, 1993).

The radiation use efficiency (RUE) on the day before the last cut was calculated as the ratio of the simulated daily gross photosynthesis to the simulated light absorption (mol C mol^{-1} photons) on this day. At low N supply, the simulated RUE of the subordinate grass was greater than that of the dominant clover (Fig. 9.4C) and, under severe shading (60 days' regrowth), the grass RUE reached 0.078 mol C mol^{-1} photons, a value close to the maximal quantum yield (0.093 mol C mol^{-1} photons (Long *et al.*, 1993)) of Rubisco. In contrast, at high N supply, the simulated RUE of the dominant grass dropped down to 0.02 mol C mol^{-1} photons, while that of the subordinate clover increased above 0.04 mol C mol^{-1} photons (Fig. 9.4C).

Hence, this modelling approach confirms that shading increases the radiation use efficiency of the subordinate species within mixed stands (Willey, 1990). Trade-offs between light capture and light use would therefore contribute to the stabilization of the botanical composition of crowded stands, and subordinate grass or legume populations could delay competitive exclusion by acquiring shade plant characteristics (Soussana and Lafarge, 1998).

Inorganic N uptake and N_2 fixation in cut mixtures

Recent simulation models of grass–legume mixtures (Thornley *et al.*, 1995; Schwinning and Parsons, 1996a) consider a nitrogen-based competitive trade-off between grass and clover. When soil mineral nitrogen is low, clover achieves a greater relative growth rate than grass, since it can supplement mineral N uptake with N_2 fixation. When soil N is high, the grass would have the greater relative growth rate because mineral N uptake is more efficient than the combination of N uptake and N fixation. Changes in the balance between the grass and the legume will cause fluctuations in soil mineral N and this may result in the complex dynamics of the grass–legume populations that were demonstrated through modelling (Thornley *et al.*, 1995; Schwinning and Parsons, 1996a). In contrast to these plant–soil models, the soil N mineralization is a fixed parameter input with Canopt. Therefore, the effects of the legume content on the soil mineral N cannot be estimated, which prevents the use of the model for long-term predictions.

In the model by Thornley *et al.* (1995), the balance between N uptake and N_2 fixation in clover is fixed, which appears to be in contradiction with the experimental evidence from both field and controlled environment studies (Böller and Nösberger, 1987; Brophy *et al.*, 1987; Faurie and Soussana, 1993). In contrast, the model by Schwinning and Parsons (1996a) assumes that, as soil mineral N increases, clover obtains an increasing amount of N from soil nitrate, although some N_2 fixation may remain engaged even at high mineral N. This conclusion is also supported by the Canopt model, which predicts, in good agreement with field data (Böller and Nösberger, 1987; Brophy *et al.*, 1987), that the percentage N derived from symbiosis by clover in mixed swards declines with N supply from 90 down to 40–50% (Fig. 9.6A). This occurs as a consequence of an increased concentration of substrate N in the simulated legumes (data not shown), thereby reducing the nitrogen fixation activity per unit root weight, a concept supported by data from controlled environment studies, which have demonstrated that nitrogenase activity in forage legumes is fine-tuned to the nitrogen demand of the plant (Faurie and Soussana, 1993; Hartwig *et al.*, 1994).

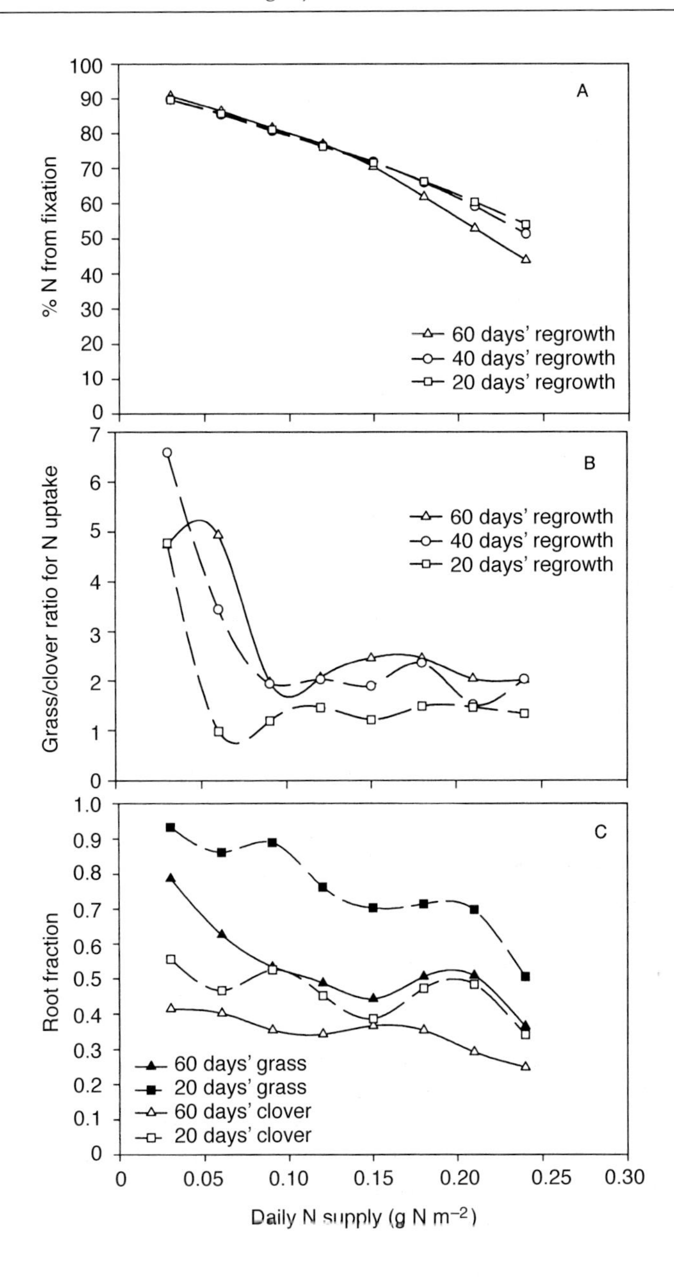

Fig. 9.6. Effects of the daily N supply (g N m^{-2}) and of the cutting frequency (20, 40 and 60 days' regrowth) on the percentage N derived from symbiosis in clover (A), on the ratio of the specific nitrogen uptake rates (in mg N m^{-1} root day^{-1}) of grass and clover roots (B) and on the root fraction (C) of simulated grass and clover in cut mixtures. In A and B, the results are means calculated for the last regrowth. In C, for clarity, the 40 days' regrowth treatment was omitted and the results correspond to the last day of the simulation.

In most replacement experiments, mixing a grass with a legume improves the nitrogen nutrition of the grass (Cruz and Soussana, 1997). This does not necessarily imply a transfer of biologically fixed N from the legume to the grass, as soil N is, in relative terms, more available to the grass plants in a mixture, due to the reduction of the grass density and to the lower competitive ability for N uptake of the legume (Vallis *et al.*, 1977; Cruz and Lemaire, 1986; Soussana *et al.*, 1989). This lower below-ground competitive ability of the legume is also predicted by the Canopt model, especially for a low N supply, since: (i) the grass : clover ratio for the specific root uptake activity is above 1 (Fig. 9.6B); and (ii) the simulated contribution of clover to the total N uptake by the mixture is always lower than its contribution to the total root mass (see Fig. 9.5B).

With a highly mobile nutrient like nitrate, the uptake rate is usually not affected by nitrate concentration over a large range (e.g. between 14 μM and 14 mM for perennial ryegrass (Clement *et al.*, 1978)) and mechanistic models of uptake have shown that the main parameters controlling the flux are the root uptake capacity, the root surface area and the total amount of nutrient in the soil (Barber, 1995). The first two of these parameters are controlled by the plant, and fast-growing grasses tend to respond to N deprivation by a derepression of uptake (Imsande and Touraine, 1994) and/or by an increase in root length or area (see Dawson *et al.*, Chapter 4, this volume). These two levels of plasticity are considered by the simulation model.

First, when the N supply is reduced, the simulated grass increases its root fraction to a greater extent than the simulated clover (Fig. 9.6C). Therefore, in agreement with controlled environment studies using hydroponics (Faurie, 1994), the root fraction of the grass becomes approximately twice that of the legume at a low N supply, whereas it is only 25–30% higher for a high N supply (Fig. 9.6C). Such differences in the relative size of the roots and shoots between grass and clover, which were observed with hydroponically grown mixtures (Faurie, 1994), have clearly strong implications for their respective above- and below-ground competitive ability. For example, a reduced root fraction of the grass under a high N supply and/or a long regrowth duration (Fig. 9.6C) increases its ability to overtop and partly shade the clover, which results in the decline of the legume in the mixture (see Fig. 9.3C).

Secondly, a reduction in N supply causes a greater decline in N uptake per unit root length for clover than for grass, thereby increasing the grass : clover ratio for N uptake (Fig. 9.6B). This simulation result is consistent with previous conclusions from NO_3^- uptake studies showing that N deprivation increases the nitrate uptake capacity of perennial ryegrass roots, but not of white clover roots (Soussana and Faurie, 1998). As the K_m values of these species are similar for both nitrate and ammonium uptake (Høgh-Jensen *et al.*, 1997), their specific uptake rate is reduced in the same proportion by a limiting substrate concentration and, therefore, the species balance for uptake can be modelled without knowing the exact concentration of nitrate or ammonium at the root surface.

Taken together, these simulation results illustrate how a grass and a legume can compete for a below-ground resource, such as inorganic N, while avoiding competitive exclusion through both morphological (root fraction and root diameter) and physiological (specific N uptake and N_2 fixation capacities) adjustments of their root systems.

The role of leaf turnover in cut mixtures

A previous sensitivity analysis of the Canopt model, run with two mixed-grass populations under optimal conditions (Soussana *et al.*, 1999b), has shown that the model is most sensitive to the value of three parameters, ruling: (i) the allometry between the weight, area and length of a leaf; (ii) the phyllochron; and (iii) the fraction of the C and N substrates that is available in each plant for branching. Therefore, the highest sensitivity was found for parameters governing leaf turnover, leaf size and branching, respectively. This shows that, when competition for resources between neighbours is modelled explicitly, as in Canopt, the shoot morphogenesis appears to be a major determinant of plant growth.

To further illustrate this point, we shall focus on the role of leaf turnover for the dynamics of mixed grass and clover swards. Lemaire and Chapman (1996) have shown that leaf turnover is one of the key parameters for the dynamics of mixed plant populations subjected to grazing. When defoliation is infrequent, a slow leaf turnover may be beneficial, leading to the development of longer leaves and to a reduction in the nutrient losses that are caused by leaf senescence. In contrast, under frequent defoliation, a fast leaf turnover could be advantageous, resulting in shorter leaves, which may partly escape defoliation, and in an increased number of axillary buds, which may develop into new shoot axes (Soussana and Lafarge, 1998). The consequences of a 25% increase in the phyllochron of grass and clover for their dynamics in cut mixtures were simulated with the Canopt model (Fig. 9.7).

With infrequent defoliation (60 days' regrowth), a slow leaf turnover conferred a vertical dominance to clover at all N supplies (Fig. 9.7B) and, as a result, the percentage of clover in shoot biomass rose above 50% at high N supply (Fig. 9.7D). Conversely, a slow leaf turnover of the grass increased its leaf length relative to that of the legume (Fig. 9.7B) and the percentage of clover in shoot biomass dropped down to 20% at high N supply (Fig. 9.7D). However, under low N supply, since competition for light was reduced (mean LAI 2.4 during the 60 days' regrowth), there was no effect of a change in leaf turnover on the grass–clover balance (Fig. 9.7D).

Under frequent defoliation (20 days' regrowth), a 25% change in leaf turnover had only small effects on the ratio of the length of clover petioles and grass leaves (Fig. 9.7A). Yet, at a low N supply, the model predicted an increase in the clover content of the mixture when the grass was assumed to have a slow leaf turnover (Fig. 9.7C). This last simulation result can be better understood by considering that, in frequently cut and nutrient-limited swards, species that tend initially to produce longer leaves will lose more nutrients through defoliation and hence have a reduced competitive ability and a reduced final leaf length.

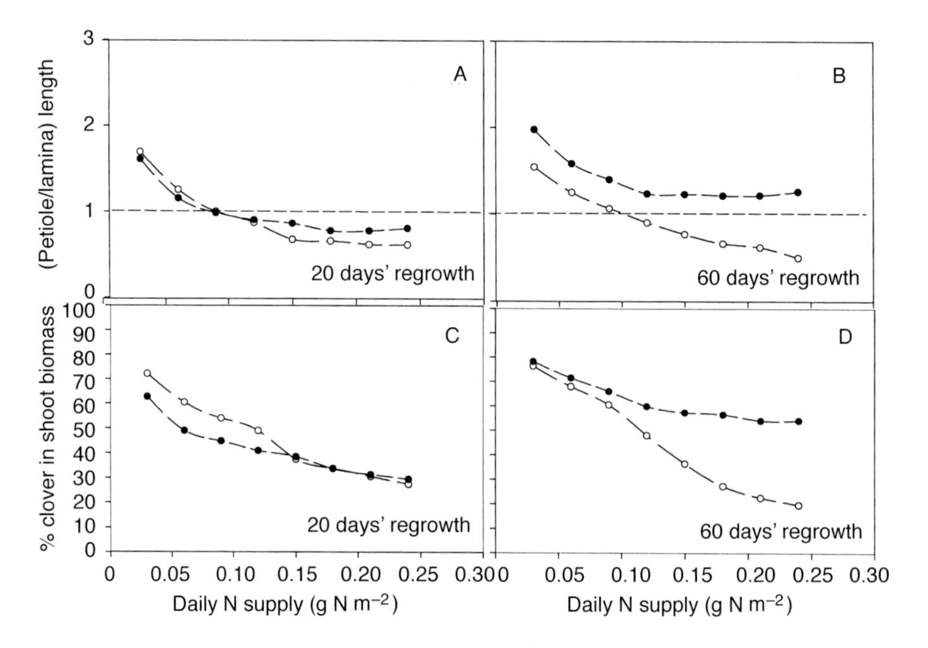

Fig. 9.7. Effects of the phyllochron, of the daily N supply (g N m^{-2}) and of the cutting frequency (A and C, 20, and B and D, 60 days' regrowth) on the ratio of the extended length of clover petioles and grass leaves (A, B) and on the clover content in the shoots (C, D) of simulated grass clover mixtures. (○), grass and clover phyllochron of 125°- and 100°-days, respectively. (●), grass and clover phyllochron of 100°- and 125°-days, respectively. The results are means calculated for the last regrowth.

Efficient nutrient conservation due to a long lifespan of the leaves and roots has often been suggested as an explanation for the long-term success of plant species in nutrient-poor and relatively undisturbed environments (Chapin, 1980; Berendse and Aerts, 1987). Our modelling approach emphasizes the suggestion that, in grass–clover mixtures, where competition interacts continuously with the disturbance induced by grazing or mowing, a longer leaf lifespan (resulting from a higher phyllochron) may confer an advantage only when there is ample nutrient supply and when defoliation is infrequent. In contrast, under frequent defoliation and moderate to low nutrient supply, a faster leaf turnover would be beneficial for mixed grasses and legumes. Similar conclusions were reached previously by modelling the competition between mixed grass populations (Soussana and Lafarge, 1998). Hence, in the broader context of semi-natural grasslands, predictions from the Canopt model suggest that pasture species with a fast leaf turnover are better adapted to frequent cutting or grazing and, conversely, that species with a slow leaf turnover may become dominant in fertile and infrequently defoliated swards.

However, the modelling results presented here should not be generalized too quickly to grazed conditions. Grazing provides differential defoliation (see Lascano, Chapter 13, this volume) and the degree to which grass and clover are allowed to retain their leaves, or whether these are removed by hervivores, has been shown to be a major determinant of their relative abundance under continuous grazing (Parsons *et al.*, 1994; Louault *et al.*, 1997). Parsons *et al.* (Chapter 15, this volume) show that the response of the vegetation to grazing should be conceptualized at the very fine scale where competition for resources between neighbours occurs and where herbivores make their individual bites. By using rules for defoliation by herbivores within the Canopt model, some detailed predictions concerning the dynamics of mixed pasture species subjected to grazing could also be made at this patch scale.

Evaluation of the Model

The field data used to evaluate the model concern the first-year growth of cut grass–clover mixtures in a free-air carbon dioxide enrichment facility (FACE) in Switzerland and were reported by Hebeisen *et al.* (1997). Mixtures between perennial ryegrass and white clover were grown by these authors at two atmospheric CO_2 concentrations (350 and 600 $\mu l\ l^{-1}$), at two N fertilizer supplies (100 and 420 kg N ha^{-1}) and at two cutting frequencies (three and seven cuts per year). The average air temperature during the growing season was 13°C and the average irradiance reached 600 μmol PAR $m^{-2}\ s^{-1}$ during 14 h (H. Blum, Zurich, 1999, personal communication). Water and nutrients other than N were non-limiting.

Simulations were run with the Canopt model over 210 days, using exactly the same parameter values as above and assuming: (i) the average environmental conditions and the sward management reported by Hebeisen *et al.* (1997); and (ii) a net soil N mineralization of 60 kg N ha^{-1} $year^{-1}$. The results show that the simulated and measured values for the clover content in the shoot dry matter are significantly correlated ($n = 8$, $r^2 = 0.72$, $P < 0.05$) with no significant bias, since the slope of the regression line is equal to 1.02 ± 0.26 (Fig. 9.8).

Therefore, this comparison shows that, when nitrogen is the only limiting below-ground resource, a detailed and mechanistic individual-based model, such as Canopt, may be used to predict the relative abundance of grass and clover in cut swards. However, it should be acknowledged that this model does not yet account for the effects of an elevated atmospheric CO_2 concentration on the grass–legume balance in field conditions. As shown in Fig. 9.8, the clover proportion of the cut mixtures studied by Hebeisen *et al.* (1997) was enhanced under elevated CO_2, whereas the simulated clover content was not affected by the atmospheric CO_2 concentration. This discrepancy may be attributed to the role of plant–soil interactions that are not currently depicted by the model, since elevated CO_2 has been shown to cause a temporal immobilization of soil N, which favours the growth and N_2 fixation of mixed clover (Soussana and Hartwig, 1996).

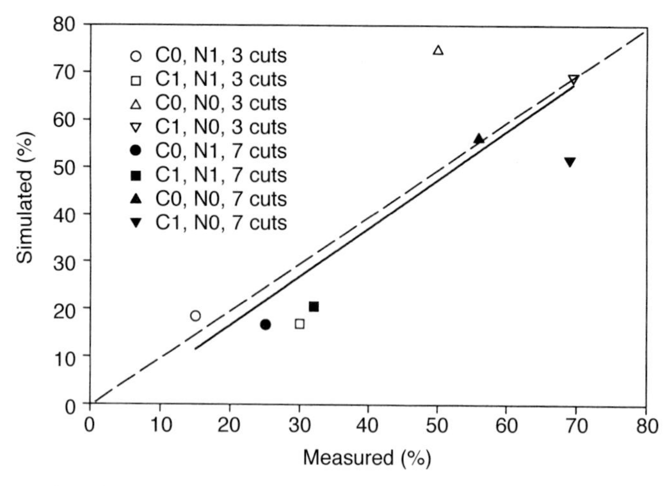

Fig. 9.8. Simulated versus measured clover content in shoots from cut perennial ryegrass–white clover mixtures, grown by Hebeisen *et al.* (1997) at two N supplies (N0, 100 and N1, 420 kg N ha^{-1} $year^{-1}$), at two atmospheric CO_2 concentrations (C0, 350 and C1, 600 p.p.m.) and at two cutting frequencies (three and seven cuts per year). The simulations were run during 210 days with the same average environmental conditions (13°C, 600 μmol PAR m^{-2} s^{-1} during 14 h) and management conditions as in the experiment by Hebeisen *et al.* (1997). The solid line indicates the linear regression between simulated and measured values ($n = 8$, $r^2 = 0.72$, $P < 0.05$).

Conclusion

We have shown here that a mechanistic model at the patch scale can be used to test the consequences of the adaptive responses of individual pasture plants to competition by neighbours and to management factors. Some plant traits, such as the morphology and demography of the leaves and axis, are best analysed at the individual plant level. To understand and predict the competitive interactions or the interferences between legumes and associated grasses, it is therefore necessary to describe the local modifications of the environment at the scale of the individual plant (Lemaire, 1994). By analysing resource partitioning and use at this very fine scale, a mechanistic understanding of the mixed population dynamics and of its consequences for the C and N fluxes can be gained. However, before this can be achieved, further work is still needed in order to account, also at the patch scale, for some of the major plant–soil and plant–herbivore interactions, which strongly affect the dynamics of mixed grasses and legumes in temperate and tropical pastures.

References

Barber, S.A. (1995) *Soil Nutrient Bioavailability: A Mechanistic Approach*, 2nd edn. John Wiley & Sons, New York, 440 pp.

Berendse, F. and Aerts, R. (1987) Nitrogen-use efficiency: a biologically meaningful definition. *Functional Ecology* 1, 293–296.

Böller, B.C. and Nösberger, J. (1987) Symbiotically fixed nitrogen from field-grown white and red clover mixed with ryegrasses at low levels of ^{15}N-fertilization. *Plant and Soil* 104, 219–226.

Bouman, B.A.M., Schapendonk, A.H.C.M., Stol, W. and van Kraalingen, D.W.G. (1996) Description of the growth model LINGRA as implemented in CGMS. *Quantitative Approaches in Systems Analysis* 7, 1–56.

Brophy, L.S., Heichel, G. H. and Russelle, M.P. (1987) Nitrogen transfer from forage legumes to grass in a systematic planting design. *Crop Science* 27, 753–758.

Brouwer, R. (1963) Some aspects of the equilibrium between overground and underground plant parts. In: *Jaarboak*. Instituut voor Biologish en Scheukendig Onderzoek van Landbouwgewassen, Wageningen, pp. 31–39.

Chapin, F.S. (1980) The mineral nutrition of wild plants. *Annual Review of Ecology and Systematics* 11, 233–260.

Chen, J.L., Reynolds, J.F., Harley, P.C. and Tenhunen, J.D. (1993) Coordination theory of leaf nitrogen distribution in a canopy. *Oecologia* 93, 63–69.

Clement, C.R., Hopper, M.J., and Jones, L.H.P. (1978) The uptake of nitrate by *Lolium perenne* from flowing nutrient solution. *Journal of Experimental Botany* 29, 453–464.

Cruz, P. and Lemaire, G. (1986) Analyse des relations de compétition dans une association de luzerne et de dactyle: effets sur la nutrition azotée des deux espèces. *Agronomie* 6, 735–742.

Cruz, P. and Soussana, J.F. (1997) Mixed crops. In: Lemaire, G. (ed.) *Diagnosis of the Nitrogen Status in Crops*. Springer Verlag, Berlin, Heidelberg, pp. 131–144.

Davidson, R.L. (1969) Effects of root leaf temperature differentials on root/shoot ratio in some pasture grasses and clover. *Annals of Botany* 33, 561–569.

Davies, A. (1974) Leaf tissue remaining after cutting and regrowth in perennial ryegrass. *Journal of Agricultural Science, Cambridge* 82, 165–172.

Davies, A. and Evans, E. (1990) Axillary bud development in white clover in relation to defoliation and shading treatment. *Annals of Botany* 66, 349–357.

de Wit, C.T., Tow, P.G. and Ennick, G.C. (1966) Competition between legumes and grasses. *Agriculture Research Report, Wageningen* 687, 3–30.

Farquhar, G.D., Von Caemmers, S. and Berry, J.A. (1980) A biochemical model of photosynthetic CO_2 assimilation in leaves of C_3 species. *Planta* 149, 78–90.

Faurie, O. (1994) Interactions carbone–azote dans des associations prairiales graminée (*Lolium perenne* L.) légumineuse (*Trifolium repens* L.). Etude d'associations simulées en conditions contrôlées. Thèse de doctorat, Université Blaise Pascal, Clermont-Ferrand II, 203 pp.

Faurie, O. and Soussana, J.F. (1993) Oxygen-induced recovery from short-term nitrate inhibition of N_2 fixation in white clover plants from spaced and dense stands. *Physiologia Plantarum* 89, 467–475.

Faurie, O., Soussana, J.F. and Sinoquet, H. (1996) Radiation interception, partitioning and use in grass–clover mixtures. *Annals of Botany* 77, 35–45.

Field, C. (1983) Allocating leaf nitrogen for the maximization of carbon gain: leaf age as a control on the allocation program. *Oecologia* 56, 341–347.

Frame, J. and Newbould, P. (1986) Agronomy of white clover. *Advances in Agronomy* 40, 1–88.

Gautier, H. and Varlet-Grancher, C. (1996) Regulation of the leaf growth of grasses by blue light. *Physiologia Plantarum* 98, 424–430.

Harley, P.C., Thomas, R.B., Reynolds, J.F. and Strain, B.R. (1992) Modelling photosynthesis of cotton grown in elevated CO_2. *Plant, Cell and Environment* 15, 271–282.

Harper, J.L. (1977) *Population Biology of Plants*. Academic Press, London, 892 pp.

Hartwig, U.A., Heim, I., Lüscher, A. and Nösberger, J. (1994) The nitrogen-sink is involved in the regulation of nitrogenase activity in white clover after defoliation. *Physiologia Plantarum* 92, 375–382.

Haynes, R.J. (1980) Competitive aspects of the grass–legume association. *Advances in Agronomy* 33, 227–256.

Hebeisen, T., Lüscher, A., Zanetti, S., Fisher, B.U., Hartwig, U.A, Frehner, M., Hendrey, G.R., Blum, H. and Nösberger, J. (1997) The different responses of *Trifolium repens* L. and *Lolium perenne* L. grassland to free air CO_2 enrichment and management. *Global Change Biology* 3, 149–160.

Hilbert, W. and Reynolds, J.F. (1991) A model allocating growth among leaf proteins, shoot structure, and root biomass to produce balanced activity. *Annals of Botany* 68, 417–425.

Hirose, T. and Werger, M.J.A. (1987) Maximising daily canopy photosynthesis with respect to the leaf nitrogen allocation pattern in the canopy. *Oecologia* 72, 520–526.

Hirose, T., Pons, T.L. and van Rheenen, J.W.A. (1988) Canopy structure and leaf nitrogen distribution in a stand of *Lysimachia vulgaris* L. as influenced by stand density. *Oecologia* 77, 145–150.

Høgh-Jensen, H., Wollenweber, B. and Schloerring, J.K. (1997) Kinetics of NH_4^+ and NO_3^- absorption and accompanying H^+ fluxes in roots of *Lolium perenne* and N_2 fixing *Trifolium repens* L. *Plant, Cell and Environment* 20, 1184–1192.

Imsande, J. and Touraine, B. (1994) N demand and the regulation of nitrate uptake. *Plant Physiology* 105, 3–7.

Johnson, I.R. and Thornley, J.H.M. (1987) A model of shoot : root partitioning with optimal growth. *Annals of Botany* 60, 133–142.

Kessler, W. and Nösberger, J. (1994) Factors limiting white clover growth in grass/clover systems: grassland and society. In: t'Mannetje, L. and Frame, J. (eds) *Proceedings of the 5th General Meeting of the European Grassland Federation*.Wageningen Pers, Wageningen, pp. 525–538.

Laidlaw, A.S. (1980) The effects of nitrogen fertilizer applied in spring on swards of ryegrass sown with four cultivars of white clover. *Grass and Forage Science* 35, 295–299.

Ledgard, S.F. and Steele, K.W. (1992) Biological nitrogen fixation in mixed legume/grass pastures. *Plant and Soil* 141, 137–153.

Lemaire, G. (1994) Ecophysiological approaches to intercropping. In: Sinoquet, J. and Cruz, P. (eds) *Ecophysiology of Tropical Intercropping*. INRA, Versailles, pp. 9–25.

Lemaire, G. and Chapman, D. (1996) Tissue flows in grazed plant communities. In: Hodgson, J. and Illius, A.W. (eds) *The Ecology and Management of Grazing*. CAB International, Wallingford, UK, pp. 3–35.

Long, S.P., Postl, W.F. and Bolhar Nordenkampf, H.R. (1993) Quantum yields for uptake of carbon dioxide in vascular plants of contrasting habitats and taxonomic groupings. *Planta* 198, 226–234.

Louault, F., Carrère, P. and Soussana, J.F. (1997) Grass and clover herbage use efficiencies in mixtures continuously grazed by sheep. *Grass and Forage Science* 52, 388–400.

Nijs, I., Behaeghe, T. and Impens, I. (1995) Leaf nitrogen content as a predictor of photosynthetic capacity in ambient and global change conditions. *Journal of Biogeography* 22, 177–183.

Parsons, A.J., Thornley, J.H.M., Newman, J. and Penning, P.D. (1994) A mechanistic model of some physical determinants of intake rate and diet selection in a two-species temperate grassland sward. *Functional Ecology* 8, 187–204.

Porter, J. R. (1984) A model of canopy development in winter wheat. *Journal of Agricultural Science, Cambridge* 102, 383–392.

Schwinning, S. and Parsons, A.J. (1996a) Analysis of the coexistence mechanisms for grasses and legumes in grazing systems. *Journal of Ecology* 84, 799–813.

Schwinning, S. and Parsons, A.J. (1996b) A spatially explicit population model of stoloniferous N-fixing legumes in mixed pastures with grass. *Journal of Ecology* 84, 815–826.

Sheehy, J. E., Gastal, F., Mitchell, P.L., Durand, J.L., Lemaire, G. and Woodward, F.I. (1996) A nitrogen-led model of grass growth. *Annals of Botany* 77, 165–177.

Simon, J.C. and Lemaire, G. (1987) Tillering and leaf area index in grasses in the vegetative phase. *Grass and Forage Science* 42, 373–380.

Soussana, J.F. and Faurie, O. (1998) The regulation of N_2 fixation and NO_3^- uptake in grass–clover mixtures. In: Emelrich, C., Kondorosi, A. and Newton, W.E. (eds) *Biological Nitrogen Fixation for the 21st century*. Kluwer Academic Publishers, Dordrecht, p. 662.

Soussana, J.F. and Hartwig, U. (1996) The effects of elevated CO_2 on symbiotic N_2 fixation: a link between the C and N cycles in grassland ecosystems. *Plant and Soil* 187, 321–332.

Soussana, J.F. and Lafarge, M. (1998) Competition for resources between neighbouring species and patch scale vegetation dynamics in temperate grasslands. *Annales de Zootechnie* 47, 371–382.

Soussana, J.F., Arregui, C. and Hazard, L. (1989) Assimilation du nitrate et fixation symbiotique dans les associations trèfle blanc – ray-grass anglais. In: *Proceedings of the XVIth International Grassland Congress*. AFPF, Nice, pp. 147–148.

Soussana, J.F., Vertes, F. and Arregui, M.C. (1995) The regulation of clover shoot growing points density and morphology during short-term decline in mixed swards. *European Journal of Agronomy* 4, 205–215.

Soussana, J.F., Teyssonneyre, F. and Thiéry, J. (1999a) Un modèle dynamique d'allocation basé sur l'hypothèse d'une co-limitation de la croissance végétale par les absorptions de lumière et d'azote. In: Bonhomme, R. and Maillard, P. (eds) *Fonctionnement des peuplements végétaux et environnement*. INRA, Versailles, France, pp. 87–116.

Soussana, J.F., Teyssonneyre, F. and Thiéry, J. (1999b) Un modèle simulant les compétitions pour la lumière et pour l'azote entre espèces herbacées à croissance clonale. In: Bonhomme, R. and Maillard, P. (eds) *Fonctionnement des peuplements végétaux et environnement*. INRA, Versailles, France, pp. 325–350.

Thompson, L. and Harper, J.L. (1988) The effect of grass on the quality of transmitted radiation and influence on the growth of white clover (*Trifolium repens* L.). *Oecologia* 75, 343–347.

Thornley, J.H.M., Bergelson, J. and Parsons, A.J. (1995) Complex dynamics in a carbon–nitrogen model of a grass–legume pasture. *Annals of Botany* 75, 79–94.

Vallis, I., Henzel, E.F. and Evans, T.R. (1977) Uptake of soil nitrogen by legume in mixed swards. *Australian Journal of Agricultural Research* 28, 413–425.

van Loo, E.N. (1993) On the relation between tillering, leaf area dynamics and growth of perennial ryegrass (*Lolium perenne* L.). Doctoral thesis, Wageningen Agricultural University, The Netherlands, 169 pp.

Varlet-Grancher, C. and Bonhomme, R. (1979) Application aux couverts végétaux des lois de rayonnement en milieu diffusant. II. Interception de l'énergie solaire par une culture. *Annales Agronomiques* 26, 1–26.
Varlet-Grancher, C. and Gautier, H. (1995) Plant morphogenetic responses to light quality and consequences for intercropping. In: Sinoquet, H. and Cruz, P. (eds) *Ecophysiology of Tropical Intercropping*. INRA, Versailles, France, pp. 231–256.
Varlet-Grancher, C., Moulia, B. and Jacques, R. (1989) Phytochrome mediated effects on white clover morphogenesis. In: *Proceedings of the XVIth International Grassland Congress*, Vol. 1. AFPF, Nice, pp. 477–478.
Willey, R.W. (1990) Resource use in intercropping systems. *Agricultural Water Management* 17, 215–231.

10

Plant–Animal Interactions in Complex Plant Communities: from Mechanism to Modelling

I.J. Gordon

Macaulay Land Use Research Institute, Craigiebuckler, Aberdeen AB15 8QH, UK

Introduction

It is widely acknowledged that ruminant herbivores have a major impact on the floristic composition and stability of grassland and shrub vegetation (McNaughton, 1979; Miles, 1981; Crawley, 1983; Hodgson and Illius, 1996). If overexploitation by grazing animals or undesired vegetation changes are to be avoided, protocols need to be developed for vegetation management through the manipulation of the herbivore population. To do this, it is first necessary to identify and quantify the processes by which vegetation dynamics are modified by the activities of herbivores in time and space (e.g. Fig. 10.1). The centre of Fig. 10.1 represents the hierarchical levels within vegetation – the individual plant, the population of individuals, mixed species associations (i.e. the plant community) and mixed community associations (Senft *et al.*, 1987; Birske, 1989). An understanding of events at the level of the individual plant and the plant populations is required for a full appreciation of the pattern and process at the vegetation community level (Bullock, 1996). This diagram demonstrates the need to understand the responses of plants to the intensity and frequency of herbivore impact in relation to environmental conditions and the animal factors, which affect not only the intensity and frequency of impact but also the distribution of that impact. While this review is primarily concerned with the activities of herbivores (i.e. defoliation, excretion and treading) as they influence vegetation, it is necessary to be aware of the importance of the abiotic factors and the possibility that interactions might occur such that individual plant and plant community responses to herbivores may be modified by variation in these factors.

(eds G. Lemaire, J. Hodgson, A. de Moraes, C. Nabinger and P.C. de F. Carvalho)

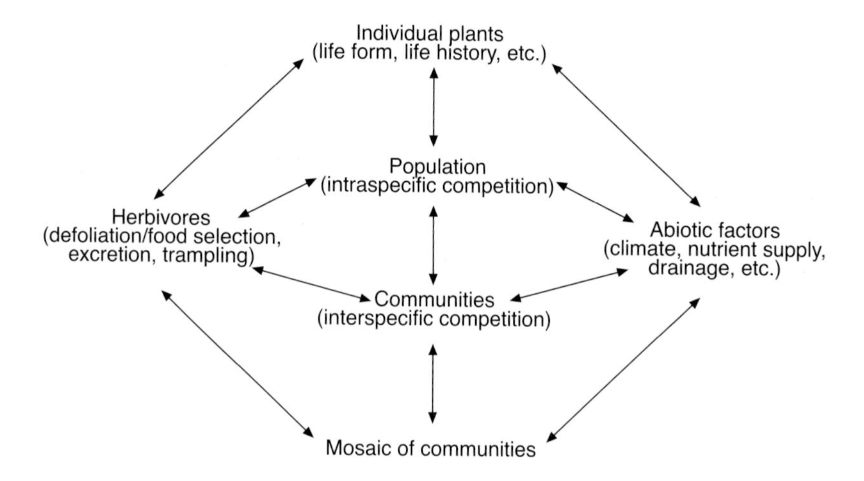

Fig. 10.1. A hierarchical representation of vegetation showing the interrelationships requiring investigation to aid identification and quantification of the processes by which vegetation composition in time and space are modified by the activities of herbivores and abiotic factors.

Plant–animal interactions

To understand the approach proposed here, it is helpful to examine the plant–animal interactions of grazing systems on grasslands in terms of the flow and partitioning of material along alternative pathways through the system. The main flow pathways, shown in Fig. 10.2, are: the growth of herbage, which is partitioned into that consumed and that which matures and senesces; herbage consumed, which is partitioned into a digested fraction and an excreted fraction; the pathways of decomposition leading to the return of nutrients; and finally the loss pathways due to loss of plant nutrients by leaching and to the removal of animal products. Circled at either side of the diagram are the factors influenced by grazing management that have a direct impact on the various rate processes.

Taking two of these rate processes, herbage growth and herbage consumption, and the factors influencing these rates, the importance of numerous sward attributes, collectively described as sward conditions, soon becomes clear. For example, herbage growth is influenced by sward attributes such as leaf area index, canopy structure and inherent species characteristics, which affect light interception, the ability to expand leaves and the photosynthetic efficiency of those leaves (Wright, 1981; Osmond, 1987; Grime *et al.*, 1988). Herbage consumption is also influenced by a variety of sward attributes, including the nature of plant material (green or dead, species, chemical composition), the amount (height, biomass) and the canopy structure (Arnold and Dudzinski, 1967; Hodgson, 1981; Gordon and Iason, 1989; Gordon and Lascano, 1993). Demand for food is determined by several animal factors, for example the species, sex and physiological state of the

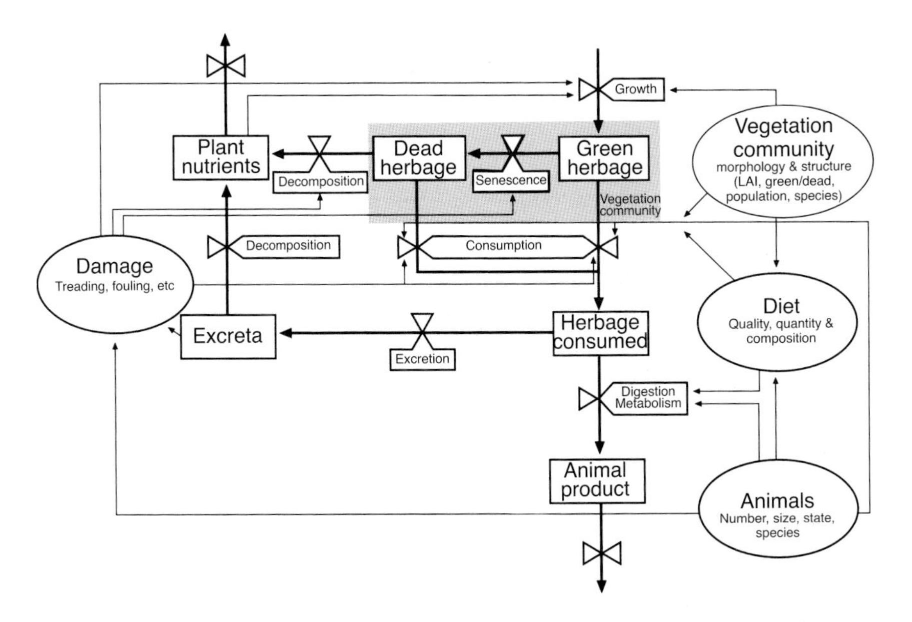

Fig. 10.2. Plant and animal interrelationships in grazing systems. The bold lines and arrows indicate the flows and partitioning of material through the system; the factors that are altered by the grazing management decisions and that influence the various rate processes (growth, consumption, etc.) are circled; faint lines are used to indicate impacts and relationships. (From Grant and Maxwell, 1988.)

animal (Prache *et al.*, 1998). The interaction between the animal and the sward factors influences diet selection and diet quality, which together determine the rate of consumption and herbage intake of the animal. It is clear that the effect of stocking rate on herbage production and animal performance is largely mediated through its effect on the sward state factors. How should the natural-resource manager interpret the message of this examination in relation to management of grasslands?

On grasslands, among the factors that require consideration are the life-form, seasonality of growth and quality and quantity of herbage produced by the different plant species. Variation in these attributes between species results in uneven distribution of grazing activity in time and space (Hunter, 1962; Gordon, 1989), which has significant consequences for vegetation dynamics. In order to define the scope for manipulating species composition of native grasslands, information and understanding in three specific research areas is required. First, it is necessary to know how the plant species of major interest respond to different seasonal patterns and levels of use. Secondly, an understanding of the factors affecting grazing choice of different plant species, both between vegetation types and between plant species within types, is essential. Finally, it is important to know how the feeding value of different vegetation types varies within season (Grant and Maxwell, 1988).

In order to predict vegetation responses to grazing, a basic knowledge of the biological characteristics and ecology or habitat requirements of the individual plant species is required. Likewise, knowledge and understanding of the factors influencing the competitive relationships of intraspecific and interspecific neighbours in plant communities, and of the ways in which different patterns and levels of defoliation might modify the competitive relationships, are needed. Manipulative experiments in the laboratory, glasshouse and field are all likely to be useful. Given the importance of sward conditions in influencing plant growth and animal behaviour, for grazing studies to be really useful they should at some stage include experiments in which treatments are based on the controlled manipulation of sward conditions (Hodgson, 1985).

On plant communities that contain few species with similar digestibilities and preference ranking (for example, on sown swards), the simple provision and control of a range of sward heights by frequent adjustments of stock numbers has been used as the experimental variable. The key role of sward conditions in influencing herbage growth, efficiency of herbage utilization and animal performance has been examined, and ways of determining the stock numbers can be maintained throughout the whole year to maximize the use of grass for sheep and beef production have been suggested (Maxwell *et al.*, 1988).

For plant communities containing many species with a wide range of digestibility and/or preference ranking, it is first necessary to acquire a basic knowledge of factors affecting grazing choice or foraging behaviour. Foraging behaviour is the area of the interface between plant and animal scientists (Gordon and Illius, 1993). From the animal scientists' viewpoint, it deals with the ways in which sward conditions influence diet selection and the nutrient intake rate of the grazing animals. From the plant scientists' viewpoint, it deals with the ways in which sward conditions and grazing animals interact to influence patterns and levels of offtake of different plant species.

An Experimental Approach to Understanding Plant–Animal Interactions in Complex Plant Communities

The experimental approach involves a number of phases of study. First, there is a descriptive phase, in which detailed measurements are made of sward biomass, composition and structure and of diet composition and levels of offtake of particular species by ruminants at a series of points in time (or at the same point in time across a range of stocking rates). In this way, sward factors of importance in determining offtake can be identified and hypotheses set up. The second phase involves experimentally manipulating swards, in which the hypotheses are tested. The third phase involves the development of grazing systems for vegetation management, in which the effects of grazing pressure on vegetation dynamics, animal performance and biodiversity are measured. The fourth phase is to draw the experimental information together into models of the plant–animal interactions in order to assess the consequences of management manipulation on vegetation dynamics and animal performance. The final phase is to test these models in the field. This approach will be illustrated, drawing on work with *Nardus* grassland and associated vegetation communities in the Scottish hills and uplands.

Context of offtake studies on Nardus *grassland*

Nardus grassland is dominated by the grass *Nardus stricta*, which is a coarse, tufted grass with bristle-like leaves, with the main between-tussock species being *Agrostis capillaris*, *Agrostis canina*, *Anthoxanthum odoratum* and *Festuca ovina*. *Nardus stricta* is tolerant of infertile nutrient-stressed conditions and, though it is found on acidic gleys, it is most common on freely drained acidic soils (Grime *et al*., 1988). Its feed quality, in terms of digestibility, is broadly similar to that of other fine-leaved grasses in the vegetation community, although the feed quality declines more with maturity than other fine-leaved grasses (Armstrong *et al*., 1986), but its grazing value is much lower, because of the high silica content and roughness of its cuticle, which results in reluctance of sheep to graze it (Thomas and Fairbairn, 1956; Armstrong *et al*., 1986). Sheep in particular avoid *Nardus*; as a consequence, where acidic native grasslands are grazed by sheep alone, *Nardus* becomes dominant and the carrying capacity of the pasture declines. This has increased in the UK over the past century, as the balance between cattle and sheep on hill pastures has increasingly favoured sheep (Fenton, 1937; Chadwick, 1960; Armstrong, R.H. *et al*., 1997).

Descriptive phase – generating hypotheses

Initial work on this community was linked with research to describe the overall intake and digestibility of material harvested from this community relative to other communities typical of the hills and uplands of Scotland. Indoor feeding experiments with sheep showed that, across vegetation communities, daily voluntary food intake (VFI) was related to the overall organic matter digestibility (OMD) of the forage cut from the community (Armstrong *et al*., 1986). The material derived from the *Nardus* community varied seasonally in intake and digestibility, with the highest digestibility and intake in the early growing season and lowest digestibility and intake in the winter. The experiments were used to assess the overall nutritive value of the communities on offer and link this with the likely agricultural potential of the communities, with the *Nardus* community demonstrating relatively high nutritive value in springtime but soon losing its nutritive value as the leaves of the *Nardus* plant mature (Armstrong and Hodgson, 1986). However, because of the unselective nature of the mechanical harvesting of plant material from communities and the limited opportunity for animals to select out preferred plant species and components, the nutritive values derived from indoor feeding trials are unlikely to reflect the diets consumed by animals feeding on these complex plant communities. As a consequence, field experiments were initiated to estimate the intake and diet composition of various domestic species grazing from the natural vegetation community.

Experimental phase – manipulating grazing pressure

When experimental field studies were initiated on this grassland in the late 1970s, research was focused on comparing the diet selection of sheep and cattle within plant communities (Grant *et al.*, 1985, 1987; Hodgson *et al.*, 1991). The experimental animals were grazed simultaneously (i.e. sheep and cattle together in sequence across a range of sites) with a 2-week period of observation at each site. During each grazing period, the sward biomass, species composition (leaf, stem, flower, live or dead) and canopy structure were characterized. Diet samples were collected from three to five oesophageally fistulated animals of each species. Offtake was not measured in this study, as there was no way of knowing which of the two species had grazed the various plants. The sites were grazed so that, over a period of 3 years, each plant community was sampled at different seasons of the year.

The *Nardus* community had had little previous grazing for several years prior to the start of the experiment. Intake, diet composition and diet digestibility estimates were made on the treatments on six occasions, ranging from May to November. The sward biomass was determined separately for the *Nardus* tussock and the between-tussock areas of the sward. Examination of the total data set for diet composition and sward biomass showed that the amount of *Nardus* in the diets of cattle and sheep, respectively, was strongly influenced by the live herbage biomass (or height) in the between-tussock areas. Figure 10.3, which shows the relationship between *Nardus* as a proportion of total diet and the biomass (dry matter (DM)) of the live herbage between tussocks, clearly demonstrates, first, that cattle consistently ingest more *Nardus* than do sheep and, secondly, that the proportion of *Nardus* in the diets of both species is inversely related to the biomass of the preferred between-tussock vegetation (Grant *et al.*, 1985).

The knowledge derived from the first phase of the field research was used to design the next phase, still conducted at a within-community level, in which the aim was to investigate the scope to manipulate the *Nardus* content and feeding value of *Nardus* dominant pasture by controlled grazing (Grant *et al.*, 1996; Armstrong, R.H. *et al.*, 1997). Plots were set up in which the sward height of the preferred between-tussock vegetation was controlled, maintaining it at predetermined heights throughout the growing season by making twice-weekly measurements of between-tussock sward height and adjusting animal numbers as necessary. Different animal species and a range of sward heights were maintained on different plots. The treatments were imposed for a period of 5 years. Offtake of *Nardus* was measured in terms of percentage of leaves grazed and the extent to which leaves were grazed. The floristic composition of the sward and the diet composition, intake and performance of the experimental animals were also recorded.

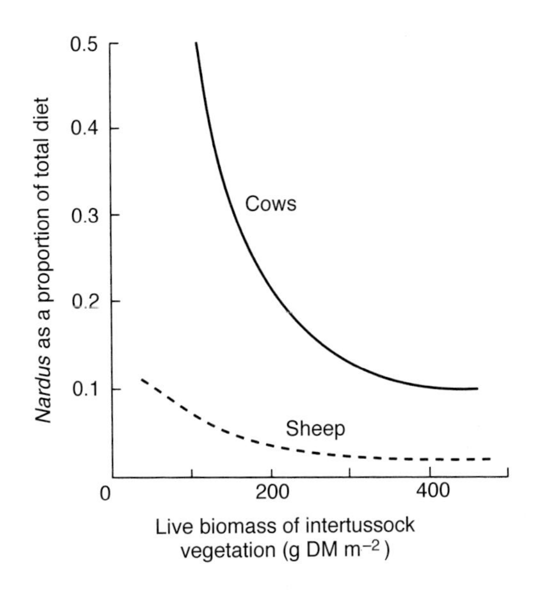

Fig. 10.3. Relationship between the amount of *Nardus* in the diets of grazing sheep and cattle and the biomass (g DM m^{-2}) of the live between-tussock vegetation (Grant *et al.*, 1985).

Selected results are presented for three treatments: cattle grazing between-tussock vegetation maintained at 4.5 cm height, a height at which the earlier phase of the work would lead to substantial grazing of *Nardus*; a sheep treatment, also with a between-tussock sward at a height of 4.5 cm; and, given the known reluctance of sheep to graze *Nardus*, a sheep-grazed treatment of 3.5 cm between-tussock height was imposed. The data in Table 10.1 are the 5-year averages for percentage of *Nardus* tillers grazed (with percentage for leaves in brackets) and the leaf length remaining after grazing as a measure of the extent to which leaves are grazed. The results comparing cattle with sheep grazing at sward heights maintained at 4.5 cm confirms the greater readiness of cattle compared with sheep to graze *Nardus* – not only were more leaves grazed, they were grazed more closely

Table 10. 1. Effect of species of grazer and control of height of the preferred between-tussock vegetation on offtake of *Nardus* in terms of proportion of tillers (% of leaves) grazed and the intensity of grazing on *Nardus* leaves (the length of lamina remaining after grazing). The data are the means of 5 years' record for July (Grant *et al.*, 1996).

Species of grazer: Between-tussock height:	Cattle 4.5 cm	Sheep 4.5 cm	Sheep 3.5 cm
Proportion of *Nardus* leaves grazed (%)	88 (67)	23 (12)	47 (30)
Grazed leaf length (cm)	3.9	10.8	7.8

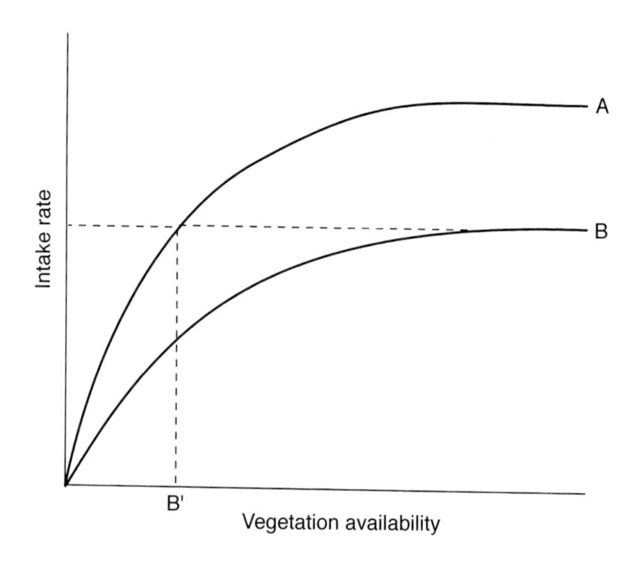

Fig. 10.4. Schematic representation of the relationship between intake rate and vegetation availability in two vegetation communities (A and B). B′ denotes the point at which the resource availability on community A results in an intake rate equal to that on community B.

(Grant *et al.*, 1996; Armstrong, R.H. *et al.*, 1997). The comparison of sheep grazing swards maintained at 4.5 cm with sheep at 3.5 cm confirms the importance of height or mass of the preferred sward component in determining the level of offtake of the unpreferred *Nardus*. The sheep diets were also of higher digestibility than were those of the cattle.

Similar results were obtained in a second short-term between-community study on adjacent areas of *Agrostis–Festuca* (the most preferred and best-quality native grassland in Scotland) and *Nardus* pasture (Gordon *et al.*, 1995). A graphical model of the relationship between the intake rate from a community and the resource availability on the community was developed (Fig. 10.4), which describes the functional response for two communities (A and B). From it predictions about the use that the herbivore would make of different communities in relation to the resource availability can be made. When resource availability is high, the herbivore is expected to spend most time feeding from the community that offers the highest intake rate (A). This preference for vegetation patches offering the highest intake rate has been shown between grass species (Illius *et al.*, 1999). However, where grazing pressure is higher than the productivity of vegetation in the preferred plant community, the resource availability in the preferred community will decline and the herbivore will increase the proportion of the less preferred community (B) in its diet. At some stage, the herbivore will consume a greater proportion of the less preferred community, because the intake rate on the preferred community is low. To test this hypothesis, sheep were grazed for short periods of time on 0.33 ha plots, which comprised 50% by area of each community type. The sward height on the *Agrostis–Festuca* community was maintained at either 3 cm or 4.5 cm. Measurements were made of the distribution of sheep across the two communities, using time-lapse photography. Measurements were also made of *Nardus* utilization. The results are summarized in Table 10.2, which shows that, when the preferred *Agrostis–Festuca* community was maintained at 3 cm, the sheep spent a greater proportion of their time on the *Nardus* community, and also that the proportion of the *Nardus* leaves grazed and the intensity of grazing on *Nardus* leaves was increased compared with where the *Agrostis–Festuca* community was maintained at a height of 4.5 cm.

In the longer-term within-community experiment previously described (Grant *et al.*, 1996), it is worth examining the trend in *Nardus* offtake over time. A graph comparing the proportion of *Nardus* tillers grazed on the sheep and cattle plots in successive years (Fig. 10.5) suggests that, in the case of sheep, the acceptability of *Nardus* declined as the study advanced. At the start of the experiment, surface dead material had been removed by fire; the *Nardus* tillers were tall and long-leaved, reflecting previous light grazing. With cattle grazing, the tussocky nature of the sward disappeared in 2 years and the *Nardus* tillers changed in morphology, becoming shorter; the proportion of tillers grazed and the intensity of grazing on *Nardus* leaves changed little over time. With sheep grazing, there was a steady decline in both the proportion of tillers and leaves grazed and in the intensity of grazing on *Nardus* leaves. The changes in *Nardus* cover since the start of the experiment (Table 10.3) clearly indicate the potential of cattle grazing to control or reduce *Nardus* and increase the proportion of live and more nutritious

Table 10.2. Effect of height control on the *Agrostis–Festuca* area on use of the adjacent *Nardus* community by sheep (proportion of sheep sightings) and offtake of *Nardus* (proportion of leaves grazed and length of lamina remaining after grazing). Data for July 1988 (Gordon *et al.*, 1995).

	Agrostis–Festuca height (cm)	
	3.0	4.5
Sheep sightings on *Nardus* area (%)	37.0	25.5
Proportion of *Nardus* leaves grazed (%)	60.0	17.0
Nardus grazed leaf length (cm)	3.0	7.5

Table 10.3. Effect of species of grazer and control of between-tussock sward height on *Nardus* cover (%) (Armstrong, R.H. *et al.*, 1997).

Species of grazer:	Cattle	Sheep	Sheep
Between-tussock height:	4.5 cm	4.5 cm	3.5 cm
1984	55.4	58.1	54.5
1989	30.0	86.5	72.2
Change	−25.4	+28.4	+17.7

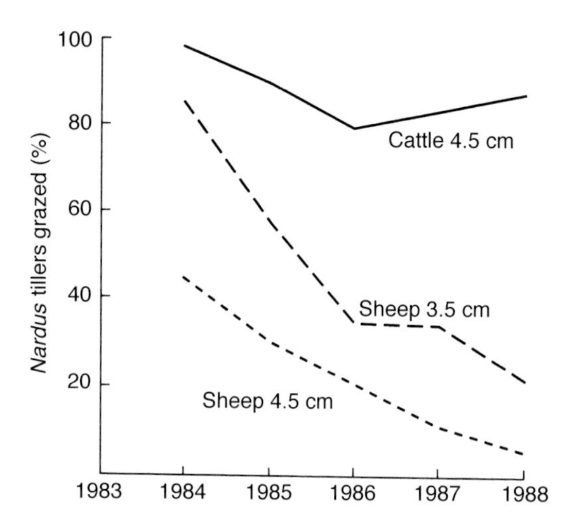

Fig. 10.5. Trends over time in the percentage of *Nardus* tillers grazed by sheep and cattle grazing separate plots; the between-tussock sward height was maintained at the heights shown by making twice-weekly measurement of sward height and adjusting animal numbers as necessary (Grant *et al.*, 1996).

grasses in the community (Armstrong, R.H. *et al.*, 1997). In contrast, *Nardus* cover increased under sheep grazing – even on the treatment where the between-tussock sward height was maintained at 3.5 cm, i.e. where the preferred grasses were in very short supply. In the case of the sheep, it was postulated that avoidance of *Nardus* increased as the dead material built up in the tussock surface. The different effects of grazing by cattle and sheep resulted in the stock-carrying capacity of the *Nardus* grassland increasing more in successive years on the cattle than on the sheep treatments and it was greater on the sward maintained at 3.5 cm than at 4.5 cm by sheep (Armstrong, R.H. *et al.*, 1997).

Further work has examined the influence of variation in *Nardus* tussock state on offtake of *Nardus*. In a within-community experiment, the different *Nardus* tussock states had been achieved by grazing the sward to maintain the between-tussock vegetation at a range of heights, using goats or sheep during the previous 5 years. All plots were subsequently grazed by sheep with the between-tussock sward height maintained at 4.5 cm. The results are summarized in Table 10.4; the tussock state data show that the height on the *Nardus* tussock was lowest and the proportion of live to dead highest on the plots previously grazed most closely by goats. The data for the proportion of *Nardus* leaves grazed and leaf length remaining after grazing clearly show that the sheep grazed the *Nardus* to a greater extent where the *Nardus* was shorter with least dead plant material.

In a second experiment, two pretreatments of the *Nardus* community were created on the two community plots (*Nardus* vs. *Agrostis–Festuca*) by either light grazing or heavy grazing by sheep in the previous year (Gordon *et al.*, 1995). This created two contrasting sward structures in the *Nardus* community. The lightly

Table 10.4. Effect of tussock condition (as influenced by previous grazing regime) of *Nardus* on the *Nardus* tussock characteristics and utilization of *Nardus* by sheep under conditions in which the between-tussock sward height was maintained at 4.5 cm. Data for June 1989 (I.J. Gordon, unpublished results).

		Previous regime		
	Sheep 4.5 cm	Goats 4.5 cm	Goats 5.5 cm	Goats 6.5 cm
Nardus tussock height (cm)	23.6	8.7	13.8	22.2
Proportion of *Nardus* state:				
Live leaf (%)	42.9	94.3	53.3	50.0
Flower (%)	7.2	1.9	16.0	11.0
Dead (%)	43.5	3.8	30.0	39.0
Proportion of *Nardus* leaves grazed (%)	49.1	71.9	61.8	43.6
Grazed leaf length (cm)	4.8	2.3	2.5	2.9

Table 10.5. Effect of previous level of grazing of *Nardus* community on the use of the subsequent *Nardus* (proportion of sheep sightings) and offtake of *Nardus* (% leaves grazed and length of lamina remaining after grazing) under conditions in which the adjacent *Agrostis–Festuca* community was maintained at a uniform height of 4.5 cm. Data for June 1989 (Gordon *et al.*, 1995).

	Nardus area previously heavily grazed	*Nardus* area previously lightly grazed
Sheep sightings on *Nardus* community (%)	39.6	24.5
Proportion of *Nardus* leaves grazed (%)	38.3	24.6
Nardus grazed leaf lengths (cm)	6.4	10.3

grazed treatment had a high proportion of mature *Nardus* tussocks and a relatively greater density of tussocks, whereas the heavily grazed *Nardus* treatment had fewer *Nardus* tussocks, which had a high proportion of new leaf relative to mature leaf and stem. During the experiment, the sward height on the *Agrostis–Festuca* community was maintained at 4.5 cm across all plots. The results from this experiment showed that the sheep spent a greater proportion of their time grazing *Nardus* on the heavily grazed pretreatment than on the lightly grazed pretreatment (Table 10.5). Associated with this, they also grazed the tussocks to a greater extent. These between-community experiments demonstrate that the proportion of time that the sheep spent grazing on the preferred *Agrostis–Festuca* community was related to the resource availability on that community, and that the sward state on the less preferred *Nardus* community had an influence on the proportion of time that the sheep spent grazing that community. This implies that *Nardus* offtake will be greater, and hence *Nardus* control more effective, in systems that integrate sheep and cattle grazing (or sheep and goats), as compared with systems where the different animal species are grazed separately. However, cattle might suffer a production penalty if they were grazed late into the growing season, when the *Nardus* leaves are mature and of low nutritive value.

This hypothesis was tested, with productive cattle (spring-calving beef cows and their calves) being grazed at two grazing pressures (between-tussock sward heights of 4–5 cm or 6–7 cm) on *Nardus* grassland for six consecutive summer grazing seasons (Common *et al.*, 1998). Again, the cattle grazed significant amounts of *Nardus* tussock in both treatments. However, the intake and diet digestibility of the cattle grazing the 4–5 cm treatment were significantly lower than on the 6–7 cm treatment. As a result, the cattle lost live-weight and calf live-weight gains were significantly lower on the 4–5 cm between-tussock height grazing treatment. There was a substantial seasonal effect, with the greatest live-weight gains for both cows and calves occurring during the first 5–6 weeks of the grazing season and the differences between treatments increasing as the grazing season went on. These results clearly demonstrate that any benefit in terms of reduction in *Nardus* cover due to cattle grazing is at the cost of reduced cattle performance if the cattle are grazed on the community late into the grazing season.

Systems phase – putting it all together

A final systems experiment was established in which the effects of grazing pressure (between-tussock sward height 4–5 cm or 6–7 cm) and mixed (cattle and sheep) or monospecies (sheep) on *Nardus* tussock cover and structure, vegetation dynamics, animal performance and invertebrate biodiversity were measured over a period of 3 years. The cattle, 12–18-month-old steers, were only grazed for the period from June until the end of July, whilst the sheep were grazed from June until late September. During the course of the experiment, the *Nardus* cover declined most (by 22% units) on the cattle-plus-sheep treatment grazed at a between-tussock sward height of 4–5 cm, with little change occurring on the other treatments (I.A. Wright and C.L. Howard, unpublished results). As with the previous experiments, the diets of cattle contained a higher proportion of *Nardus* than those of sheep at both levels of grazing pressure. There was an enhanced performance of both the ewes and the lambs when grazing in mixtures with cattle at both grazing pressures (Howard and Wright, 1994; Table 10.6). Due to the differences in the structure of the *Nardus* grasslands created by grazing, there were differences in the abundance of different species of invertebrate. For example, as cattle grazing produced rounded, relatively short tussocks, there was a lower abundance of spiders, which rely on structural vegetation for their webs, in cattle-grazed treatments than in the sheep-only-grazed treatment (P. Dennis, R.J. Aspinall and I.J. Gordon, unpublished results). On the other hand, where the cattle grazing at high grazing pressure reduced *Nardus* tussock cover and increased the area of open *Agrostis–Festuca*-dominated grassland, there was a greater abundance of the cursorial predators, such as rove beetles (Dennis *et al.*, 1997).

Table 10.6. Sheep and cattle live-weight changes (g day^{-1}) when grazing either alone (sheep) or together (cattle and sheep) on a *Nardus* vegetation community at two different between-tussock sward heights (Howard and Wright, 1994).

	Sheep alone		Sheep and cattle	
Between-tussock sward height (cm):	4–5	6–7	4–5	6–7
Year 1 (1991)				
Ewe: Preweaning (June–Aug.)	−72	−6	−16	69
Ewe: Postweaning (Aug.–Sept.)	−80	5	−13	43
Lamb: (June–Aug.)	83	143	139	217
Steer: (June–July)	–	–	318	613
Year 2 (1992)				
Ewe: Preweaning (June–Aug.)	−102	−72	−83	2
Ewe: Postweaning (Aug.–Sept.)	−10	72	17	36
Lamb: (June–Aug.)	100	160	119	174
Steer: (June–July)	–	–	322	698

Developing offtake models

One of the ultimate goals of the research on plant–animal interactions in grasslands is the development of decision support tools to help land managers assess the consequences of their management decisions for animal production and the impact on the vegetation resource. For example, a computer-based decision support tool (Hill Grazing Management Model) has been developed by the Macaulay Land Use Research Institute, which provides a means of estimating the consequences of sheep stocking density on vegetation utilization, individual animal intake and diet composition (Armstrong, H.M. *et al.*, 1997). This model is being developed further to include the effects of grazing impact on vegetation dynamics and animal performance (both sheep and cattle). This model, HillPlan, uses the theoretical frameworks described earlier, quantitative and qualitative rules and relationships derived from experiments for a number of other vegetation communities, as well as *Nardus*, at a number of sites across Scotland and also expert knowledge to provide a comprehensive computer-based program relevant to sheep and cattle management in the Scottish context (Milne and Sibbald, 1998). HillPlan helps farmers and policy-makers assess the consequences of different grazing management regimes on the livestock and the vegetation and hence allows improved management of the fragile ecosystems for agricultural and environmental benefits.

Conclusions

With these few examples of manipulative experiments, this chapter has tried to show how a quantitative understanding of plant–animal interactions can be built up over time. By collecting data on floristic change and on diet selection, intake and animal performance in the same, or similar, experiments, a body of information can be amassed which ultimately will allow particular grazing systems to be modelled. By varying sites for the grazing experiments and by carrying out supporting cutting studies, in which nutrient supply as well as defoliation regimes are manipulated, allowances can be made for the modifying effects of site characteristics, such as nutrient stress, on the response of swards to herbivory. In this way, objective management decision rules can be identified and the options for management explored.

To sum up, this approach recognizes the key role of sward conditions in influencing both plant growth and offtake, i.e. the diet selection, intake and foraging behaviour of grazing animals. The work on mixed-species pasture involves several phases of study. The first aim is to identify the sward factors of importance in determining offtake of the species of interest. Hypotheses are generated by measuring sward biomass, composition and structure over short periods of time, in which information on levels of offtake and/or diet composition is collected. This can be done at a series of points in time on individual sites, or at the same point in time across plots covering a range of grazing intensities. Having generated a hypothesis, the second phase of study tests the hypothesis and quantifies the nature of the relationships identified. The approach is to use experimentally manipulated swards, in which sward conditions are controlled by frequent (i.e. weekly or twice-weekly) adjustments of stock numbers. The approach, though labour-intensive, is very rewarding. The third phase is to put together the information derived from the experimental phase to develop systems and determine the agricultural and environmental costs and benefits of the grazing management. Finally, decision support tools can be developed and field-tested in order to allow land managers to test scenarios applicable to their local situation. As well as advancing scientific understanding, the knowledge gained is in a form that is of maximum practical use to managers of land resources.

Acknowledgements

I would like to thank colleagues at the Macaulay Land Use Research Institute for their discussions on the grazing management of *Nardus*-dominated grasslands over the years. These include Dick Armstrong, Peter Dennis, Claire Howard, Sheila Grant, John Milne and Iain Wright. Murray Beattie, Gordon Common, Elaine Sim, Iain Thomson and Lynne Torvell all helped with fieldwork. John Milne provided valuable comments on the manuscript. The research described in this chapter was funded by the Scottish Executive Rural Affairs Department.

References

Armstrong, H.M., Gordon, I.J., Sibbald, A.R., Hutchings, N.J., Illius, A.W. and Milne, J.A. (1997) A model of grazing by sheep on hill systems in the UK. II. The prediction of offtake by sheep. *Journal of Applied Ecology* 34, 186–207.

Armstrong, R.H. and Hodgson, J. (1986) Grazing behaviour and herbage intake in cattle and sheep grazing indigenous hill plant communities. In: Gudmundsson, O. (ed.) *Grazing Research at Northern Latitudes*. Plenum Publishing, New York, pp. 211–218.

Armstrong, R.H., Common, T.G. and Smith, H.K. (1986) The voluntary intake and *in vivo* digestibility of herbage harvested from indigenous hill plant communities. *Grass and Forage Science* 41, 53–60.

Armstrong, R.H., Grant, S.A., Common, T.G. and Beattie, M.M. (1997) Controlled grazing studies on *Nardus* grassland: effects of between-tussock sward height and species of grazer on diet selection and intake. *Grass and Forage Science* 52, 219–231.

Arnold, G.W. and Dudzinski, M.L. (1967) Studies on the diet of grazing sheep. III. The effects of pasture species and pasture structure on the herbage intake of sheep. *Australian Journal Agricultural Research* 18, 657–666.

Birske, D.D. (1989) Vegetation dynamics in grazed systems: a hierarchical perspective. In: *Proceedings of the XVIth International Grassland Congress, Nice, France*. AFPF, Versailles, France, pp. 1829–1833.

Bullock, J.M. (1996) Plant competition and population dynamics. In: Hodgson, J. and Illius, A.W. (eds) *The Ecology and Management of Grazing Systems*. CAB International, Wallingford, UK, pp. 69–100.

Chadwick, M.J. (1960) *Nardus stricta*. *Journal of Ecology* 48, 255–268.

Common, T.G., Wright, I.A. and Grant, S.A. (1998) The effect of grazing by cattle on animal performance and floristic composition in *Nardus*-dominated swards. *Grass and Forage Science* 53, 260–269.

Crawley, M.J. (1983) *Herbivory: The Dynamics of Animal–Plant Interactions*, Blackwell Scientific Publications, Oxford, 437 pp.

Dennis, P., Young, M.R., Howard, C.L. and Gordon, I.J. (1997) The response of epigeal beetles (Col: Carabidae, Staphylinidae) to varied grazing regimes on upland *Nardus stricta* grasslands. *Journal of Applied Ecology* 34, 433–443.

Fenton, E.W. (1937) The influence of sheep on the vegetation of hill grazings in Scotland. *Journal of Ecology* 25, 424–430.

Gordon, I.J. (1989) Vegetation community selection by ungulates on the Isle of Rhum. III. Determinants of vegetation community selection. *Journal of Applied Ecology* 26, 65–79.

Gordon, I.J. and Iason, G.R. (1989) Foraging strategy of ruminants: the significance to vegetation utilization and management. In: Anon. (ed.) *Macaulay Land Use Research Institute, Annual Report, 1988–1989*. Macaulay Land Use Research Institute, Aberdeen, pp. 34–41.

Gordon, I.J. and Illius, A.W. (1993) Foraging strategy: from monoculture to mosaic. In: Speedy, A. (ed.) *Progress in Sheep and Goat Research*. CAB International, Wallingford, pp. 153–177.

Gordon, I.J. and Lascano, C. (1993) Foraging strategies of ruminant livestock on intensively managed grasslands: potential and constraints. In: *Proceedings of the XVIIth International Grassland Congress*. SIR Publishing, Wellington, New Zealand, pp. 681–689.

Gordon, I.J., Beattie, M.M. and Thomson, I.J. (1995) Factors affecting choices between two hill plant communities by Scottish blackface sheep. In: *Proceedings of the International Symposium on Wild and Domestic Ruminants in Extensive Land Use Systems*. Instituut für Genossenschaftswesen an der Humbolt-Universität, Berlin-Mitte, pp. 140–144.

Grant, S.A. and Maxwell, T.J. (1988) Hill vegetation and grazing animals: the biology and definition of management options. In: Usher, M.B. and Thompson, D.B.A. (eds) *Ecological Change in the Uplands*. Blackwell Scientific Publications, Oxford, pp. 201–214.

Grant, S.A., Suckling, D.E., Smith, H.K., Torvell, L.F and Hodgson, J. (1985) Comparative studies of diet selection by sheep and cattle grazing individual hill plant communities as influenced by season of the year. 1. The indigenous grasslands. *Journal of Ecology* 73, 987–1004.

Grant, S.A., Torvell, L., Smith, H.K., Suckling, D.E. and Hodgson, J. (1987) Comparative studies of diet selection by sheep and cattle: blanket bog and heather moor. *Journal of Ecology* 75, 947–960.

Grant, S.A., Torvell, L., Sim, E.M., Small, J.L. and Armstrong, R.H. (1996) Controlled grazing studies on *Nardus* grassland: effects of between-tussock sward height and species of grazer on *Nardus* utilization and floristic composition in two fields in Scotland. *Journal of Applied Ecology* 33, 1053–1064.

Grime, J.P., Hodgson, J.G. and Hunt, R. (1988) *Comparative Plant Ecology: a Functional Approach to Common British Species*, Unwin Hyman, London, 724 pp.

Hodgson, J. (1981) Nutritional limits to animal production from pasture. In: Hacker, J.B. (ed.) *Proceedings of International Symposium*. Commonwealth Agricultural Bureaux, Farnham Royal, pp. 153–166.

Hodgson, J. (1985) The significance of sward characteristics in the management of temperate sown pastures. In: *Proceedings of the XVth International Grassland Congress*. The Science Council of Japan, Nishi Nasuno, Japan, pp. 63–67.

Hodgson, J. and Illius, A.W. (eds) (1996) *The Ecology and Management of Grazing Systems*. CAB International, Wallingford, UK.

Hodgson, J., Forbes, T.D.A., Armstrong, R.H., Beattie, M.M. and Hunter, E.A. (1991) Comparative studies of the ingestive behaviour and herbage intake of sheep and cattle grazing indigenous hill plant communities. *Journal of Applied Ecology* 28, 205–227.

Howard, C.L. and Wright, I.A. (1994) Effects of mixed grazing by sheep and cattle on *Nardus stricta* dominated grassland. In: Haggar, R. and Heel, S. (eds) *Grassland Management and Nature Conservation*. British Grassland Society Special Publication No. 28, British Grassland Society, Aberystwyth, pp. 292–294.

Hunter, R.F. (1962) Hill sheep and their pasture: a study of sheep-grazing in south east Scotland. *Journal of Ecology* 50, 651–680.

Illius, A.W., Gordon, I.J., Elston, D.A. and Milne, J.D. (1999) Diet selection in goats: a test of intake rate maximization. *Ecology* 80, 1008–1018.

McNaughton, S.J. (1979) Grassland–herbivore dynamics. In: Sinclair, A.R.E. and Norton-Griffiths, M. (eds) *Serengeti: Dynamics of an Ecosystem*. University of Chicago Press, Chicago, pp. 46–81.

Maxwell, T.J., Grant, S.A. and Wright, I.A. (1988) Maximising the role of grass and forages in systems of beef and sheep production. In: Anon. (ed.) *Proceedings of the 12th General Meeting of the European Grassland Federation*. Irish Grassland Association, Belclare, Ireland, pp. 84–97.

Miles, J. (1981) Problems in heathlands and grassland dynamics. In: Poissonet, P., Romane, F., Austin, M.P., van der Maarel, E. and Schmidt, W. (eds) *Vegetation Dynamics in Grasslands, Heathlands and Mediterranean Ligneous Formations*. Dr W. Junk, The Hague, pp. 61–74.

Milne, J. and Sibbald, A. (1998) Modelling of grazing systems at the farm level. *Annales de Zootechnie* 47, 407–417.

Osmond, C.B. (1987) Photosynthesis and carbon economy in plants. *New Phytologist (Supplement)* 106, 162–175.

Prache, S., Gordon, I.J. and Rook, A.J. (1998) Foraging behaviour and diet selection in domestic herbivores. *Annales de Zootechnie* 47, 335–345.

Senft, R.L., Coughenour, M.B., Bailey, D.W., Rittenhouse, L.R., Sala, O.E. and Swift, D.M. (1987) Large herbivore foraging and ecological hierarchies. *Bioscience* 37, 789–799.

Thomas. B. and Fairbairn, C.B. (1956) The white bent, its composition, digestibility and probable nutritive value. *Journal of the British Grassland Society* 11, 230–234.

Wright, C.E. (1981) *Plant Physiology and Herbage Production*. British Grassland Society, Hurley, 249 pp.

Modelling Spatial Aspects of Plant–Animal Interactions

11

E.A. Laca

Department of Agronomy and Range Science, University of California, One Shields Avenue, Davis, California 95616–8515, USA

Spatial Dimension of Plant–Animal Interactions

The goal of this chapter is to present an integrated overview of the most relevant processes and mechanisms that determine spatial grazing behaviour, which should be considered in a research model of plant–ruminant interactions. First, concepts of spatial pattern and heterogeneity are introduced in the context of dynamic ruminant–pasture interactions. Secondly, I briefly review some experimental evidence that indicates what elements should be incorporated in spatial models of grazing behaviour. Thirdly, a conceptual framework and a modelling approach are proposed and, finally, the approach is used to derive a series of hypotheses and suggest management schemes to manipulate grazing patterns.

Consideration of spatial characteristics of grazing is important for managing primary and secondary productivity, sustainability and the environmental impacts of grazing and, more globally, to generate realistic paradigms for the study of grazed ecosystems. Models indicate that the spatial placement of bites can have significant consequences for overall pasture productivity, essentially because plant growth is a non-linear function of leaf area index that is relatively continuous in time, whereas grazing is a spatially and temporally discontinuous process. Intake rate is affected by spatial distribution of forages. Other things being equal, grazers achieve higher intake rates when a given amount of forage is available in patches than when it is more uniformly and thinly spread over a given area. Presumably, animal productivity reflects this effect. Some experimental evidence and theoretical considerations indicate that the distribution of grazing impacts determines the environmental consequences of grazing. Grazing patterns that concentrate on sensitive habitats, such as riparian areas, increase the risk of negative

 Grassland Ecophysiology and Grazing Ecology (eds G. Lemaire, J. Hodgson, A. de Moraes, C. Nabinger and P.C. de F. Carvalho)

consequences for water quality and biodiversity, relative to more uniformly distributed patterns.

Finally, I argue that incorporation of spatial heterogeneity in grazing models is essential, in order to integrate small-scale studies with practical applications. Such research models may help to identify those scales that are more relevant and those that can be ignored, to construct simpler practical models and to concentrate research and management efforts at the appropriate scale.

Spatial Pattern, Heterogeneity and Their Relevance in the Grazing Process

The dimensions and scale of spatial heterogeneity

Spatial heterogeneity is a multidimensional and complex concept. For categorical variables, such as the identity of plants occupying a certain area, heterogeneity has five components: number of patch types, proportions of each type, spatial arrangement of patches, patch shape and contrast between neighbouring patches (Li and Reynolds, 1994). The importance of spatial heterogeneity stems from its pervasiveness in ecosystems (Kolasa and Rollo, 1991) and the basic postulation that it simultaneously reflects and determines ecosystem processes (Kolasa and Pickett, 1991; Turner and Gardner, 1991a).

Reviews and definitions of the terms and concepts related to spatial heterogeneity can be found in Kolasa and Rollo (1991) and Turner and Gardner (1991b). In this chapter, I consider spatial heterogeneity as the presence of different values of a given vegetation descriptor measured in different locations at the same time; grain size is the resolution or finest level of measurement in a study and it is used to define scale and operationally relate it to quadrat size.

Two aspects of spatial heterogeneity that are relevant for plant–animal interactions can be distinguished, the degree of variability in the values of the variable of interest at the smallest scale and the spatial arrangement of these values (Palmer, 1992). Consider for example the distribution of herbage height in a paddock in the mixed prairie of West Texas (E.A. Laca, unpublished data). Height was measured at nodes every 5 m in both directions. Figure 11.1a shows the observed distribution as mapped by kriging, with block size ranging between 2 m × 2 m and 16 m × 16 m. Block size may be thought of as the scale at which animals can perceive the resource or effect selective behaviour. Figure 11.1b shows exactly the same statistical distribution of heights in a hypothetical random spatial distribution. As expected, the variance of average height declines with increasing block size, and the rate of decline is determined by the spatial arrangement of the original values. A spatial distribution that is random at all scales (Fig. 11.1b) exhibits a rapid decline in variance with increasing block size (Fig. 11.2), whereas the natural distribution (Fig. 11.1a) exhibits heterogeneity and pattern at multiple scales and thus a minor decline in variance with increasing block size. Thus, heterogeneity and opportunity for a grazer to select among patches of

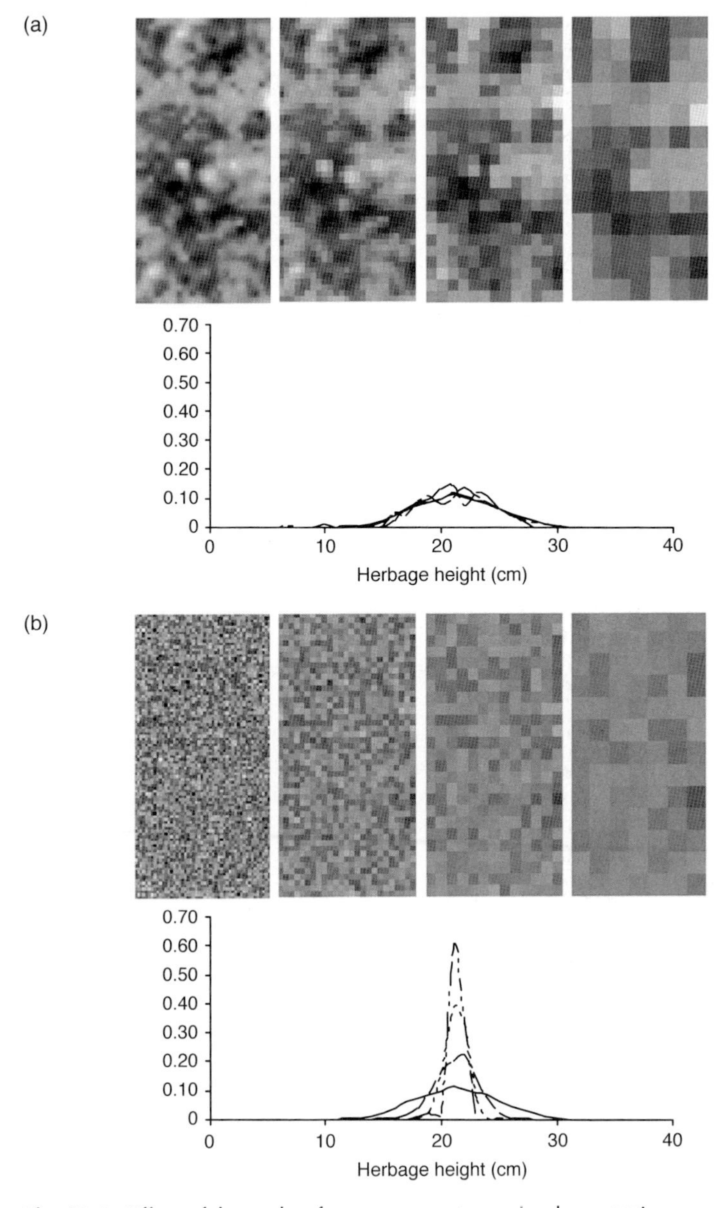

Fig. 11.1. Effect of the scale of measurement or animal perception on spatial and statistical distribution of forage height. (a) Herbage height observed in a mixed prairie site in Texas. (b) Hypothetical random spatial distribution of the same values displayed in (a). The top part of each figure has maps of average height in the same paddock, measured in 2 m × 2 m, 4 m × 4 m, 8 m × 8 m, and 16 m × 16 m quadrats, from left to right. Darker squares have taller average herbage height. The bottom part of each figure represents the probability density function of height (continuous line: 2 m × 2 m; dashed line: 4 m × 4 m; short-dashed line: 8 m × 8 m; dash-dotted line: 16 m × 16 m).

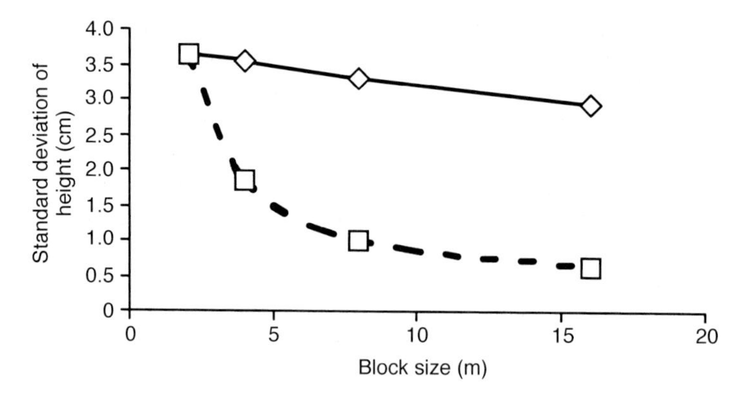

Fig. 11.2. Effect of scale of measurement on standard deviation of herbage height in Fig. 11.1. Continuous line: observed patchy distribution, dashed line: simulated random spatial distribution.

different heights is determined by the interaction between spatial distribution and scale at which animals perceive the characteristics of the forage.

Spatial heterogeneity is therefore scale-dependent (i.e. it may vary with grain size). For example, the botanical composition of a grassland can be highly heterogeneous when measured with high resolution but homogeneous when measured with a larger quadrat, because the larger quadrat may always include very similar proportions of small patches (Palmer, 1988). Natural spatial variability is characterized by different degrees of heterogeneity at different scales (Burrough, 1981; Palmer, 1988; Wiens, 1989; Robertson and Gross, 1994). Typically, spatial variance increases smoothly with increasing resolution and extent of the measurements (O'Neill *et al.*, 1991). From the point of view of the grazer, the fractal nature of resource distribution may determine a decline in resource density as scale of search increases, because larger scales will incorporate increasingly larger areas without acceptable forage.

Biodiversity, Stability and Spatial Pattern of Plant Communities

Stability of productivity and sustainability are enhanced by biodiversity because, when some species are affected (the 'susceptible' species), there are others ('resistant' species) that can expand their niche and take over the released resources and ecosystem functions (Tilman, 1996; Tilman *et al.*, 1996). However, rapid niche expansion is only possible if resistant individuals are in close proximity to susceptible ones. Given that most competition appears to take place at the scale of cm (Rees *et al.*, 1996), a minor change of distribution of plants at this scale can potentially void the compensatory potential of complementary species. Hence, pattern affects the relation between stability and diversity through the balance between intra- and interspecific competition, as well as the invadability of open patches through dispersal of species exogenous to the community under consideration. A finely grained (homogeneous, *sensu* Palmer (1988)) community should have an abundance of interspecific interactions, and dispersal into open patches should be equally likely for all species (i.e. no dispersal constraint, *sensu* Tilman (1994)). As the community becomes coarser-grained (i.e. more spatially heterogeneous at a particular scale), competition should become more dominated by intraspecific interactions, and the composition of plants dispersing into open patches should be more dependent on the specific location of the open patch. Therefore, the positive effect of biodiversity on sustainability and stability should be maximized when biodiversity is finely grained.

Contrary to the effects at the community level, populational stability within communities should be directly related to patch size. In finely interspersed communities, individuals of susceptible species are near individuals of resistant species. Because these susceptible species are at a competitive disadvantage during interspecific interactions with resistant species, they are further excluded beyond the direct effect of the initial disturbance. When monospecific patches are larger, interactions are mostly intraspecific. Susceptible individuals persist in larger patches, which act as spatial 'refuges', where most neighbours are conspecific and of similar competitive ability (Schmida and Ellner, 1984). Cellular automaton models of spatially explicit plant competition have indicated that communities with five grasses tend to lose species rapidly when the initial spatial distribution is random (Silvertown *et al.*, 1992). When species were arranged in monospecific bands, diversity was more persistent, and the trajectories of community composition over time were various and depended in a complex manner on the initial arrangement. Thus, when monospecific patches are large, gaps will have a greater probability of being recolonized by the same species as when they are small.

The relation between diversity and community stability and the relation between spatial patterns of species assemblages and the role of competitive interactions are particularly relevant for the California annual grassland, because of its high diversity and spatial heterogeneity (Rice, 1989). Annual grasslands are heterogeneous within (α diversity) and between (β diversity) habitats. While β diversity seems to be more related to abiotic patterns, such as changes in soils, topography and moisture gradients, α diversity depends on equilibrium and non-equilibrium mechanisms of competitive coexistence, including niche partitioning along nutrient and time axes, gopher disturbances, grazing by small rodents and livestock and competitive reversals due to year-to-year variations in autumn rainfall patterns (reviewed by Rice, 1989). Rice (1987) showed that the persistence of *Erodium* species depends on the elimination of small-mammal grazing. Small mammals were significantly excluded from large patches cleared of litter cover but not by small clearings. As a result, it was predicted that *Erodium* would be excluded from small patches but it would persist in the large ones. Interestingly, although sheep consume *Erodium*, areas grazed by sheep maintain stable populations of these forbs, because sheep grazing eliminates the more selective grazing by small rodents.

Further proof of the importance of spatial structure in plant communities is given by Hobbs and Mooney (1985), who found that annual serpentine grasslands have clear patches of *Bromus* and perennial bunch grasses (*Stipa* and *Sitanion*) with average dimensions of 45 and 37 cm, respectively. These dimensions roughly correspond to the diameter of newly formed gopher mounds. Gopher activity opened disturbance gaps, which were colonized by different species, depending on the distance to the gap and the timing of the disturbance, and allowed several species to persist in the community. Simulation studies of this system indicated that the main determinant of the species behaviour is their dispersal distance (Hobbs and Hobbs, 1987). In the model, larger gaps remained dominated by *Bromus mollis* and *Calycadenia glandulosa* for much longer than smaller ones, because *Bromus* and *Calycadenia* disperse further than the other two species. By using sticky seed traps and marked seeds, Hobbs and Mooney (1985) determined that, for all species, more than 75% of the seed does not disperse more than 25 cm, and virtually no seeds dispersed beyond 125 cm. This is a remarkable finding, because it points out that dispersal is limited to the neighbourhood of the parent plant. In a most complete spatially explicit simulation study of the annual serpentine grassland, Moloney and Levin (1996) found that size and spatial distribution (as determined by autocorrelation) of disturbances were only mildly important in determining the equilibrium abundance of species. However, this result may be due to the somewhat unrealistic way in which dispersal was simulated. The authors hypothesized that, with a more realistic dispersal submodel, the model would show a close link between community structure and spatial characteristics of the disturbances. Thus, both individual responses to competitive release through growth and numerical responses through seed production and dispersal are dependent on the small-scale arrangement of gaps and intact individuals.

Interactions between defoliation, competition and spatial pattern

Impacts of defoliation are complex and depend on the scale of analysis (Belsky, 1987; Brown and Allen, 1989) and on the biotic and environmental conditions around the defoliated plants. In general, individual fitness and productivity decline under grazing (Belsky, 1986; Briske and Richards, 1994; Lauenroth *et al.*, 1994), unless very restrictive conditions of timing of defoliation and resource availability are met (Paige and Whitham, 1987; Maschinski and Whitham, 1989). The ability of defoliated plants to compensate for herbivory declines as competition with other plants increases, as nutrient levels decrease and as defoliation comes later in the season (Maschinski and Whitham, 1989). On the other hand, growth and use of resources by intact plants located near defoliated plants can be significantly boosted by competitive release. When competition was reduced by clipping the vegetation in a 90 cm radius around bluebunch couch grass (*Agropyrum spicatum*) plants to ground level, couch grass plants produced *c.* 85% more herbage mass and 234% more flower stalks than controls with full competition (Mueggler, 1972). When competition was reduced by clipping the surrounding vegetation to ground level, herbage production and the number of flower stalks of plants clipped to 50% of their weight just before the emergence of flower stalks were similar to those of intact plants under full competition. The striking effect of defoliation on the utilization of resources by neighbouring plants is further illustrated by an experiment in which big sagebrush plants acquired six times as much labelled P from interspaces shared with defoliated couch grass individuals as they obtained from interspaces shared with intact grass plants (Caldwell *et al.*, 1987). These results clearly demonstrate the role of neighbourhood competition and competitive release on the response to defoliation.

The variety of responses simulated and observed in defoliation experiments has been explained on the basis of patterns of carbohydrate and nutrient allocation within plants, relative growth rate and intra- and interspecific competition (reviewed by Belsky, 1986; Briske and Richards, 1994). All of these processes have a strong spatial component: carbohydrates and nutrients can be redistributed within tillers and plants, but not between plants; movement of resources should be faster and easier between closely linked plant parts than between distant pieces of a vegetatively spreading individual. Competition requires spatial proximity of competitors. Typically, the area of influence of a grass plant is not much larger than three times its basal area (Walker *et al.*, 1989). For example, most root mass of an average blue grama plant (*Bouteloua gracilis*) with a basal cover of 320 cm^2 is found in a volume of 0.45 m^3 (Coffin and Lauenroth, 1991). Based on the average distance of 10 cm between plants, only four other blue grama individuals have roots within this volume. In spite of this strong spatial component, most experimental and theoretical studies of the impacts of defoliation have ignored its spatial distribution, particularly at the small scales at which the explanatory processes of competition and carbohydrate allocation pattern are relevant. The substantial effects of competitive release on individual and community responses to defoliation described in the previous paragraph are, therefore, completely contingent on the spatial distribution of defoliated and undefoliated individuals.

The study by Gdara *et al.* (1991) is a remarkable exception to the general exclusion of spatial considerations from the experimental study of responses to defoliation. Pairs of lucerne plants in 1 m^2 plots received clipping treatments that differed in the proportion of stems cut and the height removed. Stems were cut to the desired height in a continuous sector of the plant. Yield in plots where the top 1/3 or 2/3 of the plants was clipped was only 78 and 41%, respectively, of the yield in plots where a vertical 1/3 or 2/3 sector was cut, leaving the rest of the plants intact. Obviously, the distribution of defoliation within plants had a strong influence on the ability of plants to grow back. The physiological integration among different ramets may enable plants to average over local heterogeneity in resources and to buffer the impacts of localized and patchy defoliation, which is a characteristic of many herbivore grazing patterns (Archer and Tieszen, 1986). Yet these results show that defoliation impacts will not be buffered if no individuals or ramets are left intact in proximity to the resources released by the clipped ramets, which is central to my hypotheses that spatial patterns of defoliation can have dramatic effects on community dynamics and that population stability under selective grazing should be lowest at intermediate scales of patchiness.

Disturbances as pattern-generating processes

Disturbances can be classified along a continuum between total spatial dependence and complete independence of plant pattern. Examples of the former are frost damage or drought in a community with species that differ in their tolerance to low temperatures or to lack of moisture (ignoring potential effects of plants on their own microclimate). Regardless of the distribution of the plants, a frost will create gaps that coincide with the distribution of the frost-sensitive species, with the exception of microenvironmental refuges (Noy-Meir, 1996). A specialized arthropod with high dispersal ability would fall under the same category. An example where disturbance is largely independent of plant distribution may be found in patterns of soil turnover by gophers (Hobbs and Hobbs, 1987), although gopher activity may be correlated with certain plants or soil types.

Livestock grazing presents an interesting case because the animals' ability and motivation to forage selectively may decline as patches become smaller. Thus, grazing is an 'intermediate' situation between perfect dependence and independence between disturbance and plant pattern. At very small scales, grazing patterns may be random in relation to plant species. At larger scales, there is typically a pattern of matching or overmatching (Senft *et al.*, 1987; Senft, 1989; Distel *et al.*, 1995), by which the distribution of defoliation closely matches the distribution of preferred plants or habitats.

Herbivory is a pattern-generating process that may be strongly influenced in its distribution by the pattern in the plant community (Kareiva, 1982; Bergelson, 1990; Coughenour, 1991; Vinton and Hartnett, 1992; Vinton *et al.*, 1993; Bailey *et al.*, 1996). Herbivores respond as individuals, populations and communities to the patterns of plant communities. However, the spatial scale of the response depends on the body size and life history of the herbivore. A caterpillar moves mostly within a plant during its life stage. In this case, selection among plants, which would affect plant competitive interactions, takes place at the temporal scale of generations and involves demographics and dispersal processes at the animal population level. A larger herbivore interacts with plants at intermediate spatial scales, where it can most effectively affect the neighbourhood competitive processes that drive plant community change and determine the persistence and stability of communities as they face disturbances (Tilman, 1996; Tilman *et al.*, 1996). Thus, spatial models of plant/animal interactions should consider the potential asymmetry between scales of plant and animal processes.

Because herbivores can be a powerful force structuring plant communities, it is essential to understand how the scale of patchiness in plants determines the degree to which selective defoliation is more intense on some plant species than on others. It is equally important to integrate the effects of spatial structure on competitive interactions among plants with those on herbivore selectivity. As mentioned above, larger plant patches lend stability to community composition and diversity. On the other hand, larger patches allow more selective herbivore impacts, which should promote the extinction of preferred species or assemblages. This view is fully supported by the study of slug herbivory in mixtures of *Senecio vulgaris* and *Poa annua* (Bergelson, 1990). *Senecio* exhibited greater population growth rate in clumped than in random *Poa* distributions, because clumped distributions had gaps of low density of *Poa*, where survival of *Senecio* seedlings was greater. The benefit of the clumped distribution through locally reduced competition was opposed in the presence of herbivores, because slugs consumed more *Senecio* seedlings when the grass was clumped.

Necessary Elements for a Comprehensive Spatial Model of Grazing

The previous sections emphasize the importance of the spatial dimension in plant–animal interactions, with emphasis on plant processes. I now turn to characteristics of the forager that point to the effects of spatial behaviour on grazing systems. A comprehensive model of spatial grazing must consider a variety of characteristics of grazing behaviour, because there is empirical evidence that mammalian herbivores have complex spatial foraging abilities. In this section, I present some of these characteristics and empirical evidence to support their inclusion in realistic plant–herbivore models. Although the characteristics of grazing behaviour are presented in different sections, they are clearly interrelated.

Scale-dependent results of experiments

The relationship between grazing behaviour and forage resources depends on the scale of analysis. This is important from the point of view both of modelling and of interpreting results from empirical grazing studies. For example, in a study of the grazing pattern of steers in the mixed prairie in West Texas, the correlation between number of bites taken per unit area and the availability of forage N per unit area depended on the block size used to calculate the correlations (Fig. 11.3; E.A. Laca, unpublished data). In this experiment, steers grazed 2.5 m × 50 m 'corridors', where number of bites, botanical composition and herbage height were measured in 0.5 m × 0.5 m quadrats. An indicator of N availability in each quadrat was calculated by adding the products of N concentration, height and percentage cover for each species. Quadrats were grouped in rows and average characteristics for groups of rows spanning from 0.5 to 11 m were calculated. The effect of scale on the relation between preference and herbage characteristics was studied by calculating the correlation between average proportion of bites and average N index as a function of the number of rows used to calculate the averages. Correlation was low at small scales, peaked at scales around 2–5 m and

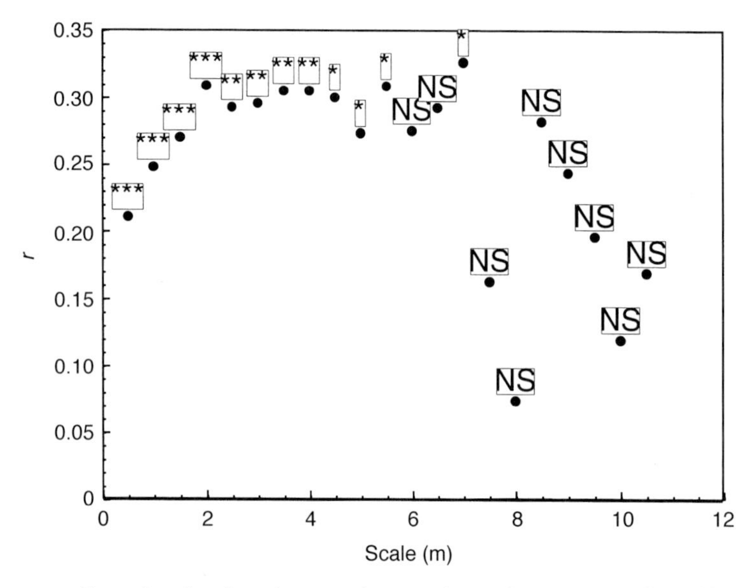

Fig. 11.3. Effect of scale of analysis on the correlation between number of bites per unit area taken by steers and an index of availability of forage nitrogen per unit area. The scale axis represents the size of quadrats used to compute number of bites and forage nitrogen. *, Significant at the 5% level; **, significant at the 1% level; ***, significant at the 0.1% level; NS, not significant.

declined at larger block sizes. I surmise that the correlations reflected the interaction between the vegetation pattern and the animal's ability to perceive and respond to pattern at various scales. It is reasonable to hypothesize that these interactions and similar results extend beyond the spatial extent of this experiment to spatial patterns over hundreds and even thousands of metres.

Flow of information across scales

Herbivores have the ability to sense and integrate information about their foraging environment and internal state (e.g. rumen fill) across scales of space and time. Laca *et al.* (1997) conducted an experiment to test whether sheep select diets based on instantaneous rewards (digestible dry matter intake rate (DDMIR) while eating), or whether they adjust instantaneous behaviour and forgo instantaneous rewards for the sake of greater daily intake of digestible dry matter (DDMI). The negative relation between diet quality and instantaneous intake rate was simulated by offering animals both a forage of high quality and low intake rate (lucerne) and one of low quality and high intake rate (timothy). Intake rate was manipulated by inserting wooden pegs in the side of the feed box containing lucerne. An intake model predicted that, if animals maximized instantaneous

Table 11.1. Effect of eating time available on selectivity by sheep offered a low-quality, high-intake rate hay (*Phleum pratense*), and a high-quality, low-intake rate hay (*Medicago sativa*). Selectivity, measured as percentage of lucerne in the diet, was also predicted on the basis of maximization of instantaneous (DDMIR) and daily (DDMI) intake rate of digestible forage dry matter.

Eating time (min)	% Lucerne observed	% Lucerne for max. DDMI	% Lucerne for max. DDMIR
30	24 (5.0)	0	0
60	32 (3.1)	17	0
90	37 (4.4)	28	0
120	33 (3.1)	29	0
180	44 (4.4)	23	0
240	35 (4.4)	16	0

Error, degrees of freedom = 33. Numbers between parentheses are standard errors of the mean.

DDMIR, they should consume only timothy, whereas maximization of DDMI should result in diets with proportions of lucerne that depend on total eating time available. Although sheep were more selective than predicted by maximization of DDMI, results strongly suggest that the response of selectivity to meal time was consistent with daily maximization (Table 11.1). This result is remarkable because it evinces sheep's ability to modify instantaneous behaviour on the basis of long-term, positive, nutritional consequences. This ability suggests a profound but fore-seeable implication for the modelling of plant–animal interactions: temporal and spatial foraging decisions at any scale are simultaneously affected by the charac-teristics of the foraging situation at multiple scales. A corollary of this is that a reductionist approach that attempts to build large-scale grazing patterns on the basis of small-scale information is severely flawed.

Use of spatial and non-spatial experience

Livestock use spatial and non-spatial experience to modify search and selectivity patterns. I illustrate this point with an example of spatial memory in cattle (Laca, 1998). Steers were exposed over time to two kinds of spatial distributions of feed in experimental arenas: constant and variable food locations. Where food loca-tions were constant, steers used spatial memory and applied a complex search strategy, by which they avoided locations recently depleted (within sessions) and revisited locations where food was expected (between sessions). This strategy resulted in high search efficiency and spatially concentrated search patterns. Where food locations were variable, steers were unable to use spatial memory, but they implemented an optimal systematic search pattern, which was widely spread in space, and had a low search efficiency relative to the constant food locations. Beyond indicating that cattle can use spatial memory, these results show that cattle are able to implement different search strategies to match the foraging challenge. Moreover, they seem to have the ability to simultaneously implement different search strategies at different spatial scales, a result that challenges modelling efforts but offers opportunity to manage spatial patterns of grazing impacts, as explained below.

The use of non-spatial experience by livestock while foraging is well established, particularly the use of gustatory and visual cues to select forages and foraging locations. In short, livestock exhibit the ability to associate flavours, colours and shapes with nutritional consequences, and these abilities are extremely plastic, allowing quick adaptation to new foraging situations (Provenza and Balph, 1987; Entsu, 1989a, b; Bazely, 1990; Distel and Provenza, 1991; Entsu *et al.*, 1992; Provenza, 1995; Edwards *et al.*, 1997).

Exploitation of patchiness at multiple scales

In a seminal paper, Senft *et al.* (1987) described herbivore foraging as a hierarchical process, ranging from landscape to bite scale. In this and more elaborated versions of the model (Bailey *et al.*, 1996), herbivores can select among options at each scale, and each decision constrains the choices at smaller scales. Experimental evidence supports the idea that animals can exploit multiple scales of heterogeneity. Furthermore, some evidence indicates that ability to select increases with the coarseness of the spatial distribution of forages and that, for a given scale of patchiness, smaller species can be more selective than larger ones. The effects of body size on the interaction between resource availability and utilization patterns are related to the allometry of search speed, metabolic requirements and intake rate (Milne *et al.*, 1992).

In order to illustrate the relationship between spatial pattern of resources and foraging pattern by cattle, I use results of two experiments that are familiar to me. First, Laca and Ortega (1996) studied search patterns and intake rate of cattle when eating pelleted feed arranged in different spatial patterns, with and without visual cues. Eight foraged individually, in a 40 m $\times$ 40 m arena with a 2 m grid painted on the ground, for 20 min in each of eight treatments. Treatments were a factorial of four spatial distributions (uniform, random, two patches and fractal (Fig. 11.4)) of 160 18 g feed pellets with or without a visual cue (one yellow flag per pellet). The fractal distribution had patches at five scales, ranging from 1 to *c.* 23 m and a constant fractal dimension. Search efficiency (intake/distance walked) increased with visual cues and degree of patchiness ($P = 0.01$ (Fig. 11.4)). In particular, search efficiency was greater in the fractal feed distribution than when there was a single level of patchiness. Regardless of visual cues, steers were able to recognize and exploit patchiness at multiple levels by ignoring the empty space between patches and concentrating search effort in areas where pellets were found. On average, intake rate and search efficiency increased ($P < 0.01$) with visual cues, but the magnitude of the effect on intake rate declined with increasing food patchiness (Fig. 11.5). This can be explained because, as food becomes

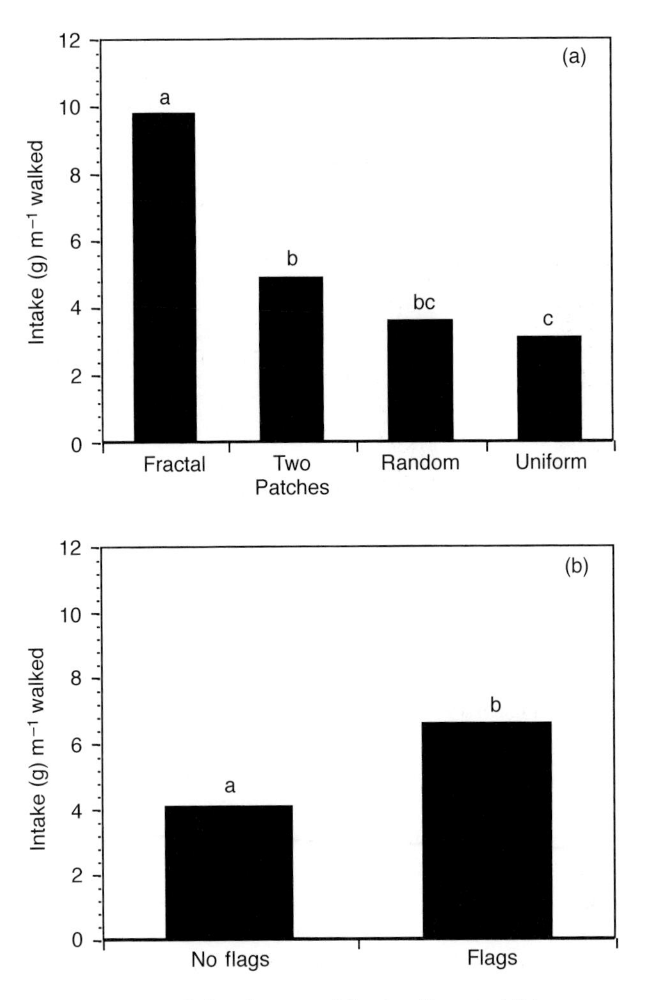

Fig. 11.4. Effect of (a) spatial distribution of feed pellets and (b) presence of visual cues on distance walked per unit intake by steers foraging in a 40 m × 40 m experimental arena. Bars with different letters differ significantly at the 5% level.

more patchy, patch detection and exploitation mechanisms not dependent on visual cues can be implemented more effectively.

The fractal dimension (D) of the search path describes its sinuosity (With, 1994) and therefore the scale of the searching effort. A high value of D for a certain scale indicates that the path has frequent turns at that scale. Fractal dimension of paths in all treatments, except for the fractal with visual cues, was greater than 1.0, and tended to peak at a scale of 2 to 7 m. In most cases, D was not constant across scales, indicating that search paths were not self-similar, presumably because the mechanisms differed between scales. This result also supports the idea that animals implemented multiple scales of search. Search paths in the fractal

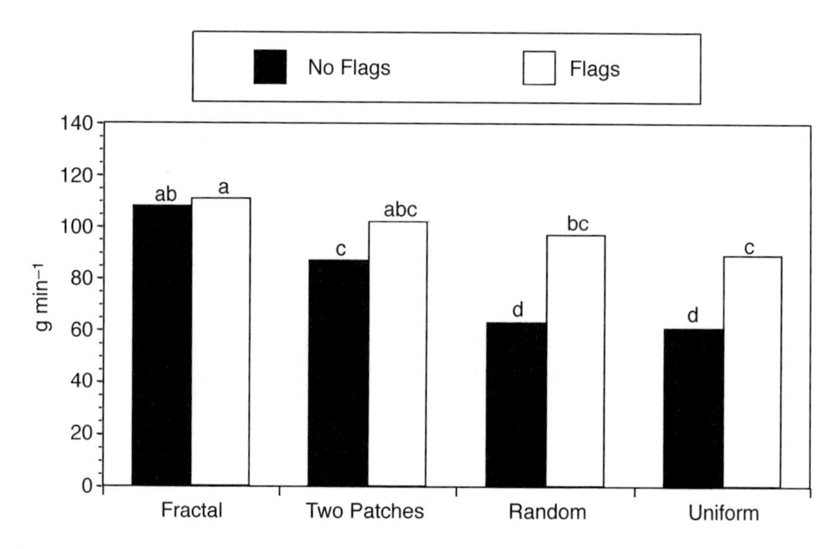

Fig. 11.5. Effect of spatial distribution of feed pellets and presence of visual cues (flags) on intake rate by steers foraging in an experimental arena (same experiment as in Fig. 11.4).

distribution with flags had a dimension between 1.0 (smooth line) and 1.1, indicating that when the perceptual range is greater, animals increase the search efficiency by walking straight to food locations and closely matching search and food patterns.

A second experiment illustrated how the ability of steers to select preferred patches varies with scale of patchiness (WallisDeVries *et al.*, 1999). Steers grazed random mosaics of short/high-quality and tall/low-quality grass patches in equal proportion at grid sizes of 2 m × 2 m and 5 m × 5 m, and foraging behaviour and search paths were monitored continuously for 30 min. Selectivity was calculated at the level of patches, feeding stations and bites. Regardless of patch size, steers selected short patches, both by concentrating the search path in short patches and by establishing more feeding stations per unit distance walked within short patches (Table 11.2). Overall, selectivity was more pronounced in large patches than in small ones, mainly because of greater selectivity at the patch and feeding station level. In contrast, number of bites per feeding station was not affected by patch size, suggesting that selection between and selection within feeding stations are essentially different processes. These results indicate that selectivity in grazers is facilitated by large-scale heterogeneity, particularly by enhancing discrimination between feeding stations and larger selection units.

Table 11.2. Effects of patch size on the index of grazing selectivity at each level of the foraging process (FS = feeding station). Each selectivity index is the ratio of values in short to tall patch types. Selectivity between patch types within treatment is present when the index deviates from 1.0. Differences between columns within each line indicate that selectivity was affected by patch size. Treatments did not allow selectivity in terms of encounters with different patch types.

	Patch size			Effect of
			Selectivity	patch size
Short/tall ratio	2 m × 2 m	5 m × 5 m	(*P* value)	(*P* value)
Steps per patch	3.39	2.03	< 0.0001	0.019
FS per step	5.60	2.63	< 0.0001	0.036
Bites per FS	0.85	0.83	NS	NS
Digestible intake per bite	0.49	0.49	< 0.0001	NS

NS, not significant.

Conceptual Model of Plant–Animal Interactions

Based on the previous experimental evidence, it is reasonable to conclude that a realistic model of plant–animal interactions should consider the interactions among multiple spatial scales of heterogeneity in the plant community and animal behaviour. The present conceptual model (Fig. 11.6), outlined by Laca and Ortega (1996), extends these concepts and presents a general framework for a simulation model that integrates mechanisms of animal behaviour over scales of time and space. Obviously, this model is not a finished product for making a quantitative prediction, but I offer it as a didactic tool and a basis for further development.

The model focuses on the acquisition and processing of information about the environment and the internal state at multiple spatial and temporal scales. At the top of the figure, I represent the processes by which animals obtain information about the foraging environment, according to the hierarchical levels of foraging. Although information may be perceived at multiple scales, attention to one scale may constrain other scales. For example, ability to see distant patches is diminished while the animal obtains information about a bite by touch or smell, because the range of visibility declines as the head is lowered. Conversely, if attention is on assessing the landscape within which the animal is moving, details of bites and feeding stations that fall within the search path are missed. To my knowledge, the effect of search rate on efficiency of detection of preferred forages has not been studied in large herbivores, but there is a wealth of literature on other animals that supports the idea that attention to certain scales or objects diminishes the efficiency to detect other objects at other scales. The interested reader is referred to Dawkins (1971), Gendron and Staddon (1983), Getty (1985), Gendron (1986) and Guilford and Dawkins (1988) and references thereof

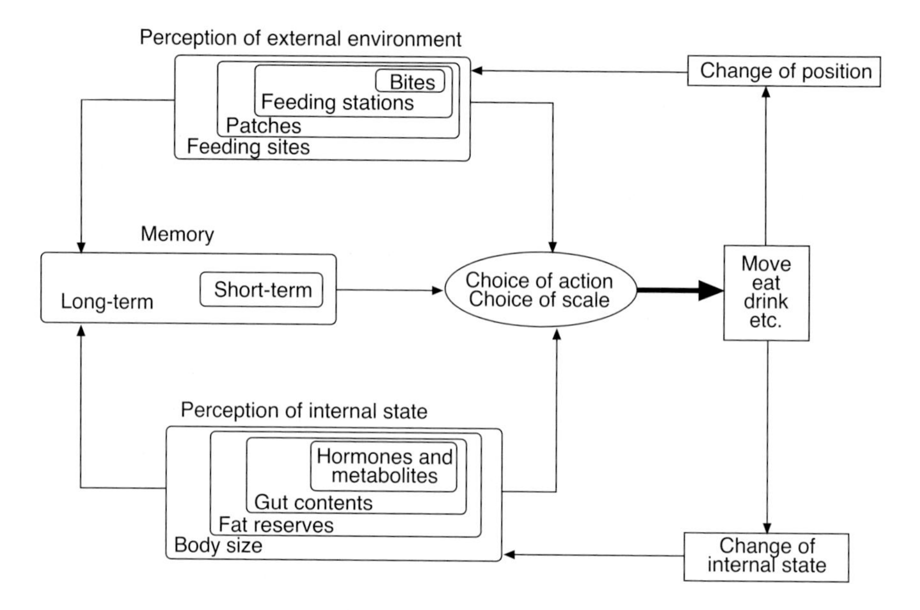

Fig. 11.6. Schematic representation of an integrative model of spatial grazing behaviour of ruminants.

for an introduction to the subject. In general, small-scale information is more accurate, but also more costly per unit of forage than large-scale information, because time costs involved in large-scale decisions are spread over a larger quantity of potential forage.

The lower part of the model represents an abstraction of the internal state and processes that impose temporal scales on animal behaviour. I have arbitrarily focused on variables that may reflect the energy balance of the herbivore, but the concepts can, and probably should, be extended to a variety of currencies that are important for fitness, survival and productivity (e.g. nitrogen, water, temperature, toxins). At the fastest temporal scale, animals may 'perceive' their levels of hormones and metabolites in the internal medium. These can vary at the scale of minutes, and these variations can elicit changes in behaviour at such scales. At longer time-scales, there are putative mechanisms that 'monitor' gut contents, fat reserves and body mass. Because animals move at a certain rate that is more or less species-specific, there is a necessary correspondence between temporal scales of internal changes and spatial scales of foraging. The nested nature of these processes makes it possible to study short-term, small-scale patterns, under the assumption that long-term, large-scale factors remain constant. This fact, however, does not contradict the idea of the simultaneous effects of factors across hierarchical levels of space and time.

On the left side of the model, I consider the role of spatial and non-spatial memory in foraging behaviour. This feature of the model allows the incorporation of effects such as learned aversions, spatial memory and area-restricted search. In the model, memory has a role similar to that of a lagged state variable, and simply reflects our lack of knowledge about and the excessive complexity of the immediate state variables through which experience affects instantaneous behaviour. Through the effect of 'memory' state variables, the model can predict different behaviours for two animals with identical internal and external states but which differ in their experience. For example, based on the experiment on spatial memory (Laca, 1998), an animal exposed to variable food locations would search systematically, whereas one exposed to constant food locations would exhibit area-restricted search.

Finally, on the right side of the model, multiscale spatial information is integrated with memory and internal state to determine behaviour and the scale of the behaviour. For example, the choice may be to move at the feeding station level. I operationally assume that the forager is continuously deciding between moving, eating or gathering information. Choice of action and location results in changes in the internal and external state of the animal. In a simplistic but potentially fruitful view, I assume that these changes are perceived as rewards (e.g. intake rate) or aversive stimuli, and are the basis for the feedback mechanisms that regulate selection of foods and foraging sites. Thus, if the animal chooses to graze a patch and as a result it obtains large mouthfuls of tasty succulent forage that stimulates rumen fermentation, the choice is reinforced and the forager continues to do more of the same. Conversely, if it obtains small bites of forage that compare poorly with what the animal remembers from recent experience, the behaviour is weakened or discontinued. One way of implementing this view is by using intake rate of nutrients as the reward.

Foraging decisions proceed from coarse to fine and from fine to coarse scales (Fig. 11.7). Upon beginning to forage, the animal selects a feeding site from a set of alternatives and then it selects a patch from those available in the site, and so on. Once a bite has been selected, grazing continues at small scales for as long as the short-term intake rate does not fall below a certain threshold. This threshold may be based on the integration of information at the next larger temporal and spatial scale. If rewards fall below the threshold, decisions shift to the next hierarchical level, where a new choice is made. When intake rate within the feeding station becomes too low, a new feeding station is selected, and, when acceptable feeding stations become scarce, a new patch is selected. Thus, consequences of foraging choices are integrated from small to large scales of time and space by various organs and tissues (e.g. rumen, fat reserves, brain), and then choices are controlled by feedback information, which cascades down from large to small scales.

Because herbivores experience information constraints and forage in vegetation with multiscale patchiness, I hypothesize that they can implement a hierarchical foraging strategy. This strategy requires the ability to assess patches and effect behaviours at multiple scales, but it increases search efficiency by eliminating the need to perform a costly small-scale search in patches that can be identified as poor at larger scales. The effectiveness of different search strategies depends on the spatial distribution of selected forages. I believe that the study of search and grazing patterns is a fertile area in the study of plant–animal interactions, and

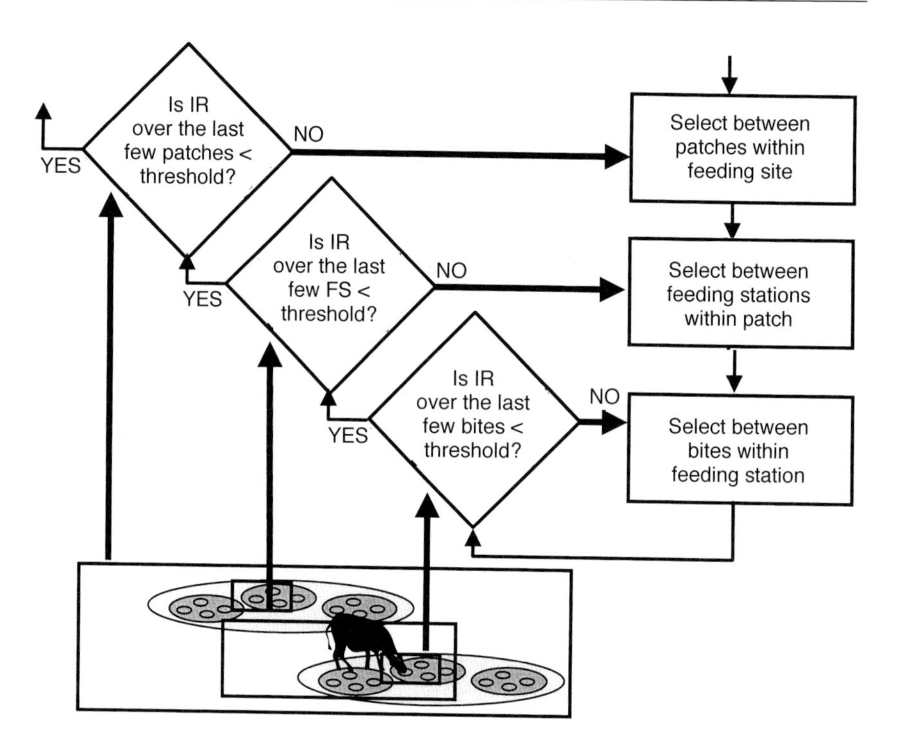

Fig. 11.7. Procedural approach to simulate spatial grazing decisions by ruminants across scales of space and time. IR, intake rate; FS, feeding stations.

it may be the next bottleneck as we try to improve our understanding and management of grazing systems. Although a discussion of search strategies is beyond the scope of this chapter, I contend that this modelling exercise and the experimental evidence presented open up a new dimension for the management of grazing distribution. In the next section, I present and discuss some hypotheses to be tested towards the exploitation of this management dimension.

Preliminary Applications of the Conceptual Model to Grazing Management

Spatial characteristics of grazing behaviour may be modified by manipulating the internal or external environment of the animal. Moreover, because animals use memory, exposure to different experiences can be used to modify future behaviour. Howery *et al.* (1999) proposed a series of potential techniques to modify the spatial behaviour of cattle, ranging from invasive (direct modification of internal state by hormones) to manipulations of the external environment, where animals develop grazing patterns.

Animals can learn to associate specific behaviours, such as searching or moving in certain directions, with the consequences of the behaviours. If the consequences are positive (or negative), the behaviour is reinforced (or weakened). Thus, within the proper context, animals will perform actions in order to obtain rewards, such as food, water or cover. The intensity and rate of a behaviour depend on the 'schedule of reinforcement' (Staddon and Ettinger, 1989). In the jargon of experimental psychology, a 'fixed-interval' schedule of reinforcement is one in which a reward (food pellet) results from an action (pressing lever) only after a fixed time has elapsed. Animals learn to respond (i.e. perform the action) intensively just prior to the expiration of the fixed time, and to completely cease responding immediately after the reward is delivered. A 'variable-interval' schedule of reinforcement involves a random time interval between rewards, which results in a more constant response by the animal over time. The results reported by Laca (1998) seem to extend the concepts of fixed and variable schedules from time to space, because steers adjusted the spatial strategy of search to both spatial and temporal predictability of food rewards. Following this rationale, it would be interesting to test whether landscape-scale patterns of search and grazing can be modified by manipulating the distribution of the pelleted supplement over space and time.

I hypothesize that cattle will develop an intense systematic pattern of search if a palatable supplement is distributed in variable locations at variable intervals, particularly if the supplement is fed on the ground, without a conspicuous visual cue, such as a feeder would provide. Conversely, if the supplement is provided at predictable times and locations, as is usually done, animals develop search patterns that are concentrated in space and time. Furthermore, it is likely that livestock can learn to simultaneously exhibit scale-dependent grazing patterns, and that these can be manipulated by careful planning of the schedule and spatial pattern of supplementation. Consider, for example, how livestock in a large paddock might respond if they were exposed to a feeding regime in which only a few large patches within the paddock receive feed on random days of the week, and if the feed is distributed in a set of randomly located subpatches within each large patch.

Conclusions

In this chapter, I argue for an integrative view of plant–animal interactions in a spatially explicit manner. Brief experimental evidence is presented to support the standpoint that realistic models of plant–animal interactions must include, at least in principle, consideration of the reciprocal feedback between the spatial pattern of the plant community and the spatial pattern of herbivory. A synthesis of basic knowledge about interaction among plants (e.g. competition), the nonlinearity of plant growth, the spatial discontinuity of defoliation relative to plant growth and herbivore foraging strategies and behaviour leads to the following conclusions:

1. The stability, productivity and invasibility of plant communities under grazing disturbances should depend on the scale of heterogeneity and spatial pattern of disturbances.
2. The defoliation pattern and degree of herbivore selectivity should depend on the scale and degree of vegetational heterogeneity relative to animal size.
3. Large herbivores have complex abilities to respond to food spatial and temporal patterns at multiple scales.
4. Spatial patterns of herbivory may be modified by manipulation of the spatial and temporal distribution of food rewards at multiple scales.

Acknowledgements

This research was funded by NSF Award Number IBN-9311463, US-Israel BARD grants US-1329-87 and IS-2331-93C, and award 9701033 from NRI Competitive Grants Program/USDA.

References

Archer, S.R. and Tieszen, L.L. (1986) Plant response to defoliation: hierarchical considerations. In: Gudmundsson, O. (ed.) *Grazing Research in Northern Latitudes*. Plenum Press, New York, pp. 349–351.

Bailey, D.W., Gross, J.E., Laca, E.A., Rittenhouse, L.R., Coughenour, M.B., Swift, D.M. and Sims, P.L. (1996) Mechanisms that result in large herbivore grazing distribution patterns. *Journal of Range Management* 49, 386–400.

Bazely, D.R. (1990) Rules and cues used by sheep foraging in monocultures. In: Hughes, R.N. (ed.) *Behavioural Mechanisms of Food Selection. NATO ASI Series G: Ecological Series*, Springer-Verlag, Heidelberg, pp. 343–367.

Belsky, A.J. (1986) Does herbivory benefit plants? A review of the evidence. *American Naturalist* 127, 870–892.

Belsky, A.J. (1987) The effects of grazing: confounding of ecosystem, community, and organism scales. *American Naturalist* 129, 777–783.

Bergelson, J. (1990) Spatial patterning in plants: opposing effects of herbivory and competition. *Journal of Ecology* 78, 937–948.

Briske, D.D. and Richards, J.H. (1994) Physiological responses of individual plants to grazing: current status and ecological significance. In: Vavra, M., Laycock, W.A. and Pieper, R.D. (eds) *Ecological Implications of Livestock Herbivory in the West*. Society for Range Management, Denver, pp. 147–176.

Brown, B.J. and Allen, T.F.H. (1989) The importance of scale in evaluating herbivory impacts. *Oikos* 54, 189–194.

Burrough, P.A. (1981) Fractal dimensions of landscapes and other environmental data. *Nature* 294, 240–242.

Caldwell, M.M., Richards, J.H., Manwaring, J.H. and Eissenstat, D.M. (1987) Rapid shifts in phosphate acquisition show direct competition between neighbouring plants. *Nature* 327, 615–616.

Coffin, D.P. and Lauenroth, W.K. (1991) Effects of competition on spatial distribution of roots of blue grama. *Journal of Range Management* 44, 68–71.

Coughenour, M.B. (1991) Spatial components of plant–herbivore interactions in pastoral, ranching, and native ungulate ecosystems. *Journal of Range Management* 44, 530–542.

Dawkins, M. (1971) Perceptual changes in chicks: another look at the 'search image' concept. *Animal Behaviour* 19, 566–574.

Distel, R.A. and Provenza, F.D. (1991) Experience early in life affects voluntary intake of blackbrush by goats. *Journal of Chemical Ecology* 17, 431–450.

Distel, R.A., Laca, E.A., Griggs, T.C. and Demment, M.W. (1995) Patch selection by cattle: maximization of intake rate in horizontally heterogeneous pastures. *Applied Animal Behaviour Science* 45, 11–21.

Edwards, G.R., Newman, J.A., Parsons, A.J. and Krebs, J.R. (1997) Use of cues by grazing animals to locate food patches: an example with sheep. *Applied Animal Behaviour Science* 51, 59–68.

Entsu, S. (1989a) Discrimination between a chromatic colour and an achromatic colour in Japanese Black cattle. *Japanese Journal of Zootechnical Science* 60, 632–638.

Entsu, S. (1989b) Shape discrimination training for cattle with a Landolt ring. *Japanese Journal of Zootechnical Science* 60, 542–547.

Entsu, S., Dohi, H. and Yamada, A. (1992) Visual acuity of cattle determined by the method of discrimination learning. *Applied Animal Behaviour Science* 34, 1–10.

Gdara, A.O., Hart, R.H. and Dean, J.G. (1991) Response of tap- and creeping-rooted alfalfa to defoliation patterns. *Journal of Range Management* 44, 22–26.

Gendron, R.P. (1986) Searching for cryptic prey: evidence for optimal search rates and the formation of search images in quail. *Animal Behaviour* 34, 898–912.

Gendron, R.P. and Staddon, J.E.R. (1983) Searching for cryptic prey: the effect of search rate. *American Naturalist* 121, 172–186.

Getty, T. (1985) Discriminability and the sigmoid functional response: how optimal foragers could stabilize model–mimic complexes. *American Naturalist* 125, 239–256.

Guilford, T. and Dawkins, M.S. (1988) Why blackbirds overlook cryptic prey: search rate or search image? *Animal Behaviour* 37, 157–164.

Hobbs, R.J. and Hobbs, V.J. (1987) Gophers and grassland: a model of vegetation response to patchy soil disturbances. *Vegetatio* 69, 141–146.

Hobbs, R.J. and Mooney, H.A. (1985) Community and population dynamics of serpentine grassland annuals in relation to gopher disturbance. *Oecologia* 67, 342–351.

Howery, L.D., Bailey, D.W. and Laca, E.A. (1999) Impact of spatial memory on habitat use. In: Launchbaugh, K.L., Sanders, K.D. and Mosley, J.C. (eds) *Grazing Behaviour of Livestock and Wildlife*. Idaho Forest, Wildlife and Range Experiment Station, University of Idaho, Moscow, Idaho, pp. 91–100.

Kareiva, P. (1982) Experimental and mathematical analyses of herbivore movement: quantifying the influence of plant spacing and quality on foraging discrimination. *Ecological Monographs* 52, 261–282.

Kolasa, J. and Pickett, S.T.A. (1991) *Ecological Heterogeneity*. Springer-Verlag, New York, 332 pp.

Kolasa, J. and Rollo, C.D. (1991) Introduction: the heterogeneity of heterogeneity: a glossary. In: Kolasa, J. and Pickett, S.T.A. (eds) *Ecological Heterogeneity*. Springer-Verlag, New York, pp. 1–23.

Laca, E.A. (1998) Spatial memory and food searching mechanisms of cattle. *Journal of Range Management* 51, 370–378.

Laca, E.A. and Ortega, I.M. (1996) Integrating foraging mechanisms across spatial and temporal scales. In: West, N.E. (ed.) *Fifth International Rangeland Congress*. Society for Range Management, Salt Lake City, Utah, pp. 129–132.

Laca, E.A., Ortega, I.M. and Soca, P. (1997) Controlling diet selection of sheep by restricting eating time. In: *50th Annual Meeting, Society for Range Management*. Society for Range Management, Denver, Colorado and Rapid City, South Dakota, p. 30.

Lauenroth, W.K., Milchunas, D.G., Dodd, J.L., Hart, R.H., Heitschmidt, R.K. and Rittenhouse, L.R. (1994) Effects of grazing on ecosystems of the Great Plains. In: Vavra, M., Laycock, W.A. and Pieper, R.D. (eds) *Ecological Implications of Livestock Herbivory in the West*. Society for Range Management, Denver, pp. 69–100.

Li, H. and Reynolds, J.F. (1994) A simulation experiment to quantify spatial heterogeneity in categorical maps. *Ecology* 75, 2446–2455.

Maschinski, J. and Whitham, T.G. (1989) The continuum of plant responses to herbivory: the influence of plant association, nutrient availability, and timing. *American Naturalist* 134, 1–19.

Milne, B.T., Turner, M.G., Wiens, A.J. and Johnson, A.R. (1992) Interactions between the fractal geometry of landscapes and allometric herbivory. *Theoretical Population Biology* 41, 337–353.

Moloney, K.A. and Levin, S.A. (1996) The effects of disturbance architecture on landscape-level population dynamics. *Ecology* 77, 375–394.

Mueggler, W.F. (1972) Influence of competition on the response of bluebunch wheat grass to clipping. *Journal of Range Management* 25, 88–92.

Noy-Meir, I. (1996) The spatial dimensions of plant–herbivore interactions. In: West, N.E. (ed.) *Fifth International Rangeland Congress*. Society for Range Management, Salt Lake City, Utah, pp. 152–154.

O'Neill, R.V., Gardner, R.H., Milne, B.T., Turner, M.G. and Jackson, B. (1991) Heterogeneity and spatial hierarchies. In: Kolasa, J. and Pickett, S.T.A. (eds) *Ecological Heterogeneity*. Springer-Verlag, New York, pp. 85–96.

Paige, K.N. and Whitham, T.G. (1987) Overcompensation in response to mammalian herbivory: the advantage of being eaten. *American Naturalist* 129, 407–416.

Palmer, M.W. (1988) Fractal geometry: a tool for describing spatial patterns of plant communities. *Vegetatio* 75, 91–102.

Palmer, M.W. (1992) The coexistence of species in fractal landscapes. *American Naturalist* 139, 375–397.

Provenza, F.D. (1995) Postingestive feedback as an elementary determinant of food selection and intake in ruminants. *Journal of Range Management* 48, 2–17.

Provenza, F.D. and Balph, D.F. (1987) Diet learning by domestic ruminants: theory, evidence and practical implications. *Applied Animal Behaviour Science* 18, 211–232.

Rees, M., Grubb, P.J. and Kelly, D. (1996) Quantifying the impact of competition and spatial heterogeneity on the structure and dynamics of a four-species guild of winter annuals. *American Naturalist* 147, 1–32.

Rice, K.J. (1987) Interaction of disturbance patch size and herbivory in *Erodium* colonization. *Ecology* 68, 1113–1115.

Rice, K.J. (1989) Competitive interactions in California annual grasslands. In: Huenneke, L.F. and Mooney, H. (eds) *Grassland Structure and Function: California Annual Grassland*. Kluwer Academic Publishers, Dordrecht, The Netherlands, pp. 59–71.

Robertson, G.P. and Gross, K.L. (1994) Assessing the heterogeneity of belowground resources: quantifying pattern and scale. In: Caldwell, M.M. and Pearcy, R.W. (eds) *Exploitation of Environmental Heterogeneity by Plants. Ecophysiological Processes Above- and Belowground*. Academic Press, San Diego, California, pp. 237–253.

Schmida, A. and Ellner, S.P. (1984) Coexistence of plants with similar niches. *Vegetation* 65, 163–173.
Senft, R.L. (1989) Hierarchical foraging models: effects of stocking and landscape composition on simulated resource use by cattle. *Ecological Modelling* 46, 283–303.
Senft, R.L., Coughenour, M.B., Bailey, D.W., Rittenhouse, L.R., Sala, O.E. and Swift, D.M. (1987) Large herbivore foraging and ecological hierarchies: landscape ecology can enhance traditional foraging theory. *BioScience* 37, 789–799.
Silvertown, J., Holtier, S., Johnson, J. and Dale, P. (1992) Cellular automaton models of interspecific competition for space – the effect of pattern on process. *Journal of Ecology* 80, 527–534.
Staddon, J.E.R. and Ettinger, R.H. (1989) *Learning. An Introduction to the Principles of Adaptive Behaviour*, 1st edn. Harcourt Brace Jovanovich, San Diego, California, 436 pp.
Tilman, D. (1994) Competition and biodiversity in spatially structured habitats. *Ecology* 75, 2–16.
Tilman, D. (1996) Biodiversity – population versus ecosystem stability. *Ecology* 77, 350–363.
Tilman, D., Wedin, D. and Knops, J. (1996) Productivity and sustainability influenced by biodiversity in grassland ecosystems. *Nature* 379, 718–720.
Turner, M.G. and Gardner, R.H. (1991a) *Quantitative Methods in Landscape Ecology*. Springer-Verlag, New York, 536 pp.
Turner, M.G. and Gardner, R.H. (1991b) Quantitative methods in landscape ecology: an introduction. In: Turner, M.G. and Gardner, R.H. (eds) *Quantitative Methods in Landscape Ecology*. Springer-Verlag, New York, pp. 3–14.
Vinton, M.A. and Hartnett, D.C. (1992) Effects of bison grazing on *Andropogon gerardii* and *Panicum virgatum* in burned and unburned tallgrass prairie. *Oecologia* 90, 374–382.
Vinton, M.A., Hartnett, D.C., Finck, E.J. and Briggs, J.M. (1993) Interactive effects of fire, bison (*Bison bison*) grazing and plant community composition in tallgrass prairie. *American Midland Naturalist* 129, 10–18.
Walker, J., Sharpe, P.J.H., Penridge, L.K. and Wu, H. (1989) Ecological field theory – the concept and field tests. *Vegetatio* 83, 81–95.
WallisDeVries, M.F., Laca, E.A. and Demment, M.W. (1999) The importance of scale of patchiness for selectivity in grazing herbivores. *Oecologia* 121, 355–363.
Wiens, J.A. (1989) Spatial scaling in ecology. *Functional Ecology* 3, 385–397.
With, K.A. (1994) Using fractal analysis to assess how species perceive landscape structure. *Landscape Ecology* 9, 25–36.

12

Defoliation Patterns and Herbage Intake on Pastures

M.H. Wade[1] and P.C. de F. Carvalho[2]

[1] *Facultad de Ciencias Veterinarias, UNCPBA, Tandil, Argentina;* [2] *Facultade de Agronomia, UFRGS, Porto Alegre, Brazil*

Introduction

This chapter is concerned with the herbage intake of grazing ruminants and the defoliation patterns that result from the interaction between the sward and the animal's mouth. In the early history of grassland research, many assumptions were made about defoliation patterns and their role in determining animal productivity. It was assumed for instance that, under continuous stocking, defoliation was so frequent as to reduce herbage production and therefore animal production. Work during the 1950s indicated that this was mostly not the case and that animal production levels were similar in rotational and continuous stocking (McMeekan and Walshe, 1963). Subsequently, when more detailed studies were carried out, it was found that intervals between defoliations under continuous stocking were not so different from those under rotational stocking (Hodgson, 1966; Hodgson and Ollerenshaw, 1969) and probably not sufficient to have a large effect upon herbage production. Meanwhile, following the paper of Mott (1960), emphasis was being placed on stocking rate, rather than grazing method, as being the principal determinant of animal productivity.

The factors affecting herbage production and animal production under humid temperate conditions continued to be studied under cutting and under rotational grazing up until the beginning of the 1980s. Since that time, there would appear to have been a predominance of grazing studies carried out under continuous stocking or under controlled conditions similar to continuous stocking conditions. This has been associated, on the one hand, with considerable advances in the technology of grazing management, particularly the use of sward height to control both sward growth and herbage intake (Hodgson, 1990). On the other hand, headway has also

 Grassland Ecophysiology and Grazing Ecology (eds G. Lemaire, J. Hodgson, A. de Moraes, C. Nabinger and P.C. de F. Carvalho)

been made in the study of the actual mechanisms involved in the act of defoliation (Ungar, 1996). By their nature, these studies tend to be very short-term, with the result that they are more relevant to a continuous than to a rotational stocking situation. While at an empirical level the responses to both grazing method and stocking rate are now apparently well known, at the level of the actual mechanisms involved there is still considerable work to be done. If the productivity of grazing systems is to be made more predictable, these mechanisms need to be better understood within the context of those systems.

The purpose of this chapter is to explore the interplay of pattern of defoliation with levels of herbage production and utilization, on the one hand, and individual daily herbage intake and animal performance, on the other. The emphasis will be placed upon intensively grazed temperate pasture systems, because most of the relevant information derives from these; nevertheless, the principles involved should be applicable to rangelands.

The Effects of Defoliation Upon Herbage Grown and Harvested

There is general agreement that the decline in individual herbage intake and animal performance with increasing stocking rate is due to a decrease in the quantities of herbage available per animal for consumption. The herbage mass, along with other associated characteristics, such as height, of the pasture that is presented instantaneously to the animal influences the current level of intake. This, in turn, is the result of the balance between herbage growth, consumption and decomposition (net accumulation or net loss) occurring at that stocking rate (Bircham and Hodgson, 1983).

The effects of cutting and grazing defoliation upon herbage production has been given great emphasis in research over the last 50 years. However, it is true to say that, although the effects of varying the interval between defoliations were clear, the effects of severity of cutting were more equivocal (Davidson, 1969). The shorter the period between defoliations, the greater the reduction in the quantity harvested (Marsh, 1976), and, in apparent agreement, many results showed reduced production with increasing severity of cutting (Jameson, 1963). Nevertheless, many others showed increased production with increased severity (i.e. lower cutting height), an apparent paradox that was difficult to explain. The principal problem was one of distinguishing between herbage accumulated and that harvested. The essential analysis that is necessary is how much stocking rate affects the production of harvested herbage and how much that contributes to the observed responses of animal production, whether per animal or per unit area.

In their review of the literature, Hodgson and Wade (1978) concluded that, in general, variation of stocking rate had relatively small effects upon herbage production, probably insufficient to directly influence the amount of herbage harvested. Nevertheless, they further suggested that circumstances might occur in which defoliation at high intensity, possibly combined with adverse environmental conditions, could be sufficient to have such an effect (Morley, 1966; Wade, 1979). This possible spectrum of responses was hypothesized by Wade (1979) to explain the apparently contradictory effects of severity of cutting referred to above.

The situation was substantially clarified for a grazing situation by the work of Parsons *et al.* (1983), using sheep continuously stocked on perennial ryegrass. They estimated rates of CO_2 assimilation, growth, senescence and herbage intake at two levels of grazing intensity, which were defined by the approximate values of leaf area index (LAI) at which they were maintained. They were thus able to make estimates of gross growth and its partition between: (i) respiration and roots; (ii) intake per unit area by sheep; and (iii) loss to senescence of material not consumed, over the range of defoliation intensities applied. Because they had obtained a wide range of LAIs, they were able to propose general relationships between herbage growth and consumption, on the one hand, and intensity of defoliation, on the other (Fig. 12.1). On the left hand part of the graph in Fig. 12.1, the quantity harvested increases steadily as intensity of defoliation increases, while growth is slightly reduced; herbage harvested is therefore independent of any effect of defoliation upon growth. However, a point is reached where the effect on growth is such that it starts to affect the quantity of herbage harvested, which becomes increasingly dependent upon growth.

Parsons *et al.* (1983) drew a parallel between the shape of their curve for herbage harvested and that obtained by Mott (1960) for animal production per unit area response to increasing stocking rate. Further modelling work by Parsons *et al.* (1988) suggested that the same principle could be applied to the effect of

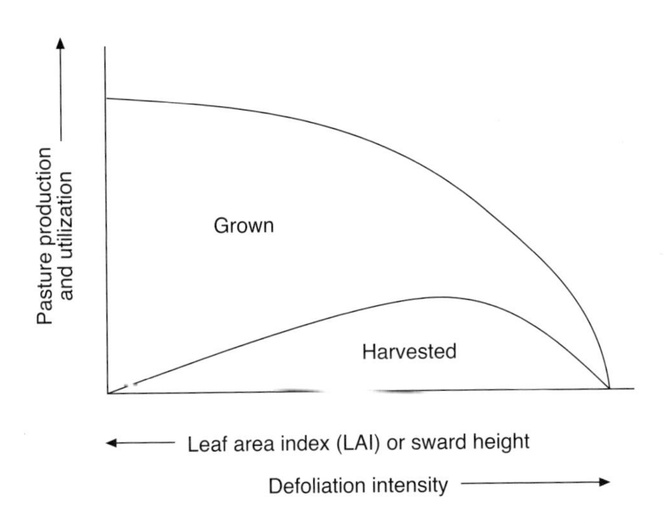

Fig. 12.1. The effect of a range of intensities of defoliation upon the net primary production of pasture and the amount removed by the grazing animal. Leaf area index (LAI) is used as an index of intensity of defoliation and decreases with increasing intensity (modified from Parsons *et al.*, 1983).

variation of severity of and interval between intermittent defoliations, thus providing a possible basis for comparing the two apparently distinct grazing methods. One important outcome of the work of Parsons *et al.* (1983), in addition to that of Bircham and Hodgson (1983), was the development of sward state, defined by LAI and sward height, as a means of controlling defoliation effects upon pasture growth in grazing management.

Herbage Intake

With the advent of new methods to estimate daily herbage intake, efforts to relate these estimates to levels of herbage on offer began. Greenhalgh *et al.* (1966) introduced an element of supply and demand by defining herbage intake and herbage on offer (daily herbage allowance) in the same terms (kg dry matter (DM) cow^{-1} day^{-1}). This was the reciprocal of the term grazing pressure, as defined by Mott (1960), but with a time element added. Greenhalgh *et al.* (1966) found an asymptotic relationship between daily herbage intake and daily herbage allowance. However, the approach was restricted in its usefulness to strip grazing and so, when, in the 1970s, interest started in continuous grazing, it could not be extended to these conditions. At the beginning of the 1980s, it was realized that LAI was related to sward height and that these parameters could be used as a common basis for managing both sward and animal performance (Hodgson, 1985a, 1990). Indeed, at this time, both stocking rate and daily herbage allowance tended to fall into disuse in research work, particularly in the UK, because they were considered too imprecise to be useful as variables, and they were replaced by 'sward state' (LAI and particularly sward height) as control variable. Since herbage intake is such a key process in the working of a grazing system, considerable emphasis will be placed upon certain aspects of the mechanics of the processes involved.

Sward factors affecting herbage intake

Defoliation at the bite level

In contrast to the treatment of the effects of defoliation upon growth and utilization of herbage, the sward and animal interactions determining herbage intake must be dealt with in terms of the individual animal. However, it should be borne in mind that, on a systems level, it is not only the individual animal that is important, but also the animal population on the area in question. It follows that the relevance of different effects upon individual animals needs eventually to be evaluated in the context of the stocking rate and the method of grazing used for an experiment or to obtain an observation.

Allden and Whittaker (1970) were the first to quantify the intake response to sward structure; they found that the longer the herbage being eaten, the larger the mass prehended per bite and the level of daily herbage intake. In short, temperate grass sward height is considered to be the major factor determining daily herbage intake; hence its usefulness as a management tool. Nevertheless, in terms of the mechanics of intake, Penning *et al.* (1994) have shown the situation to be a little more complicated than that. In a detailed study of sward factors and components of animal feeding behaviour, they found that, in sheep grazing under continuous and rotational stocking conditions, intake responded more to the green leaf mass than to sward height.

The importance of leaf in this respect had been detected in Australian work by Arnold and Dudzinski (1967). Stobbs, also in Australia, made considerable advances in trials in a tropical environment (Stobbs, 1973a, b; Chacon and Stobbs, 1976). These trials established the importance of the density of green leaf lamina in the sward in determination of bite mass, intake rate and daily herbage intake. In addition, Chacon and Stobbs (1976) established a relationship between the eating behavioural responses of dairy cows and the proportion of leaf material in the pasture. They noted that, when the amount of leaf was reduced, the amount of night grazing was also reduced and concluded that light was necessary to permit animals to see the residual leaf in order to eat it. To test for the factors influencing the response to the presence or absence of leaf, they emptied the rumens of cows on pastures with different amounts of leaf present. Those faced with a high proportion of leaf dedicated significantly more time to attempting to refill the rumen. These results imply some sort of recognition, possibly visual, of the presence or absence of leaf within tropical grass pastures, which characteristically have a low density of leaf material (Stobbs, 1973b).

A similar response to the proportion of live leaf material was not detected by Hodgson (1981) in his interpretation of the trials reported by Jamieson and Hodgson (1979a, b) for temperate grass swards, which were shorter and of higher bulk density than the tropical swards studied by Stobbs. As a result, he concluded that the height of the sward above ground level had a greater influence upon intake rate than did the proportion or density of leaf material present. Following the observation (Barthram and Grant, 1984) that sheep did not appear to graze into the sheath layer of a vegetative perennial ryegrass sward, Hodgson (1981) further speculated that bite depth may be a variable that determines the reduction in intake rate as height is reduced. Subsequent, more detailed work has confirmed that bite depth is very closely related to sward height (e.g. Mursan *et al.*, 1989; Burlison *et al.*, 1991), even when animals penetrate the sheath layer, as found by Wade *et al.* (1989). These authors found that defoliation depth was a constant 34% of the extended tiller height (ETH) during grazing down of a perennial ryegrass sward over 5 days from the top layer, where only 10% of marked tillers which were defoliated were severed at the sheath, to the last day, when the figure was 80%. Using other data, Wade (1991) concluded, with Barthram and Grant (1984), that depth of the leaf layer was important in determining daily herbage intake, but he could not explain the apparent constancy of bite depth in relation to sward height, assuming that it could be modified by the animal in response to the presence of the sheath below the leaf layer.

This method of trying to detect variation of bite depth during grazing down had been used because the literature was not conclusive on the mechanisms involved. Subsequent controlled work, such as that of Flores *et al.* (1993), could not detect any effect of vegetative sheath on intake rate in hand-constructed swards of *Paspalum dilatatum*, while cut inflorescent stems did act as a barrier. The constancy of bite depth in any given sward seems to suggest a passive response to sward height, rather than the animal's ability to control it according to the presence or absence of sheath. This would leave bite area as the key dimensional variable influencing intake per bite, but this is much more difficult to determine experimentally.

The effect of cut inflorescent stem upon bite depth, observed in cattle by Flores *et al.* (1993), would be similar to the commonly observed effect of wheat stubble: where stubble is dense no pasture below the combine cutting height is generally grazed by cattle. This is presumably due to the direct effect of the hard stubble upon the sensitive muzzle, rather than resistance to a grazing bite, as suggested by Hodgson *et al.* (1994). In the case of vegetative swards, there is evidence of penetration of the sheath layer in the field: Barthram and Grant (1984) showed a loss of approximately 30% of the sheath height during grazing down by sheep and Wade (1991) also found up to 40% of the sheath height of marked tillers reduced in strip grazing by dairy cows. In the latter case, penetration varied with both daily herbage allowance and pregrazing sward height: the lower the daily herbage allowance and the higher the pregrazing height, the greater the degree of penetration; in both cases, between about 20 and 40% was removed. This penetration would therefore be accidental and easily explained in terms of the intimate way leaf lamina and sheath are mixed at the boundary between the two layers. The mechanism that suggests itself is that the 'barrier' occasioned by the sheath in vegetative swards is due to the absence of leaf in the grazed horizon, rather than the presence of sheath at the level of incision.

Daily grazing time

Grazing time per day has a relatively small impact upon daily herbage intake. However, since Chacon and Stobbs (1976) and Stobbs (1977) have suggested that there may be a response of grazing time to the presence or absence of leaf in the grazed horizon, this component of grazing behaviour will be examined from this point of view and in the context of temperate pastures. Few studies in which grazing time was estimated have distinguished between activity during day and night. One of these was Baker *et al.* (1981a), in which cows with calves in daily strips were found to graze less at low daily herbage allowance and most of this difference was at night. Under strip grazing at low daily herbage allowance, it often happens that, before the strip is changed, nearly all of the leaf material is eaten. Jamieson and Hodgson (1979a) observed that animals grazing daily strips had marginally less grazing time under low daily herbage allowance than those at higher daily herbage allowance. In another experiment with continuous grazing, they obtained greater grazing time with lower sward heights – that is, at higher grazing pressure (Jamieson and Hodgson, 1979b). The explanation offered by

Jamieson and Hodgson (1979a) for the reduced grazing time in daily strips at high grazing pressure was that the animals preferred to wait till the next strip of fresh grass. However, in the original observation of reduced grazing time in the paddocks, studied by Chacon and Stobbs (1976), the lower level was maintained for several days, so there can be no question of such anticipation; therefore, it would appear to have been a response to sward conditions.

A superficial analysis of the phenomenon reveals that, under strip grazing, there is generally very little response to daily herbage allowance in terms of grazing time (Stobbs, 1977; Combellas and Hodgson, 1979; Le Du *et al.*, 1979; Baker *et al.*, 1981a). Typically, grazing time at low allowance was 10% less than at medium and high levels. However, under continuous stocking, where swards adapt to higher grazing pressures by dense tillering and maintenance of considerable quantities of leaf, there appears to be considerable variation of grazing time (Allden and Whittaker, 1970; Baker *et al.*, 1981b; Hodgson, 1985b; Penning *et al.*, 1991), with up to 2 to 3 h greater grazing time at higher stocking rates than at lower stocking rates (i.e. taller pastures).

Therefore, the possibility needs to be explored that this is another example of a direct behavioural response to the presence of leaf material, even at low sward heights. If it is indeed a response to a predominantly visual cue, as Chacon and Stobbs (1976) and Stobbs (1977) have proposed, it would add weight to our suggestion that the sheath 'barrier' in short, temperate, grass swards is due to an absence of leaf in the grazed horizon and not necessarily directly to differences between leaf lamina and leaf sheath in terms of shearing strength. There is ample evidence of the potential for the use of vision in feed acquisition for both sheep and cattle at a distance (Edwards *et al.*, 1997; Uetake and Kudo, 1994; Laca and Ortega, 1995), though the gustatory, olfactory and tactile senses are generally considered to be stronger overall for feed recognition and harvesting action, particularly at closer quarters (Arnold, 1966; Krueger *et al.*, 1974; Provenza, 1995). Such visual recognition of leaf material could be explained by postdigestive or postruminal effects: the material might be recognized visually as being easily comminuted and therefore to be favoured, from previous experience, over sheath or stem material, which normally is not (Provenza, 1995).

Intake and systems efficiency

At the conclusion of his review of the mechanisms involved in ingestive behaviour, Ungar (1996) questioned whether the new methodologies of microswards and links between ecological and agricultural research disciplines would translate into improvements in grazing management. Above all, he questioned the value of descending down spatial and temporal scales, if there was no benefit in a 'more agriculturally useful synthesis'. This is partly because of the complexity of the processes involved and the different scales upon which they act. But it is worth noting that his review barely touches upon the concepts of grazing pressure, stocking rate or grazing method, so that mechanisms were discussed largely in the absence of reference to limitations to intake, progressive or static, which may be considered the result of variation of grazing pressure or method.

Demment *et al.* (1987) were clear about the basis for their subsequent series of studies of sward factors affecting herbage intake mechanisms: the interface is where plant production is harvested by the animal and therefore this is the 'process that may well control the behaviour of the system'. Therefore, at the system level, efficiency cannot be evaluated without consideration of the interplay between daily herbage intake per animal and intake per unit area. A desired level of production per animal will depend upon the level of pasture production and the relative ease with which this can be harvested. From the Mott (1960) curves, we know that the possibility of optimizing the balance between animal production per hectare and per animal exists. What is needed is the means of predicting and manipulating different processes on a day-to-day basis.

Unfortunately, one of the biggest obstacles to the pursuance of these objectives is still the lack of adequate techniques to measure short-term and even long-term levels of intake: nearly all require uniformity from day to day and so only daily strips or continuous stocking of relatively uniform swards can be used. The importance of the development of techniques to estimate intake variation from day to day cannot be overemphasized: it is essential for a detailed analysis of the sward and animal factors that determine herbage intake. It is this lack of adequate techniques that makes it even more important to develop a common and workable conceptual framework within which to integrate different levels of the system.

Defoliation Patterns in Grazed Pastures

The act of defoliation has long been recognized as the probable key to understanding the plant–animal interface in relation to intake and sward responses (Voisin, 1959; Spedding, 1965; Hodgson, 1985a). Nevertheless, the detailed study of defoliation patterns has been largely descriptive, with only limited efforts to integrate the interaction of grazing methods, stocking rates, herbage intake and sward responses. Up till now, the patterns of grazing have not been considered in terms of the dynamics of the severity and frequency of individual defoliations. Intensity of defoliation was implicit, but not quantified in any detail: it was expressed either as variation of LAI or in terms of how it related to ease of ingestion and to bite dimensions. In this section, the factors controlling defoliation patterns will be explored, first of all under continuous stocking conditions and then in the context of a wider analysis of the roles of grazing method and stocking rate.

Defoliation of individual tillers

Work under continuous stocking conditions (Wade and Baker, 1979) confirmed the close relationship between defoliation interval and stocking density found by Hodgson (1966) and Hodgson and Ollerenshaw (1969). At stocking densities

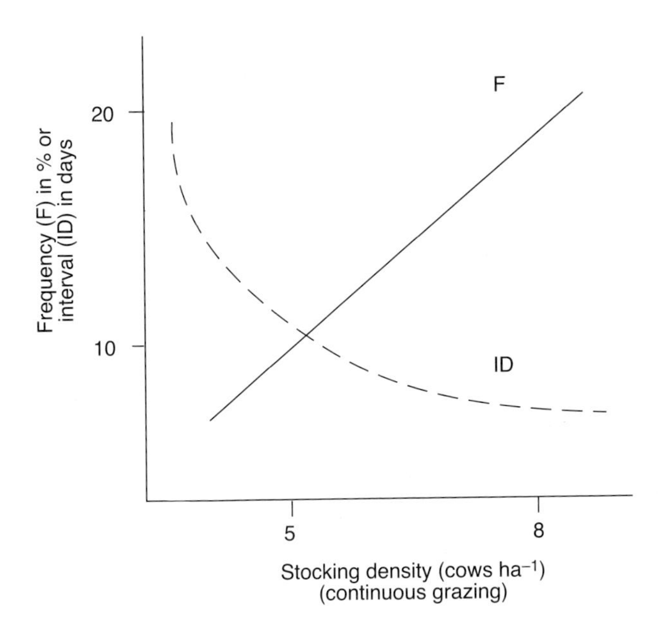

Fig. 12.2. The effect of stocking rate (and stocking density) upon the frequency of defoliation (F) expressed as the mean percentage of marked tillers defoliated per day and from which the interval between defoliations (ID) was calculated (modified from Wade, 1991).

normally found on farms, the interval was in the region of 14 to 21 days, which is comparable to many rotational systems. Further, Wade and Baker (1979) were able to collect sufficient of their own data to fit a negative exponential relationship to the relationship between defoliation interval and stocking density of cattle. The interval between defoliations estimated at continuous stocking rates between two and eight dairy cows ha^{-1} varied between about 18 and 5 days (Fig. 12.2). When data from sheep and cattle experiments were compared on a live-weight stocking density basis, a difference of about 5 days was found; however, when data were combined in a metabolic-weight expression of stock density, a similar relationship was found. This indicates that the interval was related to the level of intake per unit area, rather than to any intrinsic animal species difference. The technique used by these authors was simpler than that used by Hodgson (1966), in that the extended height of marked individual tillers (ETH) was measured, not the length of each green leaf lamina; nevertheless, defoliation events were clearly identifiable. Thus the severity of each defoliation was estimated as a reduction of ETH, which would equate to bite depth, as described previously.

The technique used by Hodgson (1966) and Hodgson and Ollerenshaw (1969) was to mark individual tillers and measure the length of each green leaf thereon at intervals of 2 or 3 days. Thus they were able to estimate not only the interval between defoliations, but also the severity of each defoliation on a leaf-by-leaf basis. They showed that the probability of defoliation tended to be greater in the higher-positioned plants and leaves in the sward. These authors used simple pastures of *Lolium perenne*, but, since that time, there have been a number of studies of defoliation patterns in more complex pastures. One of the most notable of these was that of Clark *et al.* (1984), who followed the variations in interval and severity of defoliation of three species that dominated a hill pasture in New Zealand, managed under continuous and rotational grazing during a complete year. Although stocking rates were not given, it is clear that there was considerable variation between the pasture species in terms of the interval between defoliations under continuous grazing. In general, *L. perenne* (10 to 21 days) was grazed more frequently than *Agrostis* spp. (19 to 50 days) and these, in turn, more than *Trifolium repens* (20 to 90 days). Clark *et al.* (1984) also combined the data from the two contrasting grazing methods to obtain intensity of defoliation (expressed as the total leaf grazed in millimetres per tiller) and a spectrum of its components, the frequency and severity of grazing. In this way, the intensities of the defoliations in the different species and managements could be compared, even though they were generated by large differences in frequency and severity in the different methods of grazing.

The factors that lead to such differences in defoliation interval between the different species in the same sward are far from clearly understood (Hodgson *et al.*, 1994). Relative height within the pasture may have an important role (Gammon and Roberts, 1978; Clark and Harris, 1985). However, leaf material may be removed from below the level of stem and inflorescence where contrasts in morphology are great enough (Gardener, 1980; L'Huillier *et al.*, 1984; Grant *et al.*, 1985; Forbes, 1988; Carvalho *et al.*, 1999). Thus, in cases such as that of *Cynodon dactylon* grazed by cattle (Forbes, 1988), the inflorescence is too small or does not have the characteristics that cause avoidance in other species, such as *Bromus catharticus* (Forbes, 1988) or the savannah species studied by Gammon and Roberts (1978). Selection between *T. repens* and *L. perenne* plants is also affected by horizontal distribution, as well as by the relative heights of the two species (Clark and Harris, 1985; Ridout and Robson, 1991).

There are internal animal factors that affect selection, independently of the height of the different plant species or morphological units (Provenza, 1995). For instance, there is evidence from microswards and stall feeds that animals demonstrate preference for the items that will yield the fastest intake rate. However, this is not always the case in the field: Parsons *et al.* (1994) found that sheep concentrated grazing on white clover in the morning, but on perennial ryegrass in the evening. Possible explanations included the shorter rumination time needed for clover, leaving more time for grazing during the day, when nearly all grazing is done. Another was the possible need to maintain a diverse rumen flora, as sheep can perform very well on white clover alone, if necessary. There were also differences between starved and lactating ewes in selectivity. To quote Edwards *et al.* (1997),

> Indeed, it is sobering to note that the wealth of information collected on the flexibility of the response of herbivores to their foraging environment makes it all the more difficult to predict dynamically how much animals will eat and from which species.

They add that 'this remains a major impediment to understanding the impact of herbivores on vegetation dynamics'. Nevertheless, in spite of this apparently complex situation, it may be possible to impose some level of order upon the broad effects of grazing method and stocking rate.

Defoliation patterns and grazing methods

The same technique that was used by Wade and Baker (1979) under continuous grazing was later used in France to study the process of grazing down in 5-day paddocks and in daily strips by dairy cows at one and three levels of herbage allowance, respectively (Wade, 1991; Peyraud *et al.*, 1996). When the results from these studies were combined with those from the previous continuous-stocking work (Wade and Baker, 1979), it became immediately apparent that defoliation frequency was a more appropriate index to use than its reciprocal, defoliation interval (see Fig. 12.2), because there was a continuous relationship between instantaneous stocking rate and defoliation frequency across the two sets of studies (Wade, 1991; since partially reported by Lemaire and Chapman, 1996; their Fig. 1.5). Both the UK and the French studies were carried out on perennial ryegrass swards.

Frequency of defoliation of individual tillers also gives an estimate of the area harvested in a day: it is the proportion of the total area available that is defoliated, and which, when multiplied by the area, gives an estimate of the area (m^2) covered by the cows in a day. This figure divided by the number of cows gives an estimate of the absolute area grazed per cow per day. Estimates for continuous grazing were about 200 m^2, which agree with figures obtained in New Zealand by estimating bite area and multiplying it by the number of bites per day (C.W. Holmes, 1995, personal communication). In the 5-day paddocks, the area grazed per cow per day was estimated to fall from about 200 m^2 to 100 m^2 (Wade *et al.*, 1989). As mentioned previously, the mean depth of defoliation was found to be the same at the beginning and at the end of the 5 days – grazing approximately 34% of the extended height.

These estimates of area and depth of grazing per day gave rise to the idea of estimating the total area covered in a day in the daily strips, with their much higher instantaneous stocking rate. This could be done because the initial and final ETH had been measured in the strip grazing treatments at three levels of herbage allowance: the number of times the area was grazed was calculated from the number of times the tiller had been reduced in height by 34%. This apparent continuum of combinations of approximate area grazed per day and instantaneous stock densities provides a basis for comparison between all possible combinations of grazing methods and system stocking rates.

Such a basic dynamic structure may be a useful starting-point for drawing together the animal and plant community variables that determine the efficiency of grazing management. For example, it has been shown that it is possible to quantify, to some degree, the dynamics of defoliation in a grazed grass sward. This appears to provide a functional framework for discussing many phenomena related to grazing management. Nevertheless, we are very aware that the situation in which this was developed is a simple one and that the majority of grazing situations are more complex, as shown in the previous section and discussed more fully by Gordon (Chapter 10, this volume) and Laca (Chapter 11, this volume).

Even in pastures of one or two species the formation of grazed patches at low stocking rates is often associated with selectivity (Gibb and Ridout, 1988; Parsons *et al.*, Chapter 15, this volume). The reason can be derived from the quantification of area grazed, discussed above. The areas of the sward that cannot be physically covered by the limited numbers of animals present (or which are close to dung) will develop to be areas less preferred than those grazed 20 to 40 days before. Therefore, in spite of their greater height, it is the shorter areas of the sward which will be preferred and which will be grazed. An interesting aspect of the frequency/stock density relationship is that the effective stocking density on any given part of such a heterogeneous pasture may be estimated from frequency of defoliation measurements. It is unfortunate that there has been a tendency in the literature to refer to 'grazing' without clarifying the method of grazing used or to the method of grazing without specifying grazing pressure or stocking rate, and there is often a failure to mention the size and number of animals used.

Defoliation Patterns and Animal Performance

In the previous section, the dominant factors determining defoliation patterns were seen to be method of grazing and stocking rate. In the context of intensive systems since the 1970s, there has been an increasing tendency to regard the differences between grazing methods as being minimal in terms of animal performance. This has been based upon sound empirical observations in both research and practical situations. Nevertheless, the fact remains that direct comparisons of grazing methods over a wide range of stocking rates have been extremely rare. McMeekan and Walshe (1963), with milk production, and Conway (1963) and O'Sullivan (1984), with beef production, found evidence for an interaction of method and stocking rate: in both cases, performance was reduced under continuous stocking, but only at the higher stocking rate. Bransby (1993) compared the performance of beef animals at three levels of stocking rate and again obtained considerably reduced production per animal at the highest stocking rate under continuous stocking.

These interactions may be due to the extra energy expenditure of animals under continuous stocking, as O'Sullivan (1984) suggested. An additional explanation may lie in the variation of defoliation patterns between methods. While the analysis of Parsons *et al.* (1988) suggests similarity between methods in terms of sward responses, using LAI as the index of intensity of defoliation may obscure real differences in the effects of defoliation intensity upon sward growth resulting from similar stocking rates in the differing methods. Thus part of the observed interaction between methods may be due to effects upon herbage growth, which the Parsons *et al.* (1988) model might not take into account.

The importance of these speculations is to be found principally in the approaches taken to the investigation of grazing systems. Considerable functional differences between grazing methods have been indicated in this chapter. If it is accepted that it is important to study the key mechanisms of herbage intake and production, particularly at high levels of herbage utilization, it must also be important to be clear about these differences when designing experimental programmes and interpreting and extrapolating results. In conclusion, it must be said that there still exist large gaps in our knowledge of the details of defoliation dynamics. This is an indication that, in spite of the many exhortations for more emphasis on the plant–animal interface in grazing studies, there may in fact be some way to go for a true integration of the disciplines to occur.

References

Allden, W.G. and Whittaker, I.A. (1970) The determinants of herbage intake by grazing sheep: the interrelationship of factors influencing herbage intake and availability. *Australian Journal of Agricultural Research* 21, 755–766.

Arnold, G.W. (1966) The special senses in grazing animals. I. Sight and dietary habits in sheep. *Australian Journal of Agricultural Research* 17, 521–529.

Arnold, G.W. and Dudzinski, M.L. (1967) Studies on the diet of the grazing animal. III. The effects of pasture species and pasture structure on the herbage intake of sheep. *Australian Journal of Agricultural Research* 18, 657–666.

Baker, R.D., Alvarez, F. and Le Du, Y.L.P. (1981a) The effect of herbage allowance upon the intake and performance of suckler cows and calves. *Grass and Forage Science* 36, 189–199.

Baker, R.D., Le Du, Y.L.P. and Alvarez, F. (1981b) The herbage intake and performance of set-stocked suckler cows and calves. *Grass and Forage Science* 36, 201–210

Barthram, G.T. and Grant, S.A. (1984) Defoliation of ryegrass-dominated swards by sheep. *Grass and Forage Science* 39, 211–219.

Bircham, J.S. and Hodgson, J. (1983) The influence of sward conditions on rates of herbage growth and senescence in mixed swards under continuous stocking management. *Grass and Forage Science* 38, 323–331.

Bransby, D.I. (1993) Interactions of rotational and continuous grazing with stocking rate on warm and cool-season pastures. In: *Proceedings of the 17th International Grassland Congress*. SIR Publishing, Wellington, New Zealand, pp. 1285–1286.

Burlison, A.J., Hodgson, J. and Illius, A.W. (1991) Sward canopy structure and the bite dimensions and bite weight of grazing sheep. *Grass and Forage Science* 46, 29–38.

Carvalho, P.C. de F., Prache, S., Roguet, C. and Louault, F. (1999) Defoliation process by ewes of reproductive compared to vegetative swards. In: *Proceedings of the 5th International Symposium on the Nutrition of the Herbivore, San Antonio, Texas, 11–16 April 1999*. http://cnrit.tamu.edu/conf/isnh/post-online/post0019/

Chacon, E. and Stobbs, T.H. (1976) Influence of progressive defoliation of a grass sward on the eating behaviour of cattle. *Australian Journal of Agricultural Research* 27, 709–727.

Clark, D.A. and Harris, P.S. (1985) Composition of the diet of sheep grazing swards of differing white clover content and spatial distribution. *New Zealand Journal of Agricultural Research* 28, 233–240.

Clark, D.A., Chapman, D.F., Land, C.A. and Dymock, N. (1984) Defoliation of *Lolium perenne* and *Agrostis* spp. tillers and *Trifolium repens* stolons in set-stocked and rotationally grazed hill pastures. *New Zealand Journal of Agricultural Research* 27, 289–301.

Combellas, J. and Hodgson, J. (1979) Herbage intake and milk production by grazing dairy cows. 2. The effects of variations in herbage mass and daily herbage allowance in a short-term trial. *Grass and Forage Science* 34, 209–214.

Conway, A. (1963) Effect of grazing management on beef production. II Comparisons of three stocking rates under two systems of grazing. *Irish Journal of Agricultural Research* 2, 243–258.

Davidson, J.L. (1969) Growth of grazed plants. In: *Proceedings of the Australian Grasslands Conference*, Vol. 2, pp. 125–137.

Demment, M.W., Laca, E.A. and Greenwood, G.B. (1987) Forage ingestion in grazing ruminants: a conceptual framework. In: *Proceedings of a Symposium on Feed Intake*, Agricultural Experimental Station, Oklahoma State University, pp. 208–225.

Edwards, G.R., Newman, J.A., Parson, A.J. and Krebs, J.R. (1997) Use of cues by grazing animals to locate food patches: an example with sheep. *Applied Animal Behaviour Science* 51, 59–68.

Flores, E.R., Laca, E.A., Griggs, T.C. and Demment, M.W. (1993) Sward height and vertical morphological differentiation determine cattle bite dimensions. *Agronomy Journal* 85, 527–532.

Forbes, T.D.A. (1988) Researching the plant–animal interface: the investigation of ingestive behaviour in grazing animals. *Journal of Animal Science* 66, 2369–2379.

Gammon, D.M. and Roberts, B.R. (1978) Patterns of defoliation during continuous and rotational grazing of the Matopos Sandveld of Rhodesia. *Rhodesia Journal of Agricultural Research* 16, 147–164.

Gibb, M.J. and Ridout, M.S. (1988) Application of double normal frequency distributions fitted to measurements of sward height. *Grass and Forage Science* 43, 131–136.

Grant, S.A., Suckling, D.F., Smith, H.K., Torvill, L., Forbes, T.D.A. and Hodgson, J. (1985) Comparative studies of diet selection by sheep and cattle. *Journal of Ecology* 73, 987–1004.

Greenhalgh, J.F.D., Reid, G.W., Aitken, J.N. and Florence, E. (1966) The effects of grazing intensity on herbage consumption and animal production. 1. Short-term effects in strip-grazed dairy cows. *Journal of Agricultural Science, Cambridge* 67, 13–23.

Hodgson, J. (1966) The frequency of defoliation of individual tillers in a set-stocked sward. *Journal of the British Grassland Society* 27, 258–263.

Hodgson, J. (1981) Variations in the surface characteristics of the sward and the short term rate of herbage intake by calves and lambs. *Grass and Forage Science* 36, 49–57.

Hodgson, J. (1985a) The significance of sward characteristics in the management of temperate sown pastures. In: *Proceedings of the 15th International Grassland Congress*. The Science Council of Japan, Nishi Nasuno, Japan, pp. 31–34.

Hodgson, J. (1985b) The control of herbage intake in the grazing ruminant. *Proceedings of the Nutrition Society* 44, 339–346.

Hodgson, J. (1990) *Grazing Management: Science into Practice*. Longman Scientific and Technical, Harlow, UK, 203 pp.

Hodgson, J. and Ollerenshaw, J.H. (1969) The frequency and severity of defoliation of individual tillers in set-stocked swards. *Journal of the British Grassland Society* 24, 226–234.

Hodgson, J. and Wade, M.H. (1978) Grazing management and herbage production. In: *Proceedings of the 1978 Winter Meeting of the British Grassland Society*. British Grassland Society, Hurley, UK, 1.1–1.12.

Hodgson, J., Clark, D.A. and Mitchell, R.J. (1994) Foraging behaviour in grazing animals and its impact on plant communities. In: Fahey, C.G. (ed.), *Forage Quality, Evaluation, and Utilization*. ASA–CSSA–SSSA, Madison, Wisconsin, pp. 796–827.

Jameson, D.A. (1963) Responses of individual plants to harvesting. *Botanical Review* 29, 532–594.

Jamieson, W.S. and Hodgson, J. (1979a) The effect of daily herbage allowance and sward characteristics upon the ingestive behaviour and herbage intake of calves under strip-grazing management. *Grass and Forage Science* 34, 261–271.

Jamieson, W.S. and Hodgson, J. (1979b) The effects of variation in sward characteristics upon the ingestive behaviour and herbage intake of calves and lambs under a continuous stocking management. *Grass and Forage Science* 34, 273–282

Krueger, W.C., Laycock, W.A. and Price, D.A. (1974) Relationships of taste, smell, sight, and touch to forage selection. *Journal of Range Management* 27, 258–262.

Laca, E.A. and Ortega, I.M. (1995) Integrating foraging mechanisms across spatial and temporal scales. In: West, N.E. (ed.) *Proceedings of the 5th International Rangeland Congress*. Society for Range Management, Salt Lake City, Utah, pp. 129–132.

Le Du, Y.L.P., Combellas, J., Hodgson, J. and Baker, R.D. (1979) Herbage intake and milk production by grazing dairy cows. 2. The effects of level of winter feeding and daily herbage allowance. *Grass and Forage Science* 34, 249–260.

Lemaire, G. and Chapman, D. (1996) Tissue flows in grazed plant communities. In: Hodgson, J. and Illius, A.W. (eds) *The Ecology and Management of Grazing Systems*. CAB International, Wallingford, UK, pp. 3–36.

L'Huillier, P.J., Poppi, D.P. and Fraser, T.J. (1984) Influence of green leaf distribution on diet selection by sheep and the implications for animal performance. *Proceedings of the New Zealand Society for Animal Production* 44, 105–107.

McMeekan, C.P. and Walshe, M.J. (1963) The inter-relationships of grazing method and stocking rate in the efficiency of pasture utilization by grazing dairy cattle. *Journal of Agricultural Science, Cambridge* 61, 147–163.

Marsh, R. (1976) Systems of grazing management for beef cattle. In: Hodgson, J. and Jackson, D.K. (eds) *Pasture Utilization by the Grazing Animal*. Occasional Symposium No. 8, British Grassland Society, pp. 119–128.

Morley, F.H.W. (1966) Stability and productivity of pastures. *Proceedings of the New Zealand Society of Animal Production* 26, 8–21.

Mott, G.O. (1960) Grazing pressure and the measurement of pasture production. In: *Proceedings of the 6th International Grassland Congress, Reading, UK*, pp. 606–611.

Mursan, A., Hughes, T.P., Nicol, A.M. and Sugiura, T. (1989) The influence of sward height on the mechanics of grazing in steers and bulls. *Proceedings of the New Zealand Society of Animal Production* 49, 233–236.

O'Sullivan, M. (1984) Measurement of grazing behaviour and herbage intake on two different grazing management systems for beef production. In: Holmes, W. (ed.) *Grassland Beef Production*. Martinus Nijhoff Publishers, The Hague, pp. 141–150.

Parsons, A.J., Leafe, E.L., Collett, B., Penning, P.D. and Lewis, J. (1983) The physiology of grass production under grazing. 2. Photosynthesis, crop growth and animal intake of continuously grazed swards. *Journal of Applied Ecology* 20, 127–139.

Parsons, A.J., Johnson, I.R. and Harvey, A. (1988) Use of a model to optimize the interaction between frequency and severity of intermittent defoliation and to provide a fundamental comparison of the continuous and intermittent defoliation of grass. *Grass and Forage Science* 43, 49–59.

Parsons, A.J., Newman, J.A., Penning, P.D., Harvey, A. and Orr, R.J. (1994) Diet preference of sheep: effects of recent diet, physiological state and species abundance. *Journal of Animal Ecology* 63, 465–478.

Penning, P.D., Parsons, A.J., Orr, R.J. and Treacher, T.T (1991) Intake and behaviour responses by sheep to changes in sward characteristics under continuous grazing. *Grass and Forage Science* 46, 15–28.

Penning, P.D., Parsons, A.J., Orr, R.J. and Hooper, G.E. (1994) Intake and behaviour response by sheep to changes in sward characteristics under rotational grazing. *Grass and Forage Science* 49, 476–486.

Peyraud, J.L., Comeron, E.A., Wade, M.H. and Lemaire, G. (1996) The effect of daily herbage allowance, herbage mass and animal factors upon herbage intake by grazing dairy cows. *Annales de Zootechnie* 45, 201–217.

Provenza, F.D. (1995) Postingestive feedback as an elementary determinant of food selection and intake in ruminants. *Journal of Range Management* 48, 2–17.

Ridout, M.S. and Robson, M.J. (1991) Composition of the diet of sheep grazing swards of differing white clover content and clover composition. *New Zealand Journal of Agricultural Research* 34, 89–93.

Spedding, C.R.W. (1965) The physiological basis of grazing management. *Journal of the British Grassland Society* 20, 7–14.

Stobbs, T.H. (1973a) The effect of plant structure on the intake of tropical pastures. I Variation in the bite size of grazing cattle. *Australian Journal of Agricultural Research* 24, 809–819.

Stobbs, T.H. (1973b) The effect of plant structure on the intake of tropical pastures. II Differences in structure, nutritional value, and bite size of animals grazing *Setaria anceps* and *Chloris gayana* at various stages of growth. *Australian Journal of Agricultural Research* 24, 821–829.

Stobbs, T.H. (1977) Short-term effects of herbage allowance on milk production, milk composition and grazing time of cows grazing nitrogen-fertilized tropical grass pasture. *Australian Journal of Experimental Agriculture and Animal Husbandry* 17, 892–897.

Uetake, K. and Kudo, Y. (1994) Visual dominance over hearing in feed acquisition procedure of cattle. *Applied Animal Behaviour Science* 42, 1–9.

Ungar, E.D. (1996) Ingestive behaviour. In: Hodgson, J. and Illius, A.W. (eds) *The Ecology and Management of Grazing Systems*. CAB International, Wallingford, UK, pp. 185–218.

Voisin, A. (1959) *Grass Productivity*. Crosby Lockwood, London.

Wade, M.H. (1979) The effect of grazing by dairy cows given three levels of herbage allowance on the dynamics of leaves and tillers in swards of *Lolium perenne*. MPhil thesis, University of Reading.

Wade, M.H. (1991) Factors affecting the availability of vegetative *Lolium perenne* to grazing dairy cows with special reference to sward characteristics, stocking rate and grazing method. Thèse de Doctorat, Université de Rennes.

Wade, M.H. and Baker, R.D. (1979) Defoliation in set-stocked grazing systems. *Grass and Forage Science* 34, 73–74.

Wade, M.H., Peyraud, J.L., Lemaire, G. and Comeron, E.A. (1989) The dynamics of daily area and depth of grazing and herbage intake of cows in a five day paddock system. In: *Proceedings of the 16th International Grassland Congress, Nice, France*. AFPF, Versailles, France, pp. 1111–1112.

13

Selective Grazing on Grass–Legume Mixtures in Tropical Pastures

C.E. Lascano

CIAT, AA 67-13, Cali, Colombia

Introduction

Livestock production in tropical areas of Latin America has been shown to increase significantly with the introduction of well-adapted and highly productive African grasses (e.g. *Brachiaria* spp.) (Lascano and Euclides, 1996). However, there is ample evidence that these grasses degrade over time as a result of nitrogen deficiency. This degradation process is partially reflected in loss of grass productivity and weed invasion, which affect carrying capacity and animal performance (Oldeman, 1994).

Associating grasses with nitrogen-fixing legumes offers an economic option for improving pasture quality and productivity and hence animal production in tropical regions with soils of low natural fertility (Lascano *et al.*, 1989). The rationale for this alternative is that tropical legumes have a higher nutritive value than grasses and, through symbiotic nitrogen fixation, can enhance the production and quality of the companion grass and improve soil fertility (Thomas *et al.*, 1995). However, management of grass–legume mixtures by farmers is complicated by the fact that animals exhibit selection preference for either the grass or the legume.

Research on animal selectivity in tropical grass–legume pastures aimed at establishing relationships between legume content in the pasture and in the diet selected by grazing animals has been limited. This chapter discusses results from work carried out at the Centro Internacional de Agricultura Tropical (CIAT) on the evaluation of tropical legume-based pastures and on legume selectivity by cattle. Results are discussed in terms of the implications of animal selectivity for management of tropical legume-based pastures.

 Grassland Ecophysiology and Grazing Ecology
(eds G. Lemaire, J. Hodgson, A. de Moraes, C. Nabinger and P.C. de F. Carvalho)

Development of Legume-based Pastures in Tropical America

Even though tropical America is a centre of diversity for many important forage legumes, such as *Stylosanthes*, *Centrosema* and *Arachis*, most of the scientific evaluations have been limited to a few species (e.g. *Neonotonia wightii* (Wight & Arn.) Lackey (glycine), *Macroptilium atropurpureum* (DC.) Urb. (Siratro), *Pueraria phaseoloides* (Roxb.) Benth. (kudzu) and *Centrosema pubescens* Benth. (Centro)), which are marginally adapted to the edaphic conditions prevailing in major livestock areas of tropical America (Toledo, 1985). This was recognized by the Red Internacional de Evaluación de Pastos Tropicales (RIEPT) and the former Tropical Pastures Program of CIAT (Toledo, 1982). This recognition was translated into a major effort, which began in the early 1980s, to evaluate and select grass and legume germplasm with adaptation to the acid soils found in important ecosystems, such as savannahs and humid forests, in tropical America. A tangible result of this effort was the release of legume cultivars, such as *Stylosanthes capitata* Vogel cv. Capica for the savannahs of Colombia (Anon., 1983), and *Stylosanthes guianensis* (Aublet) Sw. (stylo) cv. Pucallpa for the humid forests by a national research organization in Peru (Reyes *et al.*, 1985). More recently, *Arachis pintoi* Krapov. & D. Gregory (perennial peanut) cv. Maní Forrajero was released as a forage legume in Colombia and Costa Rica (Rincón *et al.*, 1992; Argel and Villareal, 1998). This perennial and stoloniferous forage peanut has high nutritional quality (Lascano and Thomas, 1988), is compatible with aggressive and stoloniferous grasses and is persistent under very heavy grazing (Grof, 1985).

At an early stage of forage legume dissemination, it was recognized that germplasm distribution was not sufficient to promote adoption of grass–legume technology by farmers. Therefore, in RIEPT, considerable effort was made to investigate factors associated with pasture establishment (Lascano and Spain, 1991), develop alternative methodologies for grazing experiments (Lascano *et al.*, 1986) and integrate improved grass–legume pastures into production systems. Emphasis was also given to the development of seed supply systems, which, if not in place, would limit the adoption of forage legumes by farmers.

Results from grazing trials conducted in RIEPT have shown that in grass–legume pastures live-weight gain has ranged from 200 to 400 kg ha^{-1} in locations with dry-season stress and from 500 to 600 kg ha^{-1} in areas with no dry-season stress (Tergas *et al.*, 1984; Reátegui *et al.*, 1985; Lascano *et al.*, 1989; Lascano and Thomas, 1990; Lascano, 1994). Results also indicated that, on average, live-weight gain was 30% greater in grass–legume pastures than on grass pastures. However, in some experiments conducted in the llanos of Colombia, animal weight gain on grass–legume pastures was twice as high as that recorded on grass pastures. In these cases, the larger increments in live-weight gain in the mixed pastures were related to the nutritional quality of the companion grass and to the legume content of the pasture. For example, in pastures of *Brachiaria humidicola* (Rendle) Schweick (known to be protein-deficient), animal production on grass associated with 30% perennial peanut (*A. pintoi*) was twice as high as on grass alone (Lascano, 1994). In contrast, pastures having the same grass associated with 10% of the legume resulted in only 35% more live-weight gain in the grazing animals.

In general, experimental evidence clearly indicates that, in areas with or without dry-season stress, animal live-weight gains were significantly higher in tropical grass–legume pastures than in grass-alone pastures. However, the advantage of mixed pastures over grass pastures in terms of animal production depends on the quality of the companion grass and legume content of the pasture.

The management of grass–legume mixtures by farmers is complicated by the fact that animals exhibit selection preference for either the grass or the legume. It is well known that there are differences between species in acceptability to grazing animals, as well as seasonal differences within species. These differences could be used for controlling the legume content of grass–legume pastures. For example, in Australia low preference for *Calopogonium caeruleum* (Bentham) Hemsley was associated with its persistence under grazing (Middleton and Mellor, 1982). Similarly, lower preference for M. *atropurpureum* (Mocino and Sesse ex Decandolle) Urban, *Stylosanthes hamata* (L.) Taubert and *Stylosanthes scabra* J. Vogel during part of the year (spring and early summer) has been shown to favour their survival under grazing (Stobbs, 1977; Gardener, 1980; McLean *et al.*, 1981; Walker *et al.*, 1982). A similar seasonal effect on legume selectivity was observed in the llanos of Colombia with cattle grazing associations of the erect bunch grass *Andropogon gayanus* Kunth. with *P. phaseoloides* (Roxburgh) Bentham (Kudzu) and S. *capitata* J. Vogel (Böhnert *et al.*, 1985). However, these seasonal effects on legume selection by grazing cattle have not been observed to the same extent in associations of prostrate grasses with some stoloniferous tropical legumes, such as *Desmodium heterocarpon* subsp. *ovalifolium* (l.) Decandolle and A. *pintoi*, where the proportion of legume in the diet relative to that in the forage on offer changed with legume species (Lascano and Thomas, 1988).

Legume content in tropical pastures can also be affected by fertilizer management, particularly phosphorus, in terms of both increasing legume content and improving acceptability of the legume to grazing animals (McLean and Kerridge, 1987). Thus, phosphorus management could be viewed as a way of regulating legume in the pasture, provided legume growth is not affected, which is unlikely in the case of legumes having high internal requirements for this element (Ozanne *et al.*, 1976).

An important determinant of legume content in pastures and hence legume selection by livestock is the grazing pressure used to manage the pastures. Increased stocking rate in tropical grass–legume pastures is usually accompanied by reduction of the legume component, particularly those having a twining growth habit (Jones and Jones, 1978). In contrast, creeping legumes seem to be less affected by grazing at high stocking rates (Grof, 1985; Jones and Clements, 1987). In a review by Curll and Jones (1989) on factors associated with legume persistence in Australia, it was concluded that frequent and intense defoliation can favour survival of temperate prostrate legumes (i.e. *Trifolium repens*) and of tropical legumes that are susceptible to shading (i.e. *Stylosanthes humilis* H.B.K.) by the companion grass.

In another review, by Roberts (1980), on the effect of stocking rate on grass–legume pastures, an important conclusion was that the animal gain/stocking rate relationship could be affected by the botanical composition of the pasture, which, in turn, is influenced by selective grazing. In grass–legume pastures managed at a low stocking rate, taller species can suppress low-growing species. If the dominant species have a lower feeding value than the species they replace, live-weight gain can drop. In pastures managed at a high stocking rate, if the remaining species are not tolerant of grazing, they will not persist, but, if the remaining species are less palatable and have lower nutritional value than those they suppress, animal weight gain will be less. For example, Partridge (1979) found that M. *atropurpureum* cv. Siratro did not persist in association with *Dichanthium caricosum* (L.) A. Camus (nadi bluegrass) at medium and high stocking rates. The sown legume was replaced in the pasture by *Desmodium heterophyllum* (Willd.) DC (hetero) and by naturalized legumes of low acceptability at the medium and high stocking rates, respectively. The net result of losing Siratro was little change in animal live-weight gain at the middle stocking rate, but a sharp drop in gain at the high stocking rate, when the sown legume was replaced by unpalatable, naturalized legumes.

The experience with tropical grass–legume pastures is that, in most cases, the grass suppresses the legume, particularly when pastures are grazed heavily. However, there are examples where the opposite situation occurs under grazing. For example, the stoloniferous legume D. *heterocarpon* subsp. *ovalifolium* is not readily consumed by grazing animals and, as a consequence, the animals selectively graze out the grass component (Toro, 1990). Animal production in pastures dominated by D. *heterocarpon* subsp. *ovalifolium* is low in terms of both individual live-weight gain and production per hectare. In contrast, in a grass pasture mixture with A. *pintoi*, where the legume was in higher proportion than the grass, animal performance was not affected, mainly due to high intake of the legume (Grof, 1985).

With the evidence available, it is difficult to decide on the best strategy to manage tropical grass–legume pastures in order to maximize legume persistence and animal performance. However, there are results that suggest that grass–legume balance can be influenced by the relative acceptability of the grasses and legumes being used, which, in turn, can be affected by the season of the year and grazing pressure. Thus, it would seem that, in certain tropical grass–legume pastures, some form of deferred and rotational grazing system should be utilized to favour persistence of the more palatable species (whether grass or legume).

Factors Associated with Grazing Selectivity in Tropical Pastures

It is well established that grazing animals have preferential selection of green leaves in both temperate and tropical pastures (e.g. Böhnert *et al.*, 1985). Leaf selection and the ingestive behaviour of animals grazing tropical grasses are generally affected by sward structure, which includes leaf bulk density (kg ha^{-1} cm^{-1}), total herbage mass (kg ha^{-1} cm^{-1}) and sward height (Stobbs, 1973).

In the case of grass–legume pastures, the legume proportion in the diet of grazing animals could be influenced, according to Milne *et al.* (1982), by: (i) the proportion of the legume in the sward; (ii) the relative distribution of the morphological components of grass and legume in the sward canopy; and (iii) the herbage mass, height and density of the pasture.

In grass–legume mixtures where plant species are homogeneously distributed in the sward, animal selectivity for the grass or legume component may be influenced by its relative distribution in the canopy. Results from temperate grass–legume mixtures (e.g. ryegrass–white clover) indicate that a larger proportion of the variation of legume in the diet of sheep was accounted for by considering the legume content in the upper horizons of the sward (linear, $r^2 = 0.83$) compared with the total sward (linear, $r^2 = 0.57$) (Milne *et al.*, 1982). Also, in a tropical grass–legume pasture of *Hemarthria altissima* cv. Floralta–*Aeschynomene americana*, the proportion of the legume in the diet of cattle exhibited a quadratic relationship ($r^2 = 0.79$), with the legume proportion in the upper layer of the pasture (Moore *et al.*, 1985). In both studies, the proportion of the legume in the diet was greater than the proportion in the grazed horizon of the sward, possibly as a result of canopy modification by the grazing animal prior to biting.

In other grazing studies, involving pastures of a prostrate grass (*B. humidicola*) in a mixture with *A. pintoi*, results indicate no relationship between the proportion of legume on offer in the upper layer of the canopy (> 20 cm) and the proportion in the diet (Hess, 1995). In contrast, legume proportion in the medium layer of the canopy (10–20 cm) accounted for 34% of the variation in legume selected, whereas legume in the lower strata (2–10 cm) explained 59% of the observed variation. In these pastures, a higher proportion of the variation in legume selected (67%) was accounted for when considering the legume proportion in the whole canopy. These contrasting results could be due to the differences in positioning of the legume relative to the grass in the canopy of white clover and forage peanut-based pastures, and to differences in quantity of biomass on offer, both of which are known to affect the capacity of animals to defoliate in a selective manner (Hodgson, 1981; Curll *et al.*, 1987).

Other factors that have been associated with legume selectivity in tropical grass–legume mixtures are related to the stage of maturity of the plants being grazed. In several studies, selectivity for the legume has been higher in pastures where the companion grass is mature, as is the case during the dry season of the year (Böhnert *et al.*, 1985). Secondary compounds in legumes (i.e. cyanogenic glucosides, condensed tannins) have also been mentioned as factors that can influence selectivity by grazing animals. However, according to Sheath and Hodgson (1989), selection of white clover plants growing in a perennial ryegrass sward was more related to leaf size and positioning of the legume in the sward canopy than to glucoside activity. However, the authors acknowledge that, in certain temperate legumes (e.g. *Lotus pedunculatus*), a high tannin concentration may act as a deterrent for plants being grazed, even if the companion grass is mature. In tropical legumes with high levels of tannins (e.g. *D. heterocarpon* subsp. *ovalifolium*), pastures tend to become legume-dominant, as a result of low consumption by grazing animals.

Previous exposure of animals to a diet may also influence the selective behaviour of grazing ruminants, such as calves (Hodgson and Jamieson, 1981) and sheep

(Curll and Davidson, 1983). Limited information on the effect of previous experience on selectivity by ruminants grazing tropical grass–legume pastures also indicates that cattle with previous exposure to *A. pintoi* selected more legume as compared with those with less experience (Carulla *et al.*, 1991).

Legume Selectivity

In the work carried out by CIAT and partners in tropical America, an attempt has been made to establish relationships between the content of different legume species in the pasture and animal performance. The range of data obtained in these studies in contrasting grass–legume pastures was also useful for exploring general relationships between legumes selected by grazing animals and legumes available in pastures with different growth habits or canopy structure. Little information of this type is available in the literature for tropical grass–legume pastures.

A model that allows the description of animal selectivity with a single parameter was used as a first approximation to establishing the relationship between legumes on offer and legumes selected by cattle in contrasting grass–legume pastures (Chesson, 1983). The parameter (α_i) estimated with the model represents selection of the legume component and implies a curvilinear relationship between the proportion in the diet and that in the available forage. Estimates of α_i can vary from < 0.5 (selection against the legume) to > 0.5 (selection in favour of the legume) in the range 0 to 1. The validity of α_i is dependent on two assumptions: (i) that it does not vary with time, or in other words it is an instantaneous measure of animal behaviour; and (ii) that it is not dependent on absolute quantities of biomass available.

Pastures with Prostrate Grasses and a Palatable Legume

Measurements on the proportion of legume selected by grazing cattle were carried out at the Carimagua Research Station in the llanos of Colombia for 1 year on mixtures of prostrate and semi-erect species of *Brachiaria* (*B. humidicola* CIAT 679, *B. dictyoneura* CIAT 6133, *B. brizantha* CIAT 644, *B. ruziziensis* CIAT 6291) with *A. pintoi* CIAT 17434, a stoloniferous legume. The site represents a well-drained savannah ecosystem (N 4° 34′, W 71° 20′), with an average rainfall of 2100 mm (April–November) and a well-defined dry season (December–March). Each legume-based pasture established in 1227 m^2 plots, with two replications, was grazed using a rotational scheme of 7/21 days (grazing/rest) with two oesophageal-fistulated steers (200 kg), resulting in an equivalent stocking rate of four animals per hectare. The proportion of legume in the total available biomass and in the diet selected by grazing animals was measured on a monthly basis during the first, third and last day of grazing of each association.

The relationship between legume proportion in the diet and in the total forage on offer in *Brachiaria* spp.–*A. pintoi* pastures (Fig. 13.1) was described by α = 0.52, which indicates that the legume was selected in a similar proportion to that present in the total available forage. However, results from another grazing study, also carried out in the llanos of Colombia, indicated that selection indices (α) for *A. pintoi* in association with *B. humidicola* were 0.65 and 0.79 for the wet and dry seasons, respectively (Hess, 1995), which is an indication that cattle selected in favour of the legume throughout the year (Fig. 13.2).

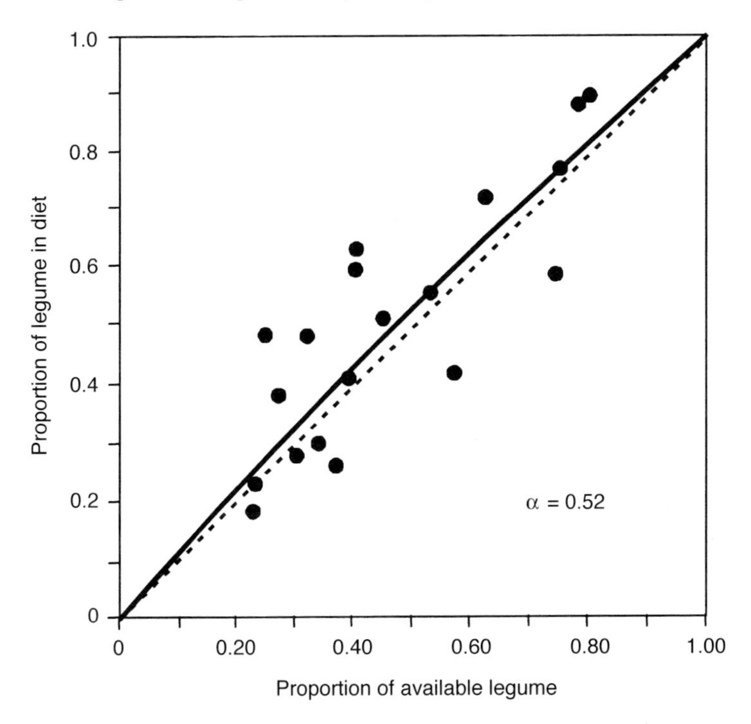

Fig. 13.1. Relationship between legume proportion in the diet and in total available forage in *Brachiaria* spp.–*Arachis pintoi* pastures.

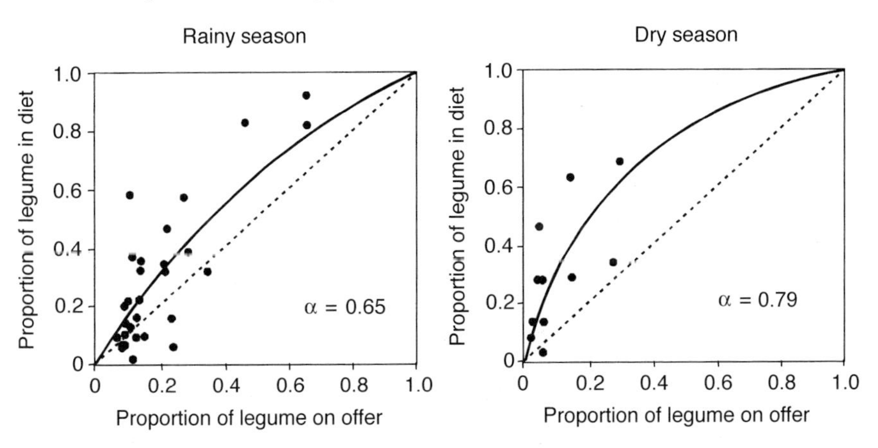

Fig. 13.2. Relationship between legume proportion in the diet and in total available forage in *Brachiaria humidicola*–*Arachis pintoi* pastures in different seasons of the year.

Pastures with Prostrate Grasses and an Unpalatable Legume

Another set of data used to analyse relationships between legumes on offer and selected was collected at the CIAT Research Station in Quilichao, Cauca Valley, Colombia in a *B. dictyoneura* (CIAT 6133)–*D. heterocarpon* subsp. *ovalifolium* (CIAT 350) (a stoloniferous legume) pasture. The site is located at latitude N 3° 6′, with an average rainfall of 1772 mm with a bimodal distribution (March–June and September–December). Pastures were again rotationally grazed by zebu steers (200 kg) at stocking rates of 3.0, 4.0 and 5.5 animals ha^{-1}, with 7 days of grazing and 21 days of rest. Dietary forage samples were obtained every 28 days in the paddocks composing the rotation, using oesophageal-fistulated steers.

During the study (2.5 years), large differences in botanical composition of the available forage were generated as a result of stocking-rate treatments and season of the year. Results showed that selection indices (α) were 0.31, 0.32 and 0.29 for the high, medium and low stocking rates, respectively (Fig. 13.3), which indicate

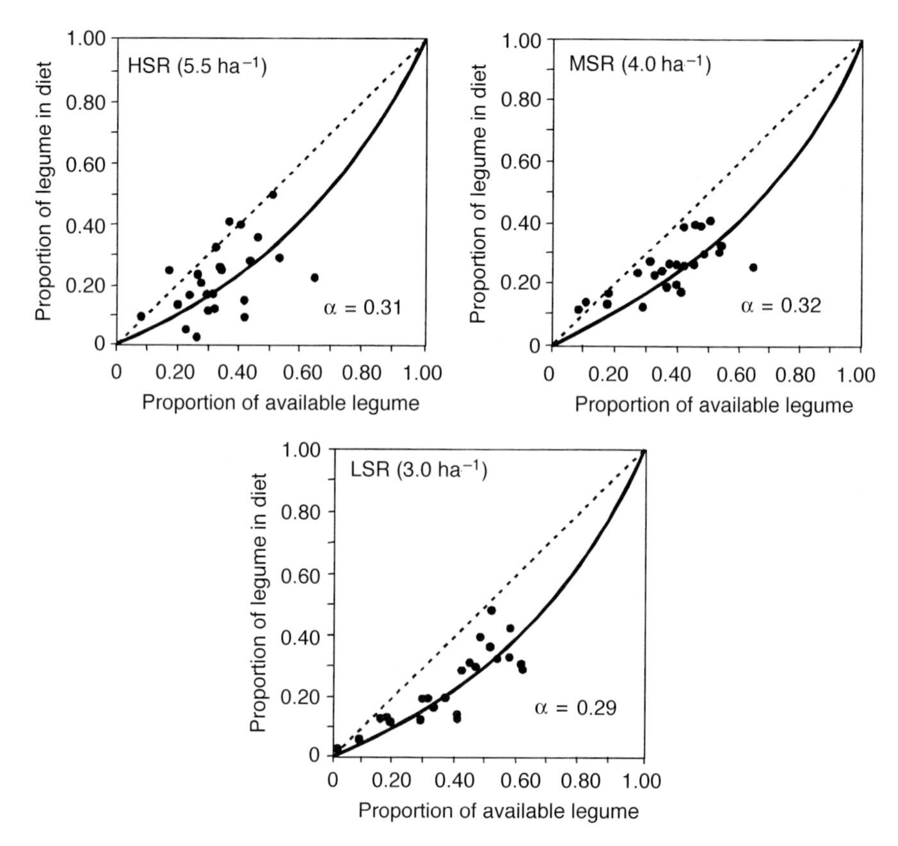

Fig. 13.3. Relationship between legume proportion in the diet and in total available forage in *Brachiaria dictyoneura–Desmodium heterocarpon* subsp. *ovalifolium* at three stocking rates (high, medium or low SR).

that, at the three stocking rates, animals selected against the legume in a similar manner. These results are interesting, since they indicate that, in the association with *D. heterocarpon* subsp. *ovalifolium* legume, selectivity was not related to legume availability resulting from the three stocking rates imposed, at least within the range of the data analysed. However, it is conceivable that, at higher levels of legume in the pasture, the proportion of legume in the diet might have been higher than the proportion available in the biomass on offer, as selection for grass becomes increasingly difficult in a strongly legume-dominant pasture. In this case, the response function would be described by a curve with a sigmoid shape (Chesson, 1983). It has been postulated that, in legume-dominant temperate white clover-based pastures, selection for the legume could decline or reverse in favour of the grass (Clark and Harris, 1985).

In general, results from the legume-based pastures with prostrate grasses were interpreted to mean that difference in the selection indices of legumes was primarily related to the palatability of the two legumes species, being high in the case of *A. pintoi* and low in the case of *D. heterocarpon* subsp. *ovalifolium*. The low palatability of this legume species has been associated with high tannin concentration, particularly when grown on sulphur-deficient soils (Lascano and Salinas, 1982; Salinas and Lascano, 1982).

Pastures with Tall Bunch Grasses

Measurements of legume selectivity were also carried out for 1 year in a grazing trial at the CIAT Carimagua Research Station, which involved mixtures of *A. gayanus* cv. Carimagua 1 (an erect, tall bunch grass) with *Centrosema acutifolium* cv. Vichada (a stoloniferous, twining legume). Legume selectivity was measured on a monthly basis through oesophageal collections from steers in pastures continuously grazed at 0.75, 1.0 and 1.5 animals ha^{-1} and rotationally grazed at 1.5 animals ha^{-1}.

The proportion of legume in the diet and in the available forage obtained over the entire year was used for the analysis, and the selection index estimated was 0.24 (Fig. 13.4A), which indicated that cattle selected against the legume. However, a closer analysis of the results showed that selection against the legume was mainly in the wet season (Fig. 13.4B), as compared with the dry part of the year (Fig. 13.4C), when the companion grass loses quality.

As mentioned before, results from other experiments had shown that, in pastures in the llanos of Colombia with the erect, tall bunch grass *A. gayanus* in mixture with *P. phaseoloides* (a twining legume) and *S. capitata* cv. Capica (a semi-erect legume), animals preferentially selected the legume in the dry season (Böhnert *et al.*, 1985). Thus, it would seem that, in tropical grass–legume mixtures with bunch-type grasses, cattle are highly selective for or against the legume component, depending on the season of the year.

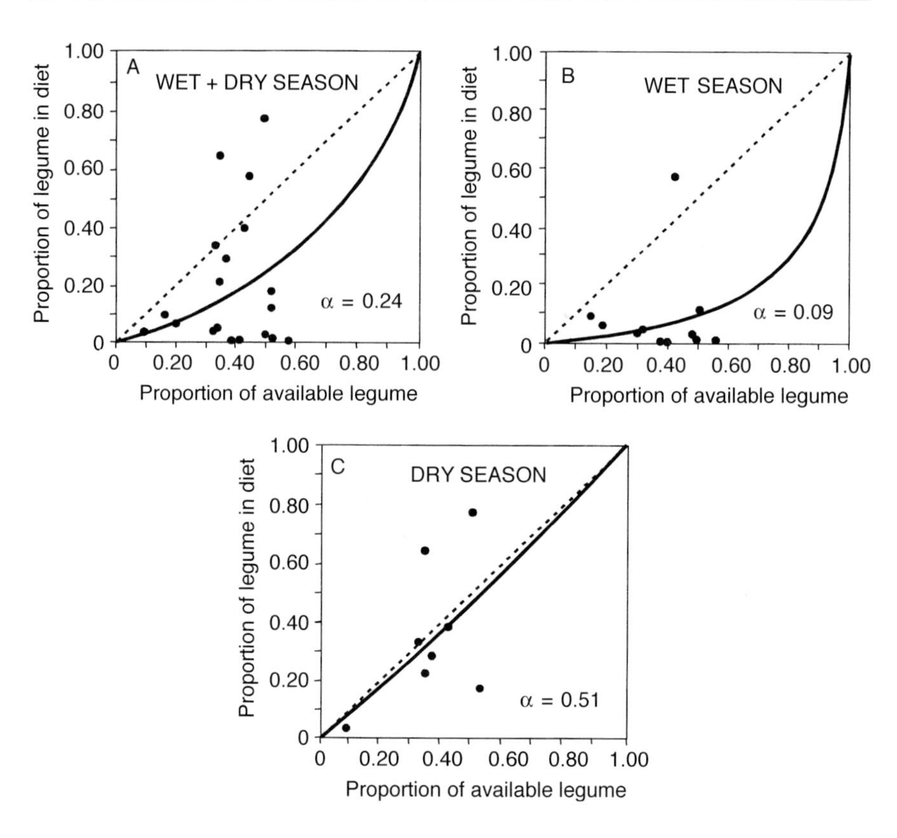

Fig. 13.4. Relationship between legume proportion in the diet and in total available forage in an *Andropogon gayanus–Centrosema acutifolium* pastures in different seasons of the year.

Guidelines for Managing Tropical Grass–Legume Pastures

Although there are decided advantages to adding a legume to a grazing system in terms of animal performance, managing a grass–legume pasture and maintaining the correct balance between the two components can be more of a challenge to farmers than keeping a grass pasture productive. This difficulty has in part been responsible for the low adoption of legumes by livestock farmers in the tropics. Thus a research priority has been to develop guidelines for managing grass–legume pastures, based on the understanding in a broad sense of how sward structure and composition interact with animal selectivity.

The relationship between legume proportion in the diet of grazing animals and in the available forage, as affected by pasture structure (within or between different sward types), can be useful for developing grazing-management guidelines applicable to different grass–legume mixtures, based on legumes well adapted to biotic and abiotic constraints.

Based on the different selection indices found for the grass–legume associations considered above, one would expect that management requirements for an adequate grass–legume balance over time would differ between ecoregions and the growth habit of the companion grass. In areas with a defined dry season (e.g. neotropical savannahs), mixtures of palatable legumes (e.g. *A. pintoi*) with prostrate grasses (e.g. *Brachiaria* spp.) should be managed so as to favour recovery of the legume, given that it is selected in a similar or higher proportion than that on offer in the pasture. This would, in turn, require frequent or continuous grazing, with seasonal adjustments of stocking rate.

In contrast, in the same environment, pastures with legumes that are unpalatable and aggressive (e.g. *D. heterocarpon* subsp. *ovalifolium*), relative to the companion grass (e.g. *Brachiaria* spp.), would require some form of deferred or rotational grazing during the wet season, together with stocking-rate adjustments to favour recovery of the grass in the mixture. This was demonstrated in a grazing trial in the Cerrados of Brazil with a *B. ruziziensis–Calopogonium mucunoides* (low-palatability) pasture that was legume-dominant (30% grass–70% legume) under continuous grazing. The pasture was reverted to a more balanced mixture (75% grass–25% legume) by resting for 4 months in the wet season, followed by a reduction of stocking rate from 2.5 to 1.5 animals ha^{-1} (CIAT, unpublished results). In areas with an undefined dry season (e.g. tropical rain forest), pastures with prostrate grasses and unpalatable legumes would certainly require long rest periods to maintain an acceptable grass–legume balance and thus animal performance.

Grazing management required for keeping an adequate balance of grass and legume in mixtures with tall bunch grasses would also depend on the aggressiveness and relative palatability of the legume, as well as on grazing pressure and rainfall pattern. Mixtures with well-adapted and aggressive legumes in areas with short dry seasons are likely to require some form of rotational grazing to prevent legume dominance. Frequency of grazing is also likely to be very critical in maintaining a desired grass–legume balance, since animals would have the ability of selecting against legumes during most of the year, particularly if they are of low palatability. With more palatable and less aggressive legumes, frequent or continuous grazing may be required to reduce competition (i.e. for light, water and soil nutrients) from the grass and to favour the legume in the mixture.

The evidence available from numerous grazing trials indicates that animal performance in tropical grass–legume pastures is as good under continuous grazing with set stocking as with any other grazing management ('t Mannetje *et al.*, 1976). However, results from pasture management studies should be viewed not only in terms of animal performance, but also in terms of stability of pasture components. One example of this was in Uganda, Africa, with a mixture of *Panicum maximum–M. atropurpureum* grazed continuously and rotationally. Animal gains were similar for continuous stocking and a three-paddock rotational grazing system, but lower for a six-paddock rotation, when grazed at the same stocking rate (Stobbs, 1969). However, after 3 years of grazing, the proportion of the grass (*P. maximum*) was much lower under continuous stocking than under rotational grazing. Pastures under continuous stocking also had more weeds than those rotationally grazed. In the long run, it is likely that animal production would have been greater in the more stable pastures under rotational grazing.

If it is accepted that different grass–legume associations have different grazing management requirements to ensure legume persistence, the question is: what research methodology should be followed to define these requirements and to give guidelines to farmers? One approach is the traditional small-plot grazing experiment, with a combination of grazing frequencies and grazing pressures in a factorial arrangement (Paladines and Lascano, 1983) or in a central composite design (Mott, 1983). Another alternative is the flexible management approach proposed by Spain *et al.* (1985). With this strategy, no fixed stocking rates or grazing frequencies are employed. Rather, stocking rate and grazing frequency are adjusted, depending on two pasture parameters: (i) stocking rate is adjusted when grazing pressure reaches selected limits; and (ii) grazing frequency is adjusted when the legume proportion reaches selected limits.

References

Anon. (1983) *Capica* (Stylosanthes capitata *Vog.*). Boletín Técnico No. 103, Instituto Colombiano Agropecuario, Bogotá, Colombia.

Argel, P.J. and Villareal, M. (1998) *Nuevo maní forrajero perenne cultivar Porvenir (CIAT 18744): Leguminosa herbácea para alimentación animal, el mejoramiento y conservación del suelo y el embellecimiento del paisaje*. Boletín Técnico del MAG, Costa Rica y CIAT, Cali, Colombia, 32 pp.

Böhnert, E., Lascano, C.E. and Weniger, J.H. (1985) Botanical and chemical composition of the diet selected by fistulated steers under grazing on improved grass–legume pasture in the tropical savannas of Colombia. I. Botanical composition of forage available and selected. *Journal of Animal Breeding and Genetics* 102, 385–394.

Carulla, J.E., Lascano, C.E. and Ward, J.K. (1991) Selectivity of resident and oesophageal fistulated steers grazing *Arachis pintoi* and *Brachiaria dictyoneura* I. The llanos of Colombia. *Tropical Grasslands* 25, 317–324.

Chesson, J. (1983) The estimation and analysis of preference and its relationship to foraging models. *Ecology* 64, 1297.

Clark, D.A. and Harris, P.S. (1985) Composition of the diet of sheep grazing swards of differing white clover content and spatial distribution. *New Zealand Journal of Agriculture Research* 28, 233–240.

Curll, M.C. and Davidson, J.C. (1983) Defoliation and productivity of a *Phalaris*–subterranean clover sward, and the influence of grazing experience on sheep intake. *Grass and Forage Science* 38, 159–167.

Curll, M.L. and Jones, R.M. (1989) The plant–animal interface and legume persistence – An Australian perspective. In: Marten, G.C., Matches, A.G., Barnes, R.F., Broughman, R.W., Clements, R.J. and Sheath, G.W. (eds) *Persistence of Forage Legumes*. American Society of Agronomy, Madison, Wisconsin, pp. 339–359.

Curll, M.L., Robards, G.E. and Langlands, J.P. (1987) Nutritive value improvement of existing forage systems by management. In: Hutchinson, K.J. (ed.) *Improving the Nutritive Value of Forage*. Standing Committee of Agriculture Technical Report No. 20, CSIRO, Melbourne, pp. 60–73.

Gardener, C.J. (1980) Diet selection and live-weight performance of steers on *Stylosanthes hamata*–native grass pastures. *Australian Journal of Agriculture Research* 31, 379–392.

Grof, B. (1985) Forage attributes of the perennial groundnut *Arachis pintoi* in a tropical savanna environment in Colombia. In: *Proceedings of the XVth International Grassland Congress, Kyoto, Japan*, The Science Council of Japan, Nishi Nasuno, Japan, pp. 168–170.

Hess, H.D. (1995) Grazing selectivity and ingestive behaviour of steers on improved tropical pastures in the Eastern Plains of Colombia. Dissertation No. 11301, Swiss Federal Institute of Technology (ETHZ), Zurich, 108 pp.

Hodgson, J. (1981) Testing and improvement of pasture species. In: Morley, F.H.W. (ed.) *World Animal Science B1: Grazing Animals*. Elsevier, Amsterdam, pp. 309–318.

Hodgson, J. and Jamieson, W. (1981) Variations in herbage mass and digestibility and the grazing behaviour and herbage intake of adult cattle and weaned calves. *Grass and Forage Science* 36, 39–48.

Jones, R.J. and Jones, R.M. (1978) The ecology of *Siratro*-based pastures. In: Wilson, J.R. (ed.) *Plant Relations in Pastures*. CSIRO, Melbourne, pp. 363–367.

Jones, R.M. and Clements, R.J. (1987) Persistence and productivity of *Centrosema virginianum* and *Vigna parkerii* cv. Shaw under grazing on the coastal lowlands of south-east Queensland. *Tropical Grasslands* 21, 55–64.

Lascano, C. (1994) Nutritive value and animal production of forage *Arachis*. In: Kerridge, P. and Hardy, B. (eds) *Biology and Agronomy of Forage* Arachis. Centro Internacional de Agricultura Tropical (CIAT), Cali, Colombia, pp. 109–121.

Lascano, C.E. and Euclides, V.P.B. (1996) Nutritional quality and animal production of *Brachiaria*. In: Miles, J.W., Maass, B.L. and do Valle, C.B. (eds) Brachiaria: *Biology, Agronomy and Improvement*. CIAT, Cali, Colombia, and CNPGC/EMBRAPA, Campo Grande, Mato Grosso, Brazil, pp. 106–123.

Lascano, C.E. and Salinas, J.G. (1982) Efecto de la fertilidad del suelo en la calidad de *Desmodium heterocarpon* subs *ovalifolium*. *Pastos Tropicales: Boletín Informativo* 6, 4–5.

Lascano, C. and Spain, J.M. (1991) Establecimiento y renovación de pasturas. In: *Sexta Reunión del Comité Asesor de la Red Internacional de Evaluación de Pastos Tropicales, Veracruz, Mexico, 1991*. CIAT, Cali, Colombia, p. 421.

Lascano, C. and Thomas, D. (1988) Forage quality and animal selection of *Arachis pintoi* in association with tropical grasses in the Eastern Plains of Colombia. *Grass and Forage Science* 43, 433–439.

Lascano, C. and Thomas, D. (1990) Quality of *Andropogon gayanus* and animal productivity. In: Toledo, J.M., Vera, R., Lascano, C. and Lenné, J.J. (eds) Andropogon gayanus *Kunth – a Grass for Tropical Acid Soils*. Centro Internacional de Agricultura Tropical (CIAT), Cali, Colombia, pp. 247–275.

Lascano, C.E., Pizarro, E.A. and Toledo, J.M. (1986) Recomendaciones generales para evaluar pasturas con Animales. In: *Evaluación de pasturas con animales: Alternativas metodológicas. Segunda Reunión del Comité Asesor de la Red Internacional de Evaluación de Pastos Tropicales, 1–5 October, 1984, Lima, Peru*. CIAT, Cali, Colombia, pp. 251–265.

Lascano, C., Estrada, J. and Avila, P. (1989) Animal production of pastures based on *Centrosema* spp. in the eastern plains of Colombia. In: *Proceedings of the XVIth International Grassland Congress, Nice, France*. AFPF, Versailles, France, pp. 1177–1178.

McLean, R.W. and Kerridge, P.C. (1987) Effect of fertilizer and sulphur on the diet of cattle grazing buffelgrass/siratro pastures. In: Rose, M. (ed.) *Herbivore Nutrition*

Research. Occasional Publication, Australian Society of Animal Production, Brisbane, pp. 93–94.

McLean, R.W., Winter, W.H., Mott, J.J. and Little, D.A. (1981) The influence of superphosphate on the legume content of the diet selected by cattle grazing *Stylosanthes*–native grass pastures. *Journal of Agriculture Science* 96, 247–249.

Middleton, C.H. and Mellor, W. (1982) Grazing assessment of the tropical legume *Calopogonium caeruleum*. *Tropical Grasslands* 16, 213–216.

Milne, J.A., Hodgson, J., Thompson, R., Souter, W.G. and Barthram, G.T. (1982) The diet ingested by sheep grazing swards differing in white clover and perennial ryegrass content. *Grass and Forage Science* 37, 209.

Moore, J.E., Sollemberger, L.E., Morantes, G.A. and Beede, P.T. (1985) Canopy structure of *Aeschynomene americana–Hemarthria altissima* pastures and ingestive behavior of cattle. In: *Proceedings of the 15th International Grassland Congress, Kyoto, Japan*. The Science Council of Japan, Nishi Nasuno, Japan, pp. 1126–1128.

Mott, G.O. (1983) Evaluación del germoplasma forrajero bajo diferentes sistemas de manejo del pastoreo. In: Paladines, O. and Lascano, C. (eds) *Germoplasma forrajero bajo pastoreo en pequeñas parcelas*. CIAT, Cali, Colombia, pp. 149–163.

Oldeman, L.R. (1994) The global extent of soil degradation. In: Greenland, D.J. and Szabolcs, I. (eds) *Soil Resilience and Sustainable Land Use*. CAB International, Wallingford, UK, pp. 99–118.

Ozanne, P.G., Howes, K.M.W. and Petch, A. (1976) The comparative phosphate requirements of four annual pastures and two crops. *Australian Journal of Agricultural Research* 27, 479–488.

Paladines, O. and Lascano, C. (1983) Recomendaciones para evaluar germoplasma bajo pastoreo en pequeños potreros. In: Paladines, O. and Lascano, C. (eds) *Germoplasma bajo pastoreo en pequeñas parcelas*. CIAT, Cali, Colombia, pp. 165–183.

Partridge, I.J. (1979) Improvement of Nadi blue grass *Dichantium caricosum* pastures on hill land in Fiji with superphosphate and siratro: effects of stocking rate on beef production and botanical composition (*Macroptilium atropurpureum*). *Tropical Grasslands* 13, 156.

Reátegui, K., Ara, M., and Schaus, R. (1985) Evaluación bajo pastoreo de asociaciones de gramíneas y leguminosas forrajeras en Yurimaguas, Perú. *Pasturas Tropicales* 7, 11–14.

Reyes, C., Ordoñez, H. and Pinedo, L. (1985) *Stylosanthes guianensis* cv. Pucallpa. In: *Leguminosa forrajera para la Amazonía*, Boletín Técnico No. 3, Instituto Veterinario de Investigaciones Tropicales y de Altura e Instituto Nacional de Investigación y Promoción Agropecuaria, Pucalpa, Peru, pp. 11–24.

Rincón, A., Cuesta, P., Pérez, R., Lascano, C. and Ferguson, J. (1992) Maní forrajero perenne (*Arachis pintoi* Krapovickas and Gregory). In: *Boletín Técnico No. 219*. Instituto Colombiano Agropecuario, Bogotá, Colombia, p. 23.

Roberts, C.R. (1980) Effect of stocking rate on tropical pastures. *Tropical Grasslands* 14(3), 225–231.

Salinas, J.G. and Lascano, C. (1982) Fertilización con azufre y mezcla de la calidad de *Desmodium heterocarpon* subs *ovalifolium*. *Pastos Tropicales: Boletín Informativo* 5, 1–2.

Sheath, G.W. and Hodgson, J. (1989) Plant–animal factors influencing legume persistence. In: Marten, G.C., Matches, A.G., Barnes, R.F., Broughman, R.W., Clements, R.J. and Sheath, G.W. (eds) *Persistence of Forage Legumes*. American Society of Agronomy, Madison, Wisconsin, USA, pp. 361–374.

Spain, J., Pereira, J.M. and Gualdrón, R. (1985) A flexible grazing management system proposed for the advanced evaluation of association of tropical grasses and legumes. In:

Proceedings of the 15th International Grassland Congress, Kyoto, Japan. The Science Council of Japan, Nishi Nasuno, Japan, pp. 1153–1155.

Stobbs, T.H. (1969) The effect of grazing management upon pasture productivity in Uganda. III. Rotational and continuous grazing. *Tropical Agriculture (Trinidad)* 46(3), 195–200.

Stobbs, T.H. (1973) The effect of plant structure on the intake of tropical pastures. III. Differences in sward structure, nutritive value, and bite size of animals grazing *Setaria anceps* and *Chloris gayana* at various stages of growth. *Australian Journal of Agriculture Research* 24, 821–829.

Stobbs, T.H. (1977) Seasonal changes in preference by cattle of *Macroptilium atropurpureum* cv. Siratro. *Tropical Grasslands* 11, 87–91.

Tergas, L.E., Paladines, O., Kleinhesterkamp, I. and Velásquez, J. (1984) Productividad animal de *Brachiaria decumbens* sola y con pastoreo complementario en *Pueraria phaseoloides* en los Llanos Orientales de Colombia. *Producción Animal Tropical* 9(1), 1–13.

Thomas, R.J., Fisher, M.J., Ayarza, M.A. and Sanz, J.I. (1995) The role of forage grasses and legumes in maintaining the productivity of acid soils in Latin America. In: Lal, R. and Stewart, B.A. (eds) *Soil Management: Experimental Basis for Sustainability and Environmental Quality*. Advances in Soil Science Series, CRC Press, Boca Raton, Florida, pp. 61–83.

't Mannetje, L., Jones, R.J. and Stobbs, T.H. (1976) Pasture evaluation by grazing experiments. In: Shaw, N.H. and Bryan, W.W. (eds) *Tropical Pasture Research: Principles and Methods*. CAB Bulletin 51, Hurley, UK, pp. 194–234.

Toledo, J.M. (1982) Objetivos y organización de la Red Internacional de Evaluación de Pastos Tropicales. In: Toledo, J.M. (ed.) *Manual para la evaluación agronómica*. Red Internacional de Evaluación de Pastos Tropicales, CIAT, Cali, Colombia, pp. 13–21.

Toledo, J.M. (1985) Pasture development for cattle production in the major ecosystems of the American lowlands. In: *Proceedings of the XVth International Grassland Congress, Kyoto, Japan*. The Science Council of Japan, Nishi Nasuno, Japan, pp. 74–78.

Toro, N. (1990) Productividad animal en pasturas de *Brachiaria humidicola* (CIAT 679) solo y en asociación con *Desmodium heterocarpon* subs *ovalifolium* (CIAT 13089) bajo sistema de manejo flexible. MS thesis, Centro Agronómico Tropical de Investigación y Enseñanza, Turrialba, Costa Rica, 111 pp.

Walker, S., Rutherford, M.T. and Whiteman, P.C. (1982) Diet selection by cattle on tropical pastures in Northern Australia. In: *Proceedings of the 14th International Grassland Congress, Lexington, Kentucky*, pp. 681–684.

14

Leaf Tissue Turnover and Efficiency of Herbage Utilization

G. Lemaire[1] and M. Agnusdei[2]

[1] *Unité d'Ecophysiologie, INRA, Lusignan 86600, France;*
[2] *INTA-EEA, Balcarce 7620, Buenos Aires, Argentina*

Introduction

Leaf tissue production is a continuous process, regulated by environmental variables and by sward state characteristics. As leaf tissue accumulates in plants, it is subjected to ageing and senescence, which lead to litter accumulation and decomposition in the soil. In grazed pastures, leaf tissues are subjected to discrete defoliation events, the frequency and intensity of which greatly affect the physiology of plants and therefore the rate at which new leaf tissues are produced. Thus, the optimization of grazing systems cannot be conceived as the independent maximization of the amount of herbage production or the intake by the animals but as the result of a compromise between the three leaf tissue fluxes that cycle in grazed pastures: growth, senescence and consumption (Parsons, 1994).

The responses of individual plants to the intensity and frequency of defoliation involve major processes at the level of the plant–animal interface. In the short term, physiological responses linked to the reduction in C supply resulting from the loss of leaf area will limit leaf tissue production; in the long term, morphological responses allow the plant to adapt its architecture and escape defoliation ('avoidance strategy' (Briske, 1996)). So the plasticity of plants in adjusting to the defoliation regime plays a central role in regulating both the rate of production of new leaf tissue and the accessibility of these leaves to the grazing animals.

The efficiency of herbage use in grazing systems has to be analysed not only with the short-term objective of maximizing the ratio between herbage produced and herbage consumed, but also with the longer-term goal of maintaining the persistence of the herbage resource. For that, the structural and botanical adaptations of the swards to the grazing regime have to be considered. The literature concerning the plant–animal interaction is dominated by studies on perennial ryegrass swards in temperate regions under relatively intensive grazing managements. At a world level, information on the behaviour of plant species with different physiologies and morphologies under more extensive environmental conditions is needed in order to analyse the agronomical and ecological sustainability of pastures under contrasting grazing systems.

The aim of this chapter is to develop a conceptual basis for modelling leaf tissue fluxes and to identify the key morphological plant parameters that determine the efficiency of herbage utilization in a grazing system. We shall first analyse the dynamics of leaf tissue production as resulting from the morphogenetic responses of plants to environmental and sward state conditions; then we shall analyse the temporal links between the dynamics of leaf production and senescence; finally, we shall analyse the role of the intensity and the frequency of defoliation in the optimization of the balance between leaf tissue consumption and senescence.

Plant Morphogenesis and Leaf Tissue Dynamics

Plant growth processes

Leaf tissue production can be analysed as the result of two interacting processes: (i) the production of assimilates by individual plants resulting from light interception and photosynthesis of leaves; and (ii) the use of assimilates by leaf meristems for the production of new growing cells and, finally, for leaf area expansion. So the rate of expansion of new leaf tissue on a plant can be considered as limited by either the production or the use of assimilates. The use of assimilates by the leaf meristems is directly determined by temperature, which governs the rates of cell division and expansion (Ben-Haj-Salah and Tardieu, 1995) and creates a demand for C and N assimilates to provide energy and material for leaf tissue expansion. When the supply of assimilates is large enough to meet the meristem demands, leaf growth can reach the potential determined by temperature and the excess assimilates can be stored as carbohydrate reserves. As the plant grows, while the assimilate supply increases, as a consequence of leaf area expansion (more light is captured), the size and number of meristems also increase and hence an approximate balance between supply of and demand for assimilates (with some fluctuations according to variations in radiation levels and temperature) is maintained. In such a situation, either a trophic approach (light interception and photosynthesis) or a morphogenetic approach (temperature and plant organ demography) can be used to model plant growth or leaf tissue production. Although both aspects have to be linked together for the analysis of growth at the whole-plant level, when the interest is focused on leaf tissue production the priority of use of assimilates for leaf meristems (Tabourel-Tayot and Gastal, 1999) supports the consideration of leaf growth as mainly governed by morphogenesis. With this simplified view of a growing plant, in the absence of water stress, leaf tissue expansion can be considered as directly determined by temperature and by nitrogen nutrition (Gastal *et al.*, 1992). If the rate of assimilate supply is lower than the demand for leaf growth, then the plant limits the number of active meristems (reduced tiller density) to maintain the potential leaf growth on the main tiller. For this reason, while leaf expansion rate on mature tillers is very little affected by the C supply, tillering is highly responsive to variations in the radiation level.

Plant morphogenesis and sward structure

Plant morphogenesis can be defined as the dynamics of generation (genesis) and expansion of plant form (*morphe*) in space (Chapman and Lemaire, 1993). According to Gillet *et al.* (1984), it can be considered that plants have a genetically programmed morphogenesis, the rate of realization of which is temperature-dependent. This morphogenetic programme determines the functioning and the coordination of meristems in terms of the production and expansion rates of new cells, which, in turn, define the expansion dynamics of the growing organs (leaf, internode, tiller) and the C and N demands necessary to fill the corresponding expanding volumes (Durand *et al.*, 1991). In this way, the plant morphogenetic programme can be coupled with the trophic processes that determine the assimilate supply.

Plant morphogenesis is expressed as the rate of appearance and expansion in size of new plant organs, as well as their disappearance rate by senescence. For a vegetative grass species, such as *Lolium perenne* or *Festuca arundinacea*, leaf tissues are produced sequentially as a chain of phytomers at the level of the individual tiller, each one following a preprogrammed series of developmental stages, from primordium initiation as meristem to the mature size and the ontogenic senescence (Silsbury, 1970). At the level of the apical meristem, the time elapsing between the initiation of two successive leaf primordia is called the plastochron. At tiller level, the time elapsing between the appearance of two successive leaves is called the phyllochron (Ph). The rate at which new leaves are produced on a tiller is measured by the leaf appearance rate (LAR), which is the reciprocal of the phyllochron:

$$\text{LAR (leaf per day)} = 1/\text{Ph} \quad (1)$$

For a given species the phyllochron appears as a relatively constant parameter when expressed in thermal time (degree-days), thus providing a time-scale basis for studying plant morphogenesis. With the production of a new leaf primordium, the apical meristem produces a new axillary bud, which can potentially grow and give a new tiller. Considering the more or less constant delay of two phyllochrons (or more, depending on species) between the appearance of a given leaf and the appearance of the corresponding new tiller, it is possible to calculate directly the maximum number of tillers that can appear during a phyllochron, which, in fact, corresponds to the concept of 'site filling' (Davies, 1974;

Neuteboom and Lantinga, 1989; Van Loo, 1992) or 'site usage' (Skinner and Nelson, 1992). As shown by Matthew *et al.* (1991), the phyllochron may also be used for analysing sequential root appearance, which gives the possibility for an integrated view of clonal plant development.

As presented in Fig. 14.1, the three main leaf growth parameters are as follows:

- The leaf elongation rate (LER), which measures the daily increase in length of the individual leaf (mm day^{-1}).
- The LAR, which measures the number of leaves appearing per unit of time.
- The leaf lifespan (LLS), which measures the period of time during which a given portion of newly appearing leaf tissue remains green.

The leaf length of fully expanded leaves (final leaf length (FLL)) can be derived directly from the rate at which the individual leaf elongates (LER) and the duration of the elongation period (LED):

$$\mathrm{FLL} = \mathrm{LER} \times \mathrm{LED} \quad (2)$$

providing that leaves elongate at a more or less constant rate during the elongation period. It has been shown for *F. arundinacea* (Robson, 1967) that the LED was proportional to the Ph:

$$\mathrm{LED} = a \times \mathrm{Ph} \quad (3)$$

The parameter a gives the number of leaves growing simultaneously on the same tiller. In consequence, it should be possible to relate the final leaf size to both the LAR and the LER:

$$\mathrm{FLL} = a \times \mathrm{LER} \times \mathrm{Ph} \quad (4)$$

or:

$$\mathrm{FLL} = a \times \mathrm{LER}/\mathrm{LAR} \quad (5)$$

Table 14.1 shows the response of LAR to temperature for different C_3 and C_4 species growing in winter and spring in the *pampa* region of Argentina. For most C_3 species, the response of LAR to temperature is linear, with an apparent threshold temperature between 3 and 5°C. For the C_4 species, the threshold temperature is much higher (8–9°C). The existence of a threshold temperature different from 0°C implies the use of base temperatures of 4°C for the C_3 and 8.5°C for the C_4 species for expressing the phyllochron in degree-days. For *F. arundinacea* in temperate conditions, Lemaire (1988) found a threshold temperature of 0°C and a constant phyllochron of 217 degree-days on a 0°C basis, while Davies and Thomas (1983) obtained a constant value of 110 degree-days for *L. perenne* and Duru *et al.* (1993) proposed a value of 160 degree-days for *Dactylis glomerata*, both without clear indications of the base temperature used. It is important to note

Table 14.1. Relationship between leaf appearance rate (LAR) and daily average air temperature for different species of a grazed community in the flooding *pampa* in Argentina (after Agnusdei, 1999).

Species	LAR = $b(\theta - \theta_0)$	R^2
Chaetotropis elongata (C_3)	LAR = 0.0070 (θ − 3.6)	0.83
Hordeum stenostachys (C_3)	LAR = 0.0053 (θ − 3.8)	0.94
Lolium multiflorum (C_3)	LAR = 0.0072 (θ − 3.6)	0.88
Stipa neesiana (C_3)	LAR = 0.0033 (θ − 3.6)	0.93
Paspalum dilatatum (C_4)	LAR = 0.0055 (θ − 8.2)	0.73
Sporobolus indicus (C_4)	LAR = 0.0041 (θ − 9.0)	0.72

θ_0 is the threshold temperature, LAR is calculated in leaf day^{-1} and the phyllochron (Ph) is equal to 1/*b* and is expressed in degree-days calculated with θ_0 as base temperature.

here that, in the literature, Ph values are often calculated using 0°C as the base temperature. For species having a very low threshold temperature, such as tall fescue, this value of Ph (base 0°C) remains constant, whatever the range of temperatures experienced by the plants. As the threshold temperature of the species becomes different from zero, a Ph value expressed on a 0°C basis will tend to decrease with the increase in the average daily temperature. Consequently, the interspecific variability in leaf appearance rate must be expressed by both the threshold temperature θ_0 and the slope of the regression line LAR vs. $\theta - \theta_0$, which represents the responsiveness of the plant to increasing temperatures. In this way, it is possible to discriminate among species in the same metabolic group.

Table 14.2 shows the LER response to temperature for various native grasses of the humid *pampa* of Argentina. In the range of average daily temperatures between 5 and 17°C for the C_3 species and 12 and 20°C for the C_4 species, the response of LER to temperature appears to be approximately exponential. It is difficult to give a clear biological interpretation for the α and β coefficients of the

Table 14.2. Response of leaf elongation rate (LER) to temperature for different species of a grazed community in the flooding *pampa* in Argentina (after Agnusdei, 1999).

Species	LER = $\alpha\ e^{\beta \times \theta}$	R^2
Chaetotropis elongata (C_3)	LER = 1.15 $e^{(0.135 \times \theta)}$	0.98
Hordeum stenostachys (C_3)	LER = 1.09 $e^{(0.128 \times \theta)}$	0.98
Lolium multiflorum (C_3)	LER = 1.20 $e^{(0.140 \times \theta)}$	0.95
Stipa neesiana (C_3)	LER = 0.25 $e^{(0.235 \times \theta)}$	0.95
Paspalum dilatatum (C_4)	LER = 0.90 $e^{(0.140 \times \theta)}$	0.95
Sporobolus indicus (C_4)	LER = 0.70 $e^{(0.120 \times \theta)}$	0.93

regression line, because of the use of only one restricted range of temperatures for statistical adjustment. Nevertheless, β represents the reactivity of LER to an increase in temperature, while α should represent the intrinsic capacity of the species for leaf elongation at low temperatures. Among the C_3 species, it is possible to distinguish the response of *Stipa neesiana*, which has a low LER at low temperatures ($\alpha = 0.25$) and a high responsiveness to temperature increases ($\beta = 0.235$) compared with the other three species evaluated. Among the C_4 species, it is possible to clearly distinguish between *Paspalum dilatatum* and *Sporobolus indicus*. Nevertheless, in the range of temperatures between 12.5 and 17.5°C, corresponding to the spring period in the *pampa* region, the LER of the C_3 grasses remains higher than that of the C_4 species. It is difficult to carry out LER measurements later in the season (summer), as a consequence of the possible interactions with water stress. Previous results suggest that no direct relationship would exist between the metabolic pathway (C_3 or C_4) and the LER capacity of a species, and this reinforces the idea that the leaf morphogenetic characteristics of plants can be analysed relatively independently of C nutrition aspects.

As discussed by Cruz and Boval (Chapter 8, this volume), plant N nutrition can affect both LER and LAR, depending on the morphological type of the species. For caespitose grasses, such as *F. arundinacea* (temperate C_3) or *Setaria anceps* (tropical C_4), while the LER and the FLL increased three to four times with N application, the LAR only decreased 20%. Conversely, in the stoloniferous grass *Digitaria decumbens* (tropical C_4), N application strongly affected LAR but had a very small effect on LER and FLL.

The LLS can be measured directly on labelled tillers by determining the delay between the appearance of a new portion of leaf tissue and its yellowing. The LLS represents the period during which, after a complete defoliation, green leaf tissues accumulate on individual adult tillers without any losses by senescence. After this delay, the accumulation of dead leaf tissues starts, matching the accumulation of new leaf material and leading to a balance between the appearance and disappearance of leaf tissues on the individual tiller. At this stage of regrowth, the maximum quantity of standing green leaf tissue (ceiling yield) per tiller is achieved and no further herbage accumulation can be obtained in the sward unless there is an increase in tiller density (see Matthew *et al.*, Chapter 7, this volume). So, as proposed by Lemaire and Chapman (1996), LLS can be used for the characterization of the differential aptitudes of the species to accumulate green leaf tissues. For *F. arundinacea*, the average LLS is 570 degree-days, which corresponds to approximately 2.5 phyllochrons. In consequence, a maximum of 2.5 fully expanded leaves can be accumulated on each adult tiller before the ceiling yield is reached (Lemaire, 1988). As shown in Fig. 14.1, LLS determines the maximum number of living leaves (NLL) an individual tiller can bear when its ceiling yield is reached. NLL can be calculated as follow:

$$NLL = LLS/Ph \tag{6}$$

or:

$$NLL = LLS \times LAR \tag{7}$$

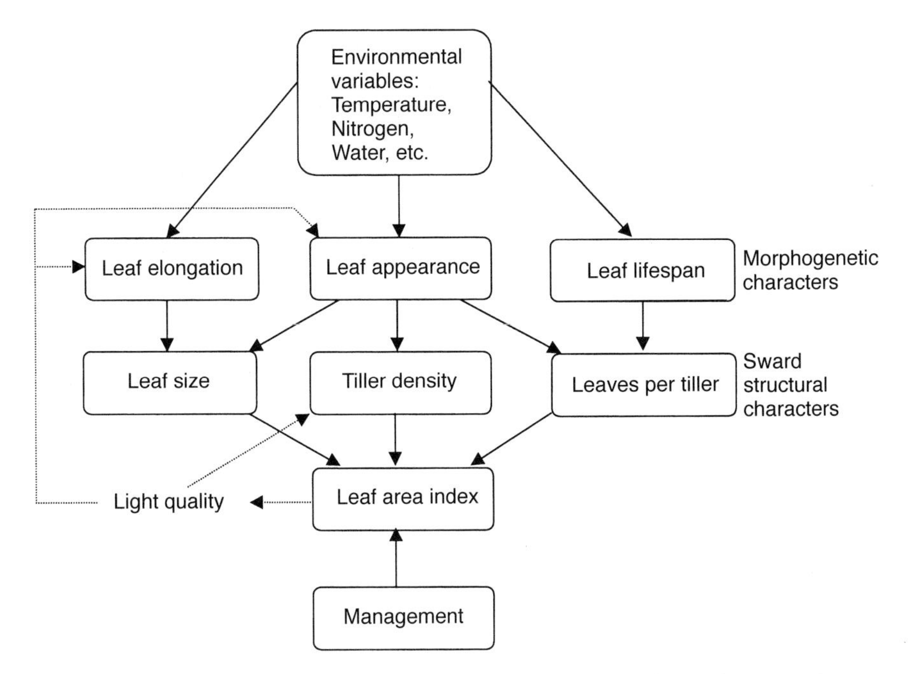

Fig. 14.1. Relationship between morphogenetic variables and sward structural characteristics (after Lemaire and Chapman, 1996).

Table 14.3 represents the LLS of different species measured at different seasons in a natural grassland community in the *pampa* region of Argentina. LLS was recorded directly on labelled tillers. In the absence of any precise information on the threshold temperature for the senescence process, a 0°C base temperature was used for expressing the values in degree-days. Results show the existence of seasonal variations in the LLS within the C_3 grasses. The reasons for this variability are not clear and might have been associated with the difficulty of precisely determining leaf senescence. Further, marked among-species differences were revealed, these variations appearing to be highly reproducible between seasons in the case of the C_3 grasses (C_4 species were only observed during spring).

When the NLL, estimated by means of equation 6 (or 7), is plotted against the observed NLL (Fig. 14.2), a unique regression line with a slope 1 : 1 holds for all species and seasons, this result being considered a validation of the model of Lemaire and Chapman (1996) presented in Fig. 14.1.

As shown in Fig. 14.1, the three main leaf morphogenetic parameters, LER, LAR and LLS, together determine the components of the sward structure: the mature leaf size (FLL), the maximum number of green leaves per tiller (NLL) and the potential number of tillers. As indicated by equations 5 and 7, and considering what was previously stated about tiller bud production, it is clear that LAR plays a central role in determining pasture structure. In this sense, a species with high LAR will tend to produce a high number of short leaves per tiller and a high

Table 14.3. Leaf lifespan (LLS) of different species of a grazed community of the flooding *pampa* in Argentina (after Agnusdei, 1999). LSS is calculated in degree-days with a base temperature of 0°C.

Species	Leaf lifespan (LLS)			
	Autumn	Winter	Spring	SE
Chaetotropis elongata (C_3)	391 cB	391 cB	541 bA	32
Hordeum stenostachys (C_3)	530 bA	425 bB	594 bA	32
Lolium multiflorum (C_3)	418 cA	335 cB	396 cAB	36
Stipa neesiana (C_3)	745 aA	745 aA	610 baA	52
Paspalum dilatatum (C_4)			550 b	
Sporobolus indicus (C_4)			700 a	

Small and capital letters indicate significant differences ($P < 0.05$) for columns and rows, respectively. SE, standard error.

tiller density, leading to a short and dense sward structure. Conversely, a species with a low LAR will tend to produce few long leaves per tiller and a lower tiller density, leading to a potentially tall sward structure. Moreover, a species with a long LLS and a low LAR, as in the case of *F. arundinacea* relative to *L. perenne* (Lemaire, 1988), can be expected to accumulate more large leaves and thus to

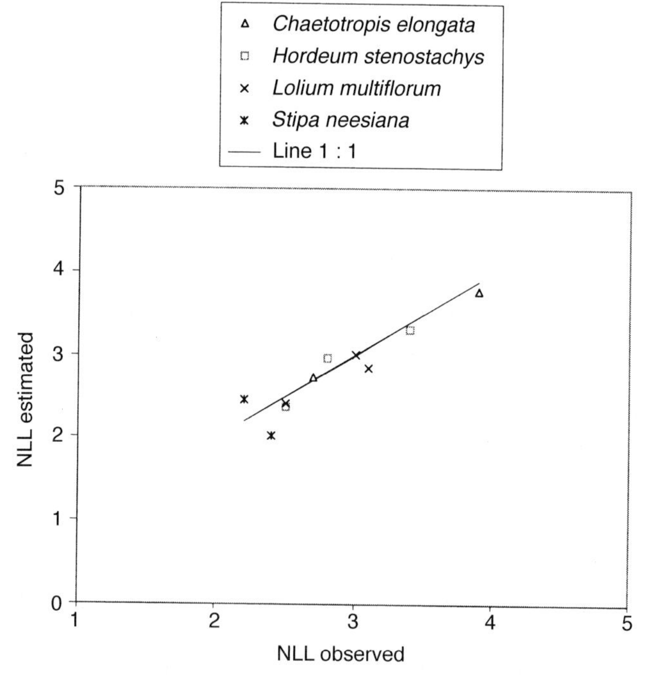

Fig. 14.2. Relationship between the number of living leaves per tiller (NLL) estimated by equation 5 (or 6) and the observed values at different seasons for different species of a grazed community of the *pampa* in Argentina (after Agnusdei, 1999).

show a greater ceiling yield capacity. A similar contrast among C_4 tropical grasses has been reported by Cruz and Boval (Chapter 8, this volume). The analysis of species or genotype productivity has been greatly confusing in the past, because no clear distinction has been made between the aptitude of a genotype for a high leaf tissue production rate and its aptitude for herbage accumulation. A precise determination of the main leaf morphogenetic attributes of species and genotypes will help to better analyse both leaf tissue and structural dynamics in grazed swards.

Morphogenetic plasticity

In Fig. 14.1, the dotted lines show that individual plants can perceive their own environmental conditions and adapt their leaf morphogenesis through a plastic response. This phenomenon, called 'phenotypic plasticity' (Bradshaw, 1965), plays an important role in the adaptation of forage species to grazing management. Phenotypic plasticity can be defined as the progressive and reversible change in the morphogenetic traits of individual plants. In these terms, pastures can be considered as highly regulated systems, where any structural change determines responses in plant morphogenesis, which, in turn, modify the sward structure itself. Within this context, LAI appears to be a major integrating sward structure characteristic.

The effect of LAI on tillering has been well demonstrated (Simon and Lemaire, 1987), a decline in tiller density in pastures maintained at high LAI levels being observed under grazed conditions (see Matthew *et al.*, Chapter 7, this volume). The role of light quality and, more precisely, of the red : far-red ratio on tillering is now well documented (Deregibus *et al.*, 1983; Casal *et al.*, 1987; Gautier *et al.*, 1999): plants perceive the changes in light quality in their surrounding environment and respond by limiting emergence from tiller buds as higher sward LAIs are maintained.

LAR can also be modified by the pasture LAI. Table 14.4 shows that the phyllochron of *P. dilatatum* and *Cynodon dactylon* increased substantially from swards maintained at a low LAI (or height) to swards maintained at a high LAI. In the same swards, *Lolium multiflorum* seemed insensible to LAI variations. As demonstrated by Gautier and Varlet-Grancher (1996), the effect of LAI on LAR might be explained by a response of the plants to a change in the quality of the light environment (blue light), large differences in the responsiveness between genotypes being reported by these authors.

Mazzanti *et al.* (1994) showed a large effect of sward height (or LAI) on the LER of tall fescue. Table 14.5 shows the effect of sward height on the LER of several species from the grasslands of the *pampa* region of Argentina. This effect of LAI on LER is not yet clearly explained. Begg and Wright (1962) observed that leaf growth is strongly reduced by full light, and suggested that this response could be phytochrome-mediated. Davies *et al.* (1983) suggested that the sheath tube, by protecting the elongating leaf from direct full light, could readily influence the LER.

Table 14.4. Effects of sward height on the phyllochron (Ph) of three different species in a grazed community in the flooding *pampa* in Argentina (after Agnusdei, 1999).

	Lolium multiflorum			*Paspalum dilatatum*			*Cynodon dactylon*		
Seasons	L	M	H	L	M	H	L	M	H
Winter	173 a	177 a	168 a						
Early spring	167 a	169 a	168 a	170 b	187 a	205 a			
Late spring				140 b	163 a	182 a	69 c	75 b	89 a

L, M and H refer, respectively, to low, medium and high sward height; Ph is calculated in degree-days, with a base temperature of 0°C, which allows comparison between sward height treatments for the same species and the same season (different letters mean significant differences between treatments).

Table 14.5. Effects of sward height on leaf elongation rate (LER) (mm day^{-1}) of different species of a grazed community of the *pampa* in Argentina (after Agnusdei, 1999).

	Lolium multiflorum			*Paspalum dilatatum*			*Cynodon dactylon*		
Seasons	L	M	H	L	M	H	L	M	H
Winter	1.0 b	1.3 b	2.0 a						
Early spring	4.7 c	6.1 b	8.9 a	5.0 b	6.5 a		6.9 a		
Late spring				10.2 b	10.6 b	14.6 a	6.2 b	7.6 b	14.1 a

L, M and H refer to low, medium and high sward height, respectively; comparison can be made between treatments for the same species and the same season (different letters mean significant differences between treatments).

If the number of growing leaves per tiller remains stable across different sward states and if its value is adequately represented by the *a* coefficient of equation 5, data from Tables 14.4 and 14.5 can be used to calculate the species FLL from equation 5. This calculated FLL has been related to observed data measured on labelled tillers from leaves that were not defoliated before reaching their mature size. Figure 14.3 shows that a general regression line of slope 1 : 1 holds for all species and seasons and for the whole range of sward heights investigated in the experiment.

Modelling leaf tissue fluxes at tiller population level

Using the different equations relating LER, LAR and LLS to temperature for a given species, it is possible to model the dynamics of leaf tissue fluxes at the level of an individual tiller. Figure 14.4 represents the predicted dynamics of growth and senescence of successive cohorts of leaves appearing through time for a given population of mature tillers of *L. perenne* in a natural grassland community of the *pampa* in Argentina. When a leaf emerges, it grows at a constant daily rate (LER). After a time period corresponding to the LED, which is proportional to the leaf

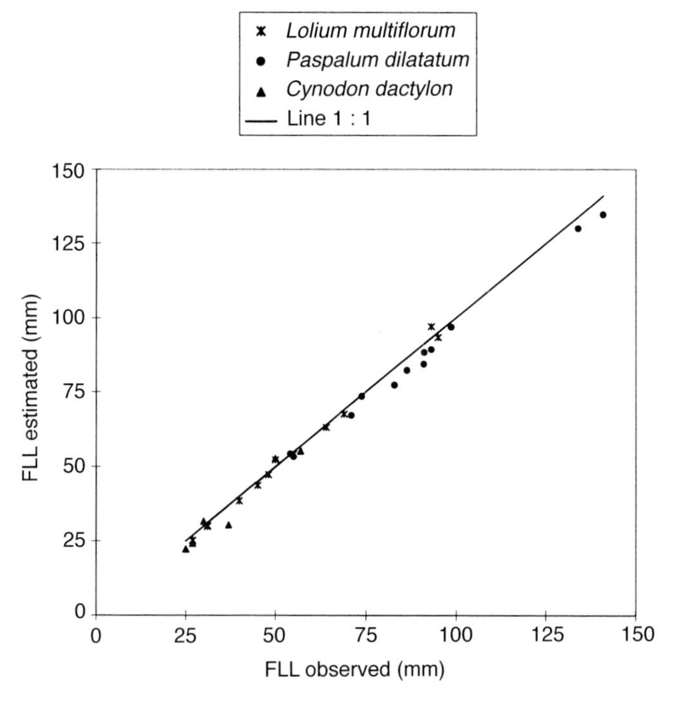

Fig. 14.3. Relationship between the final leaf length (FLL) estimated by means of equation 5 and the observed values for different species of a grazed community of the *pampa* in Argentina (after Agnusdei, 1999).

appearance interval (Ph), the elongating leaf reaches its final size (FLL) and remains green during its lifespan (LLS). Afterwards, the leaf portion (dL) that appeared on day *j* starts to senesce at day *j* + LLS, mimicking the progression of tissue senescence from the tip to the base of the leaf. Knowing the relationships of LER and LAR with temperature, it is possible to simulate the fluxes of leaf tissue production and senescence under fluctuating temperature conditions. The model simulates the fluxes for a tiller population, assuming that leaf appearance is not synchronized between tillers and that the probability of leaf appearance at tiller population level is randomly distributed each day. Thus, it is possible to simulate the effect of temperature changes on leaf size. Moreover, at each date, it is possible to calculate the average rate of leaf tissue production from the population of growing leaves and the average rate of leaf tissue senescence from the population of senescing leaves.

The model, as well as empirical observations, show that the final leaf length tends to decrease from autumn to winter as related to the general trend of decreasing temperatures, and then it increases progressively with temperature increase during spring. The average LER of the leaf population responds immediately to any change in temperature, so the leaf tissue production flux follows exactly the seasonal temperature variations. But the average senescence rate does

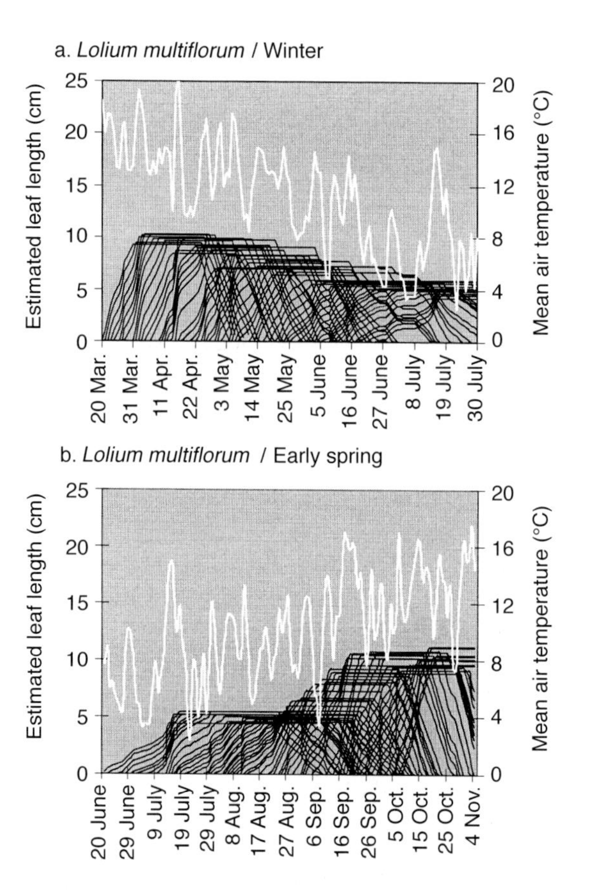

Fig. 14.4. Modelling the dynamics of tissue production and senescence at whole-leaf population level for *Lolium multiflorum*. The diagram shows the evolution of leaf lamina length of individual leaves according to temperature variations. For simplicity, only the lines corresponding to leaf cohorts appearing every 2 days are drawn.

not immediately follow fluctuations in temperature, because the portion of leaf tissue that dies each day corresponds to the portion of leaf produced an LLS before. In a period of decreasing temperatures, such as autumn, the portions of leaf tissue that senesce each day are larger than those that are currently produced, and this leads to a negative balance between growth and senescence. On the contrary, during a period of increasing temperatures, such as spring, a positive balance between both fluxes will be obtained. In other words, the leaves produced during autumn progressively senesce during winter and are replaced by shorter leaves produced under the lower temperature conditions of the season. The reverse phenomenon occurs in spring. Thus, the equilibrium between leaf growth and leaf senescence can only strictly be observed under constant temperature. In natural conditions, any event that increases the LER and the FLL, such as a temperature augmentation or N application, would increase the growth rate of the pasture without altering the current senescence and hence would modify the flux balance. A new equilibrium would be reached after a delay corresponding approximately to the LLS.

Leaf Tissue Consumption Fluxes in Grazed Swards

The rate of consumption of leaf tissues by the animals can be measured at the individual tiller level, using tiller labelling techniques (Hodgson and Ollerenshaw, 1969; Mazzanti and Lemaire, 1994). Such a method allows the recording of the components of defoliation: (i) the frequency of defoliation and (ii) the intensity of each defoliation event on either a tiller or a leaf basis.

Probability of defoliation of leaves

Defoliation events, at the individual tiller level or at the individual leaf level, can be recorded by labelling a given set of individual tillers in a grazed sward and by recording the number of defoliation events observed on the population of labelled tillers during the interval between successive measurements (2 or 3 days). If the tillers are randomly labelled, a probabilistic approach for analysing the discrete events of defoliation can be used, allowing the proportion of labelled tillers defoliated daily to be considered as a measure of the probability of defoliation of the whole tiller population of the sward. This estimation can be made for any defoliation event on an individual tiller, whatever the leaf age category or the number of leaves simultaneously removed by animal. The probability of tiller defoliation will be called Pt. A similar estimation can be made at the level of each individual leaf, with Pf_i denoting the probability of defoliation of the leaf age category i. The reciprocal of the probability of defoliation gives the average interval between two successive defoliation events on the same tiller or on the same leaf category: It or If_i, respectively. Wade *et al.* (1989) showed that the probability of defoliation of individual tillers (Pt) in a grazed sward is directly related to the stocking density through a linear function, this denoting that the proportion of tillers defoliated each day proportionally increases with the increase in stocking density.

When the probability of defoliation is analysed at the individual leaf level (Pf_i), it is possible to note that, as the age category increases (older leaves), leaves are less frequently defoliated (Mazzanti and Lemaire, 1994). Figure 14.5 shows the probability of defoliation of *P. dilatatum* for the different leaf age categories for a range of sward heights. The probabilities have been calculated for a period of one phyllochron (Pf_i(Ph)). It can be observed that it is the uppermost fully elongated leaf that is more frequently defoliated, the senescing leaves being very rarely defoliated. This result is a general picture for many grass species. The differences due to sward height can be interpreted as resulting from the differences in stocking density used for maintaining the different sward states.

a. *Paspalum dilatatum* / Early spring

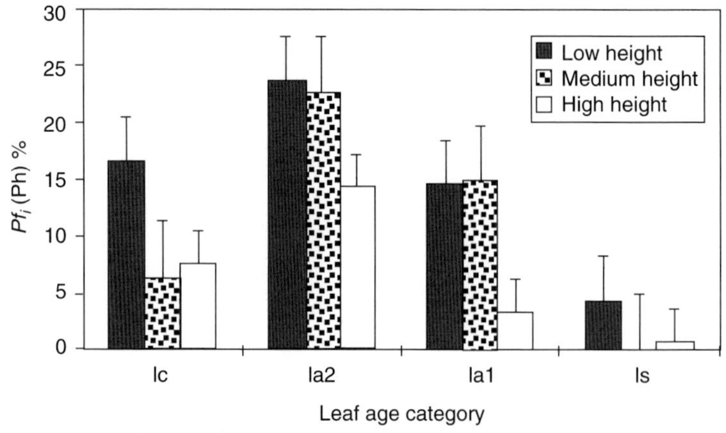

b. *Paspalum dilatatum* / Late spring

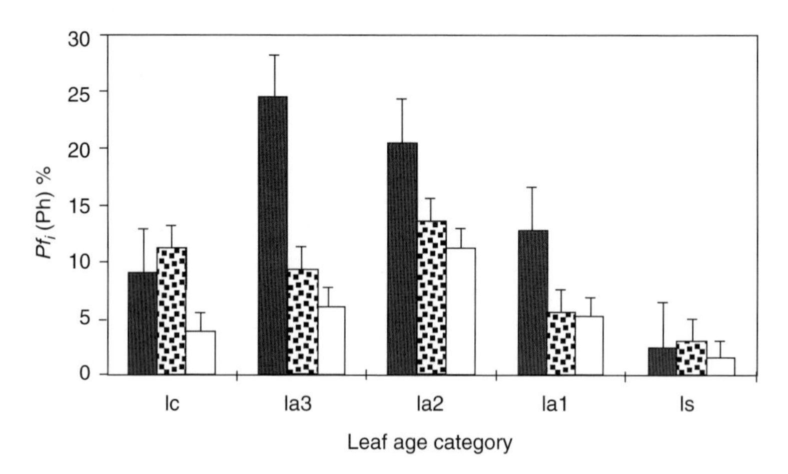

Fig. 14.5. Probability of defoliation of the different leaf age categories of *Paspalum dilatatum* calculated for one phyllochron interval: lc, growing leaves; la 1 … 3, mature leaves from oldest to youngest; ls, senescing leaves.

If we consider that a given leaf changes its category at each phyllochron (when a new leaf emerges), it is possible to cumulate the probability of defoliation of all the leaf age categories that are present on a tiller in order to calculate the average number of defoliations an individual leaf is susceptible to have during its lifespan:

$$Pf\,(\mathrm{LLS}) = \Sigma[Pf_i\,(\mathrm{Ph})] \tag{8}$$

Table 14.6 shows such an estimation for *P. dilatatum* and *L. multiflorum* growing in the same community. Even for paddocks maintained at a low sward height by

Table 14.6. Effects of sward height on the probability of defoliation (Pf_{LLS}) and on the proportion of tissue removed by grazing for an individual leaf during its whole lifespan (ID_{LLS}) for different species in a grazed community of the *pampa* in Argentina (after Agnusdei, 1999).

		Lolium multiflorum			*Paspalum dilatatum*			*Cynodon dactylon*		
Seasons		L	M	H	L	M	H	L	M	H
Winter	Pf_{LLS}	42%	17%	8%						
	ID_{LLS}	25%	9%	3%						
Early spring	Pf_{LLS}	58%	43%	27%	59%	43%	28%			
	ID_{LLS}	33%	30%	24%	36%	20%	15%			
Late spring	Pf_{LLS}				71%	46%	30%	43%	19%	15%
	ID_{LLS}				42%	22%	14%	37%	11%	7%

L, M and H refer to low, medium and high sward height, respectively; comparison can be made between treatments for the same species and the same season.

a relatively high stocking density, 40% of the individual leaves were not defoliated before they senesced. At the lowest stocking density (highest sward height), about 75% of the leaves escaped defoliation.

Intensity of defoliation

The intensity of defoliation of individual leaves can be analysed at each individual defoliation event, as well as during the LLS.

The intensity of an individual defoliation (ID_i) can be estimated by the ratio between the length of the leaf portion removed by the animal at a given defoliation event and the length of the leaf before being defoliated. As demonstrated by Mazzanti and Lemaire (1994) for *F. arundinacea*, the intensity of defoliation appears relatively constant and independent of the leaf age category. Data obtained by Agnusdei (1999) in a natural grazed community of the *pampa* in Argentina show that, for *L. multiflorum*, *P. dilatatum* and *C. dactylon*, a constant proportion of about 50–55% of the leaf length is removed at each defoliation, whatever the leaf age category or sward height. Carrère (1994) also found a constant intensity of defoliation of 55% for *L. perenne*. These results, obtained at the level of individual leaves, confirm those obtained by Wade *et al.* (1989), who demonstrated that a constant proportion of 35% of the extended tiller height of perennial ryegrass is removed at each defoliation. Taking into account the fact that the sheath height represents about 1/3 of the extended tiller height, 35% of the total tiller height corresponds approximately to 50–55% of lamina length. In this sense, the evidence indicates the existence of a remarkable adjustment of bite depth by the grazing animal to the average lamina length of the sward, which, in turn, leads to a constancy in the intensity of defoliation (see Wade and Carvalho, Chapter 12, this volume).

The intensity of defoliation during the LLS (ID_{LLS}) represents the proportion of the FLL removed by the grazing animal before leaf senescence. It can be easily approximated as the product of the probability of defoliation during the LLS and the intensity of each defoliation event:

$$ID_{LLS} = Pf\,(LLS) \times ID_i \tag{9}$$

The results of such a calculation are given in Table 14.6 for *P. dilatatum*, *L. multiflorum* and *C. dactylon*. Only 40% of the total length of each individual leaf was grazed by the animals at the highest stocking density, used to maintain a low sward height, meaning that around 60% of the leaf tissues produced in the sward escaped defoliation and returned to the soil as litter. For the less intensively stocked treatment, only 10% of leaf tissues were removed. If leaf length data are to be transformed into leaf dry weight, it has to be considered that the width and the thickness of the lamina increase from tip to base, leading to the leaf mass per unit leaf length being three times higher in the basal 1/3 portion of the lamina than in the upper 1/3 (Maurice *et al*., 1997). If expressed on a leaf mass basis, data from Table 14.6 would demonstrate that the proportion of leaf tissue removed by grazing animals is very low, even in swards maintained at relatively short heights. The data presented by Mazzanti and Lemaire (1994) on *F. arundinacea* showed that a maximum intensity of defoliation of 70% in terms of leaf length was obtained with a high N application rate and an associated high stocking density. Taking into account the mass gradient along the lamina, only 50% of the lamina dry matter produced would have been removed by animals. Moreover, in terms of total leaf tissue, the proportion of sheath vs. lamina, which is about 1/3, is indicative of the very low proportion of biomass that can be harvested by grazing animals. Nevertheless, as demonstrated by Lemaire and Culleton (1989), about 50% of C and 80% of N are recycled from senescing leaves during the senescence process and can be used by plants for supporting further leaf tissue production.

Efficiency of Herbage Utilization at Grazing

The efficiency of herbage utilization in a grazing system can be defined as the proportion of gross leaf tissue production that is removed by the animals before entering the senescent state (Lemaire and Chapman, 1996). From ecological and agronomical points of view, a rational calculation of such an efficiency should consider the maintenance of a sward state that ensures the 'sustainability' of leaf tissue production, i.e. which allows a light interception that can be maintained by the sward. Under continuous stocking, as demonstrated by Bircham and Hodgson (1983), that situation corresponds to a paddock maintained at approximately constant sward height or LAI. For a rotational grazing system, as shown by Parsons *et al*. (1988), such a state is characterized by the average LAI at which the different paddocks are maintained over time.

Under these conditions, an approximation of the efficiency of herbage utilization can be obtained by means of the average intensity of defoliation of the leaf population of the sward (ID_{LLS}). For a sward maintained at a constant LAI by frequent adjustments of stocking density during a given period, as proposed by Mazzanti and Lemaire (1994), the proportion of leaf tissue removed by the animals directly depends on the stocking density, which determines the average interval between two successive defoliations on the same leaf (the reciprocal of the probability of defoliation of individual leaves, Pf_i). Accepting that the intensity of defoliation (ID_i) can be considered to be a constant value, the probability of defoliation of an individual leaf during its lifespan (ID_{LLS}) (which measures the proportion of the leaf tissue removed by grazing) directly depends on both the stocking density and the average LLS of the species considered. If herbage production is increased by any growth factor, such as N or other mineral application, temperature and/or irrigation, a corresponding increase in stocking density has to be applied for maintaining the sward LAI, and therefore a greater proportion of leaf tissue will be removed by the animals. Data from Table 14.7 show that, as the N application rate in a grazed plant community increased, leaf senescence was lower relative to the gross leaf production in all the species evaluated. This result is not a direct effect of N on the herbage intake per animal, but an indirect effect through the corresponding increase in stocking density. Consequently, the proportion of gross leaf tissue production that escapes defoliation and returns to the soil through the senescence pathway is expected to be lower for species with long LLS, such as *S. neesiana*, as compared with species with faster leaf tissue turnover.

As stated previously, the efficiency with which the leaf tissue produced by plants is harvested by animals under continuous stocking is directly linked to the stocking density, through the average defoliation interval of tillers. However, this average includes a large intrapopulation variability, which, in addition, would probably increase with the decrease in the defoliation interval associated with

Table 14.7. Effects of N application rate in autumn and winter in a grazed community maintained at a constant sward height upon the proportion of leaf tissue produced which has been removed by grazing (after Rodriguez Palma, 1998).

Season	Species	N0	N50	N100
Winter	*Lolium multiflorum*	28%	45%	45%
	Stipa neesiana	40%	59%	59%
Early spring	*Lolium multiflorum*	49%	67%	67%
	Stipa neesiana	64%	81%	82%
	Hordeum stenostachys	45%	50%	50%

N0, N50 and N100 correspond to levels of N use of 0, 50 and 100 kg N ha^{-1}. The three N treatments were maintained at a similar sward height of 5 cm (LAI of 1.1) by contrasting stocking density: 200–270 kg LW ha^{-1} for N0, 370–420 kg LW ha^{-1} for N50, 540–900 kg LW ha^{-1} for N100 during winter and spring, respectively. LW, live-weight.

herbage growth decreases (for example, under increasingly lower N inputs in extensively managed conditions). Under these circumstances, the heterogeneity of the grazed surface increases, with frequently defoliated patches of consequently low-production tillers alternating with infrequently grazed (or refused) areas, where the relatively higher herbage production is associated with an inefficient harvest of the leaf tissues produced. As discussed by Parsons *et al.* (Chapter 15, this volume), such a discrepancy between herbage growth and herbage use from different patches creates instability in the sward, determining a progressive increase in its heterogeneity and leading, in fact, to a large decrease in the herbage use efficiency of the sward as a whole. In rotational grazing systems, the direct link between stocking density and defoliation interval can be partially broken. As demonstrated by Wade (1991), with high instantaneous stocking densities in strip grazing or in a 5-day paddock system, tillers are grazed several times per day or per grazing period. This succession of defoliations occurring during a short time interval can be considered to be similar to a unique event, whose frequency is determined by the rest period imposed by the grazing system and whose intensity depends on the instantaneous stocking density (which, in fact, determines the number of individual defoliations during any grazing period). In this way, it is possible to restrain the increase in sward heterogeneity that would occur with decreasing stocking density. Further, it is possible to adapt the duration of the rest period (and therefore the defoliation interval) to the LLS of the main species of the plant community.

A theoretical relationship between an index of leaf tissue consumption and an index of leaf tissue production in a grazed system is presented in Fig. 14.6. The point where both indexes are equal to 1 corresponds to a potential herbage production with optimum stocking rate and intake and a given herbage use efficiency. In continuously grazed swards, in a situation of herbage production extensification (as with N application reductions), the decrease in leaf tissue consumption should be more than proportional to the decrease in leaf tissue production, as a consequence of the decrease in stocking density and its unavoidable effect on herbage use efficiency (Mazzanti and Lemaire, 1994). Therefore, the corresponding trajectory AC is placed bellow the 1 : 1 line in Fig. 14.6. In a rotational grazing system, as the defoliation interval does not directly depend on stocking density, the decrease in herbage use efficiency with the decline in herbage growth could be largely avoidable and the corresponding trajectory AB should be close to the 1 : 1 line.

The graph presented in Fig. 14.6 allows the history of grazing research to be analysed in the light of present knowledge. More than 50 years ago, the average situation in the grazing area of Europe was next to point C: low herbage production and low efficiency of grazing. The first important effect of research was to move the systems from point C to point B by the application of the principles of rotational grazing. The sequential grazing of successive paddocks allowed the efficiency of herbage use to be increased with no direct increase in herbage growth. Further, the increase in the rates of N application allowed a progressive move from point B to A, with a proportional increase in both herbage production and stocking density, until the potential productivity determined by climate was

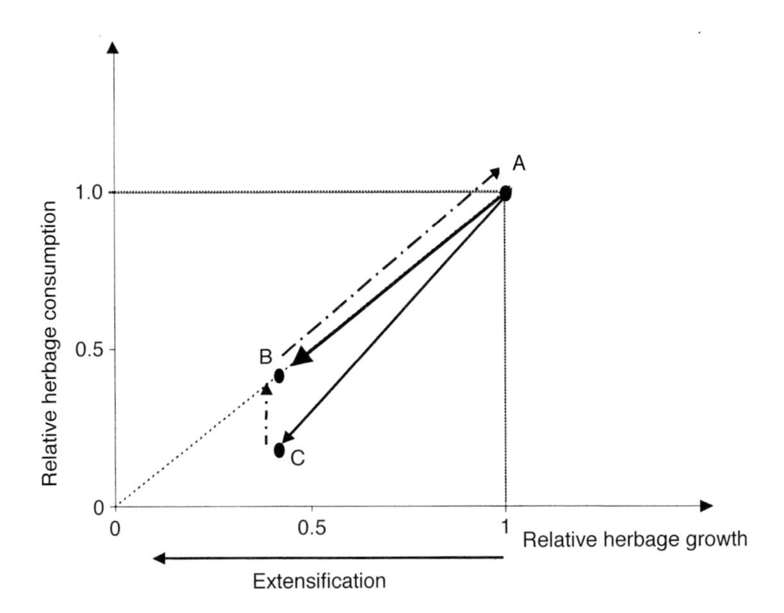

Fig. 14.6. Diagram representing the expected evolution of relative herbage growth and relative herbage consumption when herbage production is reduced by extensification. Trajectories A → C and A → B would correspond to set-stocking and rotational grazing, respectively. The slope 1 : 1 implies that herbage use efficiency would be maintained, while a steeper trajectory would correspond to a decrease in herbage use efficiency (after Lemaire, 1999).

reached. During the late 1970s, researchers progressively discovered that, when the point A was approached, it was possible to remove the fences and to obtain similar levels of herbage growth and herbage intake in both continuous stocking and rotational grazing systems (Parsons *et al.*, 1988). This is explained, in part, because, at high stocking densities, corresponding to high herbage production rates, individual tillers are grazed very frequently in continuously stocked swards (every 10 or 15 days), allowing them to achieve a similar herbage use efficiency to that in a rotational system. At present, grazing research is focusing on the sustainable use of extensively exploited systems. In this sense, it is on the new trajectories AB or AC where our research should be focused in the future. In the light of Fig. 14.6, it is possible to say that research is cycling and our contribution should be to give new answers to old questions.

Conclusions

As clearly demonstrated by Parsons *et al.* (1988), the objectives of grazing management systems are always to try to balance the maximization of herbage production and the optimization of herbage use efficiency. The approach developed above allows the grazed sward to be considered as a demography of leaves. In continuously stocked swards, the probability for an individual leaf to be defoliated before it senesces is primarily determined by the stocking density and, owing to the constant intensity of defoliation, the proportion of leaf tissue removed by the grazing animal from each individual leaf directly depends on stocking density. Consequently, as stocking density increases, the increase in the proportion of leaf tissue consumed determines a concomitant decrease in the leaf senescence flux. As demonstrated in this chapter, leaf senescence is temporally linked to leaf growth and, in the absence of defoliation, an equilibrium between both fluxes will be reached under stable environmental conditions. In a continuously stocked system, the increase in stocking density will generate a linear decrease in the senescence flux, as a result of the increase in the probability of defoliation of individual leaves. Therefore, the classical linear relationship between herbage senescence and sward height (Bircham and Hodgson, 1983), repeatedly obtained in many conditions, must be interpreted not as a sward height effect *per se*, but as an effect of the difference in stocking density associated with the different sward height treatments.

In this context, while height and LAI are relevant sward characteristics for determining the typical asymptotic herbage production response, through their control of both light interception and plant morphogenesis, they are not direct determinants of the senescence rates observed in continuously stocked swards. Each day, the quantity of leaf tissue that senesces should be equal to the quantity of leaf tissue grown one LLS period before, minus the amount of these leaf tissues that have been consumed by the animals, which is directly linked to stocking density. Therefore, the correlation between sward height and stocking density and the dynamics of leaf tissue fluxes generally observed in 'put and take' experiments may be broken when a fluctuating climate or species with contrasting growth characteristics is considered. Further, the dependence of the rate of senescence of leaf tissue on the stocking density needed to maintain a certain sward height (or LAI or biomass) under continuous stocking indicates that no general association can be expected between the sward state and both fluxes. More information on the leaf morphogenesis of a range of forage species is needed to optimize the grazing management of plant communities from contrasting ecological and agronomical conditions.

References

Agnusdei, M. (1999) Analyse de la dynamique de la morphogenèse foliaire et de la défoliation de plusieurs espèces de graminées soumises à un pâturage continu dans une communauté végétale de la *Pampa* Humide (Argentine). Thèse de Doctorat, Institut National Polytechnique de Lorraine, Nancy, France.

Begg, J.E. and Wright, M.J. (1962) Growth and development of leaves from intercalary meristems in *Phalaris arundinacea* L. *Nature* 194, 1097–1098.

Ben-Haj-Salah, M. and Tardieu, F. (1995) Temperature affects expansion rate of maize leaves without change in spatial distribution of cell length. *Plant Physiology* 109, 861–870.

Bircham, J.S. and Hodgson, J. (1983) The influence of sward conditions on rates of herbage growth and senescence in mixed swards under continuous stocking management. *Grass and Forage Science* 38, 323–331.

Bradshaw, A.D. (1965) Evolutionary significance of phenotypic plasticity in plants. *Advances in Genetics* 13, 115–155.

Briske, D.D. (1996) Strategies of plant survival in grazed systems: a functional interpretation. In: Hodgson, J. and Illius, A.W. (eds) *The Ecology and Management of Grazing Systems*. CAB International, Wallingford, pp. 37–67.

Carrère, P. (1994) Défoliation d'une association prairiale graminée (*Lolium perenne* L.) légumineuse (*Trifolium repens* L.) pâturée par des ovins. Impact sur la croissance et l'utilisation de l'herbe. Thèse de Doctorat, Université de Paris-Sud, Orsay, France.

Casal, J.J., Sanchez, R.A. and Deregibus, V.A. (1987) Tillering response of *Lolium multiflorum* plants to changes of red/far red ratio typical of sparse canopies. *Journal of Experimental Botany* 38, 1432–1439.

Chapman, D.F. and Lemaire, G. (1993) Morphogenetic and structural determinants of plant regrowth after defoliation. In: Baker, M.J. (ed.) *Grassland of Our World*. SIR Publishing, Wellington, New Zealand, pp. 55–64.

Davies, A. (1974) Leaf tissue remaining after cutting and regrowth in perennial ryegrass. *Journal of Agricultural Science (Cambridge)* 82, 165–172.

Davies, A. and Thomas, H. (1983) Rates of leaf and tiller production in young spaced perennial ryegrass plants in relation to soil temperature and solar radiation. *Annals of Botany* 57, 591–597.

Davies, A., Evans, M.E. and Exley, J.K. (1983) Regrowth of perennial ryegrass as affected by simulated leaf sheaths. *Journal of Agriculture Science (Cambridge)* 101, 131–137.

Deregibus, V.A., Sanchez, R.A. and Casal, J.J. (1983) Effects of light quality on tiller production in *Lolium* ssp. *Plant Physiology* 72, 900–912.

Durand, J.L., Varlet-Grancher, C., Lemaire, G., Gastal, F. and Moulia, B. (1991) Carbon partitioning in forage crops. *Acta Biotheoretica* 39, 213–224.

Duru, M., Justes, E., Langlet, A. and Tirilly, V. (1993) Comparison of organ appearance and senescence rates in tall fescue, cocksfoot and lucerne. *Agronomie* 13, 237–252.

Gastal, F., Bélanger, G. and Lemaire, G. (1992) A model of leaf extension rate of tall fescue in response to nitrogen and temperature. *Annals of Botany* 70, 437–442.

Gautier, H. and Varlet-Grancher, C. (1996) Regulation of leaf growth of grass by blue light. *Physiologia Plantarum* 98, 424–430.

Gautier, H., Varlet-Grancher, C. and Hazard, L. (1999) Tillering responses to light environment and to defoliation in populations of perennial ryegrass (*Lolium perenne* L.) selected for contrasting leaf length. *Annals of Botany* 83, 423–429.

Gillet, M., Lemaire, G. and Gosse, G. (1984) Essai d'élaboration d'un schéma global de croissance des graminées fourragères. *Agronomie* 4, 75–82.

Hodgson, J. and Ollerenshaw, J.H. (1969) The frequency and the severity of defoliation of individual tillers in set-stocked swards. *Journal of British Grassland Society* 24, 226–234.

Lemaire, G. (1988) Sward dynamics under different management programmes. In: *Proceedings of the 12th General Meeting of the European Grassland Federation, Dublin*. Irish Grassland Association, Belclare, Ireland, pp. 7–22.

Lemaire, G. (1999) Le déterminisme, le contrôle et l'optimisation des flux de tissus foliaires au sein des peuplements pâturés: eléments pour une conduite raisonnée du pâturage. *Fourrages* 159, 203–222.

Lemaire, G. and Chapman, D.F. (1996) Tissue flows in grazed communities. In: Hodgson, J. and Illius, A.W. (eds) *The Ecology and Management of Grazing Systems*. CAB International, Wallingford, UK, pp. 3–37.

Lemaire, G. and Culleton, N. (1989) Effects of nitrogen applied after the last cut in autumn on a tall fescue sward. 2 – Uptake and recycling of nitrogen in the sward during winter. *Agronomie* 9, 241–249.

Matthew, C., Xia, J.X., Chu, A.C.P., Mackay, A.D. and Hodgson, J. (1991) Relationship between root production and tiller appearance rates in perennial ryegrass (*Lolium perenne* L.). In: Atkinson, D. (ed.) *Plant Root Growth and Ecological Perspective*. Special Publication No. 10, British Ecological Society, Blackwell Scientific Publication, Oxford, pp. 281–290.

Maurice, I., Gastal, F. and Durand, J.L. (1997) Generation of form and associated mass deposition during leaf development in grasses: a kinematic approach for non-steady growth. *Annals of Botany* 80, 673–683.

Mazzanti, A. and Lemaire, G. (1994) Effect of nitrogen fertilisation upon herbage production of tall fescue swards continuously grazed by sheep. 2 – Consumption and efficiency of herbage utilisation. *Grass and Forage Science* 49, 352–359.

Mazzanti, A., Lemaire, G. and Gastal, F. (1994) Effect of nitrogen fertilisation upon herbage production of tall fescue swards continuously grazed by sheep. 1 – Herbage growth dynamics. *Grass and Forage Science* 49, 111–120.

Neuteboom, J.H. and Lantinga, E.A. (1989) Tillering potential and relationship between leaf and tiller production in perennial ryegrass. *Annals of Botany* 63, 265–270.

Parsons, A.J. (1994) Exploiting resource capture – grassland. In: Monteith, J.L. and Unsworth, M.H. (eds) *Resource Capture in Crops*. Nottingham University Press, Nottingham, pp. 315–322.

Parsons, A.J., Johnson, I.R. and Harvey, A. (1988) Use of a model to optimize the interaction between frequency and severity of intermittent defoliation and to provide a fundamental comparison of the continuous and intermittent defoliation of grass. *Grass and Forage Science* 43, 49–59.

Robson, M.J. (1967) A comparison of British and North African varieties of tall fescue. 1 – Leaf growth during winter and the effect on it of temperature and daylength. *Journal of Applied Ecology* 4, 475–484.

Rodriguez Palma, R. (1998) Fertilizacion nitrogenada de un pastizal de la *Pampa* Defirimada: Crecimiento y utilization del forraje bajo pastoreo de vacunos. MSc Thesis, Faculty of Agricultural Sciences, University of Mar del Plata, Argentina, 135 pp.

Silsbury, J.H. (1970) Leaf growth in pasture grasses. *Tropical Grasslands* 4, 17–36.

Simon, J.C. and Lemaire, G. (1987) Tillering and leaf area index in grasses in the vegetative phase. *Grass and Forage Science* 42, 373–380.

Skinner, R.H. and Nelson, C.J. (1992) Estimation of potential tiller production and site usage during tall fescue canopy development. *Annals of Botany* 70, 493–499.

Tabourel-Tayot, F. and Gastal, F. (1999) MecaNiCAL, a supply–demand model of carbon and nitrogen partitioning applied to defoliated grass. 1 – Model description and analysis. *European Journal of Agronomy* 9, 223–241.

Van Loo, E.N. (1992) Tillering, leaf expansion and growth of plants of two cultivars of perennial ryegrass grown using hydroponics at two water potentials. *Annals of Botany* 70, 511–518.

Wade, M.H. (1991) Factors affecting the availability of vegetative *Lolium perenne* to grazing dairy cows with special reference to sward characteristics, stocking rate and grazing method. Thèse de Doctorat, Université de Rennes, France.

Wade, M.H., Peyraud, J.L., Lemaire, G. and Cameron, E.A. (1989) The dynamics of daily area and depth of grazing and herbage intake of cows in a five day paddock system. In: *Proceeding of the 16th International Grassland Congress, Nice, France.* AFPF, Versailles, France, pp. 1111–1112.

15

Dynamics of Heterogeneity in a Grazed Sward

A.J. Parsons,[1] P. Carrère[2] and S. Schwinning[3]

[1]AgResearch Grasslands, Private Bag 11008, Palmerston North, New Zealand; [2]INRA, Unité d'Agronomie, Clermont-Ferrand, Cedex 2, France; [3]Department of Biology, University of Utah, Salt Lake City, UT 84112, USA

Introduction

Several previous chapters in this book (Chapters 10 and 11) focus on spatial heterogeneity and the way animals respond to this, making foraging decisions at a wide range of spatial and temporal scales (Senft *et al.*, 1987; Coughenour, 1991), from bites, through feeding station, to landscape (Owen-Smith and Novellie, 1982; Gordon and Lascano, 1994) and on a meal, through day, to lifetime basis (see Shipley *et al.*, 1994; Newman *et al.*, 1995; and reviews by Laca and Demment, 1996; Ungar, 1996). Although the currency for optimal foraging, and so the perceived goal of these decisions, is debated (Mangel and Clark, 1986, 1988; Stephens and Krebs, 1986), for sure animals do encounter a heterogeneous world, and emphasis on animals' response to this is of paramount importance. But in this chapter we wish to shift the focus from spatial heterogeneity *per se*, which we see as the variance in state of the vegetation and as an 'output' or emergent phenomenon of the grazing system, to focus on some of the processes and the variance in processes that have an impact on the vegetation state (see also Dutilleul and Legendre, 1993). This we see as the essential driving force for the creation and dynamics (changes over time) of heterogeneity.

We concentrate here on just one source of heterogeneity: defoliation (biting) and its contribution not only to consumption (intake) but notably also to the regeneration of resources (plant regrowth) (Possingham and Houston, 1990). In short, in this chapter we attempt to enlarge on the role of animals, through biting, not only in responding to heterogeneity but in creating and sustaining it, and its impact on yield and stability.

(eds G. Lemaire, J. Hodgson, A. de Moraes, C. Nabinger and P.C. de F. Carvalho)

There are an enormous number of possible sources of heterogeneity in vegetation, but it is beyond imagination to cover all these in one text. Some, notably biodiversity and its interaction with herbivory and nutrient 'richness' in soils, have alone been the subject of decades of research in attempts to understand what controls which species are present and their spatial distribution, abundance and succession (e.g. Silander and Pacala, 1990; Tilman and Pacala, 1993; Tilman, 1994; Tilman *et al.*, 1996; Pacala and Rees, 1998; Proulx and Mazumder, 1998; see also Huston, 1997). More directly, heterogeneity can arise in vegetation 'simply' because there is heterogeneity in the distribution of resources in soil and, for example, topography. This would be tantamount to explaining heterogeneity with heterogeneity, unless biotically induced variations in the spatial and temporal distribution of nutrients were involved, as often applies due to the local deposition of nutrients in grazed grasslands (see Jarvis, Chapter 16, this volume, and, for example, Schwinning and Parsons, 1996a, b). Rather, what we address here are some specific features of the creation and maintenance of heterogeneity – those induced by decisions made by animals about where and when to place their bites (see also Dolman and Sutherland, 1997). This approach exploits some recent advances in understanding grazing behaviour and offers a further, spatial, dimension to insights into the growth and efficiency of utilization of pastures.

Recent reviews have emphasized the need to explore the spatial dimensions of grazing (Coughenour, 1991; Illius and Hodgson, 1996; Laca and Demment, 1996) and how 'we do not know how phenomena at the bite scale express themselves at the field scale' (Marriott and Carrère, 1998). Here we attempt to add this to our understanding of the dynamics of heterogeneity in grazed pastures.

Focus on process

Perhaps the most satisfying way (see Ungar, 1996) to come to understand the origin and dynamics of spatial heterogeneity in vegetation state, and its likely consequences, is to start by imagining a totally homogeneous pasture and seeing what kind of spatial heterogeneity would arise given certain processes of defoliation. This approach helps separate 'cause' and 'effect'. For this purpose, we attempt to identify what we feel are three fundamentally contrasting examples of how animals might exploit resources spatially, in grazed pastures, and so how particular forms of heterogeneity might arise and be maintained. We look for a possible 'signature' of these different spatial ways of exploiting resources in the heterogeneity in pastures that emerges, and we look at the impact of different sources of heterogeneity on yield and stability (*sensu* Morley, 1966; Noy-Meir, 1975).

To attempt this level of clarity, we use models, both spatially implicit and spatially explicit, that can be supported with information on biological processes and observations collected from intensively managed pastoral systems. These have the benefit of having been more controllable experimentally, and having been subject to many long-standing programmes, which have provided essential insights.

Questions of scale

A lot has been said about the need to study the interaction between plants and grazing animals at a range of spatial and temporal scales (e.g. Senft *et al.*, 1987; Coughenour 1991; Laca and Ortega, 1995; review by Illius and Hodgson, 1996), but also about the pitfalls and potential confusion of extrapolating between scales (Brown and Allen, 1989; Wiens, 1989). Some important principles, perhaps, are that we should analyse both plant growth (resource regeneration) and animal intake (resource consumption) at the same scale and, where possible, at at least one scale lower than the scale at which we wish to make the prediction (Thornley and Johnson, 1990). One problem here is that, in animal science, a great deal of progress has been made by conceptualizing the response of animals to vegetation state (intake, consumption) at the scale of individual bites (see Penning, 1986; Demment *et al.*, 1987; Illius and Gordon, 1987; Ungar and Noy-Meir, 1988; Spalinger and Hobbs, 1992; Parsons *et al.*, 1994; Penning *et al.*, 1995; reviews in Hodgson and Illius, 1996), whereas the regeneration of the resource (vegetation growth) is still conceived, at least in terms of mass flux, largely at the field scale. However, the physiological understanding can readily be applied, with modifications, at the lower (bite) scale. Moreover, this offers the opportunity to integrate field-scale studies of mass flux in pastures not only with recent developments in understanding intake behaviour, but also with the detailed observations of the fate of individual leaves made on 'marked' plants or tillers (Mazzanti and Lemaire, 1994; Louault *et al.*, 1997) – the 'morphogenetic' approach advocated by, for example, Lemaire and Chapman (1996). These three approaches to studying grazing are rarely interrelated in the literature. We can imagine the studies on marked tillers as recording a temporal record of defoliation events in a grazed sward at each of a large number of 'points' in space. This is clearly a very important part of any potential signature of the spatial and temporal dynamics of grazed pastures (Coughenour, 1991).

We consider that the bite scale is appropriate for studying the origin and dynamics of spatial heterogeneity originating from defoliation, as, in essence, a homogeneous pasture becomes heterogeneous at the first bite. Less trivially, it can be calculated readily (see Parsons and Chapman, 1998) and it has been observed (Wade, 1991) that animals cannot sustainably harvest (bite from) more than some 5% of the total grazable area in 1 day. For example, animals remove approximately 35–50% of the standing biomass at that point in space in each bite (Wade *et al.*, 1989; Ungar *et al.*, 1991; Laca *et al.*, 1992; Edwards *et al.*, 1996a) and yet the daily intake in temperate pastures is only some 50 kg ha^{-1} day^{-1} from a sustainable standing biomass of some 2000 kg ha^{-1}. By the second day, then, the important issue is to what extent animals eat from the same bite-sized patches as on the day before. This involves chance events, such as the paths animals happen to take on the second day, as well as decisions made by animals whether or not to eat from all the patches encountered. To understand the creation and dynamics of heterogeneity in grazed pastures, then, one essential question to resolve is: just where and when do animals place their bites?

These simple thoughts expose some of the key features necessary for a spatial reanalysis of grazing. First, it is clear that, working at the bite scale, all defoliation must be seen as a discrete process (near-instantaneous defoliation, followed by variable periods of uninterrupted growth) and as potentially stochastic, as there will be uncertainty about when, and so at what stage or state of regrowth, a patch is revisited. Stochasticity may greatly modify the phenomena emerging from the plant–animal interaction, and several authors argue the deficiencies of a purely deterministic approach (Chesson, 1978; Saether, 1997; Stenseth and Chan, 1998). But before describing some specific examples, we must jump aside to reconsider adjustments to previous theory in order to work effectively at this lower, bite, scale.

Deconstructing functional responses

The most widely recognized functional responses between consumption and vegetation state and between growth and vegetation state are those described in the highly insightful work of Morley (1966) and Noy-Meir (1975). But these attempt to characterize responses at the field scale (growth or consumption in relation to a field-wide mean vegetation state) and are simply not valid at the bite scale. New functional responses, at the bite scale, or, more significantly, more information on processes is needed to allow us to consider the response of consumption and of vegetation growth to there being variance in the state of bite-sized patches.

When working at the bite scale, the nearest to a 'functional response' for consumption, c becomes the reportedly simple relationship (see Laca *et al.*, 1992) between the biomass of herbage present in a single bite-sized area and the proportion of standing biomass removed in a single bite. This captures the local and instantaneous response by the animal to vegetation state:

$$\begin{aligned} c &= 0 && \text{if } W \le W_{min} \\ c &= f(W - W_{min}) && \text{if } W > W_{min} \end{aligned} \tag{1}$$

where, for example, f describes the proportion of the grazable biomass removed from a patch of biomass W (prior to defoliation), above an ungrazable horizon W_{min}. One advantage of this relationship is that the biomass after defoliation (the residual biomass) can be derived from the biomass before defoliation. This makes it possible to link the functional response for consumption with that for plant growth, as the latter requires a knowledge of the patch state after defoliation (the initial conditions for the next regrowth). The bite-scale functional response (equation 1) derives the bite mass that would be expected from a given single patch state. For any one patch state, this can be related to an instantaneous intake rate (IIR), using the current understanding of the mechanistic constraints to food acquisition (prehending and masticating bites; see Spalinger and Hobbs, 1992; Parsons *et al.*, 1994), notably how small bites are 'handled' less efficiently than larger ones, hence generating a simple saturating response of IIR to a given patch state prior to defoliation. But to derive daily intake requires a knowledge of just what was the frequency distribution of patch states from which the animals grazed, and the implications for the required grazing time. This requires a knowledge of what the animals did in the rest of the field, what patch states they encountered there and how many bites in total were taken as a consequence.

In the absence of adequate knowledge of behavioural control, daily intake in models is conventionally limited either by an imposed maximum daily intake or a maximum to grazing time, and we have adopted the same approach (see Schwinning and Parsons, 1999). If the relationship between bite mass and patch state is linear, as in equation 1, the frequency distribution of patch states does not, in fact, affect the relationship between daily intake and patch state at all. This is counterintuitive (see Ungar and Noy-Meir, 1988; Demment *et al.*, 1995; Laca and Demment, 1996), given the non-linear nature of intake rate with patch state, but arises because both the sum of bite masses and the sum of biting times accumulate linearly with patch state over the day. Any non-linearity introduced into equation 1 would, however, alter the whole shape of the relationship between daily intake and vegetation state – the functional response – at the field scale. Moreover, any changes in a component of grazing behaviour, such as in the fractional bite depth (f in equation 1) or in the selectivity of grazing, will also alter the very shape of the field-scale response of consumption to vegetation state (Fig. 15.1). Should we, perhaps, then not call the field-scale relationship a functional response at all? For example, shallow biting (low f) or selection for small patches will lead to a predominance of smaller bites and this will mean that it will take more bites and so more grazing time to satisfy intake; likewise, it is more likely that intake will be depressed at a given mean vegetation state or, conversely, it will take a higher mean vegetation state to satisfy maximum daily intake. (All these cases explain the shift to the right in Fig. 15.1 at the point where intake is no longer time-constrained.) Hence, a given field-scale mean vegetation state does not dictate a given single rate of consumption, as this can depend on the frequency distribution in patch states about the mean, and no single field-scale functional response can be drawn.

Working at the bite scale, and so where the frequency distribution of patch (bite) states is known, overcomes the problems that are associated with assuming the field-scale mean vegetation state adequately represents the state of vegetation that animals are actually grazing bite by bite.

Next, it is necessary to revisit the concept of the functional response of vegetation growth to defoliation and to rework this too at the bite scale. Recall that we now see defoliation as a discrete process, as a sequence of near-instantaneous defoliations separated by variable periods of uninterrupted growth. This poses no problems, as, indeed, it has long been recognized that all systems of grazing management, even continuous grazing, are actually 'rotational grazing' at the bite (or marked plant) scale (Hodgson and Ollerenshaw, 1969; Morris, 1969; Clark *et al.*, 1984). Moreover, there is a plethora of agronomic (as well as physiological) understanding of the regrowth (regeneration of resources) of grass swards (e.g. see reviews by Robson *et al.*, 1988; Richards, 1993; Parsons and Chapman, 1999).

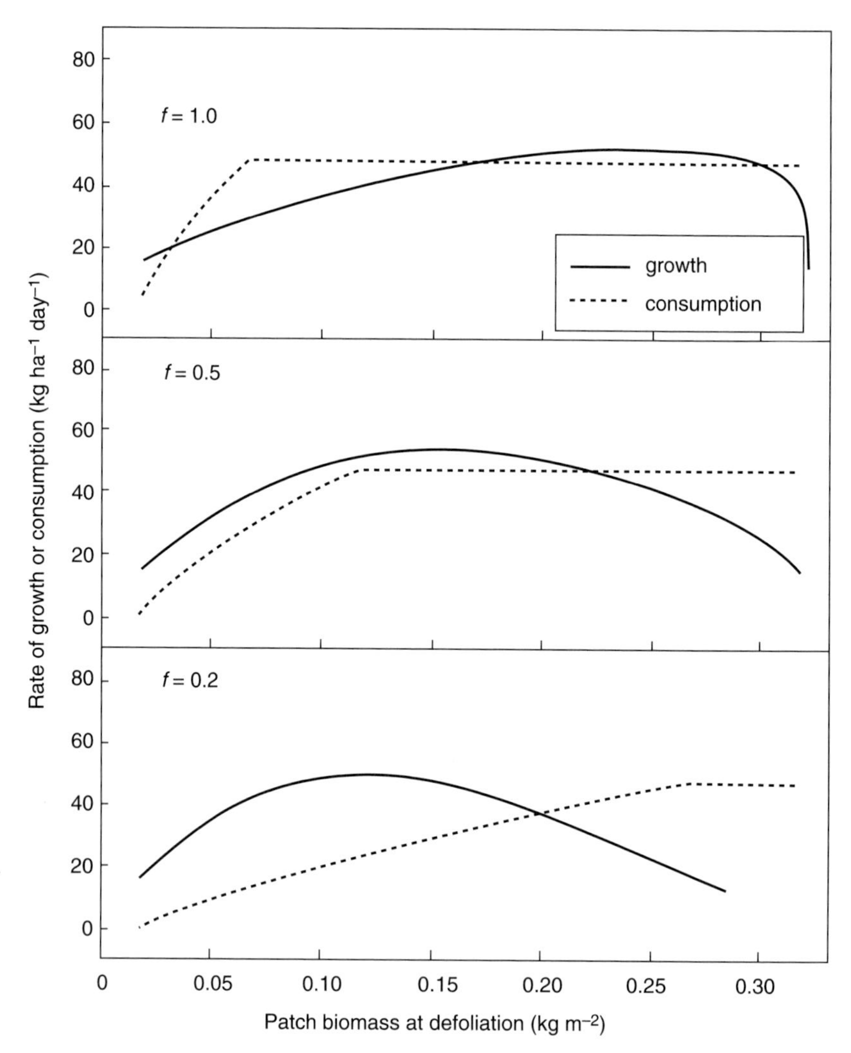

Fig. 15.1. Differences in a single component of animal grazing behaviour, here a change in bite depth (*f*, the fraction of the standing biomass removed from a patch in a single bite) can alter the whole shape of the field-scale relationship between consumption and vegetation state and between plant regrowth (resource regeneration) and vegetation state (reproduced with permission from Schwinning and Parsons, 1999).

This understanding applies equally at the field and the bite scale. But here we must emphasize how critical it is that the growth function, or mechanistic model, used to represent regrowth in each patch adequately captures the sensitivity of regrowth to initial vegetation states (see Parsons *et al.*, 1988a, b; Schwinning and Parsons, 1999), notably the effects of regrowth from high initial patch states ('lenient grazing') in depressing growth rates per unit biomass (Fig. 15.2a).

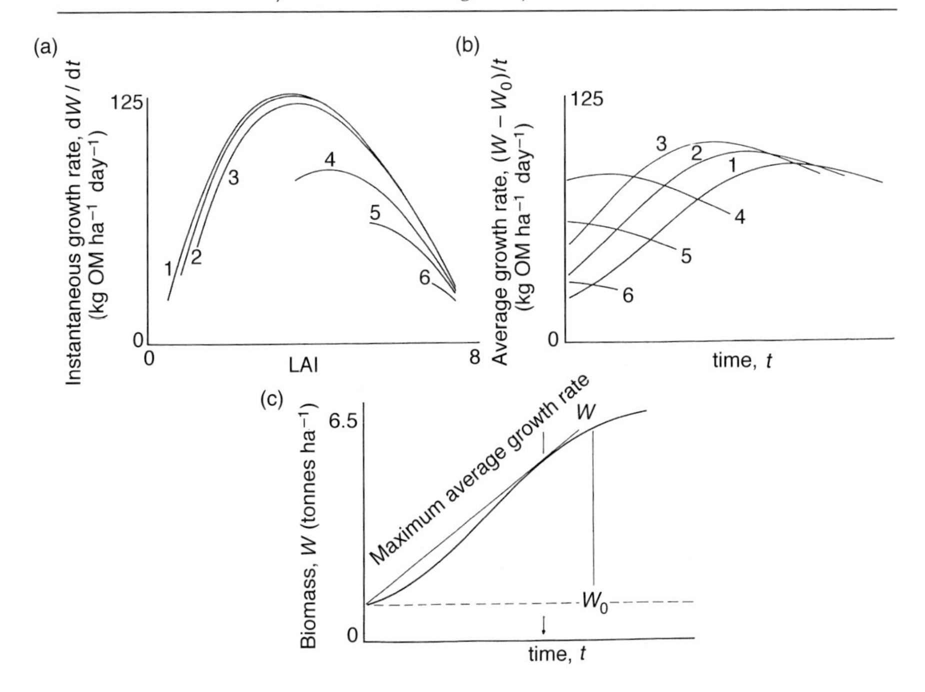

Fig. 15.2. The rate of regeneration of resources (regrowth of vegetation) depends on the residual (initial for regrowth) vegetation state and how long the vegetation is allowed to regrow (cf. defoliation interval). In (a) and (b) we compare six contrasting initial conditions for regrowth. In (a) we see how the instantaneous growth rate is depressed following lenient defoliation (e.g. line 4) and the growth rate does not retrace the corresponding part of the growth curve following severe defoliation (e.g. line 1). The effects of the duration of regrowth on the yield achieved (average growth rate) is shown in (b). The principle of the maximum average growth rate (cf. maximum sustainable yield (MSY)) is shown in (c), for line 1 only. The MSY is the steepest slope (the tangent) to the growth curve. OM, organic matter (after Parsons *et al.*, 1988a).

Whether at a patch or a field scale, the single relationship between growth and vegetation state shown by Noy-Meir (1975) is only appropriate following defoliation to a low residual mass, and it cannot be assumed that regrowth from more lenient defoliation (greater initial patch states) simply retraces the corresponding part of the same growth curve. Although this assumption has been widespread in both the agronomic (e.g. Booysen, 1966) and the ecological literature (e.g. see 'maximum sustainable yield' in Begon and Mortimer (1986) and Begon *et al.* (1990)), it has been argued that this can lead to serious misinterpretation in the context of pastures that have age-structured populations of leaves (Schwinning and Parsons 1999; A.J. Parsons, unpublished results), and failure to recognize this is at the heart of the long-standing controversy over the merits of contrasting defoliation regimes (see Brown and Blaser, 1968; reanalyses by Parsons *et al.*, 1988a, b).

At the patch scale, the 'functional response' for plant growth must consider how growth rate is affected both by the residual patch state (initial conditions for next regrowth) following grazing and by how long (and so to what state) the patch is allowed to regrow, and this is encapsulated in Fig. 15.2. Note that there may be quite independent variation in initial patch state and in the interval between defoliations. Moreover, working at the bite scale, vegetation growth is again seen to depend, somewhat counterintuitively, on animal behaviour. This is because the interval between defoliations (the time taken before animals revisit the patch) depends on the total number of bites that are taken, by all animals and in other parts of the field, in an attempt to satisfy their demand for intake. Thus, if we attempt to reconstruct a 'functional response' for vegetation growth at the field scale, we see that, once again, there is no single simple relationship between growth rate and vegetation state. Differences in animal behaviour, such as biting depth, actually alter the whole shape of the relationship between growth rate and vegetation state (by altering the size and number of bites required, and so pre- and postgrazing states and defoliation interval) (see Fig. 15.1). Likewise, different frequency distributions (variance) about the mean vegetation state do now themselves alter the whole shape of the 'functional response' at the field scale. Again, working at the patch scale overcomes the possible sampling errors associated with assuming that a spatiotemporal mean state adequately represents the state of the vegetation from which consumption and, in this case, growth are taking place.

Beyond this, all that is needed is some concept as to just how in space and time animals place their bites, and this we explore below.

Three Contrasting Methods of Spatially Exploiting a Resource

Having revised our theory for consumption and growth in relation to vegetation state, at the bite scale, we can now consider three fundamentally contrasting cases of how animals might place their bites in space and time and consider the implications:

- *Type 1: Sequential grazing.* In the first example, animals move from one bite-sized patch to the next systematically and in strict sequence (Fig. 15.3). No patches are missed and none grazed twice until all other patches are eaten. This is not unlike an example proposed by Ungar (1996), where animals always eat from the largest patches (and so have large bite masses).
- *Type 2: Random grazing.* In the second example, animals were considered to graze from the field of bite-scale patches at random (a Poisson process, or a random walk). Here the animals eat from a wide range of patch sizes (and so have variable bite masses).
- *Type 3: Selective grazing.* Here we consider that animals may make choices about where (and so when) they place their bites on the basis of the state (e.g. height or biomass) of the patches distributed before them. We consider two (of the myriad) possible examples. 'Preference for tall/large biomass' is evoked as

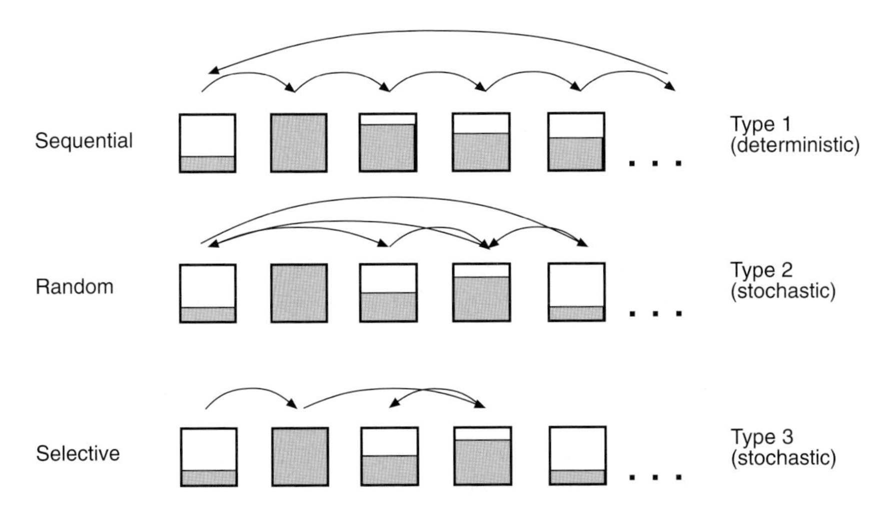

Fig. 15.3. Three proposed fundamentally contrasting processes by which animals might distribute their biting in space and time, shown schematically for a few of a huge array of bite-sized patches in a field: type 1, sequential, which leads to deterministic defoliation intervals; type 2, random (Poisson); type 3, selective, where animals choose patches on the basis of their state (biomass indicated by shaded area). Types 2 and 3 both generate stochastic defoliation intervals.

a partial rejection of short patches (a probability that not all short patches encountered are grazed from). Likewise, a 'preference for short/low biomass' is evoked as a partial rejection of tall biomass patches. There is evidence in the literature that both forms of preference (on whatever precise basis, e.g. nutritional value) are observed in reality (see later). Clearly, here too animals eat from a range of patch states, though the frequency distribution is constrained by preference.

The impact of these methods of distributing biting in space and time is shown in Fig. 15.4, along with the frequency distributions of defoliation intervals (as would be seen on marked tillers) and of patch states (the measure of heterogeneity). Other aspects of a possible signature of the form of defoliation in the heterogeneity of the pasture are discussed for each case. These allude to the relevance of the form of heterogeneity in reducing, or indeed increasing, yield and/or stability.

Type 1

Sequential defoliation leads to deterministic defoliation intervals. The defoliation interval itself will depend on stock density, and even on animal behaviour (such as bite depth), but in all cases there is no spatial or temporal variance in defolia-

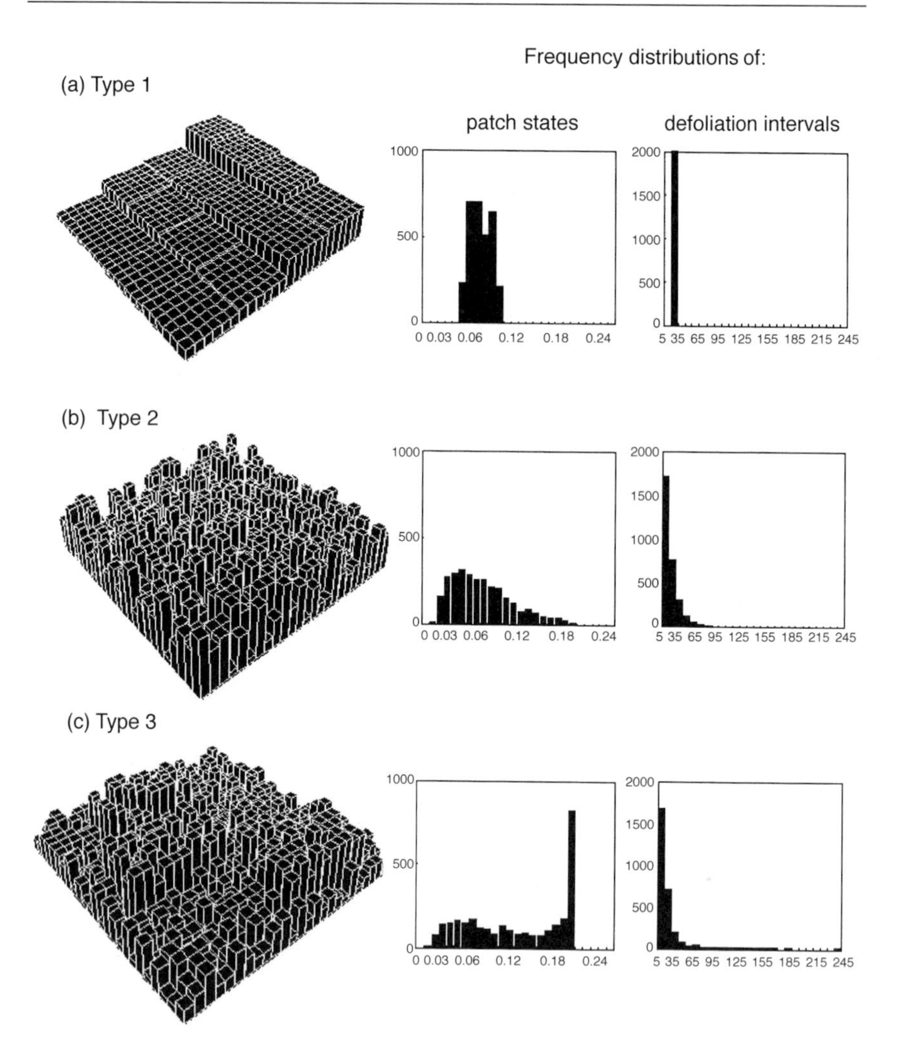

Fig. 15.4. Illustrations of the forms of heterogeneity generated by the three types of grazing process: type 1, sequential; type 2, random; and type 3, selective. Examples of the frequency distribution of patch states and the frequency distribution of defoliation intervals are shown for each case, derived for the stock density that gives close to the greatest yield per hectare (see text for further explanation). Note that in type 3, 'selective', this example shows that there may be frequently defoliated and less frequently defoliated populations of patches, which may be aggregated and visible as 'tall' and 'short' areas that do not move around.

tion interval. As stated earlier, this same result could have been achieved by cutting and, if all patches were cut simultaneously and so in phase, the pasture would remain homogeneous. However, under grazing, because patches are grazed here in sequence and so out of phase, the pasture is heterogeneous (despite the lack of variance in defoliation interval) – that is, different parts of the total grazed area are in different stages of regrowth at any one point in time. The frequency distribution of patch states (Fig. 15.4a) therefore shows that all possible patch states during regrowth (between whatever is the residual patch state and the patch state at the time of the next defoliation) are represented equally. Note that, if this part of the growth function is sigmoid, there will be an increased frequency of short and tall patch states – a U-shaped distribution – as the patches remain longer in these states during regrowth.

Because different stock densities and different bite depths will alter the residual patch states and the final patch states (by altering defoliation interval), these factors will also alter the range of patch states seen, and so will alter the coefficient of variation about the mean patch state. Hence, even a very simple source of heterogeneity can leave a bewildering signal in a measure of heterogeneity. Marked heterogeneity and even a form of bimodal distribution in patch states can arise even from this strictly sequential, and so deterministic, defoliation when stock density is low and bite depth is large, as this provides low residual states and long defoliation intervals and so 'samples' an entire sigmoid growth curve. One notable feature, however, of this source of heterogeneity is that the tall and short patches 'move around', indeed here in a sequence.

Type 2

Random defoliation clearly leads to variance in defoliation intervals, as well as in initial and final (pregrazing) patch states. Again, of course, the mean intervals and states will depend on stock density and grazing behaviour (e.g. bite depth). However, if the defoliation is totally random in space and time (Poisson), all patches will have an equal expectation of being grazed and so there will be a temporal but no spatial component to that variance. The frequency distribution of defoliation intervals is a negative exponential (see Hassell, 1978; Edwards, 1994). The pasture will show heterogeneity, as seen in the frequency distribution of patch states (Fig. 15.4b), and the shape of this distribution may be skewed, depending on stock density. At high stock density, it is skewed to the right (predominance of short patches, as shown in Fig. 15.4b; at low stock density, the mean is higher and the distribution is skewed to the left (predominance of tall patches). The frequency distribution (again, as well as the range) of patch states is an emergent property of the system and so cannot be prescribed (chosen beforehand). Again, it is a feature of this source of heterogeneity that the tall and the short patches move around.

Type 3

Selective defoliation can lead to complex sources of variance in defoliation intervals and patch states. This seems particularly so if the animals exhibit a partial rejection of tall areas. This can be shown to lead to relatively sustained bimodal distributions of patch state (Fig. 15.4c). The initial heterogeneity may arise totally by chance (as in type 2 above), but then state-dependent foraging decisions can reinforce the differences in patch states. If preference is at the single-patch scale, there is again temporal but now also some spatial variance in defoliation interval, but, if animals make choices on a scale larger than individual patches (e.g. in feeding stations), there will also be spatiotemporal covariance in defoliation intervals (see Edwards, 1994). That is, some areas of the field will be grazed frequently, while other areas of the field will be grazed less frequently, leading to 'tall' and 'short' areas, as in Fig. 15.4c. Clearly, under these circumstances, the mean vegetation state is a very poor indicator of the state (or states) of vegetation that is being grazed and which is involved in regrowth. The sward is clearly very heterogeneous, with the notable added feature that there is a stronger tendency for the tall and the short patches to remain fixed in space – they do not move around. There is increasing evidence for this form of heterogeneity to arise, notably in swards grazed continuously by cattle (Gibb *et al.*, 1989, 1997).

Role of Heterogeneity in Yield and Stability

The impact of these three contrasting sources of heterogeneity on mean vegetation state, yield (intake per hectare and per animal) and stability is shown in Figs 15.6 and 15.7 (see below), where the resulting equilibrium values are plotted against stock density (*sensu* Noy-Meir, 1975). In this way, we can consider what the consequences would be of a given number of animals choosing to distribute their grazing in the three contrasting ways, *per se*, on the performance of the grassland system, and the implication for the animals themselves.

It is important to note that, for now, we are looking at the performance of one grazing strategy relative to the others, for a given set of parameters for grazing (biting) behaviour. As such, we are comparing different levels of variance in defoliation and levels of heterogeneity. It is not until a later section that we consider how all three grazing strategies compare with the theoretical optimal defoliation regime, and so whether all forms of grazing might actually fall short of some ideal harvesting regime.

First, it is notable that there is much less evidence (Fig. 15.5), using a spatial model of grazing, of the marked 'dual stability' in grazing described by Noy-Meir (1975). We recognize that the degree of instability in grazing models depends critically on the functions used to describe consumption and growth (Illius and Hodgson, 1996). But, in a recent study (Schwinning and Parsons, 1999), we showed how the marked discontinuous stability seen in Noy-Meir's models was largely a feature of the highly deterministic nature of the underlying processes (hence there is more evidence of dual stability in the deterministic case in Fig. 15.5 than in the stochastic cases), and partly also a consequence of the vegetation growth function used by Noy-Meir (and its lesser sensitivity to residual patch states), which we have avoided here. We also illustrated how dual stability, even in this spatial model, becomes much less marked as the parameter for bite depths

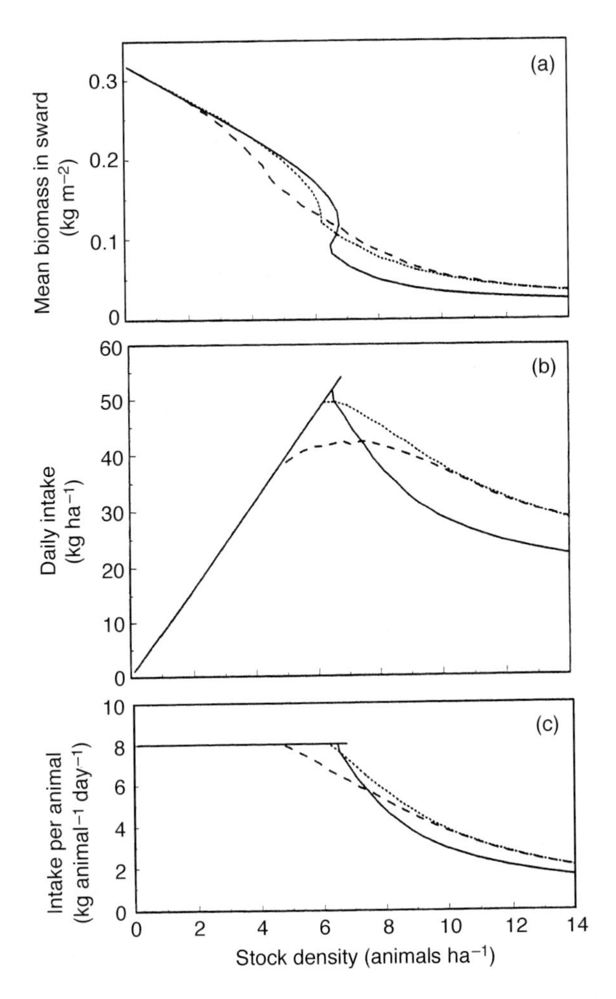

Fig. 15.5. The effect of a given numbers of animals (see range of stock density) choosing three contrasting processes for placing bites in space and time on (a) the mean vegetation state; (b) the yield (intake per hectare); and (c) intake per animal (all *sensu* Noy-Meir, 1975). The lines show the equilibria at the field-scale for type 1, sequential, solid line; type 2, random, dotted line; and type 3, selective, dashed line (partial rejection of tall areas). Note minimal evidence of dual stability (after Schwinning and Parsons, 1999).

(f in equation 1) declines from 1.0 to the observed values, for example, of *c.* 0.5 (see Laca *et al.*, 1992), used here. Overall, we concluded that dual stability is much less likely in reality (given the stochasticity and as seen in a smaller domain of parameter and state variable values) than argued by Noy-Meir (1975), but in keeping with Coughenour (1991).

At low stock densities, our analyses suggest that the foraging strategy, the defoliation regime (frequency distribution of intervals and pre- and post-defoliation patch states) and the heterogeneity that arises all have little to no effect on yield (Fig. 15.5). Consumption per hectare is the same in this example up to a stock density of about four animals per hectare (Fig. 15.5b). The lack of impact of heterogeneity on yield arises because, in situations where the animals are grazing from relatively abundant vegetation (high mean patch states at low stock density), the animals' intake is not time-constrained, and so the animals are functioning on a part of their response to vegetation that is insensitive to vegetation state. All animals are approaching maximum intake. There are, however, differences in the mean, as well as the frequency distribution, of patch states.

At higher stock densities there are interesting differences in yield per hectare under the three contrasting grazing strategies.

Sequential, systematic defoliation (type 1) leads to the greatest maximum yield per hectare (although this is partly dependent on the prediction of a highly stable branch of a region of dual stability). But, at greater stock densities, consumption per hectare is lower from this systematic defoliation than from any of the other, more stochastic, defoliation regimes (Fig. 15.5b). This seems paradoxical, as, in systematic grazing (type 1), animals are at all times biting from the largest possible patches (not unlike the case considered by Ungar (1996)), whereas, under random grazing (type 2), animals take bites from patches in a wide range of states. But this can be explained, in that, when stock density is above the optimum, sequential grazing necessarily leads to an overexploitation of all patches equally, whereas random grazing allows some patches to 'get away'. This partial escape from overexploitation under random grazing is responsible for creating a greater mean vegetation state, as can be seen in Fig. 15.5a. In short, thanks to the variance in defoliation intervals that the plants experience, the animals end up eating, albeit at random, from vegetation that is taller and so imposes fewer time constraints on intake.

This explanation is reinforced in Fig. 15.6, where intake is plotted against vegetation state, as opposed to stock density, and it is clear that random grazing, as might be expected, does indeed decrease intake for a given mean vegetation state.

Selective grazing (type 3) can have an even more marked effect on yield, as would be expected. Preference for tall (partial rejection of short) patches gives yields very similar to random grazing when compared at the same stock density (and so not shown in Fig. 15.5). It has surprisingly little effect in modifying the frequency distribution of patch states (though slightly depressing the occurrence of tall patches), and leads dynamically to a similar mean vegetation state for a given stock density. A partial rejection of tall patches has a marked effect in depressing maximum yield (see Figs 15.5 and 15.7), as animals are concentrating on taking smaller bites (from the short patches) and so both depress instantaneous intake rate and, as more bites are required to satisfy daily intake, may face time constraints as well as generating shorter defoliation intervals in the areas where grazing is concentrated.

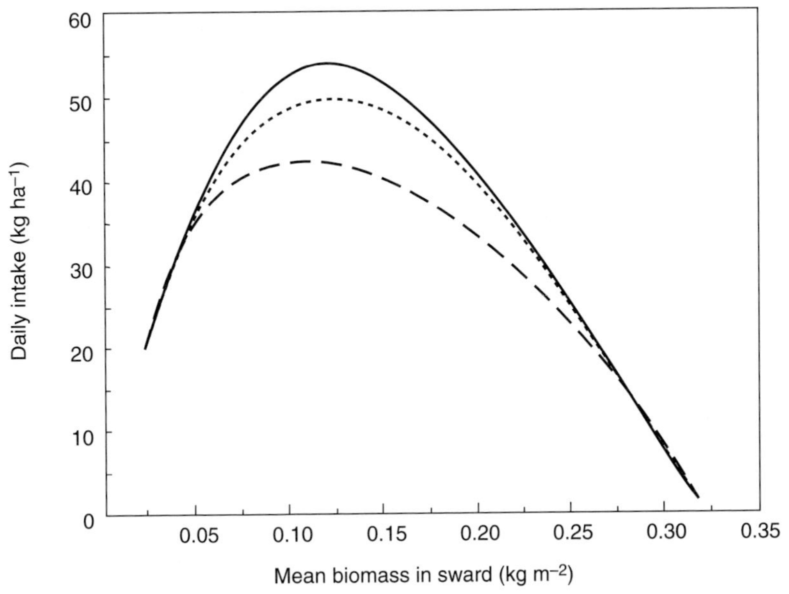

Fig. 15.6. The effect of the three contrasting processes of grazing on yield (intake per hectare) replotted from Fig. 15.4b in relation to vegetation state, as opposed to stock density. Stochastic defoliation (types 2 and 3 here), and its associated heterogeneity, always reduced yield for a given vegetation state, even though these processes of defoliation may increase yield for a given (high) stock density (see Fig. 15.4b). The lines show the equilibria at the field-scale for type 1, sequential, solid line; type 2, random, dotted line; and type 3, selective, dashed line (partial rejection of tall areas) (after Schwinning and Parsons, 1999).

Does Heterogeneity Reduce Yield?

We are now in a position to summarize the effect of heterogeneity on yield. Clearly, spatial heterogeneity can, under some circumstances, have little to no effect on yield and may even be advantageous. In the first instance, the heterogeneity that arises under sequential grazing (type 1) is inevitably of no consequence to yield, for recall that all patches are defoliated at the same deterministic interval. We can imagine how this same defoliation regime could be achieved under cutting, where all patches were defoliated at the same instant. Under cutting, then, the system would be homogeneous, whereas, under grazing, the same system is heterogeneous (but simply because patches are defoliated out of phase). Hence the yield of this heterogeneous system is identical to that of a homogeneous one, provided the defoliation intervals and mean patch states are the same. A greater component of heterogeneity arises under random grazing (type 2), but here heterogeneity is seen to be beneficial, increasing yields at high stock densities by permitting greater opportunities for plant regeneration.

Heterogeneity can also be beneficial to yield per hectare where it aids selective grazing (type 3), notably in selecting for tall patches. If the heterogeneity is perceivable by the animal, it can be advantageous, as animals may use both visual cues and spatial memory to increase their rate of encounter with desirable food items (Edwards *et al.*, 1994, 1996b, 1997).

It is the heterogeneity that arises under preferential grazing (type 3), notably when associated with the partial rejection of tall areas, that appears most deleterious. The bimodal distribution of patch states and the associated complex distribution of defoliation intervals reflect how yield is greatly reduced, because some patches are defoliated too soon and/or some patches are defoliated too late, relative to the optimal timing of harvest during regrowth. The rejection of tall patches allows some patches to regrow for too long, potentially to a ceiling yield, where their contribution to both net vegetation growth and consumption per hectare approaches zero.

Experimental Evidence

Although heterogeneity of all forms and sources is widely (perhaps ubiquitously) observed and evoked to explain all manner of uncertainty, it is rarely measured. In many cases, the variance in patch states is used only to contribute to a coefficient of variation about a mean, rather than to plot a potentially revealing frequency distribution of the patch states. Likewise, despite the enormous effort involved in making measurements on marked tillers in grazed pastures, traditionally only a mean defoliation interval is reported, and, given the complex frequency distribution of defoliation intervals the mean may be a poor summary of what plants experience (P. Carrère, F. Louault, J.F. Soussana, P.C. de F. Carvalho and M. LaFarge, unpublished results). There are therefore surprisingly few data presented in a form to support the kind of analyses proposed here to explain the creation and maintenance of heterogeneity and its impacts. However, the highly insightful studies by Gibb *et al.* (1989, 1997) have shown how markedly bimodal distributions of patch states can indeed arise under very carefully controlled continuous grazing by cattle, and in both all-grass and grass–legume pastures. These studies also showed how bimodal distributions arose under a wide range of stock densities, the stock density affecting the proportion of the total area that was 'tall' relative to the proportion that was 'short'. This is a result that can be simulated by simple models, such as those described here. Although Gibb *et al.* (1989, 1997) never actually measured defoliation intervals, we propose that they were correct in calling the tall areas 'less frequently defoliated' and the short areas 'frequently defoliated' (see Gibb *et al.*, 1997, their Table 1), and we do not feel that these populations represent the pre- and postgrazing patch states, as was proposed by Ungar (1996). There is evidence in the literature of preference both for tall and for short areas of pastures (e.g. Illius *et al.*, 1987; Harvey and Wadge, 1994; Distel *et al.*, 1995), although the real motivation for the preference is often, no doubt, the associated differences in the nutritional value and/or subsequent digestion handling costs in plant parts (e.g. Wallis de Vries and Daleboudt, 1994; Dumont *et al.*, 1995; Prache *et al.*, 1998; Shipley *et al.*, 1999).

The hypothesis that bimodal distributions of vegetation state are incurred as a dynamic consequence of a partial rejection for tall patches is difficult to test experimentally; however, some corroborating evidence is available from studies of preference in pure clover swards. Pure clover swards are characteristically far more patchy (e.g. in height) compared with pure grass swards of the same mean height, even when the variance in height in the two cases is similar (Harvey and Wadge, 1994).

But Does Any Form of Grazing Achieve the Maximum Sustainable Yield?

Finally, let us return to consider just how different foraging strategies and the resulting heterogeneity reduce yield, not only when comparing one strategy with another, but comparing all our grazing strategies relative to a theoretical maximum sustainable yield. We saw earlier how the different grazing strategies (and different stock densities) alter the means, as well as generating variance, in the two major determinants of the rate of regeneration of resources (plant regrowth), namely the residual patch states and the defoliation intervals, and that these dynamic responses are paramount in explaining the resulting yield and performance of the grazing system. Below, and in Fig. 15.7, we analyse how the complex combinations of mean and variance in residual patch state and in defoliation interval, which emerge dynamically from our contrasting foraging processes, compare with the theoretical potential of the system.

From the relationship(s) describing vegetation growth in Fig. 15.2, we can identify, for any given starting (residual) patch state an optimum timing of harvest as being the time when the average growth rate ($(W-W_0)/t$) has reached a maximum (see Parsons *et al.*, 1988a, b; Fig. 15.2c). Using an averaged growth rate as a currency for optimization is consistent with the marginal-value theorem for optimal foraging in ecology (Charnov, 1976). We can therefore solve, for all possible residual patch states, what the optimal defoliation interval would be and the corresponding yield that would be achieved in each case. These theoretical optimal solutions for grassland management (cf. calculated maximum sustainable yields) are plotted in Fig. 15.7 (solid lines). On these same axes, we can now plot the combinations of mean residual patch state and mean defoliation intervals that emerge dynamically when the pasture is grazed according to two of the three contrasting strategies for placing bites in space and time (see above) and across a wide range of stock densities.

Recall that sequential grazing gives deterministic defoliation intervals and is equivalent to a situation where there is no variance in processes and no heterogeneity in patch states. And yet, in Fig. 15.7 (open circles), it is apparent that, even at the optimal stock density, the residual patch states are too large and the defoliation interval is too short to achieve the global optimum yield. This cannot

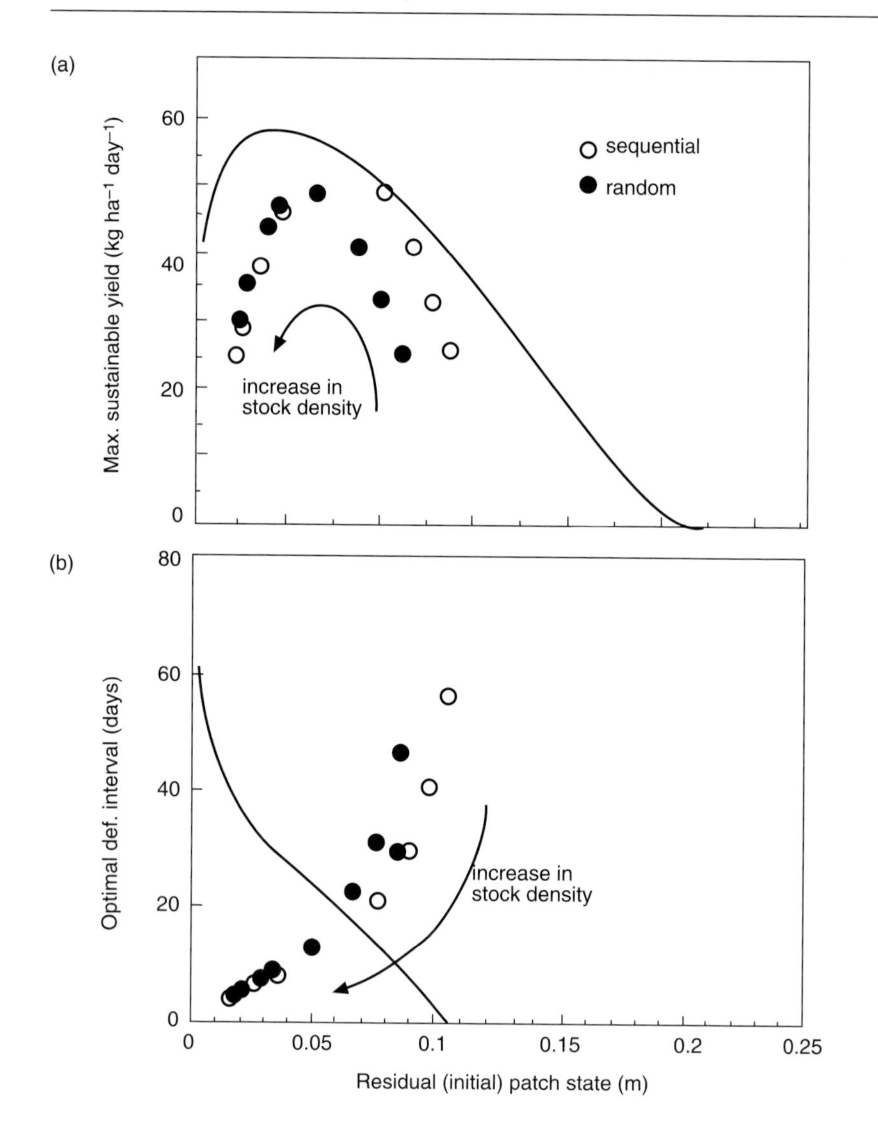

Fig. 15.7. The effect of residual patch state on (a) the maximum sustainable yield (maximum average growth rate) that may be achieved, and (b) the defoliation (def.) interval required to achieve this, as predicted by a model that seeks optimal solutions for all combinations of the residual sward state ('severity') and the timing of harvest ('duration of regrowth') in a series of discrete defoliations in all cases. The solid lines show the optimal solutions. The open circles (o) show the combinations of residual states and defoliation intervals and yields that emerge in patches when animals are assumed to graze patches in sequence (type 1, deterministically), and the solid circles (●) for when animals graze patches at random (type 2). Each circle is for a separate stock density, arrows show direction of increasing stock density. See text for explanation. (From Parsons and Chapman, 1999.)

be remedied by changing stock density. Higher stock densities give patches that are too short and are grazed too soon, and lower stock densities give patches that are too tall and grazed too late, relative to the local, let alone the global, optimum.

The situation is worse when we consider the situation where there is variance about residual patch state and defoliation interval (and so about the means plotted on each of these axes) (see solid circles for random grazing in Fig. 15.7). The variance in patch states and defoliation intervals results in a situation where the means of these must fall below even the local optimal solutions.

This reanalysis of grazing, using a spatial model incorporating recent advances in understanding intake by animals and plant growth at the bite-scale, clearly reveals some fundamental constraints to harvesting grass under grazing. Animals simply cannot reduce vegetation that has grown for the relatively long global optimal defoliation interval to the desired (small) residual patch state in a single bite (unless the proportional bite depth, f, is 1.0, which has not been observed). Harvesting patches to the desired state using a succession of bites consumes time and again constrains intake.

Heterogeneity and Complexity Abound

Even a recently sown pasture reveals heterogeneity over a range of spatial scales. At first, there may be differences only in plant size, morphology and local plant density. But subsequently invasions and local genotypic selection can give rise to increasing biodiversity (e.g. Dodd *et al.*, 1995). Animals, too, contribute to creating and maintaining heterogeneity in ways beyond those we have considered here (see Laca and Ortega, 1995; Marriott and Carrère, 1998). Treading and the spatial deposition of dung and urine can have profound effects in creating variance in plant growth, as well as in generating complex temporal and spatial patterns in species associations (see Schwinning and Parsons, 1996a, b, c). Access to water, shelter, slope and other topographic features will prevail, and social behaviour (flocking for comfort, vigilance) and diurnal and seasonal patterns of grazing add further complexity. We do not imagine that one model or one analysis can possibly address all these heterogeneity-inducing factors simultaneously. We suggest only that detailed studies might systematically target particular components of heterogeneity to explore their role in grassland ecosystems. Models may be valuable as tools to explore the impacts of numerous such effects, singly or together, always seeking some general insight, as well as providing new hypotheses (Marriott and Carrère, 1998).

Here we have aimed to show how events at the bite (or individual plant) scale are expressed in phenomena at the field-scale (as proposed by Marriott and Carrère, 1998). But reworking plant/animal interactions at a fine spatial scale, and so producing detailed spatial models, would be of questionable value, were it not that the approach reveals some new insights, as well as some of the shortcomings of working only at the field-scale (A.J. Parsons, unpublished results). One of the shortcomings we have identified here is the inevitable difficulty of attempting to define functional responses at the field scale when the field is heterogeneous and animals respond to variance in the local pasture state. In terms of insights, the present study greatly emphasizes the importance of the regeneration of resources – the regrowth of vegetation – in determining the outcome of plant/animal interactions. There have been many detailed accounts of how animals should respond optimally to heterogeneity in vegetation, but most assume no revisitation of patches and so cannot consider how resource regeneration affects plant/animal interactions (see Possingham and Houston, 1990).

Our models reveal that spatial and temporal complexity can arise even from relatively straightforward plant growth processes and simple foraging algorithms. Here we have analysed the interaction between plants and animals as a single dynamic system and have looked at the spatial consequences. Bimodal distributions in patch states arise, either because the grazing system allows patches to exist in all possible states from a low residual state to a large final state (and so the sigmoid nature of the growth curve is evident in a U-shaped distribution of patch states), or due to some basic state-dependent selective grazing. It is well established that very complex temporal behaviour, as well as elaborate spatial patterns, can arise in simple dynamic systems, and this can be enhanced, given 'noise' or stochasticity in inputs and processes (see, for example, May, 1974; Horsthemke and Lefever, 1984; Mosekilde and Mouritsen, 1995).

There is immense scope to add further realistic but potentially bewildering complexity. First, more complex state-dependent foraging decisions can be shown readily to lead to more distinctly bimodal distributions of patch states, and so greater heterogeneity. Contrasting choices that arise initially by sheer chance can be greatly reinforced by subsequent state-dependent behaviour. Secondly, there are many ways of introducing greater non-linearity in the processes and responses. For example, there is likely to be a non-linear distribution of mass with height in the vegetation canopy (e.g. Ginnett *et al.*, 1999) and so fractional bite depth will not relate directly to mass. This is simple to model (see, for example, Parsons *et al.*, 1994), but introducing such non-linearities can readily lead to complex behaviour that makes it hard to separate biological insight from possible artefact. However, some forms of non-linearity (such as density-dependent resource limitations) may stabilize interactions (see McCann *et al.*, 1998).

Managing with Heterogeneity

The analysis of grazing systems using field-scale functional responses has put forward a time-honoured paradigm for grazing management. Based on the mathematical phenomenon of discontinuous ('dual') stability (Noy-Meir, 1975), it has been argued that attempts to achieve maximum yield per hectare are associated with driving the system close to the point (e.g. with respect to stock density) where any deterioration in, for example, the rate of regrowth of vegetation would precipitate the system toward an overgrazed and unproductive state. Our spatial reanalysis of grazing exhibits much less tendency towards discontinuous stability. Since grazing does, in reality, function at the bite scale and generates heterogeneous pastures, discontinuous stability might also be much less prevalent in reality (Schwinning and Parsons, 1999). This suggests that, for any given vegetation community, gradual changes in stock density may lead to correspondingly gradual changes in grazing pressure and so offer more flexibility to grazing management.

Some authors relate the phenomenon of dual stability to that of degradation, arguing that the undegraded and degraded pastures represent the different branches (and so equilibria) of the dual stable region (see Westoby *et al.*, 1989; Laycock, 1991; review by Petraitis and Latham, 1999). If dual stability is less prevalent than previously considered, an alternative explanation for degradation may therefore be required. Note that, in a single dynamic system, dual stable solutions are alternative equilibria and cannot coexist. If a single body of animals is grazing, even sporadically, from all parts of a 'pasture', we feel that it is appropriate to look first for explanations for spatial phenomena based on a single dynamic system. In this context, degradation might instead be associated with a spatial phenomenon, such as the development of bimodal distributions of patch states, as seen here under selective grazing – that is, where two contrasting states arise spatially and do indeed coexist. But this soon evokes the vexing question as to at what point the plant/animal interaction might no longer be seen as a single dynamic system. As regions become very distinct, it could be considered that grazing should be conceived as more than one dynamic system. Certainly, we do not dispute the fact that pastures may degenerate when the grazing pressure gives rise to changes in the species composition (distribution or abundance) in the community and that less progressive reductions in grazing pressure are essential to restore some rangeland systems (see review by Tainton *et al.*, 1996).

Our studies suggest that different management options, notably 'continuous' vs. rotational (or 'paddock') grazing systems, offer alternative as opposed to fundamentally different means of utilizing grassland. In both, grazing is discrete (and so 'rotational') at the bite-scale (Clark *et al.*, 1984), and so optimizing growth and utilization revolves around the same fundamental principles for maximizing the average growth rate (see Parsons and Chapman, 1999; Fig. 15.2c). But we can add here that both can also generate marked heterogeneity. Indeed, rotational grazing purposely generates extreme heterogeneity at the between-paddock scale, albeit in an attempt to control heterogeneity (minimize it) at the within-paddock scale. The two systems can be seen, then, to provide alternative ways of distributing heterogeneity spatially. It is highly debatable if either is consistently better. The principles illustrated here for the sequential grazing of patches at the bite-scale (type 1) and the constraints to grazing this source of heterogeneity implies (Fig. 15.7) will apply equally under rotational grazing, that is for the sequential (rotational) grazing of 'paddocks' at the farm scale. Although, over and above the constraints to achieving the right combination of residual patch states and defoliation intervals, between paddocks there will be the possibility of additional heterogeneity, and so constraints within paddocks. Clearly, the more marked and deleterious heterogeneity that arises due to selective grazing (type 3 here) would be easier to control under some form of paddock (restricted-access) grazing, especially if the 'tall' areas, well above the desired mean vegetation state, were confined to single paddocks. It is in this sense that rotational (paddock) grazing is assumed to be best. However, it must be recognized, as can readily be modelled, that, while confining animals to a paddock is effective in ensuring that the paddock is reduced to the desired residual state, intake rates are depressed and time is consumed in doing so. Other areas will have an increased tendency to become 'tall' and potentially have superoptimal regrowth intervals and thus become less preferred by animals. All these issues are already recognized by grazing managers, but what the recent studies add is the basis for identifying strategic as well as tactical optimal solutions for grazing management, which can be implemented in next-generation decision support tools, which will consider the best strategy, even given uncertainties in weather and financial return (*sensu* Mathieu and Morice, 1984).

References

Begon, M. and Mortimer, M. (1986) *Population Ecology: A Unified Study of Animals and Plants*, 2nd edn. Blackwell Scientific Publications, Oxford, 220 pp.

Begon, M., Harper, J.L. and Townsend, C.R. (1990) *Ecology: Individuals, Populations and Communities*. Blackwell Scientific Publications, Oxford, 957 pp.

Booysen, P. de V. (1966) A physiological approach to research in pasture utilization. *Proceedings of the Grassland Society of South Africa* 2, 77–85.

Brown, B.J. and Allen, T.F.H. (1989) The importance of scale in evaluating herbivory impacts. *Oikos* 54, 189–194.

Brown, R.H. and Blaser, R.E. (1968) Leaf area index in pasture growth. *Herbage Abstracts* 38, 1–9.

Charnov, E.L. (1976) Optimal foraging, the marginal value theorem. *Theoretical Population Biology* 9, 129–136.

Chesson, P. (1978) Predator–prey theory and variability. *Annual Review of Ecology and Systematics* 9, 323–347.

Clark, D.A., Chapman, D.F., Land, C.A. and Dymock, N. (1984) Defoliation of *Lolium perenne* and *Agrostis* spp. tillers, and *Trifolium repens* stolons in set-stocked and rotationally grazed hill pastures. *New Zealand Journal of Agricultural Research* 27, 289–301.

Coughenour, M.B. (1991) Spatial components of plant-herbivore interactions in pastoral, ranching, and native ungulate ecosystems. *Journal of Range Management* 44, 530–542.

Demment, M.W., Laca, E.A. and Greenwood, G.B. (1987) Intake in grazing ruminants: a conceptual framework. In: Owens, F.N. (ed.) *Feed Intake by Cattle*. Symposium Proceedings, Agricultural Experimental Station, Oklahoma State University, pp. 208–225.

Demment, M.W., Peyraud, J.L. and Laca, E.A. (1995) Herbage intake at grazing: a modelling approach. In: Journet, M., Grenet, E., Farce, M.H., Theriez, M. and Demarquilly, C. (eds) *Recent Developments in the Nutrition of Herbivores: Proceedings of the 4th International Symposium on the Nutrition of Herbivores*, INRA Editions, Paris, pp. 121–141.

Distel, R.A., Laca, E.A., Griggs, T.C. and Demment, M.W. (1995) Patch selection by cattle: maximisation of intake rate in horizontally heterogeneous pastures. *Applied Animal Behaviour Science* 45, 11–21.

Dodd, M., Silvertown, J., McConway, K., Potts, J. and Crawley, M. (1995) Community stability: a 60 year record of trends and outbreaks in the occurrence of species in the Park Grass Experiment. *Journal of Ecology* 83, 277–285.

Dolman, P.M. and Sutherland, W.J. (1997) Spatial patterns of depletion imposed by foraging vertebrates: theory, review and meta-analysis. *Journal of Animal Ecology* 66, 481–494.

Dumont, B., Petit, M. and D'hour, P. (1995) Choice of sheep and cattle between vegetative and reproductive cocksfoot patches. *Applied Animal Behaviour Science* 43, 1–15.

Dutilleul, P. and Legendre, P. (1993) Spatial heterogeneity against heteroscedasticity: an ecological paradigm versus a statistical concept. *Oikos* 66, 152–171.

Edwards, G.R. (1994) The creation and maintenance of spatial heterogeneity in plant communities: the role of plant–animal interactions. PhD thesis, University of Oxford, 180 pp.

Edwards, G.R., Newman, J.A., Parsons, A.J. and Krebs, J.R. (1994) Effects of scale and spatial distribution of the food resource and animal state on diet selection: an example with sheep. *Journal of Animal Ecology* 63, 816–826.

Edwards, G.R., Parsons, A.J., Penning, P.D. and Newman, J.A. (1996a) Relationship between vegetation state and bite dimensions of sheep grazing contrasting plant species and its implications for intake rate and diet selection. *Grass and Forage Science* 50, 378–388.

Edwards, G.R., Newman, J.A., Parsons, A.J. and Krebs, J.R. (1996b) The use of spatial memory by grazing animals to locate food patches in spatially heterogeneous environments: an example with sheep. *Applied Animal Behaviour Science* 50, 147–160.

Edwards, G.R., Newman, J.A., Parsons, A.J. and Krebs, J.R. (1997) Use of cues by grazing animals to locate food patches: an example with sheep. *Applied Animal Behaviour Science* 51, 59–68.

Gibb, M.J., Baker, R.D. and Sayer, A.M.E. (1989) The impact of grazing severity on perrenial ryegrass/white clover swards stocked continuously with beef cattle. *Grass and Forage Science* 44, 315–328.

Gibb, M.J., Huckle, C.A., Nuthall, R. and Rook, A.J. (1997) Effect of sward surface height on intake and grazing behaviour by lactating Holstein Friesian cows. *Grass and Forage Science* 52, 309–321.

Ginnett, T.F., Dankosky, J.A., Deo, G. and Demment, M.W. (1999) Patch depression in grazers: the roles of biomass distribution and residual stems. *Functional Ecology* 13, 37–44.

Gordon, I.A. and Lascano, C. (1994) Foraging strategies of ruminant livestock on intensively managed grasslands: potentials and constraints. In: *Proceedings of the 17th International Grassland Congress, Palmerston North, New Zealand*, SIR Publishing, Wellington, New Zealand, pp. 681–689.

Harvey, A. and Wadge, K.J. (1994) Clover heterogeneity: preference of ewes and lambs for tall and short areas when grazing clover monocultures. In: *Proceedings of the 4th British Grassland Society Research Conference, University of Reading, UK*. British Grassland Society, Reading, pp. 153–154.

Hassell, M.P. (1978) *The Dynamics of Arthropod Predator–Prey Systems*. Princeton University Press, Princeton, New Jersey, 237 pp.

Hodgson, J. and Illius, A.W. (eds) (1996) *The Ecology and Management of Grazing Systems*. CAB International, Wallingford, UK, 466 pp.

Hodgson, J. and Ollerenshaw, J.H. (1969) The frequency and severity of defoliation of individual tillers in set-stocked swards. *Journal of the British Grassland Society* 24, 226–234.

Horsthemke, W. and Lefever, R. (1984) *Noise Induced Transitions:Theory and Applications in Physics, Chemistry and Biology*. Springer Series in Synergetics Vol. 15, Springer-Verlag, Berlin, 318 pp.

Huston, M.A. (1997) Hidden treatments in ecological experiments: re-evaluating the ecosystem function of biodiversity. *Oecologia* 110, 449–460.

Illius, A.W. and Gordon, I.J. (1987) The allometry of food intake in grazing ruminants. *Journal of Animal Ecology* 56, 989–999.

Illius, A.W. and Hodgson, J. (1996) Progress in understanding the ecology and management of grazing systems. In: Hodgson, J. and Illius, A.W. (eds) *The Ecology and Management of Grazing Systems*. CAB International, Wallingford, UK, pp. 429–457.

Illius, A.W., Wood-Gush, D.G.M. and Eddison, J.C. (1987) A study of the foraging behaviour of cattle grazing patchy swards. *Biology of Behaviour* 12, 33–44.

Laca, E.A. and Demment, M.W. (1996) Foraging strategies of grazing animals. In: Hodgson, J. and Illius, A.W. (eds) *The Ecology and Management of Grazing Systems*. CAB International, Wallingford, UK, pp. 137–158.

Laca, E.A. and Ortega, I.M. (1995) Integrating foraging mechanisms across spatial and temporal scales. In: *Proceedings of the 5th International Rangeland Congress, Salt Lake City, Utah, July 1995*. Society for Range Management, Denver, Colorado, pp. 129–132.

Laca, E.A., Ungar, E.D., Seligman, N. and Demment, M.W. (1992) Effects of sward height and bulk density on bite dimensions of cattle grazing homogeneous swards. *Grass and Forage Science* 47, 91–102.

Laycock, W.A. (1991) Stable states and thresholds of range condition on North American rangelands: a viewpoint. *Journal of Range Management* 44, 427–433.

Lemaire, G. and Chapman, D.F. (1996) Tissue flows in grazed plant communities. In: Hodgson, J. and Illius, A.W. (eds) *The Ecology and Management of Grazing Systems*. CAB International, Wallingford, UK, pp. 3–36.

Louault, F., Carrère, P. and Soussana, J.F. (1997) Efficiencies of ryegrass and white clover herbage utilisation in mixtures continuously grazed by sheep. *Grass and Forage Science* 52, 388–400.

McCann, K., Hastings, A. and Huxel, G.R. (1998) Weak trophic interactions and the balance of nature. *Nature* 395, 794–798.

Mangel, M. and Clark, C.W. (1986) Towards a unified foraging theory. *Ecology* 67, 1127–1138.

Mangel, M. and Clark, C.W. (1988) *Dynamic Modelling in Behavioural Ecology*. Princetown University Press, Princetown, New Jersey, 302 pp.

Marriott, C. and Carrère, P. (1998) Structure and dynamics of grazed vegetation. *Annales de Zootechnies* 47, 359–369.

Mathieu, J. and Morice, G. (1984) Climate changes and dairy farm economics. In: Riley, H. and Skjelvag, A.O. (eds) *The Impact of Climate on Grass Production and Quality. Proceedings of the 10th General Meeeting of the European Grassland Federation, Norway*. Norwegian State Agricultural Research Stations, Ås, Norway, pp. 520–524.

May, R.M. (1974) *Stability and Complexity in Model Ecosystems*. Monographs in Population Biology 6, Princeton University Press, Princeton, New Jersey.

Mazzanti, A. and Lemaire, G. (1994) Effect of nitrogen fertilization on herbage production of tall fescue swards continuously grazed by sheep. 2. Consumption and efficiency of herbage utilisation. *Grass and Forage Science* 49, 352–359.

Morley, F.H.W. (1966) Stability and productivity of pastures. *Proceedings of the New Zealand Society of Animal Production* 26, 8–21.

Morris, R.M. (1969) The pattern of grazing in 'continuously' grazed swards. *Journal of the British Grassland Society* 24, 65–70.

Mosekilde, E. and Mouritsen, O.G. (1995) *Modelling the Dynamics of Biological Systems: Non-linear Phenomena and Pattern Formation.* Springer Series in Synergetics Vol. 65, Springer-Verlag, Berlin, 294 pp.

Newman, J.A., Parsons, A.J., Thornley, J.H.M., Penning, P.D. and Krebs, J.R. (1995) Optimal diet selection by a generalist grazing herbivore. *Functional Ecology* 9, 255–268.

Noy-Meir, I. (1975) Stability of grazing systems: an application of predator–prey graphs. *Journal of Ecology* 63, 459–481.

Owen-Smith, N. and Novellie, P. (1982) What should a clever ungulate eat? *American Naturalist* 119, 151–178.

Pacala, S.W. and Rees, M. (1998) Models suggesting field experiments to test two hypotheses explaining successional diversity. *American Naturalist* 152, 729–737.

Parsons, A.J. and Chapman, D.F. (1998) Principles of grass growth and pasture utilisation. In: Cherney, J.H and Cherney, D.J.R. (eds) *Grass for Dairy Cattle.* CAB International, Wallingford, pp. 283–310.

Parsons, A.J. and Chapman, D.F. (1999) The principles of pasture growth and utilisation. In: Hopkins, A. (ed.) *Grass*, 3rd edn. Blackwell Scientific Publications, Oxford.

Parsons, A.J., Johnson, I.R. and Williams, J.H.H. (1988a) Leaf age structure and canopy photosynthesis in rotationally and continuously grazed swards. *Grass and Forage Science* 43, 1–14.

Parsons, A.J., Johnson, I.R. and Harvey, A. (1988b) Use of a model to optimise the interaction between the frequency and severity of intermittent defoliation and to provide a fundamental comparison of the continuous and intermittent defoliation of grass. *Grass and Forage Science* 43, 49–59.

Parsons, A.J., Thornley, J.H.M., Newman, J.A. and Penning, P.D. (1994) A mechanistic model of some physical determinants of intake rate and diet selection in a two-species temperate grassland sward. *Functional Ecology* 8, 187–204.

Penning, P.D. (1986) Some effects of sward conditions on grazing behaviour and intake by sheep. In: Gudmundsson, O. (ed.) *Grazing Research at Northern Latitudes.* Plenum Publishing Corporation, New York, pp. 219–226.

Penning, P.D., Parsons, A.J., Orr, R.J., Harvey, A. and Champion, R.A. (1995) Intake and behaviour responses by sheep in different physiological states, when grazing monocultures of grass or white clover. *Applied Animal Behaviour Science* 45, 63–78.

Petraitis, P.S. and Latham, R.E. (1999) The importance of scale in testing the origins of alternative community states. *Ecology* 80, 429–442.

Possingham, P.H. and Houston, A.I. (1990) Optimal patch use by a territorial forager. *Journal of Theoretical Biology* 145, 343–353.

Prache, S., Roguet, C. and Petit, M. (1998) How degree of selectivity modifies foraging behaviour of dry ewes on reproductive compared to vegetative sward structure. *Applied Animal Behaviour Science* 57, 91–108.

Proulx, M. and Mazumder, A. (1998) Reversal of grazing impact on plant species richness in nutrient poor vs. nutrient rich ecosystems. *Ecology* 79, 2581–2592.

Richards, J.H. (1993) Physiology of plants recovering from defoliation. Session: Plant Growth. In: *Proceedings of the 17th International Grassland Congress 1993, Palmerston North, New Zealand,* Vol. I. SIR Publishing, Wellington, New Zealand, pp. 85–94.

Robson, M.J., Ryle, G.J.A. and Woledge, J. (1988) The grass plant – its form and function.

In: Jones, M.B. and Lazenby, A. (eds) *The Grass Crop – the Physiology Basis of Production*. Chapman and Hall, London, pp. 25–83.

Saether, B.-E. (1997) Environmental stochasticity and population dynamics of large herbivores: a search for mechanisms. *Trends in Ecology and Evolution* 12, 143–149.

Schwinning, S. and Parsons, A.J. (1996a) Analysis of the coexistence mechanisms for grasses and legumes in grazing systems. *Journal of Ecology* 84, 799–814.

Schwinning, S. and Parsons, A.J. (1996b) A spatially explicit population model of stoloniferous N-fixing legumes in mixed pasture with grass. *Journal of Ecology* 84, 815–826.

Schwinning, S. and Parsons, A.J. (1996c) Interaction between grasses and legumes: understanding variability in species composition. In: Younie, D. (ed.) *Legumes in Sustainable Farming Systems. Occasional Symposium of the British Grassland Society no. 20*. British Grassland Society, Reading, pp. 153–163.

Schwinning, S. and Parsons, A.J. (1999) The stability of grazing systems revisited: spatial models and the role of heterogeneity. *Functional Ecology* 13, 737–747.

Senft, R.L., Coughenour, M.B., Bailey, D.W., Rittenhouse, L.R., Sala, O.E. and Swift, D.M. (1987) Large herbivore foraging and ecological hierarchies. *Bioscience* 37, 789–799.

Shipley, L.A., Gross, J.E., Spalinger, D.E., Hobbs, N.T. and Wunder, B.A. (1994) The scaling of intake rate in mammalian herbivores. *American Naturalist* 143, 1055–1082.

Shipley, L.A., Illius, A.W., Dannell, K., Hobbs, N.T. and Spalinger, D.E. (1999) Predicting bite selection of mammalian herbivores: a test of a general model of diet optimization. *Oikos* 84, 55–68.

Silander, J.A. and Pacala, S.W. (1990) The application of plant population dynamic models to understanding plant competition. In: Grace, J.B. and Tilman, D. (eds) *Perspectives on Plant Competition*. Academic Press, London, New York, pp. 67–91.

Spalinger, D.E. and Hobbs, N.T. (1992) Mechanisms of foraging in mammalian herbivores: new models of functional response. *American Naturalist* 140, 325–348.

Stenseth, N.C. and Chan, K.-S. (1998) Non-linear sheep in a noisy world. *Nature* 394, 620–621.

Stephens, D.W. and Krebs, J.R. (1986) *Foraging Theory*. Princeton University Press, Princeton, New Jersey, 247 pp.

Tainton, N.M., Morris, C.D. and Hardy, M.B. (1996) Complexity and stability in grazing systems. In: Hodgson, J. and Illius, A.W. (eds), *The Ecology and Management of Grazing Systems*. CAB International, Wallingford, UK, pp. 275–299.

Thornley, J.H.M. and Johnson, I.R. (1990) *Plant and Crop Modelling*. Clarendon Press, Oxford, 669 pp.

Tilman, D. (1994) Competition and biodiversity in spatially structured habitats. *Ecology* 75, 2–16.

Tilman, D. and Pacala, S.W. (1993) The maintenance of species richness in plant communities. In: Ricklefs, R.E. and Schluter, D. (eds) *Species Diversity in Ecological Communities*. University of Chicago Press, Chicago, pp. 13–25.

Tilman, D., Wedin, D. and Knops, J. (1996) Productivity and sustainability influenced by biodiversity in grassland ecosystems. *Nature* 379, 718–720.

Ungar, E.D. (1996) Ingestive behaviour. In: Hodgson, J. and Illius, A.W. (eds) *The Ecology and Management of Grazing Systems*. CAB International, Wallingford, UK, pp. 185–218.

Ungar, E.D. and Noy-Meir, I. (1988) Herbage intake in relation to availability and sward structure: grazing processes and optimal foraging. *Journal of Applied Ecology* 25, 1045–1062.

Ungar, E.D., Genizi, A. and Demment, M.W. (1991) Bite dimensions and herbage intake by cattle grazing short, hand-constucted swards. *Agronomy Journal* 83, 973–978.

Wade, M.H. (1991) Factors affecting the availability of vegetative *Lolium perenne* to grazing dairy cows with special reference to sward characteristics, stocking rate and grazing method. Doctoral thesis, University of Rennes, France.

Wade, M.H., Peyraud, J.L., Lemaire, G. and Cameron, E.A. (1989) The dynamics of daily area and depth of grazing and herbage intake of cows in a five-day paddock system. In: *Proceedings of the 16th International Grassland Congress, Nice, France*. AFPF, Versailles, France, pp. 1111–1112.

Wallis de Vries, M.F. and Daleboudt, C. (1994) Foraging strategy of cattle in patchy grassland. *Oecologia* 100, 98–106.

Westoby, M., Walker, B. and Noy-Meir, I. (1989) Opportunistic management for rangelands not at equilibrium. *Journal of Range Management* 42, 266–274.

Wiens, J.A. (1989) Spatial scaling in ecology. *Functional Ecology* 3, 385–397.

16

Soil–Plant–Animal Interactions and Impact on Nitrogen and Phosphorus Cycling and Recycling in Grazed Pastures

S.C. Jarvis

Institute of Grassland and Environmental Research, North Wyke Research Station, Okehampton, Devon EX20 2SB, UK

Introduction

Sustainable pasture management depends upon an adequacy and balance of supply of nutrients to meet the requirements of an appropriate level of dry matter production of sufficient nutritional quality for livestock. Whilst legumes have provided the nitrogen (N) supplying resource in many circumstances, fertilizer inputs of all nutrients have been the key to sustainable production. Much research in the past has been targeted at defining crop needs and responses to nutrients and in many parts of the world this is still an important need in order to promote productivity levels that meet demands for food. Increasingly, however, there has been at least an equal interest in the impact of nutrient management on environmental quality and effects on air and water composition, as well as in the maintenance of soil systems that can sustain production or ecosystems for current and long-term needs.

All components of grassland production systems are 'leaky' with respect to nutrients to a greater or lesser extent, depending upon the intensity of management, the physicochemical nature of the particular nutrient and the local management/agronomy that is applied. As socio-economic and political pressures have increased, the ability to increase efficiency of nutrient use and reduce potential for environmental impact has also increased. This has consequences for, and important interactions with, grazing; to make effective overall changes requires a knowledge of all the pathways and transfers of each nutrient throughout the particular management system. Whilst there has been a long-term acknowledgement that this level of understanding is important, interest in understanding flows at the systems level has increased with the realization that this offers opportunities

 Grassland Ecophysiology and Grazing Ecology (eds G. Lemaire, J. Hodgson, A. de Moraes, C. Nabinger and P.C. de F. Carvalho)

to exploit the potential to maximize both internal and external nutrient resources.

Within grassland-based livestock production systems, the role of the animal has an overriding influence on nutrient fluxes. Grasses and forages are generally highly effective at maximizing uptake and incorporation of available forms of nutrients into their biomass. It is often the next stages in the production cycle, associated with poor utilization by livestock, which generate much potential for inefficiency, especially for N (because of its mobility and capacity to change chemical and physical forms) but also for other nutrients as well. In those managements where animals are housed, this presents much opportunity for loss, but the grazing phase is also very important. The effects that grazing may have on nutrient cycling have been well described and can be summarized as follows:

- Removal of sink capacity (plant shoots) for nutrients and influence on internal transfers within the plant.
- Relocation of nutrients from grazed areas to houses and into manure during milking, overwintering and relocation of animals.
- Local concentration of nutrients in dung and urine patches and at water, feeding and camping areas in the field.
- Conversion of nutrients from stable, often relatively immobile, forms in plant materials into mobile forms, which have potential for transfer away from the system.
- Alteration of soil physical conditions (through treading and other effects) with impact on soil water, aeration, stability (and enhanced potential for physical transfer of particulate materials and their associated nutrients) and forms and availability of nutrients.

These factors are important in all systems, regardless of whether they are intensively or extensively managed, and contribute to the development or degradation of the long-term sustainability of the system. Extensively managed systems may often be more fragile than those that are managed under more intensive regimes, and will require consideration and research effort to a comparable level to that received by intensive pastures over the last 10 years or so, in which the impact of the animal has been clearly defined (see Jarvis, 1998). In the present chapter, the implications of the effects of the grazing animal on the behaviour of N and phosphorus (P) are considered, with particular reference to the consequences for sustainable supplies for production and losses to the environment.

Nitrogen and Phosphorus in Grassland Soils

The requirements, roles and functions of N and P in pasture management for both plants and animals are well known and have been reviewed extensively over the years (Whitehead, 1995; Tunney *et al.*, 1997). The behaviour of the two nutrients within ecosystems is quite different. Substantial amounts of the N, on the one hand, are either mobile or have the potential to be converted into mobile forms, and N is therefore considered to be 'non-conservative'. Most of the P, on the other hand, is immobile and regarded as being a 'conservative' nutrient in most circumstances. Their functions and requirements in plant and animal biomass are well defined and will not be discussed further here. Recent research has centred on their environmental impact 'downstream' from farming systems and these issues have been widely discussed and debated (see Jarvis and Pain, 1997; Tunney *et al.*, 1997; Jarvis, 1998). This interest has arisen because of concern over air (e.g. ammonia (NH_3), nitrous oxide (N_2O) and NO_x emissions and transfer) and water (nitrate (NO_3^-), nitrite (NO_2^-) and various forms of mobile P) quality. Increased levels of NH_3 in the atmosphere have resulted, after deposition, in changing N status and reduced soil pH in poorly buffered ecosystems. Nitrous oxide influences atmospheric quality and contributes to global warming and climate change and/or the controls over ozone concentrations. Although, for both NH_3 and N_2O, those livestock systems which involve housed livestock are particularly important, extensive managements make a substantial impact because of the large number of animals that such systems support at the larger scale. These issues will become more and more important as nations align themselves with international agreements and policies to reduce environmental impacts.

Directives already exist within European and other countries to hold constituents of potable and other waters to within defined limits. Nations within the European Union have implemented a range of policies to meet the requirements of directives, such as that designed to reduce NO_3^- in waters (EC, 1991). The role of P as the controlling nutrient for eutrophication in inland surface waters has long been recognized, whilst in marine waters it is N (Rekolainen *et al.*, 1997), and, because industrial sources have reduced their outputs, and are continuing to do so, diffuse agricultural sources have become more and more significant (Tunney *et al.*, 1997). These effects, as discussed later, may be as relevant to extensive managements as to intensive ones, depending upon how grasslands are positioned within the landscape.

There is also increasing interest in how nutrient supplies influence pasture composition of swards from at least two other points of view: first, with respect to maintaining floristic diversity and, secondly, in maintaining appropriate mixtures involving legume species, which allow a sustainable level of production. Enhanced levels of P in the soil have a major restrictive control on attempts to return intensively managed pastures to the diversity that was present under previous extensive management (Tallowin *et al.*, 1998). Another particular issue has been the effects of supplies of both N and P in the control and maintenance of legumes in mixed swards. There have been recent modelling developments for prediction of the changing structure and degree of patchiness within grazed mixed grass–clover swards, which have highlighted the role of supplies of soil mineral N in defining sward structure and, in this case, the role of white clover (Schwinning and Parsons, 1996). The present chapter concentrates on non-legume-based swards.

Agents of N and P Transfer within Grazed Grasslands

Nutrient sources and turnover

Inputs of N and P to the soil arise from fertilizer, atmospheric deposition and any manures 'imported' into a particular field either from another component of the system or from another enterprise altogether. If animals are provided with supplementary feeding (e.g. as concentrates), there may be a substantial import of N and P to the total farming system, some of which may find its way to the grazing areas. In some temperate managements, this input may be substantial: for example a typical dairy farm in the UK will receive > 15% of its annual N input through bought-in feeds, which is then cycled and recycled through animal excreta (Jarvis, 1993). For P, the proportion is 65% and it has been calculated that the surplus that is accumulating within a typical dairy farm almost exactly equates to that contained within purchased feeds (Haygarth *et al.*, 1998a). Other, internal supplies of N and P are regulated by the factors that control release from the soil organic matter (SOM) and from dung and urine. Without inputs, grazing pressures have been seen, in long-term studies, to jeopardize the sustainability of ecosystems – for example, by reducing fertility of rough fescue pastures in Alberta, USA (Dormaar and Willms, 1998). Similarly, exclusion of cattle from grazed native grasslands of the flooding *pampa* in Argentina induced some substantial changes in nutrient cycling, with a significant decrease in extractable P where grazing took place (Lavado *et al.*, 1996).

Available soil P supplies are regulated, in the main, by physicochemical processes and these have been the subject of much research over the years because of the widespread low P status of many soils throughout the world. Interactions between the physical and chemical status of the soil and supplies of P, from whatever original source, are therefore very important in regulating the flows and transfers within a system. Much P is also held in SOM, and the actions of soil microbial biomass (SMB) are particularly important. Mycorrhizae play an important role in the transfer of P to associated plants from sources that may not be directly accessible for uptake by roots; this mechanism plays an essential role in the acquisition of P for many legumes. Localized returns of P in dung will play an important role in determining microbial distributions and activities.

In the case of N, much is present in organic forms in the soil and, in many situations, these provide the resource that ultimately controls most of the annual flow through the system. The controlling processes are those of gross mineralization and gross immobilization, both of which are dependent upon the activities of the SMB. In practice, what is important is the net effect of the balance between the two processes, i.e. net mineralization, the outcome of which is the appearance of NH_4^+ in the soil. Recent advances in methodology mean that both the gross and the net rates of supply can now be accurately determined under practical conditions. Studies have shown how important net mineralization can be in intensive grassland, where annual releases of > 300 kg N ha^{-1} have been measured (Gill *et al.*, 1995). This rate and quality of supply is often not properly

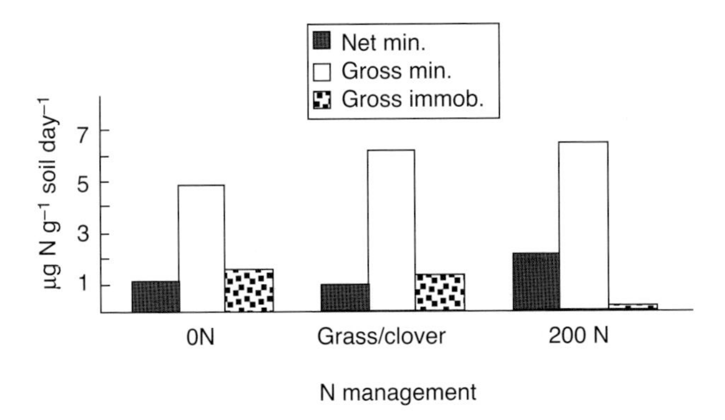

Fig. 16.1. Gross and net mineralization/immobilization rates in soils and under unfertilized (0) or fertilized (200) grass and grass–clover swards (from Ledgard *et al.*, 1998).

taken into account in fertilizer recommendation schemes, and therefore can contribute to inefficiency and loss. Although it is possible to make measurements of release accurately over short periods, prediction is difficult, because the process is tremendously variable: in the studies by Gill *et al.* (1995), < 40% of the variability could be explained by the range in the two main controlling variables i.e. soil temperature and moisture.

Progress in further understanding this release will come from two directions; first, from an improved mechanistic knowledge of the gross processes involved and, secondly, through improved practical diagnostics for field measurements. Techniques are now available which allow gross mineralization and immobilization to be measured with some accuracy, using ^{15}N pool dilution techniques, and this is allowing differences to be observed that would not be apparent from measurements of net rates. This is well illustrated by Fig. 16.1, which shows that, although the net rate of release may not differ between different background managements, the balance of gross mineralization and immobilization does differ considerably. Appreciation of these differences is essential to the further development of mechanistic models of the processes and predictive capability. The balance between these two gross processes becomes more and more important as the system becomes more extensive by providing the control over immediate supplies of N to plant roots.

A practical requirement is methods to predict rates of supply of N on a day-to-day basis in order to ensure that, where fertilizers are used, crop requirements are met precisely. Approaches to this that have been taken in the past include the use of: (i) plants as biometers; (ii) soil characteristics, such as C : N ratios; (iii) chemical extractions (e.g. hot KCl); and (iv) laboratory incubations (Jarvis *et al.*, 1996a). Whilst some of these have been successful on some occasions, many have not and there is little general applicability of any of the methods. One recent approach has been to use soil thermal units (accumulated soil temperatures), which relate well to measured release of mineral N (Table 16.1), with different

Table 16.1. Summary of the relationships ($y = ax + b$) between soil thermal units (x) and cumulative soil N turnover (y) for a number of soil/sward combinations (from Clough *et al.*, 1998).

Soil/sward	a ($\times 10^{-3}$)	b	r^2
1	1.7	0.54	0.97
2	1.6	0.33	0.98
3	0.7	0.09	0.98
4	1.6	0.22	0.93
5	2.8	0.26	0.92
6	2.3	0.13	0.93
7	0.8	0.27	0.86
8	3.7	−0.24	0.99
9	2.1	0.16	0.95
10	2.1	−0.06	0.99
11	5.1	0.15	0.99
12	6.6	−0.41	0.99
13	4.9	0.04	0.99

slopes to the relationships for different soils/swards. Once the means has been developed to provide a measure of the slope for a particular soil/sward, then day-to-day N release should be easily predicted from a knowledge of soil temperatures.

Immobilization/mineralization processes are controlled by SMB activities and these are sensitive to management changes. Over the short term, there may be little effect on total biomass content, but there are longer-term effects that have an impact on the C : N ratio of SOM materials, which influences biomass size and composition (Lovell *et al.*, 1995). Although SMB size may not be influenced immediately by change, it seems likely that its activities are, and this will control process rates and consequent pool sizes. The annual balance between mineralization and immobilization can, in principle, fluctuate from positive to negative, according to ambient conditions and the quality and quantity of the substrates. In most situations in productive pastures, there is net mineralization over the growing season, indicating a positive balance of mineralization against immobilization into SMB. This does not, however, include the 'immobilization' or accumulation of soil organic N, which is returned in plant debris, root turnover and excreta. For this reason, a positive net mineralization is compatible with a positive accumulation of organic N, provided that the N input was greater than offtake plus that in organic matter. Pastures continue to accumulate SOM over long periods, but rates decrease with time and move towards an equilibrium in permanent swards. Cultivation has an immediate impact on aeration and consequent mineralization, and this is reflected by an immediate release of N into drainage (Fig. 16.2). As reseeded swards become established, there is a return to an accumulation phase for SOM, as shown in Fig. 16.2, where there is a reversal in the trend for NO_3^- to be leached in the years after cultivation of pastures.

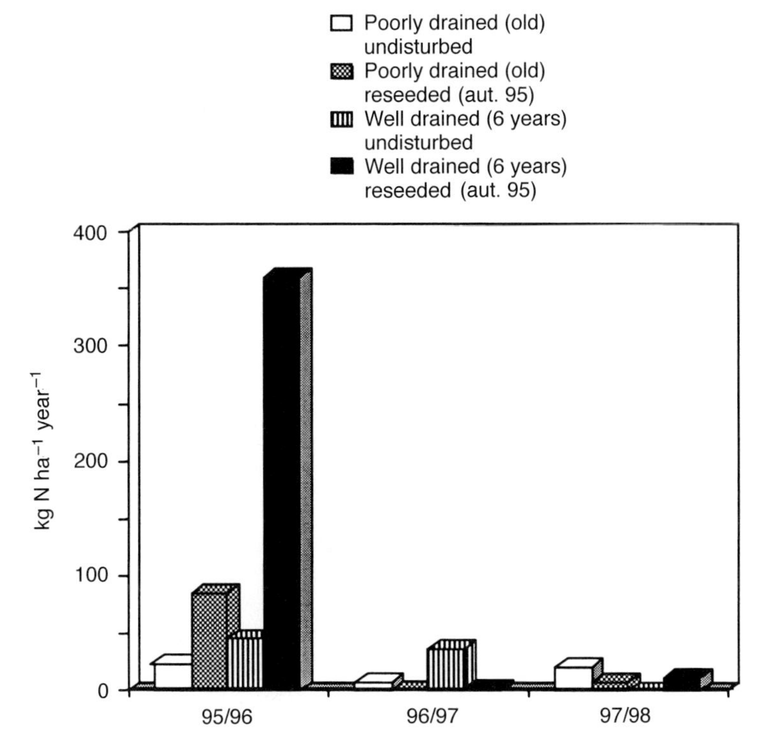

Fig. 16.2. Effects of cultivation and reseeding on leached NO_3^--N from two soil types (Jarvis, 1999a).

Role of the grazing animal

Grazing has an immediate impact on soil N pools. First, the physiological controls over uptake of N by roots are severely disrupted when shoots are removed (frequently in the case of intensively grazed swards) and uptake does not return to normal patterns for a number of days afterwards (Jarvis and Macduff, 1989), depending on the severity of defoliation. There will also be an increase in the 'leakage' of N and C compounds into the rhizosphere of the plants, with consequences for the microbial populations there and their activities (see Dawson *et al.*, Chapter 4, this volume). Over the long term, this will affect overall activities throughout the rooting zone soil, with implications for soil quality and functional sustainability. In this respect, and indeed in relation to many other soil-based processes determining fluxes of nutrients, the coupling between C and N cycles is very important. This is an area that needs further definition and research into the implications for future management of grassland soils

Other plant effects are also apparent. It has been estimated that, in grazed systems, harvested herbage represents only 50–60% of the gross dry matter (DM) production of shoot and root tissues (Parsons *et al.*, 1991) and somewhat less in extensively managed grasslands (see Lemaire and Agnusdei, Chapter 14, this volume). Senescing leaves and roots with their nutrient contents therefore return directly to the soil during the whole of the production cycle, but are not usually taken into account in discussions of either apparent recoveries or the transfers within the grazed system. An evaluation of grazing effects on vegetation structure and nutrient transfers (N and P) made in native grassland in the flooding *pampas* of Argentina (Chaneton *et al.*, 1996) showed that grazing: (i) generated a relocation of N and P in plant pools (80–90% and 63–75%, respectively) in belowground biomass in grazed and ungrazed pastures; (ii) resulted in less P in graminaceous plants than in forbs; and (iii) resulted in greater nutrient uptake (by 30–50%), concomitant with enhanced mineralization rates and generally accelerated cycling rates.

Perhaps of even greater significance to nutrient cycling within grazed systems is the return of nutrients (especially N) in excreta. Ingested N is poorly utilized by ruminants, with only small proportions of intake being incorporated into body tissues or products. 'Harvest' of N into product is therefore lower than in arable agriculture and much is excreted into dung and urine and deposited in localized, discrete areas in pastures. The amounts of N excreted depend very much on the N inputs to the system and the content in the diet; whereas excretion in dung remains relatively constant, that in urea is responsive to N intake (Jarvis *et al.*, 1989). The distribution between dung and urine has implications for turnover and losses from the soil: much more N is lost when excretion outputs are diverted to the urinary pathway. For cattle grazing grass swards with fertilizer inputs of 0, 210, 251, 420 and 545 kg N ha^{-1}, the annual excretion at grazing was 132, 155, 300, 321 and 458 kg ha^{-1}, respectively, of which 56, 60, 71, 74 and 77% was partitioned into urine (Jarvis *et al.*, 1989; Deenen, 1994).

Soils are inherently heterogeneous and grazing animals contribute further to this heterogeneity and to the processes influencing N and P transfer. Depending upon DM and liquid intakes, cattle tend to urinate and defecate nine and 12 times per day, respectively. Some estimates indicate that, after a single grazing in a rotational system, 4–9% of a pasture surface will be affected by urine and *c.* 1% by faeces (Richards and Wotton, 1976). Under intensive grazing regimes, for example with 750 cow grazing days ha^{-1} $year^{-1}$, 39% of the area will be affected by at least one urine patch (each with an area of influence of 0.7 m^2) with a reasonably uniform pattern of distribution (Lantinga *et al.*, 1987). Dung pats are estimated to cover an area of 0.5 m^2 with less chance of overlap and only 4% of the area affected, even in the heavily stocked example. Dung will contain much lower mobile N contents than urine, but could represent the equivalent of 2000 kg N ha^{-1} (Lantinga *et al.*, 1987). This highly concentrated, localized input also has important implications for heterogeneity and process rates, but over a longer time interval than with urine patches. Overall impact under field conditions is poorly defined, but modern techniques utilizing stable isotopes are being used to determine the dynamics of change after dung addition.

Excreta returned to the sward, especially urine, create 'hot spots' of high N (and other nutrient) content, activity and potential for immediate transfer from the system. Some of this may be volatilized as NH_3 or denitrified (see later), but it will almost always be in excess of immediate local crop demands for N. There will also be an immediate impact on soil microbial processes and activities. The latter are not especially well defined, but recent laboratory studies showed that soil processes mediated by SMB react quickly to additions of urea. Carbon dioxide production increased within 1 day of application, suggesting an immediate and significant increase in SMB metabolic activity (Lovell and Jarvis, 1996a): this may have been in response to a solubilization of soil organic C following urine application (Monaghan and Barraclough, 1993), which would stimulate metabolic processes. Nitrogen transformation processes were also quickly affected, with substantial, but short-lived, emissions of N_2O and NO.

Effects on microbial changes with dung are not as easily determined, because of the nature of the materials involved and the longer time step required before substrates enter active pools. However, laboratory-based studies (Lovell and Jarvis, 1996b) showed that, when dung was mixed with grassland soils under controlled conditions, the size of the SMB increased, as did respiration and specific respiration. This contrasted with effects in the field, when the application of dung pats had no measurable impact on SMB, presumably because the effects demonstrated under laboratory conditions were diluted in the field environment. However, over time, the materials involved would have significant effects, not only on nutrient supplies but on other facets of soil function as well.

Although microbial activities are enhanced by the transfer of nutrients and other materials in excreta, a comparison of soils under grazed, introduced pasture and indigenous broad-leaved forest showed that, despite the dependency that managed grassland in New Zealand has on inputs of P fertilizer, microbial P did not differ significantly between the different systems (Ross *et al.*, 1999). This indicates that community structures retain stability, even though there may be shifts in activities. One outcome of this may be changes in mycorrhizal effects; Scullion *et al.* (1998) have suggested that intensification of farming methods may reduce the effectiveness of indigenous arbuscular mycorrhizal populations, particularly where fertilizer inputs are high and inherent fertility is low; this has implications for the long-term sustainability of pasture systems for which future intensification is an option and where excretal impact may be significant.

A comparison of main grazing and stock camping areas (Haynes and Williams, 1999) demonstrated that an increase in both fertility and biological activity occurred in soils within the camp areas at the expense of these characteristics in the main grazing areas. Soil organic C and P (as well as pH) were all greater in the camped areas, as were soil enzymatic activities. These changes were related to a transfer of nutrients and organic matter to the camping areas via dung and urine. Williams and Haynes (1990) have indicated that, although the improvement of New Zealand's pastures over the last 150 years is the result of fertilizer application, increased SOM and increased biological activity, the grazing animal has also influenced the nutrient status by increasing the rate at which nutrients cycle between components of the system. This also, as indicated elsewhere in this chapter, has implications for loss and the additional needs for management and would be of at least as much importance in extensive as in intensive systems. Effects over long distances (i.e. 10s of kilometres and greater) have been shown in extensive systems. Thus, a recent study has demonstrated a livestock-generated gradient in P availability from outlying zones toward rainy-season pans (Turner, 1998) in rangeland in West Africa, which has important implications for grassland productivity. Cycling of nutrients through livestock into dung and urine has an important link with soil productivity in semi-arid zones (Powell *et al.*, 1998), where total amounts and proportions of nutrients voided in faeces and urine were highly influenced by the N and P contents, as well as other chemical characteristics of the feed. The estimated 45–54 kg N and 5–8 kg P ha^{-1} released from the faecal excretion derived from browsed forages made an important contribution to the annual requirements of milk in a mixed farming system (Powell *et al.*, 1998). Selection of feeds that not only satisfy the nutrient requirements of livestock but also produce excreta less susceptible to losses may be an important means of improving nutrient cycling in semi-arid mixed farming systems.

Flows and Losses from the System

Nitrogen

Nitrogen has been the key to enhanced production in many situations, but increased inputs have often generated increased losses. The extent of loss is very much related to the rate of input to the system; all ecosystems, whether managed or unmanaged, leak N to some extent (Garrett, 1991; Jarvis, 1999b) and particularly so where animals are involved. By and large, N flows are intricately coupled with, and regulated by, C fixation and fluxes: N processes and fluxes are much more tightly regulated and controlled in systems with a wide C : N ratio. Table 16.2 indicates how the extent of NO_3^--N loss varies across a number of temperate ecosystems and increases with intensity of management. Grazing animals influence the N cycle, not only in a physical sense, as noted above, but also by changing the chemical form of N in the system. This takes place both indirectly by influencing litter quality through altering plant diversity, and thereby affecting conditions for N mineralization and the release of readily available N, and by controlling mineral forms in the soil from urine and faeces. The magnitude of returns of plant N to the soil in excreta is a function of animal body mass and of plant characteristics, such as N content and tannin levels. Hobbs (1996) suggests that, in extensive grazing, the effects of these on N cycling can cascade throughout the ecosystem and can either stabilize or destabilize the composition of plant communities. At least part of the change that this represents comes about by the potential for transfer away from the system, because of the increased mobility that occurs.

Table 16.2. Rates of NO_3^--N loss from land types (from Jarvis, 1999a).

Vegetation type/predominant land use		Loss rate (kg ha^{-1} year^{-1})
Moorland		0.2–1.6
Low-intensity grassland		0.2–2.0
Forest		
Sweden		2.0–3.8
USA		4.5
Loch Leven (Scotland)	– agricultural catchment	7.5
Slapton Ley (SW England)	– agricultural catchment	5.7–6.8
River Ouse (England)	– agricultural catchment	26.4
Swiss Alps	– agriculture	36.8
Swiss Alps	– lowland agriculture	84.3
Intensive grazing (England)		56.5
Perthshire (Scotland)	– agriculture (general)	58.8
Perthshire (Scotland)	– soft fruit	79.5
Intensively managed lake catchment (Denmark)		131.1–198.8

Nitrate leaching

Grassland agriculture had been considered to be reasonably conservative up to high levels of N inputs with respect to leaching, because most measurements had been made with cut swards with no involvement of animals. Increasingly, data have shown that leaching losses from grazed swards can be substantial. Generally, leaching losses increase with increasing fertilizer or other inputs but the scatter in the relationship (see Jarvis *et al.*, 1995) is wide, even within sites, because of differences in soil drainage/aeration status, sward age, past and present management and weather conditions. The size of the mobile or potentially leachable pool of NO_3^- is a reflection of the balance of the removal processes, with lower rates of leaching under new swards (more immobilization) than under old swards (more mineralization), and after good growing conditions (more uptake) and in heavy textured soils (less drainage and mineralization and more denitrification). Annual losses from intensively managed temperate pastures can be substantial and represent at least 13% of the annual N input to dairy farming systems (Jarvis *et al.*, 1996b). Opportunities to reduce this have been demonstrated and, as shown in Table 16.3, this can be achieved by manipulation of the management to incorporate currently available technologies.

Much of the leaching loss has been associated with accumulation of NO_3^- generated in urine patches and its availability for transfer from the system when there is excess rainfall. Trials to examine N dynamics under excretal patches produced an estimate of leaching of between 150 and 320 kg N ha^{-1} (for dung the loss was between 3 and 28 kg N ha^{-1}) (Sauer and Harrach, 1996). As indicated above, the overall impact of this will vary with N intake and excretion by the animal and livestock density. Various studies have shown that NO_3^- leaching is much greater under grazed than under cut grassland (Ryden *et al.*, 1984; Macduff *et al*, 1990). Variability

Table 16.3. Predicted effects of different N managements on NO_3^--N leaching losses from a UK dairy farm (from Jarvis *et al.*, 1996b).

Management	Fertilizer N applied	NO_3^--N leached kg N ha^{-1} year^{-1}	% total N input to farm	% fertilizer N input
Conventional	250	56	17	23
Tactical N + slurry injection	155	32	13	21
Grass/clover	–	28	13	–
Grow maize silage	185	35	13	19

Table 16.4. Inorganic and organic leaching losses (kg N ha^{-1} year^{-1}) from farmlet experiments at IGER (preliminary unpublished results).

Treatment	Total N	NO_3^--N	Organic N	% loss as organic N
Permanent swards				
Conventional management (280 kg N ha^{-1})	12	7	5	42
Best management practice (*c.* 220 kg N ha^{-1})	7	3	4	5
Reseeded swards				
Conventional management (autumn: 280 kg N ha^{-1})	74	70	4	6
Best management practice (spring: *c.* 220 kg N ha^{-1})	12	8	4	32

is high and is demonstrated by Hack-ten Broeke *et al.* (1996) to be related predominantly to urine patches deposited in the last months of grazing. In contrast, after the winter deposition of urine in hill pastures grazed by sheep, immediate loss of excreta-derived nutrients was negligible (Sakadevan *et al.*, 1993).

As well as inorganic N loss in drainage, there will also be losses in soluble organic forms. Recent measurements have shown that the extent of these is significant in temperate managements and will make an additional, and as yet often unaccounted for, contribution to modelled estimates and calculated budgets, as the data in Table 16.4 demonstrate. Other data indicate that the amounts can be substantial and become proportionally more important as N inputs to the system decrease. It is likely, therefore, that in many extensive systems this form of loss in drainage predominates but will not always have been considered in N balances, etc. constructed for the system.

Denitrification

Denitrification is a largely microbially mitigated process transforming NO_3^- into gaseous forms (N_2 and N_2O): it is the long-term fate of most mineral N in soil or

Table 16.5. Denitrification rates in grazed and cut swards after receiving 60 kg N ha^{-1} during summer and during autumn after 420 kg N ha^{-1} had been applied (from Ryden, 1985; Ryden and Nixon, 1985).

	g N ha^{-1} day^{-1}		
		Grazed sward	
	Cut sward	Dung and urine patches	Areas not affected by excreta
Spring			
Day 3 after fertilizer	40	199	58
Days 3–20 after fertilizer	4–40	32–436	13–64
Autumn			
21 October	< 1	352	164
30 November	9	50	16

aquatic environments to be transformed in this way and to return to the atmosphere. The emission of N_2O from agricultural systems can be considerable – it has been estimated that at least 22% of the annual fertilizer N input to an intensive dairy farm system is lost in this way (Jarvis *et al.*, 1996b). The production of N_2O as one of the major products of the process is also of some considerable concern, through its role in greenhouse effects and ozone generation. Under optimal conditions (anaerobic soils, temperatures > 8°C, supplies of NO_3^- and energy), the potential for denitrification loss from grassland soils has been suggested to be > 30 kg N ha^{-1} day^{-1} (Colbourne, 1992). Both dung and urine patches are important sources of denitrification activity. Much of the denitrification loss in intensive pastures can be apportioned directly to fertilizer applications, but there will be background emissions from the soil/sward system whenever conditions are apt.

Data shown in Table 16.5 illustrate: (i) the immediate effect of fertilizer; (ii) overall differences in the longer-term effects in cut and grazed systems; and (iii) the difference between the specific impact of identifiable excreta patches and the background pasture. Denitrification is most prevalent in heavy textured, poorly drained soils, but can have an impact whenever soil pores become anaerobic, and this can occur even in the better-drained soils on occasion. It is therefore an important removal mechanism in many situations. Grazing animals contribute not only through the effects of their excreta but also through treading and compacting the soil, which will generate anaerobic conditions. Even in extensive pastures, denitrification may be important. Results from a semi-arid grazing system in the USA showed that denitrification can be an ecologically important flux of N in components of semi-arid landscapes and that there is a previously unsuspected regulation of this process by herbivores (Frank and Groffman, 1998).

Attention has turned recently to the particular effects of dung, urine and grazed pastures in general as sources of N_2O and there have been a number of studies to determine these fluxes (Yamulki *et al.*, 1998; for a review, see Oenema *et al.*, 1998). Nitrification as well as denitrification can contribute to these emissions. Oenema *et al.* (1998) calculated that the global contribution of the excreta of grazing animals is 1.55 Tg N_2O-N year^{-1} which is more than 10% of the total global emission. Emissions of N_2O are, however, extremely variable: Yamulki *et al.* (1998) monitored losses from dung and urine applied in a simulated grazing pattern through a year. Rates were extremely variable, with what appeared to be multiple controls operating through soil and environmental factors, differential removal of NO_3^- and NH_4^+ substrates by sward uptake and possible differences in the composition of urine and dung at different times of the year. Emissions from urine were approximately five times those from dung (Table 16.6). Other studies (Allen *et al.*, 1996) have shown that different patterns of emission occur over the same period on different soil types. The mechanisms controlling N_2O production in grazed pastures are reasonably well known, but the process rates and variables are poorly quantified. Because of this, the confidence limits that can be put on estimates of emissions from many grassland types are wide. Most current knowledge is based on temperate intensive pasture, and there are likely to be important global effects in many extensive systems as well. Mosier *et al.* (1998) indicate that even relatively subtle management changes, such as small annual additions of N fertilizer or changing cattle stocking rates, can have a significant impact in semi-arid grassland. Their measurements show that applications of N to simulate either a urine deposition or a fertilizer application continued to stimulate N_2O emissions 6–15 years after application. Because of the extent of grasslands of this nature, this will represent a long-term effect, but one that is difficult to quantify and predict. Recognizing that, as well as N_2O fluxes, there will be an associated N_2 loss as well, over the longer term this will represent a substantial movement of N from the system.

Ammonia volatilization

The other major transfer route for N is through NH_3 volatilization. Excreted N in the form of urea from domesticated animals provides the main source of the increasing NH_3 concentration in the atmosphere. Within a grassland-based livestock system which incorporates housed animals, the major losses of NH_3 arise when animals

Table 16.6. Summary of integrated N_2O fluxes during 100 days after dung or urine application calculated as annual flux (g N_2O-N ha^{-1} year^{-1}) (from Yamulki *et al.*, 1998).

Date of application	Dung	Urine
01/09/94	89	608
16/09/94	166	516
21/10/94	49	518
31/05/95	23	211
17/07/95	11	9
04/09/95	20	186

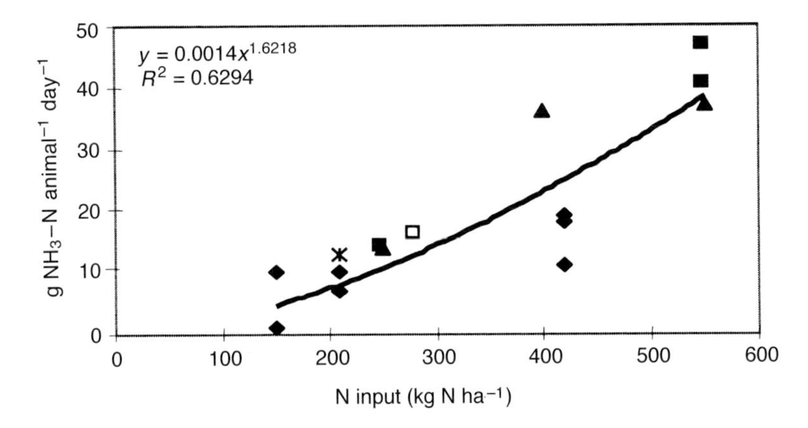

Fig. 16.3. Relationship between inorganic N input on to grassland and rate of ammonia loss from grazing cattle (T.J. van der Weerden, unpublished information). Different symbols represent measurements made at different sites.

are housed, either directly from the house itself or through the storage and dispersal of the stored manures and slurries (Pain *et al.*, 1998). Nevertheless, there are significant losses from grazing (largely as a result of urea excretion, although some may arise directly from fertilizer, especially if this is applied as urea). On average, it is estimated, for UK cattle, for example, that grazing accounts for approximately 5% of that sector's overall loss (Pain *et al.*, 1998). As with the other pathways of N loss, emissions from grazed swards are related to inputs (again reflecting the impact of input on plant N contents and subsequent effects on excretion and partitioning into urine). Figure 16.3 shows the response curve relating NH_3 loss to N inputs to temperate grass swards. The fate of NH_4 that is volatilized is still poorly quantified, although there is increasing evidence that much is deposited within a short distance from the source, but, depending upon local plant N status and environmental conditions, may be re-emitted again. In more intensively managed swards, the net effect of this NH_3 'hopping' would be for at least a partial transfer from the edge of the field. In extensive, less N-rich pastures, there will be more opportunity for plant capture, because of the lower N status of the plant (i.e. less opportunity for the N saturation status to be exceeded). Because of this, the excretal effect of concentrating N within a small area will be partially overcome by redistribution after deposition. Data for extensive grasslands are sparse, because of difficulties in providing confident measurements of NH_3 fluxes, but recent findings in grassland grazed by native ungulates indicated that the animals promoted N loss from the soil by volatilization, and leaching and/or denitrification (Frank and Evans, 1997). This occurs as the result of accelerated losses from urine- and dung-affected microsites and potentially from elevated N loss throughout the grazed landscape by grazers promoting N cycling. Ammonia volatilization generated by the impact of large grazing mammals has been shown to have an effect on nutrient (N) loss from East African grasslands (Ruess and McNaughton, 1988).

Phosphorus

Although considered to be a conservative nutrient, because of the strong retentive capacity of the soil, there are low but significant losses of P from agricultural systems. The recently revived interest in P as a research area stems from the impact that this diffuse leakage has on water quality. There are a number of extensive reviews that describe the extent, mechanisms and potential impact of leakage from agricultural managements (see contributions in Tunney *et al.*, 1997). In agronomic terms, these P losses are insignificant, but in many situations they are now the causal agent for eutrophic effects in surface waters. In many agricultural managements, the P levels within the system are rising. Table 16.7 gives examples of the balance for intensive and extensive grassland managements under UK conditions. It is the case that, even in the extensive upland grassland farm, there is a gradual accumulation of P in the soil. In some circumstances, continued inputs can be in excess of the sorption capacity of the soil, and movement away from the system can be enhanced significantly. The usual practice for fertilized grass swards, whether they are grazed or cut, is for a replacement policy for P, i.e. to balance offtake with additions. A failure to take true account of the other farm inputs and the recycled P generated by livestock leads to an increase in soil levels, as demonstrated by the case of the dairy farm in Table 16.7. In undisturbed grassland soils, the effect of this is exaggerated by the pattern of distribution in the soil, so that any input of P, whether from fertilizer or dung is, because of an overall chemical immobility, partitioned in the top few centimetres of the soil profile (Haygarth *et al.*, 1998b). This study showed that the Olsen-extractable P decreased from *c.* 18 mg kg^{-1} between 0 and 0.5 cm to *c.* 7 mg kg^{-1} at 2 cm. This distribution pattern and its potential impact on transfer would not be reflected by the normal sampling procedures for soil testing. Similar trends in distribution have been reported elsewhere through an accumulation of P not just from fertilizer, but also from excreta, manures, senescent plant material and changes in SOM accumulation. All these contribute to a reservoir of P available for transfer by surface runoff (Haygarth *et al.*, 1998b). Grazing animals help to

Table 16.7. Annual inputs, outputs and P cycling through an intensive dairy farm and an upland sheep farm (kg ha^{-1} year^{-1}) (from Haygarth *et al.*, 1998a).

	Dairy farm	Upland
Inputs		
Fertilizers	16	0.48
Feeds, etc.	27	0.20
Outputs		
Milk	15.6	–
Animals	0.5	0.14
Wool	–	<0.01
Balance	+26	+0.2

generate the potential for movement of P by physical impact and increased risk of erosion (Haygarth and Jarvis, 1999).

Poaching of the pasture surface (i.e. soil compaction and break-up due to trampling) is particularly prevalent around gateways, drinking troughs and waterways. This, in turn, reduces plant cover and soil shear strength and increases the vulnerability to erosion and the physical transfer of P. Heathewaite *et al.* (1998) showed that a heavy stocking rate resulted in an 80% reduction in infiltration capacity, which, in turn, produced a doubling of the P in runoff, compared with that which occurred from a lightly grazed area and which was at least 12 times that of ungrazed areas. Work in New Zealand (Nguyen *et al.*, 1998) has also demonstrated high predicted P losses during the winter months, when soil was vulnerable.

There is gathering acceptance that soil is not an endless sink for P uptake and that, at the landscape scale, the highest P concentrations occur in surface runoff (Nash and Halliwell, 1999). New concepts (Haygarth and Jarvis, 1999) are being developed to describe transport mechanisms and controls over P transfer from soils, which should enable strategies to be developed to improve management in order to decrease P losses from grazing systems. In more extensive systems, increased grazing pressure has been shown to decrease soil sustainability and total C and P and to increase bulk density and total N (Dormaar and Willms, 1998). The loss of P from the system is viewed with some concern, because of the long-term effect in jeopardizing the sustainability of the ecosystem. Other studies in flooding *pampa* grassland have also reported significant decreases in the extractable soil P levels (Lavado *et al.*, 1996). Part of this decrease will be related to the removal in livestock, but grazing has been shown to accelerate P cycling rates within the ecosystem, which would have enhanced the potential for transfer (Chaneton *et al.*, 1996).

Conclusion

The interactions between soil, plant and animal components and effects on nutrient supply, turnover and transfer from the system are complex but have important implications for productivity and environmental quality. Of particular current concern is the behaviour of N and P as key controlling elements of dry matter production in both extensively and intensively managed grassland agriculture and as potential pollutants. There is little doubt that the nutrient status of grassland soils is changing: the long-term effects of this may be profound in either stabilizing or destabilizing the sustainability of the system. Diagnostics are required that will allow assessments of the current and predicted effects: soil testing will have an important role to play, but tools are required to assess effects at the larger scale. Consideration of fluxes and transfers at the farm scale is increasingly being used to provide a way forward for intensive managements (Jarvis *et al.*, 1996b). Models have also been developed for the larger scale (Smaling and Fresco, 1993), which have the potential to become dynamic tools for balanced land-use policies to be developed in order to take account of both productivity and long-term sustainability. These methodologies will provide an important way forward but will need to be based on an understanding of the basic processes involved.

References

Allen, A.G., Jarvis, S.C. and Headon, D.M. (1996) Nitrous oxide emissions from soils due to inputs of nitrogen from livestock on grazed grassland in the UK. *Soil Biology and Biochemistry* 28, 597–607.

Chaneton, E.J., Lemcoff, J.H, and Lavado, R.S. (1996) Nitrogen and phosphorus cycling in grazed and ungrazed plots in a temperate sub-humid grassland in Argentina. *Journal of Applied Ecology* 33, 291–302.

Clough, T.J., Jarvis, S.C. and Hatch, D.J. (1998) Relationships between soil thermal units, nitrogen mineralisation and dry matter production in pastures. *Soil Use and Management* 14, 65–69.

Colbourne, P. (1992) Denitrification and N_2O production in pasture soil: the influence of nitrogen supply and moisture. *Agriculture, Ecosystems and Environment* 39, 267–278.

Deenen, P.J.A.G. (1994) Nitrogen use efficiency in intensive grassland farming. PhD thesis, Agricultural University of Wageningen, The Netherlands.

Dormaar, J.F. and Willms, W.D. (1998) Effect of forty-four years of grazing and fescue on grassland soils. *Journal of Range Management* 51, 122–126.

EC (1991) EC Council Directive Concerning the Protection of Waters against Pollution Caused by Nitrates from Agricultural Sources. No. 375/1. *Official Journal of the European Communities*. 91/6761 EEC.

Frank, D.A. and Evans, R.D. (1997) Effects of native grazers on grassland N cycling in Yellowstone National Park. *Ecology* 78, 2238–2248.

Frank, D.A. and Groffman, P.M. (1998) Denitrification in a semi-arid grazing ecosystem. *Oecologia* 117, 564–569.

Garrett, M.K. (1991) Nitrogen losses from grassland systems under temperate climatic conditions. In: Richardson, M.L. (ed.) *Chemistry, Agriculture and the Environment*. Royal Society of Chemistry, Cambridge, pp. 121–122.

Gill, K., Jarvis, S.C. and Hatch, D.J. (1995) Mineralization of nitrogen in long-term pasture soils: effects of management. *Plant and Soil* 172, 153–162.

Hack-ten Broeke, M.J.D., De Groot, W.J.M. and Dijkstra, J.P. (1996) Impact of excreted nitrogen by grazing cattle on nitrate leaching. *Soil Use and Management* 12, 190–198.

Haygarth, P. H. and Jarvis, S.C. (1999) Transfer of phosphorus from agricultural soils. *Advances in Agronomy* 66, 196–249.

Haygarth, P.M., Jarvis, S.C., Chapman, P. and Smith, R.F. (1998a) Mass balances for P in grassland management systems and potential for transfer to waters. *Soil Use and Management* 14, 160–167.

Haygarth, P.M., Hepworth, L. and Jarvis, S.C. (1998b) Forms of phosphorus transfer in hydrological pathways from soil under grazed grassland. *European Journal of Soil Science* 49, 65–72.

Haynes, R.J. and Williams, P.H. (1999) The influence of stock camping behaviours on the soil microbial and biochemical properties of grazed pastoral soils. *Biology and Fertility of Soils* 28, 253–258.

Heathewaite, A.L., Griffiths, P. and Parkinson, R.J. (1998) Nitrogen and phosphorus in run-off from grassland with buffer strips following application of fertilizers and manures. *Soil Use and Management* 14, 142–148.

Hobbs, N.T. (1996) Modification of ecosystems by ungulates. *Journal of Wildlife Management* 60, 695–713.

Jarvis, S.C. (1993) Nitrogen cycling and losses from dairy farms. *Soil Use and Management* 9, 99–105.

Jarvis S.C. (1998) Nitrogen management and sustainability. In: Cherney, J.H. and Cherney, D.J.R. (eds) *Grass for Dairy Cattle*. CAB International, Wallingford, pp. 161–192.

Jarvis, S.C. (1999a) Comparisons between nitrogen dynamics in natural and agricultural ecosystems. In: Wilson, W., Ball, A.S. and Hinton, R.H. (eds) *Managing Risks of Nitrates to Humans and the Environment*. Royal Society of Chemistry, Cambridge, pp. 2–20.

Jarvis, S.C. (1999b) Nitrate leaching from grassland and potential abatement strategies. In: MAFF (ed.) *Tackling Nitrate from Agriculture: Strategy from Science*. MAFF, London, pp. 18–26.

Jarvis, S.C. and Macduff, J.H. (1989) Nitrate nutrition of grasses from steady-state supplies in flowing solution culture following nitrate deprivation and/or defoliation. *Journal of Experimental Botany* 40, 965–975.

Jarvis, S.C. and Pain, B.F. (1997) *Gaseous Nitrogen Emissions from Grasslands*. CAB International, Wallingford, 452 pp.

Jarvis, S.C., Hatch, D.J. and Roberts, D.H. (1989) The effects of grassland management on nitrogen losses from grazed swards through ammonia volatilization: the relationship to excretal N returns from cattle. *Journal of Agricultural Science, Cambridge* 112, 205–216.

Jarvis, S.C., Scholefield, D. and Pain, B.F. (1995) Nitrogen cycling in grazing systems. In: Bacon, P. (ed.) *Nitrogen Fertilization and the Environment*. Marcel Dekker, New York, pp. 381–419.

Jarvis, S.C., Stockdale, E.A., Shepherd, M.A. and Powlson, D.S. (1996a) Nitrogen mineralization in temperate agricultural soils: processes and measurement. *Advances in Agronomy* 57, 157–235.

Jarvis, S.C., Wilkins, R.J. and Pain, B.F. (1996b) Opportunities for reducing the environmental impact of dairy farm managements: a systems approach. *Grass and Forage Science* 51, 21–31.

Lantinga, E.A., Keuning, J.A., Groenwold. J. and Deenen, P.J.A.G. (1987) Distribution of excreted nitrogen by grazing cattle and its effects on sward quality, herbage production and utilisation. In: van der Meer, H.G., Unwin, R.J., van Dijk, T.A. and Enruk, G.C. (eds) *Animal Manure on Grassland and Fodder Crops: Fertilizer or Waste?* Martinus Nijhoff, Dordrecht, pp. 103–117.

Lavado, R.S., Sierra, J.O. and Hashimoto, P.N. (1996) Impact of grazing on soil nutrients in a Pampean grassland. *Journal of Range Management* 49, 452–457.

Ledgard, S.F., Jarvis, S.C. and Hatch, D.J. (1998) Short-term N fluxes in grassland soils under different long-term N managements. *Soil Biology and Biochemistry* 30, 1233–1241.

Lovell, R.D. and Jarvis, S.C. (1996a) Effects of urine on soil microbial biomass, methanogenesis, nitrification and denitrification in grassland soils. *Plant and Soil* 186, 265–273.

Lovell, R.D. and Jarvis, S.C. (1996b) The effect of cattle dung on the soil microbial biomass C and N in a permanent pasture soil. *Soil Biology and Biochemistry* 28, 291–299.

Lovell, R.D., Jarvis, S.C. and Bardgett, R.D. (1995) Soil microbial biomass and activity in long-term grassland: effects of management changes. *Soil Biology and Biochemistry* 27, 969–975.

Macduff, J.H., Jarvis, S.C. and Roberts, D.H. (1990) Nitrates leached from grazed grassland systems. In: Calvet, R. (ed.) *Nitrates – Agriculture – Eau.* INRA, Paris, pp. 405–410.

Monaghan, R.M. and Barraclough, D. (1993) Nitrous oxide and dinitrogen emissions from urine-affected soil under controlled conditions. *Plant and Soil* 151, 127–138.

Mosier, A.R., Parton, W.J. and Phongpan, S. (1998) Long term large N and immediate small N addition effects on trace gas fluxes in the Colorado shortgrass steppe. *Biology and Fertility of Soils* 28, 44–50.

Nash, D.M. and Halliwell, D.J. (1999) Fertilizers and phosphorus loss from productive grazing systems. *Australian Journal of Soil Research* 37, 403–429.

Nguyen, N.L., Sheath, G.W., Smith, G.M. and Cooper, A.B. (1998) Impact of cattle treading on hill land. 2 Soil physical properties and contaminant run-off. *New Zealand Journal of Agricultural Research* 41, 279–290.

Oenema, O., Gebauer, G., Rodriguez, M., Sapek, A., Jarvis, S.C., Corre, W.J. and Yamulki, S. (1998) Controlling nitrous oxide emissions from grassland livestock production systems. *Nutrient Cycling in Agroecosystems* 25, 141–149.

Pain, B.F., van der Weerden, T.J., Chambers, B.J., Phillips, V.R. and Jarvis, S.C. (1998) A new inventory for ammonia emissions from UK agriculture. *Atmospheric Environment* 32, 309–313.

Parsons, A.J., Orr, R.J., Penning, P.D. and Lockyer, D.R. (1991) Uptake, cycling and fate of nitrogen in grass–clover swards continuously grazed by sheep. *Journal of Agricultural Science, Cambridge* 116, 47–61.

Powell, J.M., Ikpe, F.N., Somda, Z.C. and Fernandez-Rivera, S. (1998) Urine effects on soil chemical properties and the impact of urine and dung on pearl millet yield. *Experimental Agriculture* 34, 259–276.

Rekolainen, S., Ekholm, P., Ulen, B. and Gustafson, A. (1997) Phosphorus losses from agriculture to surface waters in the Nordic Countries. In: Tunney, H., Carton, O.Y., Brookes, P.C. and Johnston, A.E. (eds) *Phosphorus Loss from Soil to Water.* CAB International, Wallingford, pp. 77–93.

Richards, I.R. and Wotton, K.M. (1976) The spatial distribution of excreta under intensive grazing. *Journal of British Grassland Society* 30, 187–188.

Ross, D.J., Tate, K.R., Scott, N.A. and Feltham, C.W. (1999) Land use change: effects on soil carbon, nitrogen and phosphorus pools and fluxes in three adjacent ecosystems. *Soil Biology and Biochemistry* 31, 803–813.

Ruess, R.W. and McNaughton, S.J. (1988) Ammonia volatilization and the effects of large grazing mammals on nutrient loss from East African grasslands. *Oecologia* 77, 382–386.

Ryden, J.C. (1985) Denitrification loss from managed pasture. In: Golterman, H.L. (ed.) *Denitrification in the Nitrogen Cycle.* Plenum Press, New York, pp. 121–134.

Ryden, J.C. and Nixon, D.J. (1985) Denitrification from grazed and cut swards. In: *Grassland Research Institute Final Report, 1984–5.* GRI, Hurley, pp. 21–24.

Ryden, J.C., Ball, P.R. and Garwood, E.A. (1984) Nitrate leaching from grassland. *Nature* 311, 50–53.

Sakadevan, K., Mackay, A.D. and Hedley, M.J. (1993) Influence of sheep excreta on pasture uptake and leaching losses of sulphur, nitrogen and potassium from grazed pastures. *Australian Journal of Soil Research* 31, 151–162.

Sauer, S. and Harrach, T. (1996) Leaching of nitrogen from pastures at the end of the grazing season. *Zeitschrift für Planzenernahrung und Bodenkunde* 159, 31–35.

Schwinning, S. and Parsons, A.J. (1996) A spatially explicit population model of stoloniferous N-fixing legumes in mixed pasture with grass. *Journal of Ecology* 84, 815–826.

Scullion, J., Eason, W.R. and Scott, E.P. (1998) The effectivity of arbuscular mycorrhizal fungi from high input conventional and organic grassland and grass–arable rotations. *Plant and Soil* 204, 243–254.

Smaling, E.M.A. and Fresco, L.O. (1993) A decision-support system for monitoring nutrient balances under agricultural land-use (NUTMON) *Geoderma* 60, 235–256.

Tallowin, J.R.B., Kirkham, F.W., Smith, R.E.N. and Mountford, J.O. (1998) Residual effects of phosphorus fertilization on the restoration of floristic diversity to grassland. In: Joyce, C.J. and Wade, P.M. (eds) *European Wet Grasslands: Biodiversity, Management and Restoration*. John Wiley & Sons, Chichester, pp. 249–263.

Tunney, H., Carton, O.Y., Brookes, P.C. and Johnston, A.E. (eds) (1997) *Phosphorus Loss from Soil to Water*. CAB International, Wallingford, 467 pp.

Turner, M.D. (1998) Long-term effects of daily grazing orbits on nutrient availability in Sahalian West Africa: 2 Effects of a phosphorus gradient on spatial patterns of annual grassland production. *Journal of Biogeography* 25, 683–694.

Whitehead, D.C. (1995) *Grassland Nitrogen*. CAB International, Wallingford, 397 pp.

Williams, P.H. and Haynes, R.J. (1990) Influence of improved pastures and grazing animals on nutrient cycling within New Zealand soils. *New Zealand Journal of Ecology* 14, 49–57.

Yamulki, S., Jarvis, S.C. and Owen, P. (1998) Nitrous oxide emissions from excreta applied in a simulated grazing pattern. *Soil Biology and Biochemistry* 30, 491–500.

Sustainable Management of Pasture and Rangelands

17

J. Stuth[1] and G.E. Maraschin[2]

[1]*Department of Rangeland Ecology and Management, Texas A&M University, College Station, TX 77843-2126, USA;*
[2]*Universidade Federal do Rio Grande do Sul, Porto Allegre, Brazil*

Introduction

Over the past 50 years, many grazing systems have been devised for grazing lands worldwide (Heitschmidt and Taylor, 1991). Rotational grazing strategies were devised to improve efficiency of harvest, limit patch grazing, enhance nutrient acquisition and utilization and improve management's ability to control frequency and intensity of grazing in a relatively simple vegetation matrix on small landscapes. The primary goal was to manage inputs into the system in a cost-efficient manner, given the level of market demand and price subsidies.

During this same period, rangeland scientists in the USA, South Africa and Zimbabwe began a series of grazing studies designed to convert plant communities from a low ecological state to one possessing more productive and palatable species of a higher ecological state (as reviewed by Heitschmidt and Taylor, 1991; Heady and Childs, 1994). Regular resting via some form of rotational grazing has become a standard part of the rangeland discipline in South Africa and North America. Recent reviews reveal the lack of evidence behind some of the dogma associated with such grazing systems and have shifted the emphasis to the overriding impact of stocking rate, spatial distribution of grazing, animal type, fire and tactical resting (Pieper and Heitschmidt, 1988).

In recent times, much attention has been given to the feasibility of using high stocking rates on rangelands, in conjunction with intensive rotational grazing systems, commonly referred to as 'short-duration' or 'time-controlled' grazing systems (Pieper and Heitschmidt, 1988). Evidence to support these systems is equivocal, although several authors have contended that modest increases in stocking rate can be attained over moderate stocking levels in years with average

 Grassland Ecophysiology and Grazing Ecology (eds G. Lemaire, J. Hodgson, A. de Moraes, C. Nabinger and P.C. de F. Carvalho)

or above-average rainfall, due to improved harvest efficiency of leaf turnover and improved livestock distribution. Greater control and flexibility are provided by these multipasture systems, but greater risk of management errors is also incurred (Roberts, 1993). Therefore, managerial ability is as much a part of grazing system design as biology, husbandry and economics.

This chapter attempts to address some of the critical concepts driving 'outcome' from grazing lands managed in a sustainable manner. The framework used to address these concepts will be in the context of hierarchical relationships embodied in the concept of SWAPAH (soil, water, atmosphere, plants, animals, humans) (Stuth *et al.*, 1993). Figure 17.1 provides a generalized view of the

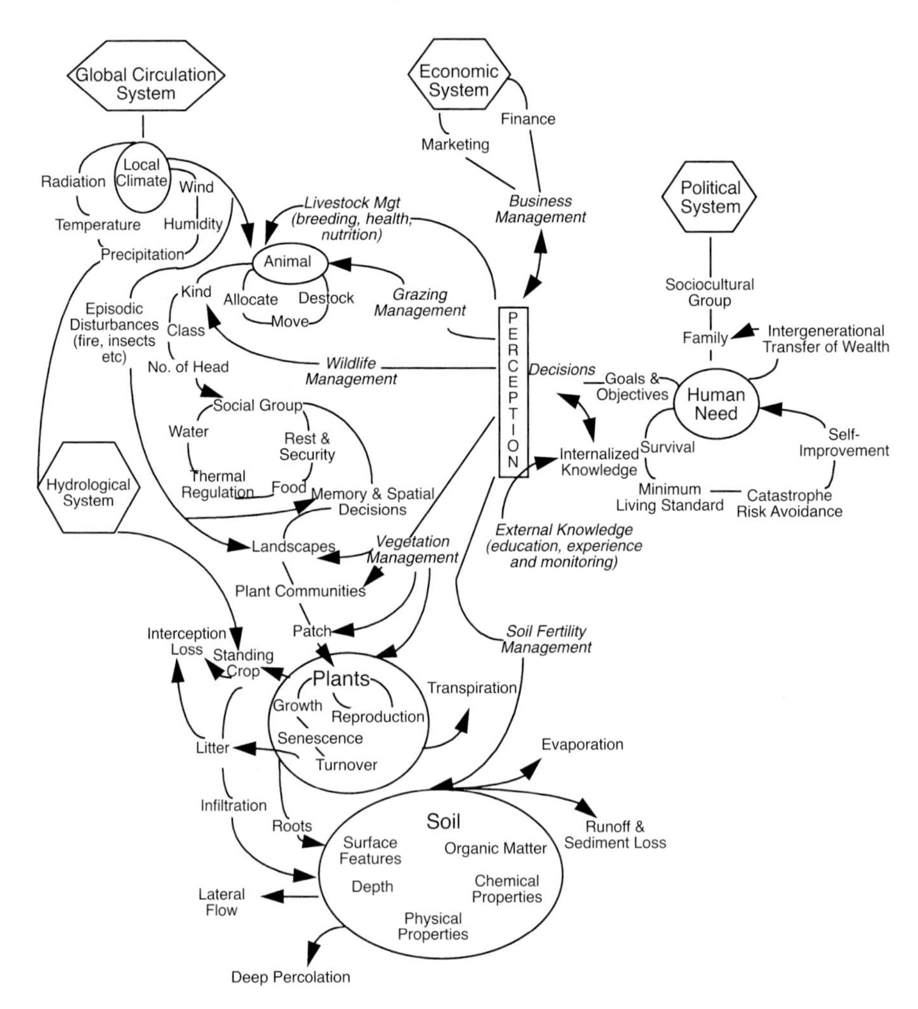

Fig. 17.1. Spatiotemporal complexity of the grazing system decision environment, which encompasses a hierarchical view of the SWAPAH concept of soil, water, atmosphere, plants, animals and humans. Mgt, management.

SWAPAH, depicting a hierarchical representation of various driving forces affecting structure and function of grazed ecosystems managed by humans.

Human Actions in Management of Grazing Lands

Management programmes applied to grazing lands are a product of human thoughts, decisions and actions and therefore humans must be considered an integral part of system function. Management implies control, whether informed or ignorant of economic or ecological processes. Decisions are based on the goals and objectives of management relative to a hierarchy of needs, either for the individual or for the firm, and on the perception of external and internal processes (ecological, weather, market) which shape those decisions (Stuth *et al.*, 1991; Foran and Howden, 1999). We believe that the process of 'landscape visioning' is critical to the success of any grazing programme, i.e. developing a perspective of the desired future landscape relative to current conditions and placing a time-line on the emergence of the envisioned landscape.

Worldwide, there is large variation in the potential degree of human interference in the diet selection process among the large array of livestock production systems (Table 17.1). Animals set-stocked within large paddocks are constrained more by the spatial configuration of land types relative to water locations, patterns of vegetation and configurations of shelter, rather than by diet preference *per se*. With increasing intensity of use comes greater subdivision of land into smaller paddocks, providing opportunities for controlling access of livestock to plant communities or land systems. Grazing impacts of animals are controlled even more through herding and tethered grazing practices. The ultimate control over the

Table 17.1. Six broad categories of managed grazing environments. The degree of human interference in the diet selection process increases from the top of the list to the bottom.

Grazing production system	Example
Free-ranging within a static set of landscape conditions	Set stocking of paddocks
Free-ranging across multiple landscape conditions, extensive herding	Rotation among several paddocks
Across multiple landscapes	Transhumance herding practices
Intensive herding within a static set of landscapes conditions	Village agriculture
Tethered grazing, rotated among patches	High-density cropping systems
Confined feeding/zero-based grazing/cut and carry	All land in crops and human shelter or feed lots

diet selection process comes from zero-based or confined feeding of animals, where the human essentially selects the animals' diet. As population density of humans increases, there is increasing pressure on grazing lands, and therefore greater needs for ecologically based planning of grazing management.

Management Must Set Goals

Traditional emphasis on the design of grazing programmes has focused on manipulating rest and grazing intensity for maximum livestock production per unit of land area. However, we find little acceptance and application of the narrow set of grazing systems that have been developed in the research community (Danckwerts *et al.*, 1993). Each property or landholding offers such a complex decision-making environment that fixed grazing strategies often employed in research systems have limited extension to livestock producers (Walker and Hodgkinson, 1999). The list below outlines common motivations for designing and implementing grazing programmes. In many cases, all of the reasons listed below are driving the decision process, and many are interdependent:

- Improve profitability.
- Sustain operations.
- Drive successional change in a desired direction.
- Facilitate implementation of other management practices.
- Facilitate other enterprises (profit centres).
- Enhance wildlife habitat or recreational experience.
- Respond to environmental issues (water quality and quantity, biodiversity).
- Provide ecological services for society.

We believe that planned grazing strategies represent the interface between grazing managers and their understanding of grazing behaviour, vegetation ecology, ecophysiology, hydrology, animal nutrition and range economics. Therefore, the land manager must consider all elements of the planning process, of which the issue of grazing method (continuous, rotational or tactical spelling) is only one consideration (Table 17.2).

Not all resource managers are interested in just livestock production. Recently, environmental organizations have expressed interest in using livestock grazing to manage biodiversity and increased water yield of high quality. In some areas, grazing can alter competitive relationships to enhance diversity through differential disturbance of preferred species (Milchunas *et al.*, 1988). Production-orientated managers usually want to maximize distribution of livestock across the landscape, while managers who focus on plant diversity may want livestock grazing to have a heterogeneous distribution, creating a variety of patches, from heavily grazed to ungrazed (Fuhlendorf, 1996).

Thus, planned grazing management operates in a range of production systems and with a diversity of management goals. In all cases, however, there are only a limited number of means of controlling the grazing process.

Table 17.2. The elements of effective grazing management planning.

Time dimension	Period of influence	Typical decisions	Decision profile
Strategic	For next few years	• Animal type(s) • Production system • Base stocking rate (carrying capacity) • Grazing plan • Brush management plan • Wildlife management plan	• Longer-term reassessment of goals and strategies • Specific • Measurable
Tactical	For next 6–18 months	• Stocking decisions – adjust numbers for whole property – adjust numbers across paddocks – spelling • Burning • Feeding • Brush and habitat treatments	• Tailoring management to accommodate variability • Key decision points and contingency plans
Operational	For next month or so	• Work plan – applying tactical decisions • Respond to unforeseen circumstances	• Focused • Flexible

Control and Influence of the Grazing Process

Set base stocking rate

Establishing a base stocking rate or carrying capacity is the most critical decision affecting the success of a grazing system and its subsequent impact on ecological processes (Illius *et al.*, 1999). The importance of overall grazing pressure far exceeds that of grazing method (e.g. continuous vs. rotational). As most enterprises have only limited ability to adjust seasonal or annual stock numbers to match forage supply, the base stocking rate represents the expectation of the number of animals that can be safely carried in most years. As such, the base stocking rate integrates the manager's understanding of climate, land capability, land condition, animal distribution and animal demand with his/her attitude to risk. Risk, in this context, should refer to the degree of security desired with respect to: (i) adequacy of forage supply to meet nutrient demand (a feed-budgeting perspective); and (ii) maintenance of the land's capability to grow useful forage (an ecological perspective). Although these aspects of risk are interdependent, the more immediate urgency of the former in relation to short-term economic performance tends to result in the latter aspect of risk being overlooked or ignored, often leading to overstocking and, subsequently, to land degradation in the medium to longer term.

Match animals with the environment

Productivity from a given grazing environment and consequences for ecological processes are strongly influenced by the degree of matching of animal species to the vegetation structure and composition, to the climate and to the array of endemic predators, parasites and diseases. Animal scientists have often emphasized the selection of species, breeds and genotypes that require minimal husbandry input, while still satisfying product specifications. However, this has not always been matched with concern for effects on the environment.

How does one determine the optimal mix of different animal species for a given landscape? Animal species have an evolutionary predisposition to particular foraging strategies and diet preferences. The functional nature of diet selectivity, its ultimate consequences for genetic fitness and its proximate causes are well documented (e.g. Provenza, 1995). However, the impact of this knowledge on natural resource management is equivocal. There exists a hierarchy of physical and vegetative factors that interact with the foraging strategies of different herbivores to determine forage demand at the species level. Recent development of decision support systems has provided planners with the capability to characterize landscape configurations and the composition of plant communities, and compute appropriate stocking rates for various combinations of wildlife and livestock species (Ranching Systems Group, 1994; Quirk, 1995; Quirk and Stuth, 1995).

Develop a grazing plan

Implicit in control of the grazing process is development of planned strategies for meeting management objectives. Land managers must: (i) determine which and how many paddocks/pastures are allocated to each class of livestock, e.g. cows vs. young steers; (ii) decide on a grazing method, e.g. continuous +/– tactical resting vs. rotational resting; (iii) plan the use of fire; (iv) assess the need for conserved fodder or other energy supplement; and (v) control forage demand via timing of parturition and purchase/sale policies.

Modify spatial distribution of grazing pressure

Control of grazing pressure occurs spatially and temporally. The degree of spatial control depends on the arrangement of water, terrain, cover and forage across a landscape relative to the hierarchy of needs of the animal. Because water is essential for life processes, location of water and its distribution across the landscape dictate the frequency of grazing and occupancy by animals in a gradient from water sources (Stuth, 1991). Animals with greater water use efficiency have more effective grazing capacity per water source. Other animal species vary with respect to mobility across terrain varying in obstacles (dense brush), roughness and slope, which, in turn, affect accessibility to forage resources.

Location of water sources relative to thermal regulation sites creates differential domains of attraction (Stuth, 1991). A recent study on landscape use patterns of cattle by Erickson *et al.* (1996) noted greater use of landscapes that were in line between multiple watering sources and where roads or trails connected watering points with adequate shade to accommodate multiple social groups of 25–35 head each. When water and thermal foci are highly associated with each other, spatial inefficiencies of grazing increase, especially when grazing capacity of the land is low, and result in a small number of social groups in the livestock population grazing that pasture.

Accommodate temporal variability

Tactical adjustment of stocking number and distribution to accommodate seasonal and annual fluctuations in forage supply is essential in all grazing environments, but is especially important in rangeland environments with highly variable rainfall. In north-east Australia, seasonal forecasts based on the El Niño–southern oscillation may contribute to a better tactical assessment of risk and opportunities (McKeon *et al.*, 1990). In more intensive grazing environments, feed budgeting and regulation of rest/graze periods help accommodate temporal variability. Monitoring of pastures and animals is required to optimize harvest efficiency across the whole production cycle (McCall and Sheath, 1993). The greater the degree of match between intake profiles of animals and growth cycles of forages grazed, the greater the harvest efficiency of annual production. However, the greater the mismatch between dietary preferences of the animal and species on offer, the lower the efficiency of harvest by the grazing animal. The critical issue is to understand what drives intake in livestock, what species they like to eat and the growth patterns of vegetation on offer relative to weather events. The advent of decision support systems is helping managers to deal with this complexity (Stuth, 1996).

Managing Competitive Interactions within Plant Communities

Changes in rangeland vegetation have been associated with many factors, including grazing by livestock, variable weather patterns and altered fire regimes. These changes occur at multiple spatial and temporal scales and are often interactive, with feedback mechanisms that confound the identification of specific driving processes. Traditionally, rangeland management and ecological theory have been based on the climax community concept, where grazing pressure dictates the movement toward or away from a single stable state (climax) through fairly linear dynamics (Dyksterhuis, 1949). This theory is based on the view that succession on rangelands can be largely driven and controlled by grazing, and range condition can be manipulated by domestic livestock to direct succession toward or away from climax. Recently, several alternatives have been proposed, including state and transition models (Westoby *et al.*, 1989; Laycock, 1991) and the concept of ecological thresholds (Archer, 1989; Archer and Smeins, 1991).

The success of grazing management regimes depends on the ability to recognize those vegetation conditions where grazing processes can move vegetation structure and function in a direction desired by management, either directly or through facilitation of other practices. Early researchers in rangeland grazing systems recognized the need for proper stocking and strategic rest periods to enhance the positive influences and limit the negative influence of grazing on the competitive relationships between desirable and undesirable plant species (Pieper and Heitschmidt, 1988).

To understand rangeland processes, vegetation should be viewed as a hierarchy of spatiotemporal scales or levels of organization, such as individuals, patches, communities and landscapes (Archer and Smeins, 1991). Interpretation of vegetation dynamics and identification of stable states, as well as the influence of grazing, are dependent upon the level or scale of observation (Fuhlendorf and Smeins, 1996). Temporal patterns observed at small scales can be different from large-scale patterns and driven by different processes. Grazing animals make decisions at each of these levels and the influence of grazing is different at each scale. Management must decide on the appropriate scale(s) for monitoring condition and trend (e.g. plant community, paddock or catchment), and develop management strategies that recognize the primary driving mechanisms at each scale (e.g. grazing, fire, drought).

When analysis of vegetation change includes an increasing woody component, the direct influence of grazing is more difficult to determine. Grazing has been reported to increase, decrease or not affect the abundance of woody plants on rangelands, depending upon the kind/class of animal and the plant community (Stuth and Scifres, 1982; Scifres *et al.*, 1987; Archer, 1989). The primary influence of grazing is reduction of fuel loads, which reduces both the probability of naturally occurring fires and the potential use of prescribed fire for control of woody plants (Smeins *et al.*, 1994; Fuhlendorf and Smeins, 1996).

Grazing may reduce the competitive ability of grasses and allow woody plants to invade at faster rates. However, the alternative has also been demonstrated, where increase in woody plant density was slower under grazing because of harsher environmental conditions at the soil surface when less vegetation was present (O'Connor, 1985, 1995). Thus, there is no universal agreement concerning the influence of grazing on the rate of increase of woody plants, but apparently woody plant density can increase under all grazing conditions, suggesting that grazing is not the primary driving variable. Altered fire regimes appear to be the primary factor in this example and grazing systems must incorporate prescribed burning to maintain a herbaceous-dominated system.

Management of changes in vegetation, therefore, requires a multiscale approach and an understanding of the influence of several processes, which can be interactive (Fig. 17.2). The traditional method for analysis of range condition and trend was developed in grassland communities and best describes the grazing-

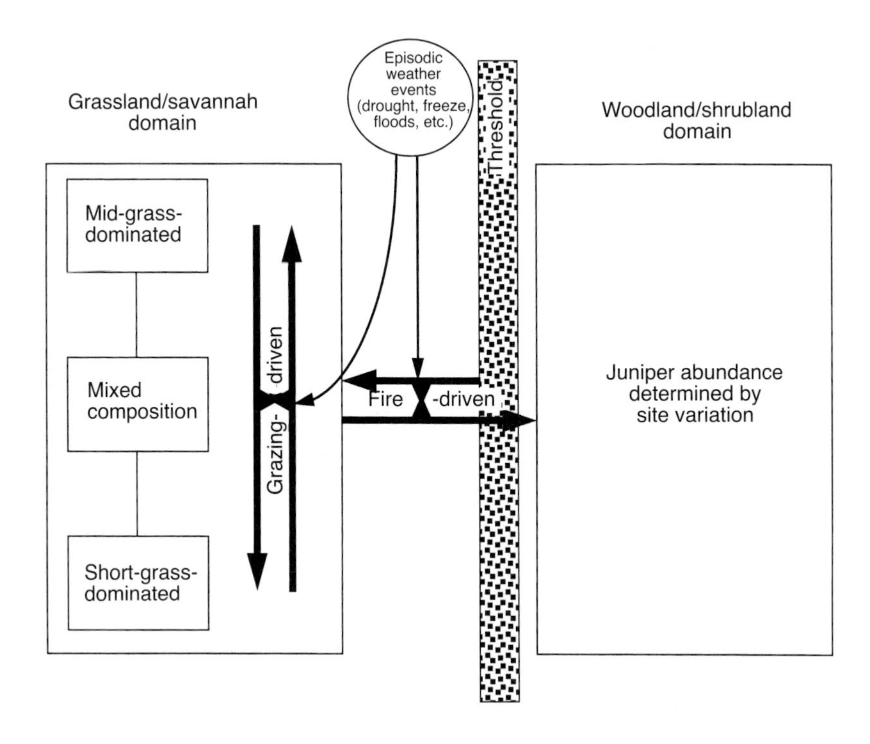

Fig. 17.2. Interactions of grazing, fire and episodic weather events driving successional change within grassland/savannah and woodland/shrubland domains and across thresholds in an oak-dominated savannah ecosystem in the Edwards Plateau of central Texas (adapted from Fuhlendorf and Smeins, 1996).

induced changes that occur in herbaceous-dominated communities and in the herbaceous interspaces of shrublands and woodlands. Population structure and morphology of individuals are directly influenced by differential grazing within the herbaceous plant community. These structural changes in populations result in competitive interactions between species that are highly preferred by grazing animals and those that are less preferred (Briske and Richards, 1995). For the herbaceous component of this system, weather patterns and/or altered fire regimes should be considered as interactive forces with grazing intensity, where the greatest influences would occur under heavy stocking rates. These changes could be considered linear and frequently reversible over reasonable management time frames. This suggests that traditional range condition theory is reasonable for many herbaceous populations, patches or communities within a landscape.

However, when a landscape perspective is taken, primary change in many rangelands involves the interaction between woody and herbaceous layers. At this larger scale, grazing is less important and variables such as altered fire regimes and long-term directional shifts in climate become more important. As woody plants increase, they can cause positive feedback through increased seed availability, nutrient redistribution and reduction in herbaceous vegetation necessary for fires, resulting in an ecological threshold where increases are exponential and often irreversible (Archer, 1989; Fuhlendorf and Smeins, 1996). These changes occur at a decade to century time-scale and can result in major decreases in livestock-carrying capacities. As these large-scale changes continue, stocking rates must be reduced, because of relative increases in less palatable woody species, and eventually a threshold could be crossed where return to a grassland-dominated state is not likely under typical management conditions. Grazing-system ecology requires integration of management strategies that focus on changes driven by processes other than grazing with more traditional grazing management practices.

Defining the Strategic Role of 'Derived' Pastures

Conversion of rangelands and woodlands to 'derived' pastures requires considerable input in human and petroleum energy to establish. Given the costs of conversion, careful consideration needs to be placed on defining the strategic role of these highly productive pastures in terms of species, acreage and spatial placement. Typically, these forage resources are planted to allow increased grazing capacity, provision of nutrients during dormant or low-quality periods of natural pastures, harvesting of stored roughages for periods of forage shortfall in native pastures and integration of land management practices on native pastures to cover shortfall in seasonal grazing capacity due to required deferments.

The primary issue of pastures focuses on returns on investments (fence, water, land preparation) and high annual costs due to agronomic inputs of fertilizer and weed/brush control. High harvest efficiencies are required to ensure adequate animal productivity to cover input costs. The primary issues of 'sustainability' of these forage resources are centred on long-term adaptation of the species used, financial stability to maintain soil fertility and reduce competition from weeds and the degree of 'match' with other ranch forage resources and enterprises.

Often, the mistake is made to establish species over a large landscape, apply minimal amounts of fertilizer across all sites and extensively graze these high-input forage resources. Careful selection of high-potential sites within the configuration of the ranch and concentration of fertilizer on those sites where the plants can exploit the nutrients allow maximum response of the production and ability of the animals to capture forage produced in a more efficient manner. The remainder of the landscape can be managed more extensively, using the derived pastures as a strategic tool to allow rest of the native pastures.

Establishing Grazing Systems

Given that each landscape is unique and the needs of managers vary, can a given set of management principles emerge to help guide the decision process and have the desired impact on ecological processes? Critical managerial considerations are set out in a hierarchical manner below:

- Landscape visioning.
- Objectives of animal production.
- Understanding climate and land capability for determining base stocking rate.
- Understanding weather and market patterns and risk to determine required flexibility in decision-making and enterprise mix.
- Understanding interactions with other enterprises (wildlife, recreation, mixed livestock).
- Matching management skills and capability with the complexity of the decision environment.
- Identifying relevant spatial and temporal scales for monitoring.
- Targeting plant species, forage residue levels, soil surface conditions and nutritional parameters for monitoring.
- Understanding critical growth periods of the target plant species.
- Understanding how to give competitive advantage to the target species and limit the impact of unwanted plant species.
- Establishing analytical capability to make strategic and tactical decisions and interpret monitoring information for operational adjustment.

It is essential for individuals to have a realistic vision of what they would like the landscape to look like in terms of species and their distribution and stature. Land managers should give the landscape condition as much attention as the traditional focus on animal condition.

An understanding of risk in terms of climate, future weather (precipitation) and market conditions is critical for the design of grazing strategies in terms of base stocking rate, flexibility in animal numbers and destock/restock decisions. The greater the coefficient of variation in annual rainfall, the greater the need for flexibility in the decision-making process and the less predictable the outcome of the plan. Monitoring of conditions is essential for timely decision-making.

Diversity of enterprises creates opportunities for risk aversion in terms of market and weather flux, but also creates higher-order complexity in decision-making. Landscape diversity (spatial and temporal events) adds to the complexity when more than one enterprise is utilizing the area, but it creates more options for management, which may, in turn, lessen the risk of adverse ecological change. The key is provision of analytical tools to help comprehend this diversity and the complexity of the dynamics involved.

Decision-making in complex environments is often limited by managerial skill levels and understanding of the management environment. Outcome can be enhanced both ecologically and economically when grazing systems are designed to match skill levels of the landholder/decision-maker. The outcome can be greater, in many cases, with less complex systems, simply because the human factor is considered an integral part of the grazed ecosystem. Greater intervention does not always equal more benefits.

A key to the success of grazing strategies is the establishment of proper monitoring systems at the appropriate scale and temporal frequency relative to landscape size and complexity. Much debate exists as to the appropriate parameters to measure, where the measurements should take place in the landscape and how often information should be gathered (Stuart Hill, 1989; Stafford Smith and Pickup, 1990; Danckwerts *et al.*, 1993; Friedel, 1994; Brown and Ash, 1996). The challenge for scientists, agency personnel and technical advisers is to identify the appropriate variables to measure to best serve the decision-making process in terms of timeliness, accuracy and perceived value to the land manager. Building the appropriate information infrastructures to best serve the sustainability of grazing lands is the challenge facing all agricultural professions in the 21st century (Stuth, 1996).

Role of Monitoring

Drought and rainfall variability

Newly emerging early warning systems, which couple livestock monitoring via faecal profiling (Stuth *et al.*, 1999a) and the use of hydrology-based forage production models with new high-definition satellite imagery (Brook and Carter, 1994), are on the verge of providing a new generation of tools for grazing managers to monitor plant response and detect early deviations in forage production that forewarn of impending shortfalls of forage supply. Livestock producers who can make early adjustments in animal numbers can sustain a higher overall livestock production throughout the life cycle of the ranch firm. Strategic planning for drought and high rainfall variability is further improved with these tools.

Body condition scoring

Perhaps one of the most important management skills of livestock producers is the ability to score the body condition of their animals and track progress toward meeting a desired degree of fatness to meet a given reproductive goal in a herd. Body condition scoring is an index to the degree of fatness expressed in the anatomy of the animal that can be viewed by the human eye. Essentially, body condition scoring is a systematic process of attempting to visualize the degree of underlying skeletal features that can be detected by observing the animal.

To attain high pregnancy rates, one must manage for scores that achieve above-average fatness to ensure sufficient fatness for reproductive fitness. Generally, scoring of cows is most convenient at weaning, since the cows are being gathered and calves separated from the cows. However, this production phase is the least sensitive to predicting the likely pregnancy rate of a cow. The most sensitive time to condition-score animals is at calving, followed by scoring at breeding. Effective nutritional management of cows on rangelands requires that the livestock producer establish a reasonable, attainable goal for the level of pregnancy rates that they desire and then establish a nutritional management system that meets these goals.

Nutrition

A major breakthrough in improving the ability of livestock producers to monitor the concentration of protein and energy in the diet of animals, coupled with a computer programme to predict performance and determine least-cost solutions to mediating shortfalls in the nutrient balance of the animal, is changing the ability of livestock producers to anticipate and understand the linkages of the animal and the land (Stuth *et al.*, 1999a). Near infrared reflectance spectroscopy faecal profiling technology provides the estimates of protein and energy content of the diet under free-ranging conditions (Lyons and Stuth, 1992). The Nutritional Balance Analyzer system provides an analytical tool to predict the response of the animal, given nutritional quality of the diet, breed characteristics, environmental conditions, feeding regime, metabolic modifiers used and level of grazing pressure relative to forage supply (Norman *et al.*, 1999; Stuth *et al.*, 1999b)

Ecological Processes

The health of the land is expressed in those ecological processes that lead to the stability and resiliency of the vegetation subjected to variations in grazing pressure. Although hotly debated, there is a recognized need for livestock producers to begin the process of monitoring key ecosystem attributes that are indicative of health of the land, such as soil organic matter, litter levels, residual standing crops, degree of soil erosion, levels of runoff and evidence of biological invasion by noxious plants. Technologies are slowly emerging that will guide livestock producers in the future in monitoring those processes that are indicative of the health of the land and the subsequent long-term sustainability of the ranch firm.

Conclusion

Throughout this discussion, it has been stressed that grazed ecosystems are generally complex, and the human decision-making process is an integral part of the ecosystem function and structure. Grazing systems have been redefined to direct primary attention away from grazing methods, enabling attention to focus on all elements of planned grazing management. Grazing systems have also been integrated in the context of new ecological frameworks. Monitoring of processes was stressed as essential for rational decision-making and the realization of the manager's landscape vision. Sustainability of grazed ecosystems into the future will require greater linkages between the human decision-making process, ecological processes, economic systems and political systems across several temporal and spatial scales. The only way we can hope to achieve this goal is to begin developing information infrastructures that link the knowledge generator with the knowledge purveyor and knowledge seeker.

References

Archer, S. (1989) Have southern Texas savannas been converted to woodlands in recent history? *American Naturalist* 134, 545–561.

Archer, S. and Smeins, F.E. (1991) Ecosystem-level processes. In: Heitschmidt, R.K. and Stuth, J.W. (eds) *Grazing Management: An Ecological Perspective*. Timber Press, Portland, Oregon, pp. 109–139.

Briske, D.D. and Richards, J.H. (1995) Physiological response of individual plants to grazing: current status and ecological significance. In: Vavra, M., Laycock, W.A. and Pieper, R.D. (eds) *Ecological Implications of Livestock Herbivory in the West*. Society for Range Management, Denver, Colorado, pp. 146–176.

Brook, K.D. and Carter, J.O. (1994) Integrating satellite data and pasture growth models to produce feed deficits and land degradation alerts. *Agricultural Systems and Information Technology* 6, 54–56.

Brown, J.R. and Ash, A.J. (1996) Managing resources: moving from sustainable yield to sustainability in tropical rangelands. *Tropical Grasslands* 30, 47–57.

Danckwerts, J.E., O'Reagain, P.J. and O'Connor, T.G. (1993) Range management in a challenging environment: a southern African perspective. *Rangeland Journal* 15, 133–144.

Dyksterhuis, E.J. (1949) Condition and management of rangeland based on quantitative ecology. *Journal of Range Management* 2, 104–115.

Erickson, D., Stuth, J.W. and Pinchak, W. (1996) Spatially explicit modeling of landscape use by cattle. In: *Proceedings of the ESRI Conference on GIS Applications, San Jose, California, USA*, pp. 23–29.

Foran, B. and Howden, M. (1999) Nine global drivers of rangeland change. In: Eldridge, D. and Freudenberger, D. (eds) *People and Rangelands: Building the Future. Proceedings of the VI International Rangeland Congress, Townsville, Queensland, Australia, 19–23 July*, pp. 7–13.

Friedel, M.H. (1994) How spatial and temporal scale affect the perception of change in rangelands. *Rangeland Journal* 16, 16–25.

Fuhlendorf, S.D. (1996) Multi-scale vegetation responses to long-term herbivory and weather variation on the Edwards Plateau, Texas. PhD dissertation, Texas A&M University, College Station, Texas, USA.

Fuhlendorf, S.D. and Smeins, F.E. (1996) Spatial scale influence on long-term temporal patterns of a semi-arid grassland. *Landscape Ecology* 11, 107–113.

Heady, H.F. and Childs, R.D. (1994) *Rangeland Ecology and Management*. Westview Press, Boulder, Colorado, 519 pp.

Heitschmidt, R.K. and Taylor, C.A. (1991) Livestock production. In: Heitschmidt, R.K. and Stuth, J.W. (eds) *Grazing Management*. Timber Press, Portland, Oregon, USA, pp. 181–178.

Illius, A.W., Derry, J.F. and Gordon, I.J. (1999) A re-assessment of the value of strategies for tracking climatic variation in semi-arid grazing systems. In: Eldridge, D. and Freudenberger, D. (eds) *People and Rangelands: Building the Future. Proceedings of the*

VI International Rangeland Congress, Townsville, Queensland, Australia, 19–23 July, pp. 504–505.

Laycock, W.A. (1991) Stable states and thresholds of range condition on North American rangelands: a viewpoint. *Journal of Range Management* 44, 427–433.

Lyons, R.K. and Stuth, J.W. (1992) Fecal NIRS equations predict diet quality of free-ranging cattle. *Journal Range Management* 45, 614–618.

McCall, D.G. and Sheath, G.W. (1993) Development of intensive grassland systems: from science to practice. In: Baker, M.J., Crush, J.R. and Humphreys, L.R. (eds) *Proceedings of the XVIIth International Grassland Congress, New Zealand.* SIR Publishing, Wellington, New Zealand, pp. 1257–1265.

McKeon, G.M., Day, K.A., Howden, S.M., Mott, J.J., Orr, D.M., Scattini, W.J. and Weston, E.J. (1990) Northern Australian savannas: management for pastoral production. *Journal of Biogeography* 17, 355–372.

Milchunas, D.G., Sala, O.E. and Lauenroth, W.K. (1988) A generalized model of the effects of grazing by large herbivores on grassland community structure. *American Naturalist* 132, 87–106.

Norman, A., Eilers, J., Stuth, J. and Tolleson, D. (1999) Enhancing conservation management via nutritional profiling of livestock on grazing lands: the NRCS National Evaluation of Forage Quality and Animal Well-being Programme. In: Eldridge, D. and Freudenberger, D. (eds) *People and Rangelands: Building the Future. Proceedings of the VIth International Rangeland Congress, Townsville, Queensland, Australia, 19–23 July*, pp. 378–379.

O'Connor, T.G. (1985) *A Synthesis of Field Experiments Concerning the Grass Layers in the Savanna Regions of Southern Africa. South Africa National Scientific Progress Report No. 114, October*, 119 pp.

O'Connor, T.G. (1995) *Acacia karroo* invasion of grassland: environmental and biotic effects influencing seedling emergence and establishment. *Oecologia* 103, 214–223.

Pieper, R.D. and Heitschmidt, R.K. (1988) Is short-duration grazing the answer? *Journal of Soil and Water Conservation* 43, 133–137.

Provenza, F.D. (1995) Post-ingestive feedback as an elementary determinant of food preference and intake in ruminants. *Journal Range Management* 48, 2–17.

Quirk, M.F. (1995) Field testing and enhancement of the diet selection algorithm of POPMIX: a decision support system for multispecies stocking of rangeland. PhD dissertation, Texas A&M University, College Station, Texas, 140 pp.

Quirk, M.F. and Stuth, J.W. (1995) Preference-based algorithms for predicting herbivore diet composition. *Annuals of Zootechnique* 44 (Suppl.), 110.

Ranching Systems Group (RSG) (1994) *GLA–Grazing Lands Application Users' Guide.* Ranching Systems Group Document 94–1, Department of Rangeland Ecology and Management, Texas A&M University, College Station, Texas, 205 pp.

Roberts, B. (1993) Grazing management in property management planning: Consensus and conflict on recommendations. In: *Will Cells Sell? Proceedings of a Grazing Systems Seminar.* Queensland Branch, Soil and Water Conservation Association of Australia, Rockhampton, pp. 26–32.

Scifres, C.J., Stuth, J.W. and Koerth, B.H. (1987) *Improvement of Oak-dominated Rangeland with Tebuthiuron and Prescribed Burning.* Bulletin 1567, Texas Agricultural Experiment Station, College Station, Texas.

Smeins, F.E., Owens, M.K. and Fuhlendorf, S.D. (1994) Biology and ecology of ashe (blueberry) juniper. In: Taylor, C.A., Jr (ed.) *Juniper Symposium.* Technical Report 94–2, Sonora Research Center, Texas Agricultural Experiment Station, Sonora, Texas, pp. 9–24.

Stafford Smith, D.M. and Pickup, G. (1990) Pattern and production in arid lands. *Proceedings of the Ecological Society of Australia* 16, 195–200.

Stuart Hill, G.C. (1989) Adaptive management: the only practicable method of veld management. In: Danckwerts, J.E. and Teague, W.R. (eds) *Veld Management in the Eastern Cape*. Government Printer, Pretoria, South Africa, pp. 4–6.

Stuth, J.W. (1991) Foraging behaviour. In: Heitschmidt, R.K. and Stuth, J.W. (eds) *Grazing Management: An Ecological Perspective*. Timber Press, Portland, Oregon, pp. 65–83.

Stuth, J.W. (1996) Harry Stobbs Memorial Lecture–1995. Managing grazing lands: critical information infrastructures and knowledge requirements for the future. *Tropical Grasslands* 30, 2–17.

Stuth, J.W. and Scifres, C.J. (1982) Integrated grazing management and brush management strategies. In: Scifres, C.J. (ed.) *Development and Implementation of Integrated Brush Management Systems (IBMS) with Special Reference to South Texas*. Bulletin 1493, Texas Agricultural Experiment Station, College Station, Texas, pp. 41–46.

Stuth, J.W., Conner, J.R. and Heitschmidt, R.K. (1991) The decision-making environment and planning paradigm. In: Heitschmidt, R.K. and Stuth, J.W. (eds) *Grazing Management: An Ecological Perspective*. Timber Press, Portland, Oregon, pp. 201–233.

Stuth, J.W., Lyons, R.K. and Kreuter, U.P. (1993) Animal/plant interactions: nutrient acquisition and use by ruminants. In: Powell, J.M., Fernandez-Rivera, S., Williams, T.O. and Renard, C. (eds) *Livestock and Sustainable Nutrient Cycling in Mixed Farming Systems of Sub-Saharan Africa. Proceedings of an International Conference, International Livestock Centre for Africa, Addis Ababa, Ethiopia, 22–26 November*, Vol. II: *Technical Papers*, pp. 63–82.

Stuth, J.W., Freer, M., Dove, J. and Lyons, R. (1999a) Nutritional management for free-ranging livestock. In: Jung, H.G. and Fahey, G.C., Jr (eds) *Nutritional Ecology of Herbivores*. American Society of Animal Science, Savory, Illinois, USA, pp. 696–751.

Stuth, J.W., Eilers, J. and Tolleson, D. (1999b) Nutritional profiling of free-ranging herbivores: what have we learned and where do we go from here? In: Eldridge, D. and Freudenberger, D. (eds) *People and Rangelands: Building the Future. Proceedings of the VIth International Rangeland Congress, Townsville, Queensland, Australia, 19–23 July*, pp. 514–515.

Walker, J.W. and Hodgkinson, K.C. (1999) Grazing management: new technologies for old problems. In: Eldridge, D. and Freudenberger (eds) *People and Rangelands: Building the Future. Proceedings of the VIth International Rangeland Congress, Townsville, Queensland, Australia, 19–23 July*, pp. 424–430.

Westoby, M., Walker, B. and Noy-meir, I. (1989) Opportunistic management for rangelands not at equilibrium. *Journal of Range Management* 42, 266–274.

18

Campos in Southern Brazil

C. Nabinger,[1] A. de Moraes[2] and G.E. Maraschin[1]

[1]*Universidade Federal do Rio Grande do Sul, Porto Alegre, Brazil;* [2]*Universidade Federal do Parana, Curitiba, Brazil*

Introduction

The Brazilian subtropical region is located between the extreme southern border of the country (approximately 33°S) and the Tropic of Capricorn. This chapter discusses the main grazing ecosystems found in this region (states of Rio Grande do Sul, Santa Catarina and Parana), based on a tradition of beef cattle livestock production, which started at the beginning of Brazilian colonization at Rio Grande do Sul and, little by little, spread north to the grasslands of Santa Catarina and Parana.

Few places in the world present such diversity in native forage species, with almost 800 grasses and 200 legumes. *Compositae*, *Cyperaceae* and other families are also present, providing a plant biodiversity that surpasses even that found in tropical rain forests (Duncan and Jarman, 1993). Moreover, the particular climatic conditions make possible an unusual coexistence of summer C_4 and winter C_3 species. The frequency of winter species is influenced by latitude, altitude, soil fertility and pasture management. Natural pasture is the main livestock feed in the region, especially in Rio Grande do Sul, occupying 40% of the territory. In Santa Catarina and Parana, native pastures are less important and have been progressively replaced by crops and cultivated pastures in integrated crop–cattle systems.

Importance of Beef Cattle Livestock Farming in Brazilian Subtropics

The population of beef cattle in the three southern states is currently 26 million head (Table 18.1), while 667,000 animals were slaughtered in inspected premises in 1996 (Table 18.2). This represents a low productivity rate, although data can

 Grassland Ecophysiology and Grazing Ecology (eds G. Lemaire, J. Hodgson, A. de Moraes, C. Nabinger and P.C. de F. Carvalho)

Table 18.1. Evolution of beef cattle numbers in the states of the Brazilian subtropics (in thousands) (from IBGE, 1999).

Year	RS	SC	PR	Total
1960	8,810	1,202	1,665	11,677
1970	12,305	1,955	4,693	18,953
1980	13,986	2,616	7,893	24,495
1990	13,715	2,994	8,616	25,325
1996	13,221	3,097	9,901	26,219

In this and subsequent tables, RS = Rio Grande do Sul, SC = Santa Catarina, PR = Parana.

Table 18.2. Cattle slaughtered in inspected premises in the states of the Brazilian subtropics (from IBGE, 1999).

Year	RS*	SC*	PR*	Total
1985	253,992	86,986	89,055	430,033
1996	344,675	157,110	165,393	66,778
1997	1,487,214	189,449	1,004,042	2,680,505

*See footnote, Table 18.1.

be misleading, since slaughter without inspection was a common practice in the region and many animals are sent to abattoirs in other states, such as São Paulo. Data from 1997 reflect a higher offtake, as a direct consequence of enforced inspection. Productivity (slaughter/total livestock) is still low, official data indicating values between 10 and 11% in Rio Grande do Sul and Parana and 6% in Santa Catarina, though estimates (Cordova, 1997) suggest values of 17% in Santa Catarina and 18% for Rio Grande do Sul and Parana.

Major Grazing Ecosystems in Brazilian Subtropics

Natural grazing ecosystem

Natural grasslands still represent the base for cattle farming, totalling 66% of all land used for livestock production in the region, and 91% in Rio Grande do Sul (Table 18.3). Until 1950, cattle farming was located mainly on native pastures. This natural resource made possible the introduction of the first cattle by the Jesuit father Cristovao Mendonza, who brought 1500 animals from Paraguay in 1634. These animals were distributed through the different Jesuit missions in order to feed thousands of Indians that lived there, and were later dispersed by the *bandeirantes* during their attacks against the missions. The cattle population increased mainly under extensive exploitation, and in 1797 Rio Grande do Sul had a livestock population of 17,471 head (Vieira, 1965). In the natural grasslands of this state, the joint farming of cattle and sheep was a common practice, deter-

Table 18.3. Evolution of natural grasslands area in the states of the Brazilian subtropics (from IBGE, 1999).

	Native pasture (1000 ha)			
Year	RS*	SC*	PR*	Total
1970	14,078	2,089	1,809	17,976
1975	13,061	1,977	1,684	16,722
1980	12,241	1,903	1,534	15,678
1985	11,940	1,928	1,423	15,291
1996	10,524	1,779	1,377	13,680

*See footnote, Table 18.1.

mining a selective grazing pressure that conditioned the general characteristics of the natural vegetation that exists today. The current sheep population of Rio Grande do Sul is 5 million animals, a reduction from 12 million before the fall in international wool prices.

Natural pastures: main features

Natural pastures in the region present a great structural diversity, with a predominance of grasses and relatively low proportions of legumes. There is a high variability in productivity in both time and space. Temporal variations are determined by the seasonal climatic variation. The response of a plant community is determined essentially by the coexistence of C_3 and C_4 species adapted to subtropical climates. The balance of these species within a given community determines the balance of growth through the different seasons of the year, and defines the balance of annual forage production. The frequency of winter species can be 17% or more (Gomes, 1996), but this is rarely observed, because of poor land management, which includes burning and winter overstocking. Spatial variation is strongly linked to soil physical and chemical features, altitude and rainfall, factors that determine important variations in productivity related to species dominance.

Figure 18.1 illustrates seasonal variations in animal performance, as a measure of pasture production, for three different regions of Rio Grande do Sul. Vacaria is located in the tall pasture zone (altitude fields), São Gabriel represents mixed pastures near the central part of the state and Uruguaiana has short grasslands (fine fields) near the frontiers with Argentina and Uruguay.

Major grassland formations in Brazilian subtropics

A better knowledge of Brazilian flora in the subtropics started in the 19th century with the first European botanists to visit the country. The first description of natural pastures in southern Brazil was published by Lindman in 1906 and republished in Portuguese in 1974. Brasil (1973) classified the communities as tall, short

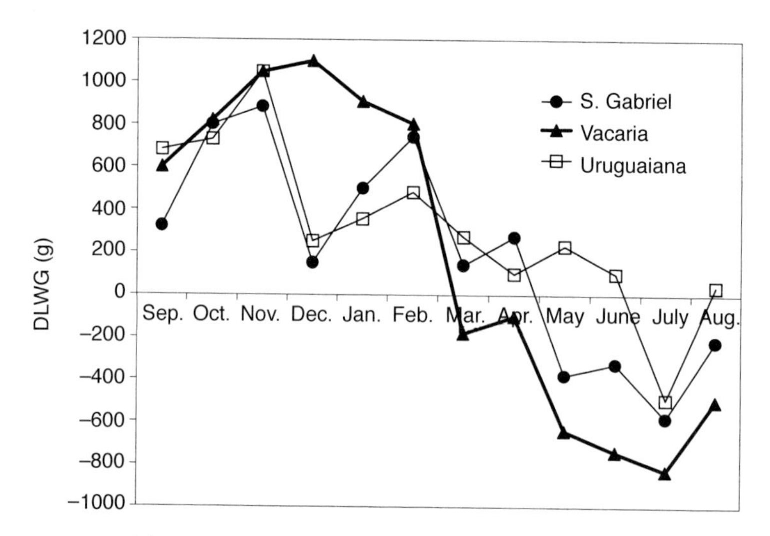

Fig. 18.1. Monthly variation on daily live-weight gain (DLWG) per animal in three regions of Rio Grande do Sul (from Grossman and Mohrdieck, 1956).

and mixed grasslands, mixed grassland/shrub communities and sea-coast formations. An adaptation of Brasil (1973) is shown in Fig. 18.2.

After Burkart (1975), these communities may be classified into two major groups: 'central Brazil' and 'Uruguayan–southern Brazil' communities (Valls, 1986). These two groups present strong internal gradients in terms of altitude, rainfall and soil texture and other characteristics that directly reflect soil moisture levels (Valls, 1986). In a more generic way, Valls (1986) considers the communities from Parana, Santa Catarina and the northern half of Rio Grande do Sul as being fit to be included in the class of 'central Brazil', representing a tall grass formation, dominated by species of *Aristida*, *Andropogon*, *Schizachyrium*, *Elyonurus* and *Trachypogon*. The so-called 'Uruguayan–southern Brazil' communities occupy the southern half of Rio Grande do Sul. In these communities, *Paspalum* species play a major role, with increasing importance of *Axonopus*, *Coelorhachis*, *Leersia* and *Luziola* species, especially on humid soils (Valls, 1986). On the sea coast, there are patches of grassland that can be included in both the above classes. These are dense communities with high quality native grasses, such as *Hemarthria altissima*, *Panicum elephantipes* and *Paspalidum paludivagum*, as well as *Luziola peruviana* and *Paspalum modestum*, which are among the species with acknowledged high quality and adaptation to humid soils.

A proper general characterization of these communities is very difficult, due to the presence of a large number of species and many ecotypes (Paim, 1983). Also, vegetation is under continuous successional development due to biotic factors (Nabinger, 1980). Throughout the centuries, subdivision, animal overstocking, the use of fire and associated factors were responsible for floral changes, resulting in a short-grass vegetation (Lindman, 1974), representing a disclimax. Climax communities reflect very low grazing pressures and are characterized pre-

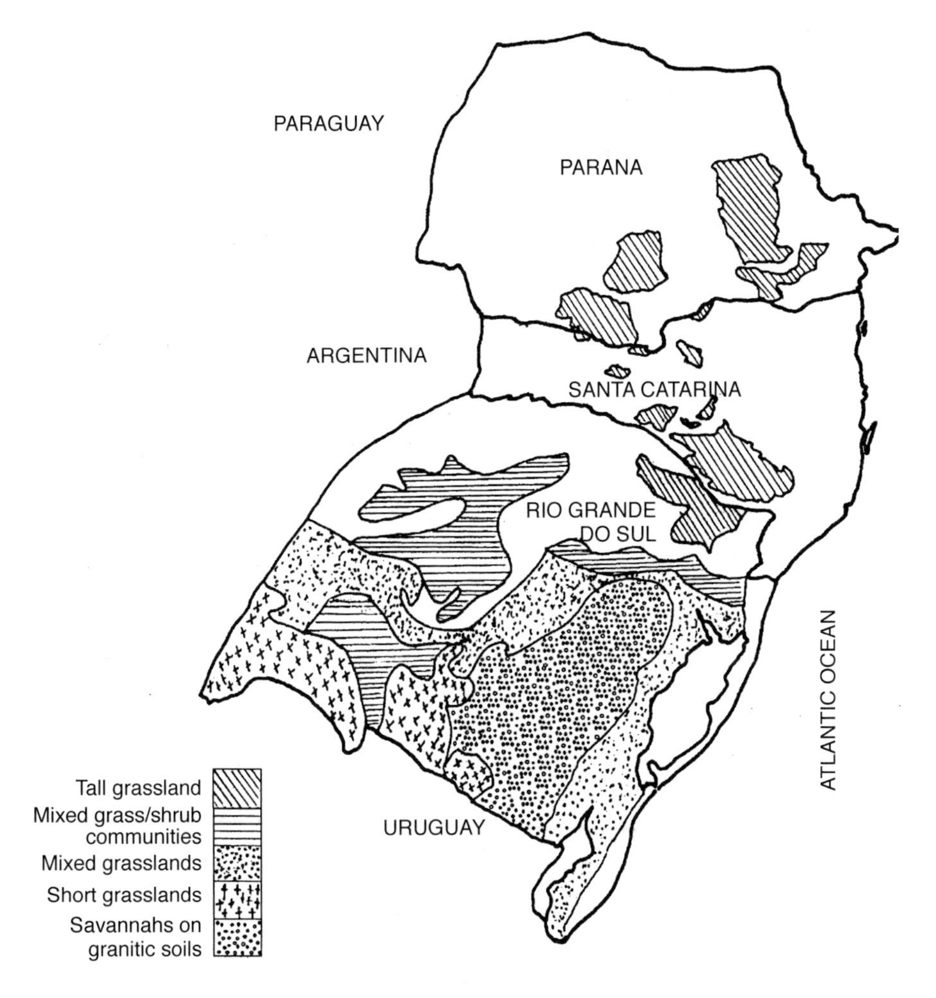

Fig. 18.2. Main natural grassland formations in Brazilian subtropics.

dominantly by few species occupying large areas, such as *caninha* grass (*Andropogon lateralis*) in the Rio Grande do Sul central depression region and goat's-beard (*Aristida jubata*) in the highlands of the same state, as well as *forquilha* grass (*Paspalum notatum*) in the Campaign region (Nabinger, 1980).

Pott's (1974) study on the dynamics of natural vegetation in the Rio Grande do Sul central depression corroborates these ideas. Areas under normal grazing utilization were compared with areas submitted to two different managements: (i) protection from grazing; and (ii) the introduction of Italian ryegrass and subterranean clover after cultivation. Figure 18.3 shows that on the protected area there was a replacement of short species by tall and less palatable species, especially from the genus *Andropogon*. The cultivated area initiated a new succession (subsere 2) towards, probably, the original situation of the grazed control treatment.

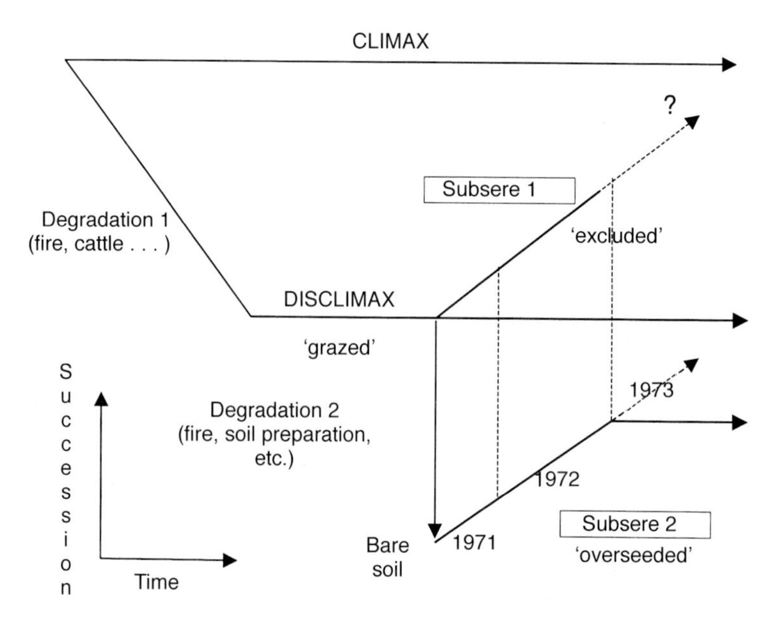

Fig. 18.3. Theoretical syngenetic–subseral scheme of vegetal succession on a natural grassland at Rio Grande do Sul, Brazil, protected from grazing (subsere 1), overseeded with winter exotic species (subsere 2) or maintained in normal grazed condition (from Pott, 1974).

These ecosystems are therefore in an extremely unstable disclimax condition, which may nevertheless be more desirable for animal production. In these cases, it is necessary to understand all the factors affecting plant succession under different climatic environments, using grazing pressure to condition vegetation to maintain the predominance of the best species for animal feeding (Nabinger, 1980). Heavy grazing pressure tends to degrade these ecosystems, resulting in low soil cover and replacement of highly productive species by less productive and generally low-quality species, or complete loss of forage species. Reduced soil cover results in erosion. On the other hand, excessively low grazing pressure can result in dominance of tall grasses of low nutritional value or by bushes and other undesirable species, mainly from the genera *Baccharis* and *Eryngeum*.

The effect of the animal on the pasture mainly reflects the influence of the frequency and intensity of defoliation of individual species on botanical composition. So, for example, Girardi-Deiro and Gonçalves (1987) noticed an increased frequency of *forquilha* grass (*P. notatum*), from 26.9% at low grazing pressure to 62.9% at high grazing pressure. Increasing frequency of this species with high levels of grazing pressure was also noticed by Martinez Crovetto (1965), Rosito and Maraschin (1984) and by Souza (1989), and is attributed to its phenotypic plasticity and rhizomatous habit. Boldrini (1993), on the other hand, even though noticing similar trends, observed that the species was well represented at any grazing pressure studied, and that soil condition was probably the most important factor influencing species balance. In this way, soil type and humidity must also be taken into account when interpreting successional trends.

Descriptions of the major grass and legume species on natural pastures in the Brazilian subtropics, together with information on geographical distribution, forage values, habits and life cycles can be found in Gomes *et al.* (1989), Barreto and Boldrini (1990), and Gonçalves (1990a, b).

Major cultivated pasture ecosystems

Recent reductions in farm size, necessitating intensification of the farming systems and the integration of cropping and animal production, has made farmers look for productive cultivated pastures in order to ensure economic returns. Despite the high productivity and nutritive potential of many native species, they are not commercially exploited, and cultivated pastures are mainly formed by exotic species.

Table 18.4 shows a substantial increase in the area of cultivated pastures in the Brazilian subtropics since the 1960s, mainly concentrated in Parana. This state went from 29% of cultivated pastures to 79% in 1996, the increase in the northern region of the state being due to replacement of permanent crops (mainly coffee) by pastures. Morais (1988) describes this process as a cattle-farming culture forming in the Parana northern region. Forage species used in this process were *colonião* (*Panicum maximum*) and *jaragua* (*Hiparrhenia rufa*). With time and especially during the last 15 years, these species were replaced by *Brachiaria*, especially *B. brizantha*, *B. decumbens*, *B. humidicola* and, more recently, *B. dictioneura*. Continual impoverishment of pasture soils has resulted in a process of degradation, with a replacement of this vegetation by low-productivity and low-quality species, such as the *mato grosso* grass (an ecotype of *P. notatum*) and even the *barba-de-bode* grass (*Aristida* sp.).

The area of cultivated pasture in south-western Parana increased by 187% between 1960 and 1970, and by a further 34% up to 1980. Basically, these cultivated pastures are similar to those in the northern region, with a massive presence of *Brachiaria bryzantina* and of *Cynodon* spp. (mainly African stargrass and coast cross). In the south-eastern region of Parana, where the proportion of cultivated pastures increased from 10% in the 1960s to 40% in 1985, the major species are

Table 18.4. Evolution of cultivated pastures area in the states of the Brazilian subtropics (from IBGE, 1999).

	Cultivated pasture (1000 ha)			
Year	RS*	SC*	PR*	Total
1960	361	233	782	1376
1970	557	379	2700	3636
1975	712	427	3299	4438
1980	1061	588	3986	5635
1985	1023	542	4577	6142
1996	1157	560	5300	7017

*See footnote, Table 18.1.

the *missioneira* grass (*Axonopus compressus*) in the Palmas region and *pensacola* (*Paspalum saurae*) in Guarapuava and more recently, *hemartria* (*H. altissima*).

In the state of Santa Catarina, *missioneira* grass is widespread from the west to the microregions of the Alto Irany, Santa Catarina Midwest, High Rio do Peixe, to the Catarinense Highlands counties and also to the east, through the microregion of the High Itajaí Valley up to the Canoinhas Valley and to the Northern Highlands (Nascimento *et al.*, 1990). In the Catarinense Highlands, this species occupies more than 80% of cultivated summer pastures (approximately 43,000 ha) (CEPA, 1984).

In Rio Grande do Sul (Valls, 1973), the area of missionary grass is about 60% of the total area of perennial summer forage species. These species are located mainly in the High Uruguay (30,839 ha), Upper Mountains Campos (13,426 ha), Middle Highlands (6839 ha), Missions (5378 ha) and Northeast Upper Slope (5373 ha). Pangola grass, Rhodes grass and *Setaria* were utilized widely, but more recently *Cynodon* (cv. Tifton) and *P. maximum* (cv. Brazilian) have increased in importance in the more subtropical regions of the state. Tables 18.5 and 18.6 illustrate the relative importance of the alternative summer and winter forage species, respectively.

Table 18.5. Major summer forage species cultivated in Brazilian subtropics.

		Present level of relative importance		
Species	Brazilian common name	PR*	SC*	RS*
Perennial summer grasses				
Panicum maximum	Colonião	++	+	+
B. brizantha	Brizanta, braquiária	+++	+	+
B. decumbens	Braquiária, decumbens	+	+	+
B. humidicula	Espetudinha, humidícula	+	+	+
Digitaria decumbens	Pangola	+	+	++
Cynodon spp.	Estrela, Coast Cross, Tifton	+++	+	++
Axonopus compressus	Missioneira, jesuíta	++	+++	+++
Pennisetum clandestinum	Quicuio	+	+	+
Pennisetum purpureum	Capim elefante	+	+	+
Paspalum saurae	Pensacola	++	++	+++
Hemarthria altissima	Hemartria	++	++	+
Hyparrhenia rufa	Jaraguá	+	–	–
Setaria sphacelata	Setaria	+	+	+
Chloris gayana	Rhodes	+	+	+
Annual summer grasses				
Pennisetum americanum	Milheto	+++	+++	+++
Sorghum spp.	Sorgo	+	+	++
Euchlaena mexicana	Teosinto	+	+	+
Brachiaria plantaginea	Papuã	++	+	+

*See footnote, Table 18.1.
+, low importance; ++, medium importance; +++, high importance.

Table 18.6. Major winter forage species cultivated in Brazilian subtropics.

Species	Brazilian common name	Present level of relative importance		
		PR*	SC*	RS*
Annual grasses				
Lolium multiflorum	Azevém	+++	+++	+++
Avena strigosa	Aveia preta	+++	+++	+++
Avena sativa	Aveia branca	+++	++	+
Secale cereale	Centeio	++	++	+
× *Triticosecale*	Triticale	++	+	+
Hordeum vulgare	Cevada	+	+	+
Perennial grasses				
Festuca arundinaceae	Festuca	+	+	+
Bromus catharticus	Cevadilha	+	+	+
Dactylis glomerata	Capim dos pomares	+	+	+
Falaris tuberosa	Falaris	+	+	+
Annual legumes				
Vicia sativa	Ervilhaca, Vica	++	++	++
Vicia villosa	Ervilhaca peluda	+	+	+
Trifolium vesiculosum	Trevo vesiculoso	++	++	+++
Trifolium subterraneum	Trevo subterrâneo	+	+	+
Ornithopus sativus	Serradela	+	+	+
Lotus subflorus	El Rincon	–	–	++
Lathyrus sativus	Chícharo	+	+	+
Perennial legumes				
Trifolium repens	Trevo branco	++	+	+++
Trifolium pratense	Trevo vermelho	++	++	++
Lotus corniculatus	Cornichão	++	++	++
Medicago sativa	Alfafa	+	+	+

*See footnote, Table 18.1.

The most important winter species are Italian ryegrass (*Lolium multiflorum*) and black oat (*Avena strigosa*). Italian ryegrass is chosen as the main option, due to its capacity for natural reseeding, disease resistance, good production potential and versatility for association with legumes. Black oat represents a more important cultivated area than Italian ryegrass in southern Brazil and is the favoured species in integrated crop–cattle systems. This is due to its early growth and short production cycle, which does not interfere with the sowing time of summer crops (Moraes, 1994). Winter perennial grasses have more limited application, due to the lack of varieties adapted to Brazilian subtropical conditions, although, in many areas of the region, climatic and soil conditions are adequate for their development. There is a good potential for the use of improved varieties developed in neighbouring countries (Machado and Machado, 1982; Salerno and Vetterle, 1984).

Of the winter legumes, arrow-leaf clover is used in Rio Grande do Sul. Red clover and some varieties of subterranean clover and more recently the annual *Lotus subbiflorus* have some importance in particular areas. Among the perennial species, white clover and bird's-foot trefoil are often used in association with Italian ryegrass.

Integrated crop–animal production systems

Agricultural areas in the Brazilian subtropics have been suffering a continuous process of degradation due to misuse. Topographical location, rainfall distribution, soil characteristics and especially agricultural practices have produced compaction, low infiltration and erosion of soil (Maraschin and Jacques, 1993). Inclusion of pastures into grain-producing regions may be a useful tool in recuperating these damaged soils, as well as a way of guaranteeing the sustainability of the system.

One of the benefits obtained through this integration is that the increase in soil nutrient status resulting from crop fertilization also benefits pasture production and quality. An example of this integrated crop–cattle system is the use of summer crop areas (e.g. soybean, maize) for winter forage production with temperate forage species, such as Italian ryegrass, oats and clovers, thus forming a supplementary forage resource to augment summer perennial pastures. The use of winter annual pastures established by direct drilling in the autumn has increased in importance in southern Brazil, representing an attractive and economical opportunity for grain producers and heifer producers.

A lowland grazing ecosystem involving integration with irrigated rice crops is common in Rio Grande do Sul, with rotation involving 1–2 crop years and 3–4 grazing years, helped by the regeneration of native species. Some farmers use with success a mixture of white clover and bird's-foot trefoil with ryegrass, oversown by aeroplane after the rice harvest. These areas are in the order of 3.5 million ha and are located mainly in the south, south-east and south-west of Rio Grande do Sul. An excellent study on the possibility of using these areas immediately after the rice harvest is given by Saibro and Silva (1999). *Trifolium subterraneum*, *Trifolium resupinatum*, *Trifolium nigrescens*, *Trifolium repens*, *Lotus subbiflorus* and *Lotus pedunculatus* are alternative legumes, to be used in mixture mainly with ryegrass. However, the invasion of agricultural areas (mainly rice) into areas of excellent natural pastures (the short-grass pastures in south-western Rio Grande do Sul) may cause an irreversible genetic erosion of forage plants exclusive to the local flora, many of them with high forage potential (Pott, 1989).

Research on Grazing Ecosystems in Brazilian Subtropics

Research on natural pasture ecosystems

Studies on natural pastures in the Brazilian subtropics are quite old, even though they became the objective of formal institutional planning only in 1961. In that year, project S3-CR-11 was initiated, involving the US Department of Agriculture, the Rio Grande do Sul State Department of Agriculture and the Federal University of Rio Grande do Sul (UFRGS). Its aim was to study the state of native pastures, with the following interests (Barreto, 1963):

1. Classification of the main native grass and legume forage species of Rio Grande do Sul.
2. Agronomic and cytological studies on the most promising species.
3. Selection and genetic improvement of the more promising species.
4. Practical studies for the improvement and best use of natural pastures.
5. Ecological studies of natural pastures in the different regions.

Studies generated from this project, describing the framework of different plant communities, improved knowledge of floristic composition related to soil and climatic variables (Brasil, 1973). They also established information on the seasonal productivity and, in some cases, qualitative aspects of these communities (Gavillon, 1963; Gavillon and Quadros, 1965; Prestes *et al.*, 1968; Freitas *et al.*, 1976; Schreiner *et al.*, 1980). Unfortunately, many of the results obtained remain unpublished and are only available in internal reports.

Few studies were made on the phenology of native species related to climatic variation and its effects on grazing selectivity, and there was no work on plant–animal relations and their interaction with the environment. The termination of project S3-CR-11 in the late 1960s forced a change of direction in forage plant research, with concentrated efforts on exotic forage species, but an important multidisciplinary and interinstitutional staff was formed and studies on vegetation dynamics and plant cytology continued. Exotic plant materials and their physiological response to management were evaluated in the absence of animal effects. After the late 1970s pasture research was focused on plant–animal relationships. This new direction made possible a better interpretation of the effects of forage allowance on individual animal performance and live-weight gain per area and the effects on the pasture. Studies initially focused on cultivated species and, more recently, on natural pastures, and this research brought new proposals for management systems, based upon the concept of grazing pressure, to provide a better validation of pasture production.

The main goals of studies on native pasture made from 1985 to 1995 in the Agronomic Experimental Station of UFRGS, located on the central depression of Rio Grande do Sul, were to acquire data and to encourage human and cultural development based on this important renewable natural resource (Moraes *et al.*, 1990). With 1300 mm of annual precipitation and on soils of low natural fertility

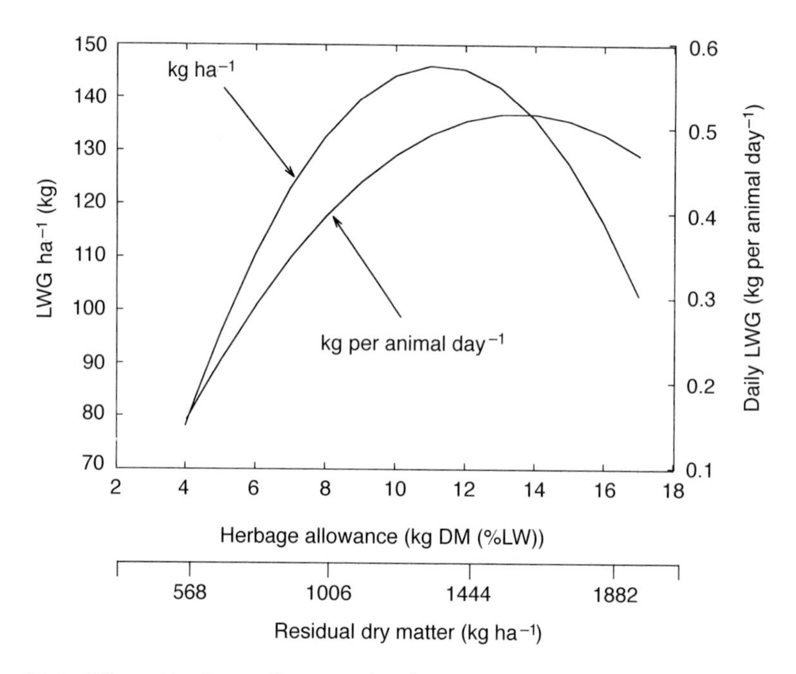

Fig. 18.4. Effect of herbage allowance levels on animal performance, and consequences for average residual dry matter (from Maraschin *et al.*, 1997). LWG, live-weight gain.

and the presence of exchangeable Al^{3+}, natural pasture grows from September to the first autumn frost (usually 220 days of growth). In these studies, forage allowance was maintained at 4.0, 8.0, 12.0 and 16.0 kg of dry matter (DM) 100 kg^{-1} live-weight (% LW) day^{-1}. Animal production was maximized at allowances between 11 and 13 kg DM 100 kg^{-1} LW (Fig. 18.4) that determines residual dry matter between 1.5 and 2.0 t ha^{-1}. Some areas remained ungrazed, indicating levels of availability above intake potential and selective grazing. Better development of plants in these ungrazed areas made possible their flowering and reseeding, thus ensuring pasture longevity and stability. Natural pasture in southern Brazil is of sufficient quality to maintain more than 500 g of daily weight gain per animal and to attain 150 to 180 kg LW ha^{-1} during the growing season. Fertilization and introduction of winter species can substantially increase production from this ecosystem.

As demonstrated early by Grossman and Mohrdieck (1956) (see Fig. 18.1), spring is the season offering the best opportunities for enhanced animal gain. Management at this time also influences production over the entire growing season (Moojen, 1992; Corrêa, 1993; Setelich, 1994). This is also true for fertilized native pasture (Maraschin and Jacques, 1993). During summer, the increase in residual dry matter follows the increase in the forage structural fraction, diluting the general quality of the available forage. This determines a grazing condition with high forage on offer and a certain level of selective grazing, so that the animal can maintain performance. Low autumn and winter temperatures reduce pasture growth rate. Accumulation of senescent herbage at this time reduces pasture quality to a maintenance diet and selective grazing becomes more evident. At such times, herbage allowances must be adjusted to available green forage and supplementation is necessary in order to maintain animal body condition (Escosteguy, 1990; Moojen, 1992).

Limited herbage production over the winter results in overgrazing at this period. A pasture management philosophy based on fixed stocking still dominates southern Brazil. As a consequence, weight gain has been kept below 0.4 kg animal^{-1} day^{-1} during the summer, and heavy weight losses occur during the winter. Under natural conditions, animals could adapt to available space and maintain a grazing pressure adequate to sustain the system. Limitations to migration and changes in animal species proportions have been responsible for an imbalance between what native pastures can offer and what animals demand. A better understanding of the dynamics of the growth of natural pastures increasingly shows the importance of grazing control (Tothill *et al.*, 1989). Research in animal production has paid little attention to the concept of forage allowance and to the complex climate–soil–plant–animal interactions. Added to this, one can observe the difficulty of researchers in using proper sampling methods to determine forage accumulation on pasture and to understand how forage availability may change grazing behaviour and its productivity. Further, results from many grazing experiments have limited value, due to poor knowledge and inadequate use of experimental methods, particularly those concerned with the measurement of herbage accumulation and selective defoliation (Maraschin and Jacques, 1993).

Herbivores are primary consumers of forage produced through capture of the sun's energy by vegetation. Using data from Maraschin *et al.* (1997), Nabinger (1998) showed the effect of optimization of energy balance on the system, which can be obtained in natural grassland through simple management techniques, such as the adjustment of animal stocking density to forage on offer (Table 18.7). An allowance of 4% of LW corresponds to a grazing pressure that forces animals to consume almost all the above-ground herbage. Under these conditions, residual leaf area is reduced and interception of solar radiation is very low, resulting in low efficiency of photosynthetically active radiation (PAR) utilization for primary production. Forage availability in these conditions makes the grazing process difficult, limiting daily forage ingestion, and results in poor animal performance, diminishing still further the efficiency of the system. Residual leaf area and solar energy capture will increase with increase in forage on offer, resulting in augmentation of pasture growth rate and optimizing the grazing process and animal performance. Greater forage availability improves forage intake and diet selection by the grazing animal, and the efficiency of conversion of PAR to animal production increases almost 100% as herbage allowance increases from 4.0% to 12.0% of LW. A simple practice like the adequate adjustment of stocking density to forage availability can increase animal production by more than 100% at very limited cost.

Further low-cost improvements are possible in natural systems. Improved

Table 18.7. Effect of grazing intensity on efficiency of solar radiation utilization in a natural pasture in Rio Grande do Sul (from Nabinger, 1998).

	Forage allowance ($kg\ DM\ 100\ kg^{-1}\ LW\ day^{-1}$)			
System components	4.0%	8.0%	12.0%	16.0%
Energy ($MJ\ ha^{-1}$)				
Global incident solar energy		48,000,000		
Incident PAR		20,600,000		
Primary production*	40,877	68,714	73,343	66,842
Secondary production*	1,835	3,144	3,415	2,738
Annual DM production ($kg\ ha^{-1}$)	2,075	3,488	3,723	3,393
LW gain ($kg\ ha^{-1}$)	78.1	132.5	145.3	116.5
Efficiency of PAR conversion				
PAR/primary production	0.20	0.33	0.36	0.32
PAR/secondary production	0.009	0.015	0.017	0.013
Primary/secondary production	4.48	4.53	4.66	4.10

*Considering energy concentration in plant and animal tissues of 19.7 and 23.5 $MJ\ kg^{-1}$, respectively. PAR, photosynthetically active radiation.

subdivision of the land according to natural soil fertility, distribution of water sources and natural protection for animals and the provision of deferment of paddocks are tools for adequate pasture management. Research results have demonstrated the importance of deferment practices for improving botanical composition and soil conditions in natural pastures (Fontaneli and Jacques, 1988; Moojen, 1992). Production may also be improved by the use of soil fertilization (Rosito and Maraschin, 1984; Perin, 1990; Moojen, 1992). Gomes (1996) demonstrated that the responses of natural pasture to soil fertilization are reduction in bare soil, reduced incidence of tall grasses of low quality, fewer invader species and less dead material. Prostrate grasses increase their participation up to intermediate levels of fertilization and native legumes show an impressive response, especially at the higher levels of fertilization (Table 18.8).

Overseeding practices with winter species in native pastures have played a double role. First, they increase pasture production during winter – a basic goal with this practice. Secondly, but no less important, there is the effect of associated fertilization practices on the quality and yield of native species (Estivalet, 1997; Dürr *et al.*, 1998; Vidor and Jacques, 1998). Overseeding with summer species, mainly legumes, also gives promising results (Silva and Jacques, 1998; Perez, 1999).

More recently, the need for a better understanding of the functioning and potential production of natural grazing ecosystems in southern Brazil encouraged research on the ecophysiological responses of some important native species. Studies designed to model the use of solar radiation in *Desmodium incanum* (Spannenberg *et al.*, 1997a, b) and in ecotypes of *P. notatum* (Costa *et al.*, 1997a, b) indicate the

Table 18.8. Effect of fertilization of native pasture on the participation of different groups of species in total dry matter (DM) during summer – at central depression of Rio Grande do Sul, 1993–1994 (Gomes, 1996).

	Fertilization levels (kg NPK ha^{-1})				
	0	90	180	360	720
Component			(% in DM)		
Tall grasses	56.4	53.0	50.3	48.6	43.1
Prostrate grasses	27.1	27.0	30.6	34.4	22.8
Native legumes	0.5	2.4	3.0	7.4	24.3
No forage species	6.5	8.7	9.0	4.0	5.6
Dead material	9.5	8.9	7.1	5.6	4.2
Bare soil (% by area)	3.4	3.0	2.9	2.1	0.7

potential of these species when nutrient and water limitation is removed. In these conditions, forage DM production was greater than 100 kg DM ha^{-1} day^{-1}. Studies on animal production responses to nitrogen fertilization on natural pasture with a predominance of *P. notatum* indicate a potential of about 700 kg LW ha^{-1} during spring–summer (P.R. Boggiano, 1999, personal communication).

The need for a better understanding of the tissue flows in pasture in order to explain the effect of grazing on the leaf area index (LAI) and the consequent use of solar energy and on succession dynamics produced studies on morphogenesis and tissue turnover of important species, such as *A. lateralis* (Cruz, 1998), *P. notatum* (Eggers, 1999), *D. incanum* (Silva *et al.*, 1998), *Bromus auleticus* (Soares *et al.*, 1998), *Coelorhachis selloana* (Eggers, 1999) and *Adesmia* spp. (Scheffer-Basso, 1999). Other equally important native species, such as *Paspalum urvillei*, *Paspalum paniculatum*, *Paspalum dilatatum*, *Briza subaristata* and *Piptochaetium montevidense*, are also under study.

Research on cultivated grazing ecosystems

Research on cultivated pastures of exotic species is not new (Grossman, 1963; Müller and Primo, 1969), but analytical studies that consider both plant and animal responses are still scarce in the Brazilian subtropics. Studies on the initial phases of introduction of exotic forage species in the three Brazilian southern states are often limited, but larger-scale evaluation is being tested in Parana with a network of tests in 15 different centres by the Parana Forage Evaluation Committee (CPAF). Little information is available on animal production from winter grasses in Parana and Santa Catarina, in spite of their importance in meat and milk production systems in these two states. There are several reasons for this, but it is unthinkable that this practice should continue in one of the most important pastoral regions of the country in total ignorance of plant productivity and persistence and animal production potential. Even though beef cattle farming in

Rio Grande do Sul is based upon natural pastures, much research has been done on cultivated pastures. Winter cultivated species received more attention, fully justified by the animal yield potential. Other studies from the Rio Grande do Sul Department of Agriculture remain unpublished, while others were written as internal reports without authorship, making their listing as references difficult.

Much of the basic research on ecophysiology and morphogenesis has been done with winter exotic species. Producers consider the common variety of Italian ryegrass to grow too late in the autumn. However, Viegas (1998) demonstrated that nitrogen is the main factor limiting leaf development and radiation use efficiency. If this limitation is removed, common Italian ryegrass has a high efficiency, is very precocious and can be utilized in autumn with very high forage production levels, because incident solar radiation and temperature do not limit growth, but a low mineralization rate limits N uptake. Similarly to the data of Costa *et al.* (1997a, b) and Spannenberg *et al.* (1997a, b), these results demonstrate the enormous climatic potential of the region, which is scarcely used, due to nutritional or water limitations. Studies on morphogenesis have also been carried on with other winter species, such as *Lotus corniculatus*, with different levels of water availability and light competition (Morales *et al.*, 1997; Morales, 1998). Water demand from forage species is a subject that has received increasing interest in recent years. For example, Cunha *et al.* (1998) identified water requirements of lucerne and generated the necessary parameters to identify the regions that are more suitable for the utilization of this species in Rio Grande do Sul (Bergamaschi *et al.*, 1997), making possible proper recommendations in terms of irrigation requirements.

Summer annual grasses, such as pearl millet (*Pennisetum americanum*) and sorghum, were also evaluated under grazing conditions in Rio Grande do Sul, due to their high quality and potential in intensive systems of animal production. One example of the potential of perennial cultivated grasses was given by Maraschin *et al.* (1993) with pangola grass (*Digitaria decumbens*). High animal performance under continuous grazing (0.76 kg animal day^{-1} and 757 kg LW gain ha^{-1}) were obtained with 9.3% LW of forage allowance, which reflected the maintenance of a residual dry matter between 2.1 and 2.5 t ha^{-1}. This level of performance is substantially greater than that from the usual management of this species by producers, who maintain less than 1.5 t DM of residue and generally do not achieve more than 300 kg of LW gain ha^{-1} year^{-1}. Maraschin and Jacques (1993) listed other summer perennial grasses as being promising for the Brazilian subtropics, such as those from the genus *Cynodon*, the new lines of Pensacola bahia grass (*P. saurae*), obtained through a recurrent selection in the USA, and elephant grass (*Pennisetum purpureum*). The dwarf variant of elephant grass (cv. Mott) was evaluated on grazing in the Ituporanga Experimental Station of Santa Catarina with different levels of forage availability and produced 1.0 kg LW animal^{-1} day^{-1} over 200 days (Almeida, 1997). Responses to nitrogen fertilization in terms of animal production have also been studied in this species (Setelich *et al.*, 1989a, b) and joint results from these two studies are being used to build up a dairy production programme in the area. Also, studies on morphogenesis have made it possible to explain some effects of management on LAI and on herbage consumption (Setelich *et al.*, 1998a), providing the basis for extrapolation to other environmental conditions.

Suggested Research on Grassland Ecosystems in the Brazilian Subtropics

In order to better understand the dynamics of biomass elaboration and renewal of natural and cultivated pastures in the Brazilian subtropics, it is necessary to adopt a more analytical and explanatory approach. To achieve this goal, one has to perform an 'ecophysiological' analysis of the dynamics of primary production from different plant communities. These studies must focus mainly on the growth dynamics (rates of appearance and senescence and expansion of leaves, leaf lifespan, dynamics of tillering) of the most important species. These morphogenetic characteristics should be related to major environmental factors: temperature, radiation, water, nitrogen and phosphorus.

This kind of approach must also bring in new knowledge and open possibilities of forecasting seasonal variation in terms of primary productivity from different species. In a second phase, one must try to use these basic data in order to build up models of productivity at a community level. Also, studies must be undertaken on the intake dynamics of different plant species by grazing animals related to morphogenetic characteristics, and to quantify the process of selective defoliation in terms of frequency and intensity at the level of organs (leaves) and species present in the plant community. In such a way, one could directly relate primary production dynamics on a given plant community with the dynamics of herbage consumption and transformation to animal products. These studies must focus on communities representative of the region. A detailed analysis of species plasticity in response to grazing intensity should also be performed on species with contrasting morphology. Studies on animal responses are seldom related to the dynamics of morphogenesis of the different species that form a plant community. The majority of these studies have been done on monospecific pastures, and their extension to the complex natural pastures in the region provides a methodological challenge still to be overcome. This kind of approach to forage research, fundamental to an understanding of the interactions between pasture and animal as part of an ecosystem, can only be implemented through the use of solid knowledge of morphogenesis and ecophysiology, and integration of different areas of knowledge is mandatory. Edaphoclimatic diversity in southern Brazil and, indeed, the whole southern part of South America, with its richness in grazing flora, makes it essential to formalize collaborative action by the excellent research groups in the region for the education of new researchers and the development of comprehensive projects independent of national borders. Prioritization of research needs should be based on interactions with farmers and extension services. This interaction is also essential when validating new technology in terms of economic and environmental viability.

References

Almeida, E.X. (1997) Oferta de forragem de capim elefante anão (*Pennisetum purpureum* Schum. cv. Mott), dinâmica da pastagem e sua relação com o rendimento animal no Alto Vale do Itajaí, Santa Catarina. Doctoral thesis, Universidade Federal do Rio Grande do Sul, Porto Alegre, Rio Grande do Sul, Brazil, 112 pp.

Barreto, I.L. (1963) Estudo da pastagem nativa no Rio Grande do Sul. In: Associação Gabrielense de Melhoramento e Renovação de Pastagens (ed.) *Anuário*. AGMRP, São Gabriel, pp. 81–88.

Barreto, I.L. and Boldrini, I. (1990) Aspectos físicos, vegetação e problemática das regiões do Litoral, Depressão Central, Missões e Planalto do Rio Grande do Sul, Brasil. In: Puigneau, J.P. (ed.) *Introduccion, Conservacion y Evaluacion de Germoplasma Forrajero en el Cono Sul: primer taller de trabajo dela red de forrageras del Cono Sur*. IICA–PROCISUR, Montevideo, Uruguay, pp. 199–210.

Bergamaschi, H., Aragonés, R.S. and Santos, A.O.S. (1997) Disponibilidade hídrica para a cultura da alfafa nas diferentes regiões ecoclimáticas do Estado do Rio Grande do Sul. *Pesquisa Agropecuária Gaúcha* 3, 99–107.

Boldrini, I.I. (1993) Dinâmica de vegetação de uma pastagen natural sob diferentes niveis de oferta de forragem e tipos de solos, Depressão Central, Rio Grande do Sul. Doctoral thesis, Universidade Federal do Rio Grande do Sul, 262 pp.

Brasil (1973) *Levantamento de reconhecimento dos Solos do Estado do Rio Grande do Sul. Boletim Técnico 30*, Departamento Nacional de Pesquisa Agropecuária, Divisão de Pesquisa Pedológica, Ministério da Agricultura, Recife, 431 pp.

Burkart, A. (1975) Evolution of grasses and grasslands in South America. *Taxon* 24, 53–66.

CEPA (1984) *Síntese anual da agricultura de Santa Catarina*, 2 vols. Instituto de Planejamento e Economia Agrícola de Santa Catarina, Florianópolis, Santa Catarina.

Cordova, U.A. (1997) O agroecossistema campos naturais do Planalto Catarinense: origens, caracteristicas e alternativas para evitar sua extinção. MSc thesis, Universidade Federal de Santa Catarina, 214 pp.

Corrêa, F.L. (1993) Produção e qualidade de uma pastagem nativa do Rio Grande do Sul sob níveis de oferta de forragem à novilhos. MSc thesis, Universidade Federal do Rio Grande do Sul, Porto Alegre, Rio Grande do Sul, Brazil, 167 pp.

Costa, J.A.A., Nabinger, C., Spannenberg, P.R.O., Jacques, A.V.A. and Rosa, L.M.G. (1997a) Eficiência de uso da radiação e ajuste de um modelo de produção potencial para biótipos de *Paspalum notatum* Flügge var. *notatum*. In: *Proceedings of X Congresso Brasileiro de Agrometeorologia*. Sociedade Brasileira Agrometeorologia, Piracicaba, São Paulo, pp. 155–157.

Costa, J.A.A., Nabinger, C., Spannenberg, P.R.O. and Rosa, L.M. (1997b) Parâmetros básicos para o ajuste de modelos de previsão da produtividade potencial de *Paspalum notatum* Flügge var. *notatum*. In: *Proceedings of X Congresso Brasileiro de Agrometeorologia*. Sociedade Brasileira Agrometeorologia, Piracicaba, São Paulo, pp. 158–160.

Cruz, F.P. (1998) Dinâmica de crescimento, desenvolvimento e desfoliação em *Andropogon lateralis* Nees. MSc thesis, Universidade Federal do Rio Grande do Sul, Porto Alegre, Rio Grande do Sul, Brazil, 105 pp.

Cunha, G.R., Bergamaschi, H., Paula, J.R.F. and de Saibro, J.C. (1998) Resposta da alfafa a diversas disponibilidades de água. *Pesquisa Agropecuária Brasileira* 33, 1113–1119.

Duncan, P. and Jarman, P.J. (1993) Conservation of biodiversity in managed rangelands, with special emphasis on the ecological effects of large grazing ungulates, domestic and wild. In: *Proceedings of the XVIIth International Grassland Congress, Palmerston North, New Zealand*, SIR Publishing, Wellington, New Zealand, pp. 2077–2084.

Dürr, J.W., Castilhos, Z.M.S., Flores, A.I.P., Freitas, J.M.O. and Jacques, A.V.A. (1998) Tratamentos de Limpeza e Adubação como Modificadores da Composição Florística do Campo Nativo. In: *Proceedings of XII Reunião do Grupo Técnico Regional do Cone Sul (Zona Campos) em Melhoramento e Utilização de Recursos Forrageiros das Áreas Tropical e Subtropical.* FAO/EMBRAPA–CPPSul, Bagé, p. 115.

Eggers, L. (1999) Morfogênese e desfoliação de *Paspalum notatum* Fl. e *Coelorhachis selloana* (Hack.) Camus em níveis de oferta de forragem. Doctoral thesis, Universidade Federal do Rio Grande do Sul, Porto Alegre, Rio Grande do Sul, Brazil, 147 pp.

Escosteguy, C.M.D. (1990) Avaliação agronômica de uma pastagem natural sob níveis de pressão de pastejo. MSc thesis, Universidade Federal do Rio Grande do Sul, Porto Alegre, Rio Grande do Sul, Brazil, 231 pp.

Estivalet, C.N.O., Jr (1997) Efeitos da ceifa, queima e diferimento sobre a disponibilidade de forragem e composição botânica de uma pastagem natural. MSc thesis, Universidade Federal do Rio Grande do Sul, Porto Alegre, Rio Grande do Sul, Brazil, 97 pp.

Fontaneli, R.S. and Jacques, A.V.A. (1988) Melhoramento de pastagem natural–Ceifa, queima, diferimento e adubação. *Revista da Sociedade Brasileira de Zootecnia* 17, 180–189.

Freitas, E.A.S., López, J. and Prates, E.R. (1976) Produtividade de matéria seca, proteína digestível e nutrientes digestíveis totais em pastagem nativa do Rio Grande do Sul. *Anuário Técnico do IPZFO, Porto Alegre* 3, 454–515.

Gavillon, O. (1963) Levantamento da composição mineral das pastagens do Rio Grande do Sul. II – Os minerais maiores e a composição imediata. In: Associação Gabrielense de Melhoramento e Renovação de Pastagens (ed.) *Anuário.* AGMRP, São Gabriel, pp. 58–64.

Gavillon, O. and Quadros, A.T. (1965) Levantamento da composição mineral das pastagens nativas do Rio Grande do Sul: o cobre, o cobalto e o molibdênio. In: *Proceedings of IXth International Grassland Congress, São Paulo, Brazil*, pp. 709–712.

Girardi-Deiro, A.M. and Gonçalves, J.O.N. (1987) Estrutura da vegetação de um campo natural submetido a três cargas animais na região sudoeste do Rio Grande Do Sul. In: EMBRAPA–CNPO (ed.) *Coletânea das pesquisas: forrageiras*, Vol. 1. CNPO, Bagé, pp. 33–62.

Gomes, K.E. (1996) Dinâmica e produtividade de uma pastagem natural do Rio Grande do Sul após seis anos de aplicação de adubos, diferimentos e níveis de oferta de forragem. Doctoral thesis, Universidade Federal do Rio Grande do Sul, Porto Alegre, Rio Grande do Sul, Brazil, 223 pp.

Gomes, K.E., Quadros, F.L.P., Vidor, M.A., Dall'Agnol, M. and Ribeiro, A.M.L. (1989) Zoneamento das pastagens naturais do Planalto Catarinense. In: *Proceedings of XI Reunião do Grupo Técnico Regional do Cone Sul em Melhoramento e Utilização dos Recursos Forrageiros das Áreas Tropical e Subtropical.* EMPASC, Lages, Santa Catarina, pp. 304–314.

Gonçalves, J.O.N. (1990a) Ecossistema da zona temperada quente–Estado do Rio Grande do Sul. In: Puigneau, J.P. (ed.) *Introduccion, Conservacion y Evaluacion de Germoplasma Forrajero en el Cono Sul: primer taller de trabajo dela red de forrageras del Cono Sur.* IICA–PROCISUR, Montevideo, Uruguay, pp. 183–186.

Gonçalves, J.O.N. (1990b) Informações básicas sobre solos, clima, vegetação, áreas agroecológicas homogêneas e centros de pesquisa, na região sul do Brasil. In:

Puigneau, J.P. (ed.) *Introduccion, Conservacion y Evaluacion de Germoplasma Forrajero en el Cono Sul: primer taller de trabajo dela red de forrageras del Cono Sur*. IICA–PROCISUR, Montevideo, Uruguay, pp. 187–198.

Grossman, J. (1963) Pastagens e seu manejo. In: Associação Gabrielense de Melhoramento e Renovação de Pastagens (ed.) *Anuário*. AGMRP, São Gabriel, pp. 9–16.

Grossman, J. and Mohrdieck, K.H. (1956) Experimentação forrageira do Rio Grande do Sul. In: Rio Grande do Sul, Secretaria da Agricultura, Diretoria da Produçào Animal (ed.) *Histórico da Diretoria da Produção Animal*. Secretaria da Agricultura, Porto Alegre, pp. 115–122.

IBGE (1999) Censo Agropecuário do Brasil. http:www.sidra.ibge.gov.br/sidra/agro/agro/htm

Lindman, C.A.I. (1974) *A vegetação do Rio Grande do Sul*. Livraria Universal, Porto Alegre, 358 pp.

Machado, M.L. da S. and Machado, N.M. (1982) Forrageiras temperadas perenes. In: Fundação Instituto Agronômico do Paraná (ed.) *Forrageiras para o primeiro planalto do Paraná*. Circular 26, IAPAR, Londrina, pp. 11–14.

Maraschin, G.E. and Jacques, A.V.A. (1993) Grassland opportunities in the subtropical region of South America. In: *Proceedings of XVIIth International Grassland Congress, Palmerston North, New Zealand*, SIR Publishing, Wellington, New Zealand, pp. 1977–1981.

Maraschin, G.E., Moraes, A., da Silva, L.F.A. and Riboldi, J. (1993) Cultivated pasture, forage on offer and animal response. In: *Proceedings of XVIIth International Grassland Congress, Palmerston North, New Zealand*, SIR Publishing, Wellington, New Zealand, pp. 2014–2015.

Maraschin, G.E., Moojen, E.L., Escosteguy, C.M.D., Correa, F.L., Apezteguia, E.S., Boldrini, I.J. and Riboldi, J. (1997) Native pasture, forage on offer and animal response. In: *Proceedings of XVIIIth International Grassland Congress, Association Management Centre, Calgary, Canada*, Vol. II, Paper 288.

Martinez Crovetto, R. (1965) Estudios ecologicos en los campos del sur de Misiones. I. Efecto del pastoreo sobre la estructura de la vegetación. *Bonplandia* 2, 1–13.

Moojen, E.L. (1992) Dinâmica e potencial produtivo de uma pastagem nativa do Rio Grande do Sul, submetida a pressões de pastejo, épocas de diferimento e níveis de adubação. Doctoral thesis, Universidade Federal do Rio Grande do Sul, Porto Alegre, Rio Grande do Sul, Brazil, 172 pp.

Moraes, A. (1994) Culturas forrageiras de inverno. In: *Proceedings of Simpósio Brasileiro de Forrageiras e Pastagens*. CBNA, Campinas, São Paulo, pp. 67–78.

Moraes, A., Moojen, E.L. and Maraschin, G.E. (1990) Comparação de métodos de estimativa de taxas de crescimento em uma pastagem submetida a diferentes pressões de pastejo. In: *Proceedings of XXVII Reunião Anual da Sociedade Brasileira de Zootecnia*. Sociedade Brasileira. Zootecnia/FEALQ, Campinas, p. 332.

Morais, N.A. (1988) A pecuária e a pecuarização no estado do Paraná. MSc thesis, Universidade Federal de Pernambuco, Recife, Pernambuco, Brazil, 631 pp.

Morales, A.A.S. (1998) Desenvolvimento inicial de *Lotus corniculatus* L. cv. São Gabriel sob efeio de restrições hídricas e luminosas. MSc thesis, Universidade Federal do Rio Grande do Sul, Porto Alegre, Rio Grande do Sul, Brazil, 74 pp.

Morales, A., Rosa Nabinger, L.M. and Maraschin, G.E. (1997) Efeito da limitação hídrica sobre a morfogênese e repartição da biomassa de *Lotus corniculatus* L. cv. São Gabriel. In: *Proceedings of XXXIV Reunião da Sociedade Brasileira de Zootecnia*. Sociedade. Brasileira, Vol. 2. Zootecnia, Juiz de Fora, Minas Gerais, pp. 124–126.

Müller, L. and Primo, A.T. (1969) Efeito da suplementação com pastagem cultivada de inverno, durante os períodos da 'desmama' e 'sobre-ano', em novilhos de corte. In: *Relatório técnico: Estação Experimental de São Gabriel.* Secretaria da Agricultura, DPA Estação Experimental de São Gabriel, São Gabriel, pp. 75–78.

Nabinger, C. (1980) Técnicas de melhoramento de pastagens naturais no Rio Grande do Sul. In: *Proceedings of Seminário Sobre Pastagens–'de que Pastagens Necessitamos'*. FARSUL, Porto Alegre, pp. 28–58.

Nabinger, C. (1998) Princípios de manejo e produtividade de pastagens. In: Gottschall, C.S., da Silva, J.L.S. and Rodrigues, N.C. (eds) *Proceedings of the III Ciclo de Palestras em Produção e Manejo de Bovinos de Corte. Êqnfase: Manejo e Utilização Sustentável de Pastagens.* ULBRA, Canoas, Rio Grande do Sul, pp. 54–107.

Nascimento, J.A.L, Freitas, E.A.G. and Duarte, C.M.L. (1990) *A grama missioneira no planalto catarinense.* EMPASC, Florianópolis, 65 pp.

Paim, N.R. (1983) Pastagens nativas da Região Sul do Brasil. In: *Proceedings of 1st Congresso Brasileiro de Forrageiras e Pastagens Nativas.* Sociedade Brasileira. Zootecnia, Olinda, Pernambuco, pp. 1–27.

Perez, N.B. (1999) Métodos de estabelecimento do amendoim forrageiro perene (*Arachis pintoi* Krapovickas & Gregory) (leguminosae). MSc thesis, Universidade Federal do Rio Grande do Sul, Porto Alegre, Rio Grande do Sul, Brazil, 83 pp.

Perin, R. (1990) Rendimento de forragem de uma pastagem nativa melhorada sob pastejo contínuo e rotativo. MSc thesis, Universidade Federal do Rio Grande do Sul, Porto Alegre, Rio Grande do Sul, Brazil, 130 pp.

Pott, A. (1974) Levantamento ecológico da vegetação de um campo natural sob três condições: pastejado, excluído e melhorado. MSc thesis, Universidade Federal do Rio Grande do Sul, Porto Alegre, Rio Grande do Sul, Brazil, 223 pp.

Pott, A. (1989) O papel da pastagem na modificação da vegetação clímax. In: *Proceedings of the Simpósio sobre Ecossistema de Pastagens.* FUNEP, Jaboticabal, pp. 43–68.

Prestes, P.J.Q., Freitas, E.G. and Barreto, I.L. (1976) Hábito vegetativo e variação estacional do valor nutritivo das principais gramíneas da pastagem nativa do Rio Grande do Sul. Anuário Técnico do Instituto de Pesquisas Zootécnicas Francisco Osório. IPZFO, Porto Alegre, vol. 3, pp. 516–531.

Rosito, J.M. and Maraschin, G.E. (1984) Efeito de sistemas de manejo sobre a flora de uma pastagem. *Pesquisa Agropecuária Brasileira* 19, 3311–3316.

Saibro, J.C. and Silva, J.L.S. (1999) Integração sustentável do sistema arroz×pastagens utilizando misturas forrageiras de estação fria no litoral norte do Rio Grande do Sul. In: Gottschall, C.S., da Silva, J.L.S. and Rodrigues, N.C. (eds) *Proceedings of IV Ciclo de Palestras em Produção e Manejo de Bovinos de Corte.* Ulbra, Canoas, pp. 27–55.

Salerno, A.R. and Vetterle, C.P. (1984) Avaliação de forrageiras de inverno no Baixo Vale do Itajaí, Santa Catarina. *Comunicado Técnico 76*, EMPASC, Florianópolis, 26 pp.

Scheffer-Basso, S. (1999) Caracterização morfofisiológica e fixação biológica de nitrogênio de espécies de *Adesmia* DC. e *Lotus* L. Doctoral thesis, Universidade Federal do Rio Grande do Sul, Porto Alegre, Rio Grande do Sul, Brazil. 247 pp.

Schreiner, H.G., Andriguetto, J.M. and Minardi, I. (1980) Características agrostológicas dos campos naturais do Paraná. Etapa I. Áreas não pastejadas. *Revista do Setor de Ciências Agrárias da Universidade Federal do Paraná*, Curitiba, 2, pp. 105–111.

Setelich, E.S.A., Almeida, E.X. and Maraschin, G.E. (1998a) Adubação nitrogenada e variáveis morfogênicas em Capim elefante anão cv. Mott, sob pastejo. In: *Proceedings of XXXV Reunião Anual da Sociedade Brasileira de Zootecnia*, Vol. II. Sociedade Brasileira Zootecnia, Botucatu, São Paulo, pp. 152–154.

Setelich, E.S.A., Almeida, E.X. and Maraschin, G.E. (1998b) Resposta à adubação nitrogenada de Capim elefante anão cv. Mott, sob pastejo. In: *Proceedings of XXXV Reunião Anual da Sociedade Brasileira de Zootecnia*, Vol. II. Sociedade Brasileira Zootecnia, Botucatu, São Paulo, pp. 155–157.

Setelich, E.S.A. (1994) Potencial produtivo de uma pastagem natural do Rio Grande do Sul, submetida a distintas ofertas de forragem. MSc thesis, Universidade Federal do Rio Grande do Sul, Porto Alegre, Brazil, 169 pp.

Silva, J.L.S. and Jacques, A.V.A. (1998) Disponibilidade estacional de matéria seca do campo nativo melhorado com sobressemeadura de leguminosas perenes de verão. In: *Proceedings of XII Reunião do Grupo Técnico Regional do Cone Sul (Zona Campos) em Melhoramento e Utilização de Recursos Forrageiros das Áreas Tropical e Subtropical.* EMBRAPA/CPPSul, Bagé, Rio Grande do Sul, p. 124.

Silva, N.C., Franke, L.B. and Nabinger, C. (1998) Morfogênese de *Desmodium incanum* DC. em resposta à disponibilidade de fósforo. In: *Proceedings of XVII Reunião do Grupo Técnico em Forrageiras do Cone-Sul–Zona Campos*. Epagri/UDESC/AEASC/PMLages, Lages, Santa Catarina, p. 156.

Soares, G.C., Dall'agnol, M., Nabinger, C., Costa, J.Q.F.F., Tonelotto, L.A. and Rosa, L.M.G. (1998) Estudo da morfogênese em uma população de *Bromus auleticus* Trin. In: *Proceedings of XVII Reunião do Grupo Técnico em Forrageiras do Cone-Sul–Zona Campos*. Epagri/UDESC/AEASC/PMLages, Lages, Santa Catarina, p. 101.

Souza, A.G.(1989) Evolução da produção animal da pastagem nativa sob pastejo contínuo e rotativo. MSc thesis, Universidade Federal do Rio Grande do Sul, Porto Alegre, Rio Grande do Sul, Brazil, 160 pp.

Spannenberg, P.R.O., Nabinger, C., Costa, J.A.A. and Rosa, L.M. (1997a) Determinação do coeficiente de extinção (K) e ajuste de modelo de evolução de IAF em *Desmodium incanum* (S.W.) D.C. In: *Proceedings of Xth Congresso Brasileiro de Agrometeorologia.* Sociedade Brasileira Agrometeorologia, Piracicaba, São Paulo, pp. 149–151.

Spannenberg, P.R.O., Nabinger, C., Costa, J.A.A., Jacques, A.V.A. and Rosa, L.M. (1997b) Modelagem do crescimento de *Desmodium incanum* (S.W.) D.C. In: *Proceedings of Xth Congresso Brasileiro de Agrometeorologia.* Sociedade Brasileira Agrometeorologia, Piracicaba, São Paulo, pp. 152–154.

Tothill, J.C., Dzowela, B.H. and Diallo, A.K. (1989) Present and future role of grassland in inter-tropical countries with special reference to ecological and sociological constraints. In: *Proceedings of XVIth International Grassland Congress.* AFPF, Versailles, France, pp. 1719–1724.

Valls, J.F.M. (1973) As entidades taxonômicas da série *Axonopus* do gênero *Axonopus* Beauv. no Rio Grande do Sul. Doctoral thesis, Universidade Federal do Rio Grande do Sul, Porto Alegre, Rio Grande do Sul, Brazil.

Valls, J.F.M. (1986) Principais gramíneas forrageiras nativas das diferentes regiões do Brasil. In: *Proceedings of 3rd Simpósio sobre Produção Animal.* Fundação Cargill, Campinas, p. 130.

Vidor, M. and Jacques, A.V.A. (1998) Comportamento de uma pastagem sobressemeada com leguminosas de estação fria e avaliada sob condições de corte e pastejo. 1. Disponibilidade de matéria seca, matéria orgânica digestível e proteína bruta. *Revista Brasileira de Zootecnia* 27, 267–271.

Viegas, J. (1998) Análise do desenvolvimento foliar e ajuste de um modelo de previsão do rendimento potencial de azevém anual (*Lolium multiflorum* Lam.). Doctoral thesis, Universidade Federal do Rio Grande do Sul, Porto Alegre, Rio Grande do Sul, Brazil, 166 pp.

Vieira, G.V.N. (1965) *Ovinocultura no Rio Grande do Sul.* Secretaria da Agricultura do Rio Grande do Sul, Porto Alegre, 42 pp.

Campos in Uruguay

19

E.J. Berretta, D.F. Risso, F. Montossi and G. Pigurina

Instituto Nacional de Investigación Agropecuaria (INIA), Estación Experimental del Norte, Ruta 5, km 386, 45000 Tacuarembó, Uruguay

Introduction

Beef and sheep meat, wool and dairy productions are the most important sources of goods in the country. Animal production systems are based on all-year-round grazing, mostly mixed (cattle and sheep together) and generally dependent on natural grassland, corresponding to 83% of the total 16,000,000 ha available for production.

Traditionally, animal husbandry practices based on native pastures have shown comparative advantages, but, in order to improve the economic profits of farmers, it has become increasingly necessary to improve such productive levels. The complementary use of such *campos*, with variable proportions of improved *campos* or highly productive cultivated pastures, makes it possible to increase farm productivity.

The general productive conditions of the country and the description of the main pasture and animal research proposals according to regions are discussed in this chapter.

Productive Conditions

Uruguay is located between 30 and 35° latitude south, with a temperate to subtropical climate and wide seasonal variations.

Mean annual temperature ranges from 16°C in the south-east to 19°C in the north. For the warmest month, January, temperature varies between 22 and 27°C for the same regions, while, for the coldest month, July, the variation is between 11 and 14°C. Frost number varies among years, with an average of 30, and a frost-free period of 240 days.

 Grassland Ecophysiology and Grazing Ecology (eds G. Lemaire, J. Hodgson, A. de Moraes, C. Nabinger and P.C. de F. Carvalho)

Average annual rainfall is 1000 mm in the south, increasing to 1300 mm in the north. Rainfall is evenly distributed, the wettest seasons are autumn and spring. In spite of this, rainfall is highly variable, resulting in the occurrence of either periods of drought or excess rain (flood) at any time throughout the year.

Evapotranspiration is fairly constant over years, varying from 800 mm in the south to 900 mm in the north, in both cases being higher during summer, resulting in soil water deficits. Consequently, the Uruguayan climate can be considered as subhumid (Corsi, 1978). The gently rolling topography is seldom higher than 300 m, with a maximum of 503 m, so altitude is not an important factor in the climatic characterization. According to such climatic conditions, the growth rate of both native and cultivated pastures is mainly affected by water deficits and, to a lesser extent, by temperature.

Although it is a small country, Uruguay has a wide variety of soil types. The most important groups are presented in Fig. 19.1. Group I is mainly composed of shallow soils (< 20 cm), associated with deeper and heavier soils with good fertility. The soils developed from basaltic and granitic layers are devoted to extensive cattle and sheep grazing. Predominant in group II are poorly drained deep soils, associated with solonetz. Rice production and extensive mixed grazing are the main activities. Group III is predominantly composed of deep heavy soils that are highly leached (medium to low fertility), associated with a lower proportion of shallow soils, mainly developed from granitic materials. Cattle and sheep farming is the most important activity, while extensive cropping and dairy production are less important. In group IV, dominant soils are deep (> 1.5 m), light (mostly sandy) and with low fertility potential. Forestry production and extensive mixed grazing are the main activities. Finally, group V comprises the soils with the highest potential in the country. They are deep (> 1 m) and heavy-textured, showing a moderate to slow permeability and high fertility. They are located in the west part of the country, stretching from south to north. Most of the intensive cropping takes place in this region, i.e. wheat, barley, oats, maize, sorghum, etc., alternating with intensive cattle and sheep fattening on cultivated pastures; also, the highest proportion of the dairy industry is developed on such soils. At present, in the north, rice cropping has become an important activity.

The dominant native vegetation of Uruguay (*campos*) is composed of numerous grass species, both C_4 and C_3, forbs and shrubs, the frequency of trees being very low (Berretta and do Nascimento, 1991). This vegetation varies according to soil type and, to a lesser extent, with climate.

Extensive Animal Production Systems

Native pastures

Prior to the introduction of farm animals (cattle and horses) in 1611, only small herbivores were present and the vegetation had a greater proportion of shrubs (mainly of the genus *Baccharis*) and tall bunch grasses, except in shallow stony

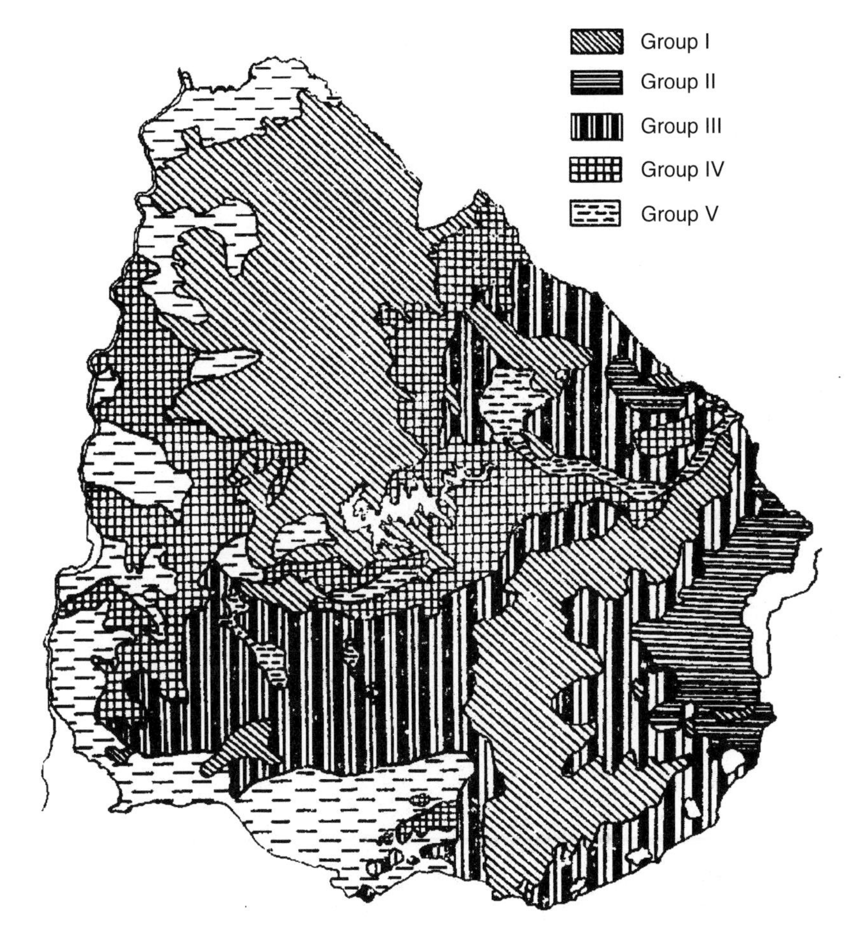

Fig. 19.1. Most important soil groups in Uruguay (CIDE, 1962).

soils. Since then, grazing has induced changes in the vegetative types, being the main factor in maintaining *campos* in a herbaceous pseudoclimax phase (Veira da Silva, 1979).

Studies in different communities and seasons show a predominance of summer-growing species (C_4) (Berretta, 1990, 1991, 1996; Formoso, 1990; Olmos, 1992). Even though, during autumn and winter, the relative frequency of cold-season species (C_3) increases, it is still lower than that of the summer species. Continuous grazing is responsible for the decrease in winter species by preventing their flowering and seed dispersal. Approximately 50% of winter species are herbs and coarse grasses, while fine ones are scarce. The predominance of warm-season species explains the higher forage production during spring and summer. In all vegetation types, the frequency of native legumes is very low.

Different soil types determine the composition of the plant communities, influencing the proportion of productive types (Rosengurtt, 1979) and varying their botanical composition, density and productivity, so annual forage yield has a wide range. In shallow or low fertility soils production is close to 2500 kg dry matter (DM) ha^{-1} and in intermediate soils it reaches 3500 kg DM ha^{-1}, while in the deep, highly fertile soils yield is greater than 5000 kg DM ha^{-1} (Formoso, 1990; Berretta and Bemhaja, 1991). For most of the native communities, such annual production concentrates mainly on spring and summer (70–85% on light soils, reaching 60–70% in vegetation with autumn growth). Winter forage production varies between 6–7% on sandy soils (Bemhaja, 1991) and 10–15% in other *campos*. On light, low-fertility soils, there is a dominance of coarse and tough species, with high annual yield but low quality and palatability; in better, fertile, structured soils, there is a higher frequency of good-quality and palatable grasses.

Forage production is mainly explained by rainfall, which shows high variability among years and seasons. As an example, a 15-year variation in forage yield for three predominant types of basaltic soils is shown in Fig. 19.2.

Crude protein (CP) content of the native vegetation varies between 6 and 15%, according to season, botanical composition and the amount of dead material. Maximum values of CP occur during winter and early spring, while the minimum occurs in summer, with cold-season grasses having a higher level of CP than warm-season grasses (Berretta, 1996). These sources of variation agree with the

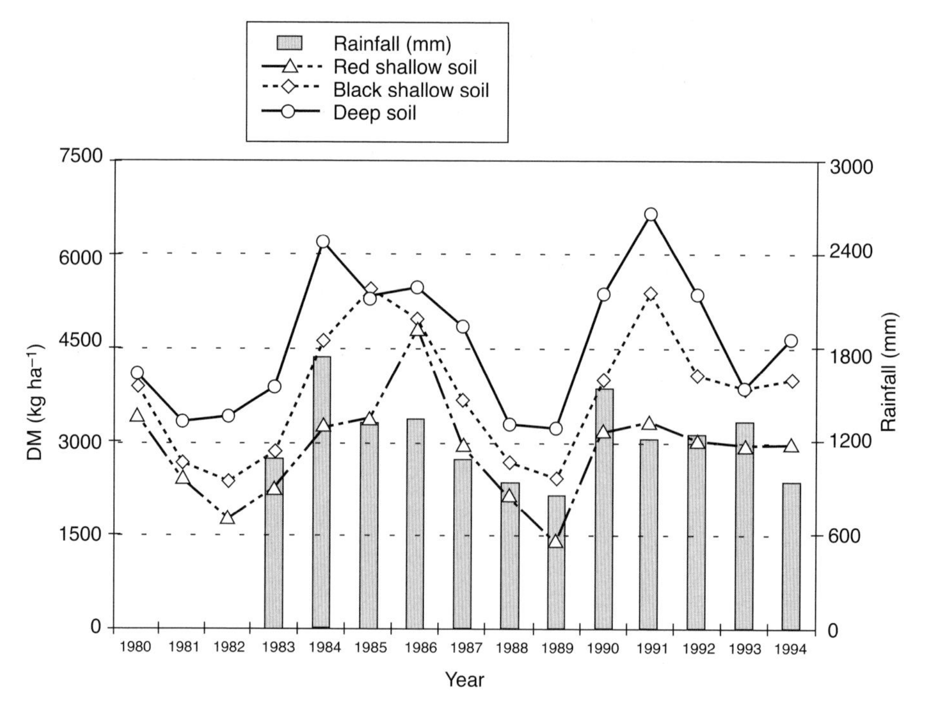

Fig. 19.2. Fifteen-years variation in forage yield for three predominant types of basaltic soils.

results obtained by Montossi *et al.* (1998c), using oesophageal fistulates. The phosphorus content of most native pastures in Uruguay is generally lower than average animal requirements. Annual average organic matter digestibility is about 58%, with fluctuations according to season and vegetation type.

The main problem of native pastures is the risk of degradation and loss of species, related to continuous stocking, high stocking rate and high sheep/cattle ratio. Degradation signs are the increase of small herbs and stoloniferous grasses, adapted to such grazing conditions, and reduced frequency of bunch grasses, as well as a reduction in species number. Such changes in botanical composition result in a 12% reduction in annual forage production, which is seldom noticeable in short periods of time (Berretta, 1998).

Beef cattle production

According to the *campos* characteristics described above, associated with mixed continuous grazing, cattle production has several limitations, mainly derived from nutritional constraints. Some of the more important are: advanced heifer mating age (3 years average); low calving rate (65%); low calf live-weight gain, with a consequently low weaning weight (130–140 kg); advanced slaughter age (4 years); and low extraction rates (18–20%). In such conditions, beef meat production in traditional extensive systems is about 65 kg ha^{-1} year^{-1}.

An example of this is shown in Table 19.1 in relation to two management systems, with or without sheep, in the basaltic region. Both systems were evaluated under continuous grazing with a stocking rate of one cow equivalent ha^{-1} during 4 years. In spite of the high stocking rate and simple management practices

Table 19.1. Reproductive performance and productivity of two management systems (from Pigurina *et al.*, 1998a).

Year	Weaning (%)	Weaning weight (kg)	Productivity (kg ha^{-1})	
			Live-weight	Wool
Cattle system				
1	77.5	141	109	–
2	55.5	141	78	–
3	75.0	137	103	–
4	70.0	143	100	–
Mean	69.5	141	98	–
Mixed system				
1	70.0	153	107	10.1
2	50.5	143	72	9.0
3	75.0	166	125	10.3
4	60.0	160	96	9.8
Mean	64.0	156	100	9.8

Table 19.2. Live-weight gain (kg animal^{-1} day^{-1}) and total productivity (kg ha^{-1}) of steers grazing *campos* in different conditions (from Pigurina *et al.*, 1998b).

Stocking rate (AU ha^{-1}):	0.6	0.8	0.8	1.06
Sheep : cattle ratio:	2 : 1	2 : 1	5 : 1	2 : 1
Season	Live-weight variation (kg day^{-1})			
Autumn	0.196	0.194	0.139	−0.076
Winter	0.089	−0.176	−0.086	−0.312
Spring	0.915	0.858	0.828	0.667
Summer	0.351	0.413	0.297	0.413
Annual mean	0.388	0.322	0.295	0.178
Beef meat production (kg animal^{-1} day^{-1})	141	118	108	65
Productivity (kg ha^{-1})	75	84	54	62

applied, results show a better animal performance than those obtained in the extensive livestock systems.

Annual differences in the rates of calving and weaning are the main factors that determine animal productivity. Weaning rates were higher in the cattle-only system, while in the mixed system the calf weaning weight and total productivity were greater.

In relation to the problems of the fattening process on *campos*, live-weight gain of steers was highly variable among seasons, according to forage availability and quality, stocking rate and presence of sheep (Table 19.2).

In order to achieve a reasonable steer performance, it is necessary to use stocking rates not higher than 0.8 animal unit (AU) ha^{-1}. Mixed grazing at low cattle/sheep ratios does not appear to be detrimental to steer live-weight gain. Deferred grazing results in an improved animal performance through a higher pasture production and better forage availability, particularly during critical periods.

Sheep production

The recent sheep research programmes developed by the Instituto Nacional Investigación Agropecuaria (INIA), covering the main extensive livestock regions of the country, have been concentrated on the evaluation of the effects of different sheep feeding, management and health control strategies on the productivity of the flock, considering the main sheep breeds used in Uruguay. The research programme has been orientated to the improvement of wool and sheep meat production efficiency, taking into consideration the influence on the quality of both products. The strategies were proved on *campos*, improved *campos* and cultivated pastures, developed on medium to deep soils. Recently, the use of supplementary feeds (concentrates and hay) has been incorporated.

Table 19.3. Summary of the productive results obtained in experiments conducted by Montossi *et al.* (1998a), using ewe CS and an autumn deferred grazing system for *campos* and improved *campos* in late pregnancy.

Pasture, ewe and lamb characteristics	Traditional system	Deferred *campos*	Deferred improved *campos*
Sward mass at lambing (kg DM ha^{-1})	400–700	1300–1500	1100*–1900
Sward height at lambing (cm)	2–3	5–8	4–7
Stocking rate (ewes ha^{-1})	4 (0.8 AU ha^{-1})	5 (1 AU ha^{-1})	10 (2 AU ha^{-1})
Ewe live-weight at lambing (kg)	35–40	42–45	45–48
Ewe CS at lambing (grades)	2–2.5	3–3.5	3.3–3.7
Lamb live-weight (kg)	2.5–3	3.6–3.8	3.8–4.6
Lamb mortality rate (%)	20–30	10–13	9–10

*Sward mass needed according to the level of legumes present in the improved *campos.*

The multidisciplinary approach of the research work applied has generated an important stock of technological options adapted to the different conditions of the production systems of the extensive livestock regions.

From the results presented in Table 19.3, it can be shown that the feeding levels of the breeding ewes in the traditional systems are deficient during late pregnancy, resulting in ewes with low body weight and condition score (CS) at lambing. This has a negative effect on lamb survival (20 to 30% of mortality), being the major cause of the low reproductive performance achieved in the national breeding flock.

In order to improve ewe reproductive performance and reduce lamb mortality to 10% with single-pregnant ewes, it is necessary to use a deferred grazing system, accumulating sward masses in the range of 1300 to 1500 kg DM ha^{-1} (5–8 cm) at the beginning of the last 5–6 weeks of pregnancy. Also, CS at lambing has to be in the range of 3 to 3.5 grades for Corriedale ewes (Montossi *et al.*, 1998a). In the case of improved *campos* at double the stocking rate (ten ewes ha^{-1}) of unimproved *campos*, with the same CS at lambing, it is possible to achieve the goal of losing just 10% of the lambs born. The deferred sward masses or sward heights recommended are in the range of 1900–1100 kg DM ha^{-1} (7–4 cm). The values of sward masses or heights needed will depend on the proportion of legumes present in the improved *campos*.

Considering the average autumn pasture growth rates of the *campos* and improved *campos* of the basaltic region (Berretta and Bemhaja, 1998) and the normal lambing season, it will be necessary to start the deferring period between 70 and 50 days and between 40 and 30 days before lambing for *campos* and improved *campos*, respectively. These figures will depend on the prevalent climatic conditions of each year and their effects on pasture growth and the sward mass present in the pasture at the start of the period of accumulation.

In Uruguay, the majority of the hoggets are mated at 2.5 years (four teeth), because a high proportion of these animals (40–60%) do not reach the minimum weight required for mating at 1.5 years. This fact has clear economic and productive consequences for the industry, reducing the number of lambs produced by the ewe in her lifetime, the genetic improvement of the flock and the level of efficiency of the resources used by the farmer. In the context of positive market signals for sheep meat, it will be very important to increase hogget reproductive performance.

Several winter feeding and management strategies have been defined to improve hogget live-weight gain in the extensive livestock enterprises of the basaltic region (San Julián *et al.*, 1998). The use of improved *campos* and annual-grazing forage crops allowed the achievement of adequate winter live-weight gains (60–90 g animal^{-1} day^{-1}). Those rates resulted in high percentages (80–90%) of hoggets reaching the first mating live-weight recommended at two teeth (greater than 32 and 35 kg for merinos and Corriedales respectively), resulting in acceptable reproductive performances. In order to achieve recommended hogget winter live-weight gains, it is necessary to manage minimum sward masses or sward heights of 1500 (5 cm) and 1000 (3.5 cm) kg DM ha^{-1} for *campos* and improved *campos*, respectively (San Julián *et al.*, 1998).

The beginning of the present decade has been characterized by low wool prices, resulting in an important reduction of the Uruguayan sheep population. Under these circumstances, it has been necessary to look for new alternatives for the sheep industry. Covering the main five sheep production regions of the country, several macroresearch programmes are being conducted by INIA on fine and ultrafine merino (Montossi *et al.*, 1998d), sheep milking (Ganzábal, 1996) and lamb meat (Montossi *et al.*, 1997). In relation to the positive market signals for lamb meat prizes, the farmers' demands for technological information associated with the factors affecting lamb production and meat quality increased. The technologies proposed by INIA to increase lamb production and quality were based on the use of cultivated pastures, annual forage and improved *campos*, taking into consideration the agroecological particularities of each region of the country.

A summary of some of the experimental work carried out by INIA in lamb fattening is presented in Table 19.4 (Montossi *et al.*, 1998b; San Julián *et al.*, 1998; Scaglia *et al.*, 1998).

The experimental results obtained in the different regions (Table 19.4) demonstrate an important production potential for high-quality lamb meat, achieving the requirements of the European market. In order to reach this goal, it is necessary to use pastures with high production capability and nutritive value, high stocking rates, proper animal health status and controlled rotational grazing systems. The small and medium farmers are potential users of this technology, resulting in positive economic returns of the investments made in the application of the fattening systems proposed by INIA (Ferreira and Pittaluga, 1998).

Native pastures improvement

Since the mid-1960s, efforts have been made to overcome the problem of the low winter forage production and the medium to low quality of native pastures. The

Table 19.4. Summary of the level of lamb production and quality achieved in the different experiments conducted by INIA in the main sheep production regions of Uruguay.

Type of soil	Type of pasture	Stocking rate (lambs ha^{-1})	Final LW (kg)	Production (kg LW ha^{-1})	Carcass weight (kg)	Fat cover (GR) (mm)
Basaltic soils	Annual forage crops	20–35	42–34	414–670	14–21	7–11
Basaltic soils	Cultivated pasture	15–25	43–45	392–446	19–20	15–16
Sandy soils	Annual forage crops	20–30	44–39	553–709	16–18	7–11
Granitic soils	Improved *campos*	8–10	43–40	351–432	18–19	11–13

LW, live-weight.
GR, the total tissue thickness between the surface of a lamb carcass and the rib at a point 11 cm from the midline, in the region of the 12th rib.

Table 19.5. Changes in vegetation attributes due to N + P fertilization policy (from Berretta *et al.*, 1998).

	Campos	*Campos* + NP
Autumn growth rate (kg ha^{-1} day^{-1})	11.1	16.9
Winter growth rate (kg ha^{-1} day^{-1})	4.8	9.1
Annual total production (kg ha^{-1})	4320	6420
N content in forage (%)	1.73	2.15
P content in forage (%)	1.62	2.18
Winter grasses frequency (%)	29	39

use of fertilizer, particularly P, and the introduction of legumes have been the main technology alternatives considered to make a direct contribution to increasing forage production and nutritive value. In addition, legumes introduce N into the ecosystem, diminishing costs. Moreover, P fertilization and legume introduction prevent *campos* degradation and improve in a sustainable manner this existing natural resource without destroying the canopy. This also allows strategically developed improved areas to be utilized by efficient ruminant classes, enhancing global productivity at the farm level.

In contrast to cultivated pastures that are introduced to restore soil properties and to make the cropping phase of the intensive pasture sustainable – cropping rotational systems after removing the native vegetation – the *campos* improvement aims to capitalize upon and improve the good soil and plant conditions already existing. In vegetation presenting winter perennial species, the combined application of N and P increases the frequency and vigour of these species and additionally extends the length of the growing season of C_4 plants. This is the case of the typical vegetation of basaltic soils. Table 19.5 shows experimental information from a study being conducted by INIA, with an annual split-fertilization policy (early autumn and late winter), where fertilizer doses are 92 kg N ha^{-1} and 44 kg P_2O_5 ha^{-1}. These favourable changes were obtained under rotational grazing by steers (14 + 42 days). At the same stocking rate (0.9 AU ha^{-1}), the daily gain was 0.297 g $animal^{-1}$ and 0.454 g $animal^{-1}$ for *campos* and *campos* + NP, respectively.

As mentioned above, the most generalized technology alternative to improve the production and nutritive value of *campos* is legume introduction with P fertilization. The response to this technology is variable, according to region, soil and vegetation types. In granitic soils (16% of the country), forage production resulting from the overseeding of adapted legumes is more than a 100% higher than the average production of 3400 kg DM ha^{-1} of the predominant *campos* of the region (Risso and Scavino, 1978; Risso, 1991). In the case of the basaltic region (21% of the country), the magnitude of the response is lower (60–70%) but starting from higher levels of productivity – 4500 kg DM ha^{-1} on medium and deep soils (45% of the basaltic area) (Risso *et al.*, 1997).

The legume species for such pasture improvements are white clover (*Trifolium repens* cv. Zapicán) and bird's-foot trefoil (*Lotus corniculatus* cv. San Gabriel and

Table 19.6. Animal output of improved *campos*, under controlled grazing of steers in different regions of Uruguay (adapted from: Ayala and Carámbula, 1996; Bemhaja *et al.*, 1996; Risso and Berretta, 1996).

Soil types	Stocking rate (AU ha^{-1})	Live-weight gain (kg ha^{-1})	Productivity (kg LW ha^{-1})
Medium granitic	1.55	533	406
Medium–deep basaltic	1.85	680	485
Lixiviated, low fertility	1.53*	700	473

*Includes mixed grazing with wethers, 2 : 1 ratio.
LW, live-weight.

Ganador). For the granitic soils, another annual bird's-foot trefoil species, *Lotus subbiflorus* cv. El Rincón, has proved to be highly adaptable, productive and persistent, even though grazing management is practised in extensive conditions (Risso and Berretta, 1996). Additionally, INIA has been concerned to develop other plants for light acid soils (sandy), as is the case of *Ornithopus compressus* cv. INIA Encantada and *Triticale* cv. INIA Caracé. For lixiviated, wet soils, INIA has made available, after evaluating its good performance, *Lotus pedunculatus* cv. Maku. In all cases, in spite of differences in efficiency of P use, it is necessary to maintain a regular policy of periodic applications of P fertilizer. Another important management practice, regardless of species being perennial or annual, is to prevent grazing from late spring to early summer, to allow flowering and seed-setting, in order to ensure seed availability in the following autumn (Berretta and Risso, 1995).

Even though this technology requires low inputs, it is environmentally friendly, promoting the sustainable development of the native vegetation and placing it in a higher productive level. Animal output (individual performance and carrying capacity) of this improved *campos* is higher, when utilized to accelerate the fattening process in different regions of the country, as shown in Table 19.6. These results were obtained under controlled grazing (five to eight paddocks, with a grazing period of 7–12 days and 30–40 days of resting), covering a grazing season of approximately 300 days.

The complementation of an area of *campos* with another improved area may constitute an intensive unit in an extensive livestock production system. As an example of this, for the basaltic region, INIA has been evaluating for more than 5 years a system comprising 50% of each type of pasture. A group of weaned calves (Hereford and zebu crosses) is introduced every autumn, remaining for 2 years, when they are sold for slaughter at 2.5 years old. The unit is divided into five paddocks, the improved area being concentrated in three of them. Year-round stocking rate averages 1.2 AU ha^{-1}, constituted by equal numbers of calves and steers. In general, both classes are managed together, except in some critical periods, when steers (closer to slaughter) are favoured.

In this framework, the average results for the first 4 years indicate that it is possible to obtain daily gains of 445 g $head^{-1}$ (with an advantage for the crossbred) for the group of animals included. Consequently, total average animal output was 265 kg live-weight (LW) ha^{-1} $year^{-1}$ for the same period (Risso *et al.*, 1998).

Intensive Animal Production Systems

Under Uruguayan conditions, the most intensive beef and sheep production systems take place in the western region, based on the highly fertile soils. Due to such high production potential, this area has a long cropping tradition. The continued and intensive cultivation resulted in the substitution of the best native pasture species with a progressive weed grass invasion (mainly *Cynodon dactylon*) and adversely affected the chemical and physical properties of the prevalent soils. Following research results obtained in the early 1970s, the application of the alternation of crops and cultivated pastures (ley-farming systems) became important. Because the cultivated pastures have such a high productive potential, they have made it possible to overcome the difficulties emerging from the low annual and winter production and forage quality of the degraded native vegetation, so resulting in improvements in the efficiency of the breeding and fattening processes in both cattle and sheep. Since the introduction of cultivated pastures, there has been an intensification of such processes in a context that ensures the bioeconomic sustainability of the predominant production systems.

Traditionally, cultivated pastures may be annual, short rotation or perennial, lasting approximately 4 years. The main species utilized in these pastures are presented in Table 19.7. For each of these species, INIA has developed a cultivar adapted to the ecological conditions and predominant production systems. This breeding strategy is a continued process with periodical releases. In the last 2 years, INIA has offered to the seed market new cultivars of oats, ryegrass, cocksfoot, red clover, bird's-foot trefoil and *Trifolium alexandrinum* as new alternatives for short rotations.

In addition to the efforts in plant improvement, particular emphasis is placed on developing technologies of pasture management and utilization, which ensure

Table 19.7. Improved annual and perennial species utilized in cultivated pastures.

	Annual	Short rotation	Perennial
Oats (*Avena byzantina, Avena sativa*)	+++	–	–
Annual ryegrass (*Lolium multiflorum*)	++	+	–
Forage wheat (*Triticum aestivum*)	++	–	–
Tall fescue (*Festuca arundinacea*)	–	–	+++
Cocksfoot (*Dactylis glomerata*)	–	++	+++
Phalaris (*Phalaris aquatica*)	–	–	++
Yorkshire fog (*Holcus lanatus*)	+	++	–
Red clover (*Trifolium pratense*)	+	+++	+
White clover (*Trifolium repens*)	–	+	+++
Lucerne (*Medicago sativa*)	–	–	+++
Bird's-foot trefoil (*Lotus corniculatus*)	–	+	+++
Chicory (*Cichorium intybus*)	–	+++	–

–, not appropriate; +, recommended.

the expression of potential productivity of all species and cultivars consistent with good persistence and nutritive value.

Beef cattle production

The results obtained with animal utilization studies on these pastures, being either analytical or physical models, show high levels of animal output (Table 19.8). The combination and integration of two or more of these alternatives with *campos*, improved or not, in different proportions, results in a highly productive nutritional basis for the various cattle processes, mainly fattening. In this case, a physical model (cropping and fattening), including weaned calves, at a stocking rate of 1.8 AU ha^{-1}, evaluated for several years at INIA, has resulted in an animal output of 440 kg ha^{-1} $year^{-1}$. In spite of this, some steer performance problems were registered during winter, mainly due to the high stocking used in relation to the low rates of pasture growth observed in that period. Consequently, the implementation of strategic supplementation with concentrates, mainly grain, allowed for higher stocking rates, resulting in better animal performance and higher outputs of approximately 550 kg LW ha^{-1} (Risso *et al.*, 1991).

Sheep production

Uruguay has a prolific history in sheep research, which started at the beginning of the 1960s at INIA in the research station La Estanzuela. Since then, with the additional contributions of the Secretariado Uruguayo de la Lana (SUL) and the Agronomy and Veterinary faculties, there has been important progress in the knowledge of those aspects related to sheep feeding, genetic improvement, management and health for the different main agroecological conditions of Uruguay. At present, there is available a series of technologies generated by the main sheep research bodies of the country, offering alternatives for producing substantial increases in the production levels of the traditional extensive farmers. Table 19.9

Table 19.8. Animal productivity obtained with several cultivated pastures under controlled grazing with fattening steers (from Risso, 1997).

Pasture type	Stocking rate (AU ha^{-1})	Daily gain (g $animal^{-1}$)	Animal output (kg LW ha^{-1})
Fescue/phalaris + white clover/red clover + bird's-foot trefoil (age: 1–4 years)	6.3	815	555
Chicory + holcus + red clover	3.5	540	569
Bird's-foot trefoil + white clover	8.1	844	602
Red clover	7.2	325	190
Forage wheat	4.5	1000	350

Table 19.9. Wool and sheep meat production levels obtained by the different sheep production intensive systems evaluated by SUL and INIA during the 1980s and 1990s.

Production system	Sheep breed	Improved area (%)	Stocking rate (AU ha^{-1})*	Wool (fleece) (kg ha^{-1})	Sheep meat (kg ha^{-1})	Beef meat (kg ha^{-1})	Lamb meat (%)†
TS‡	Corriedale (70%)	0	0.80	5.3	12	44	–
S I (SUL)§	Corriedale	0	0.96	10.9	30.5	38.9	5
S II (SUL)§	Corriedale	15	1.09	15.4	52.6	44.5	15
S III (SUL)§	Corriedale	40	1.46	25.6	93.5	66.8	23
SLI (INIA)‖	Corriedale	90	2.52	45	210	–	24
MAOI (INIA)¶	Polwarth	100	3.6	84	270	–	100

* AU (animal unit) = five breeding ewes or one breeding cow.
† Proportion of lamb meat in the total of sheep meat produced in each production system.
‡ Traditional livestock systems (G. Ferreira, Tacuarembó, 1999, personal communication).
§ Production systems I, II and III of SUL (Oficialdegui and Gaggero, 1990).
‖ Intensive sheep production system of INIA La Estanzuela (Castro and Ganzábal, 1988).
¶ Sheep–cropping intensive production system of INIA La Estanzuela (Ganzábal *et al.*, 1999).

shows the influence on production of the application of the different intensive schemes proposed by the research institutes, which integrate complementary sheep production with other productive alternatives (e.g. crops, beef, etc.). These systems proposed to integrate the knowledge generated by the national investigation with aspects of pasture management and utilization, forage and concentrate supplementation and with those aspects associated with sheep management, feeding and animal health.

The technologies proposed and evaluated during the decades of the 1980s and early 1990s were orientated to wool production, in response to the good wool prices observed during this period. These systems were based on the integration of the following factors:

- Adequate use of improved pastures with high production and nutritive value during the greater part of the year, giving high animal responses in key moments of the productive cycle, resulting in good production coefficients.
- Matching the pasture production curve with animal requirements for the different sheep classes and phases of production (flushing, pregnancy, lactation and fattening).
- Use of conserved forage (principally pasture hay resulting from the surplus of the spring pasture growth), particularly during wintertime or to enhance pasture persistency with adequate management during summer time.
- High stocking rate, adequate paddock numbers and size and intensive use of electric fences to realize proper pasture utilization and management in relation to the different physiological requirements of pasture species.
- Frequent monitoring of forage availability and animal performance.

- Strategic practices of preferential animal feeding with the use of supplements.
- Strict control of sheep health constraints (principally foot-rot and internal parasites).

With the historic collapse of wool prices, the main focus of the sheep research projects (Ganzábal *et al.*, 1999) was concentrated on proposals to increase sheep meat production, with particular emphasis on the production of high-quality lamb meat all year round. Coinciding with this objective, different lambing schemes were designed to supply the sheep meat markets during the different seasons of the year with different products – heavy lambs (34–45 kg) and light lambs (22–25 kg). These schemes reduce the seasonality and add value to sheep meat production, such as supplying 'prime lamb' to the winter–early spring sheep market. Additionally, a great challenge was to integrate sheep meat production into the intensive animal–cropping systems of the south-west of Uruguay. Sheep production has been demonstrated to be an excellent complementary and diversifying option, giving comparative advantages to: (i) legume seed production; (ii) weed control during the cropping phase of the rotation; (iii) grazing of double purpose crops, such as barley and wheat; (iv) utilization of crop residues; and (v) acting as a tool, together with the use of cultivated pastures, to recover the chemical and physical soil losses incurred during the cropping phase of the rotation, resulting in better productivity and economic sustainability of the production systems.

Conclusion

The production levels and quality of the products achieved by the technologies proposed by agricultural research bodies for pasture and animal production in Uruguay favour a more productive, ecologically sustainable and economically sound perspective, making the production systems more flexible and stable.

The future challenges of animal and pasture research will be primarily:

- To find better productive alternatives for those producers who are running farms on shallow soils (basaltic and granitic soils): examples are fine and ultrafine merino production and the evaluation and breeding of legumes and grasses well adapted to these ecological and productive conditions.
- To release continuously new plant material adapted to high-fertility soil conditions for intensive production systems, for grazing and forage conservation purposes.
- To evaluate the impact of irrigation on cultivated pastures and improved *campos*, particularly in periods of time when water deficit occurs.
- To produce better-quality and safer products to satisfy local and overseas consumer demands.
- To integrate the knowledge generated for plant and animal components at the production system level, offering easy tools to farmers to facilitate the adoption of the new technologies available.

- To evaluate the impact of the technologies proposed on our natural resources, particularly with the use of fertilizers, herbicides, insecticides and grazing intensity.
- To take into consideration the requirements of the different components of each industry sector, particularly the consumers, at both national and international levels.
- To consider the influence of social, ecological and economic aspects on the decision-making unit at farm level in the definition of future research and technology transfer strategies.
- To educate and train each agent of the wool and meat industry chain.
- To define and establish a proper scheme for the technology adoption processes, considering the cultural, economic and ecological particularities of each production system at farmer and regional levels.
- To establish an adequate framework to improve continuously the national and regional integration and coordination of the different public and private organizations associated with research and technology transfer.

References

Ayala, W. and Carámbula, M. (1996) Mejoramientos extensivos en la Región Este: manejo y utilización. In: Risso, D.F., Berretta, E.J. and Morón, A. (eds) *Producción y Manejo de Pasturas*. Serie Técnica 80, INIA, Montevideo, pp. 177–182.

Bemhaja, M. (1991) *Forrajeras de invierno en suelos arenosos*. Hoja Divulgación Pasturas 1, INIA, Montevideo, 2 pp.

Bemhaja, M., Risso, D.F. and Zamit, W. (1996) Efecto de la carga animal en la productividad y persistencia de un mejoramiento extensivo. In: *XVI Reunión Grupo Técnico Regional del Cono Sur, Zona Campos*. UFRGS, Porto Alegre, poster.

Berretta, E.J. (1990) Técnicas para evaluar la dinámica de pasturas naturales en pastoreo. In: *Reuniao do Grupo Técnico Regional do Cone Sul em Melhoramento e Utilizaçaô dos Recursos Forrageiros das Areas Tropical e Subtropical: Grupo Campos, XI, 1989, Lages (SC), Brasil. Relatório*. Lages, pp. 129–147.

Berretta, E.J. (1991) Producción de pasturas naturales en el Basalto. A. – Producción mensual y estacional de forraje de cuatro comunidades nativas sobre suelos de Basalto. In: Carámbula, M., Vaz Martins, D. and Indarte, E. (eds) *Pasturas y Producción Animal en Areas de Ganadería Extensiva*. Serie Técnica 13, INIA, Montevideo, pp. 12–18.

Berretta, E.J. (1996) Campo natural: valor nutritivo y manejo. In: Risso, D.F., Berretta, E.J. and Morón, A. (eds) *Producción y Manejo de Pasturas*. Serie Técnica 80, INIA, Montevideo, pp. 113–127.

Berretta, E.J. (1998) Impacto del pastoreo en el ecosistema de la pradera natural. In: *Recuperación y Manejo de Ecosistemas Degradados*. Diálogo 49, IICA–PROCISUR, Montevideo, pp. 55–62.

Berretta, E.J. and Bemhaja, M. (1991) Producción de pasturas naturales en el Basalto. B. – Producción estacional de forraje de tres comunidades nativas sobre suelos de Basalto. In: Carámbula, M., Vaz Martins, D. and Indarte, E. (eds) *Pasturas y Producción Animal en Areas de Ganadería Extensiva*. Serie Técnica 13, INIA, Montevideo, pp. 19–23.

Berretta, E.J. and Bemhaja, M. (1998) Producción estacional de comunidades de campo natural sobre suelos de Basalto de la Unidad Queguay Chico. In: Berretta, E.J. (ed.) *Seminario de Actualización en Tecnologías para Basalto*. Serie Técnica 102, INIA, Montevideo, pp. 11–20.

Berretta, E.J. and do Nascimento, D. Jr (1991) *Glosario estructurado de términos sobre pasturas y producción animal*. Diálogo 32, IICA–PROCISUR, Montevideo, 127 pp.

Berretta, E.J. and Risso, D.F. (1995) Native grassland improvement on basaltic and granitic soils in Uruguay. In: West, N.E. (ed.) *Proceedings of the V International Rangeland Congress*, Vol. 1. Society for Range Management, Salt Lake City, USA, pp. 52–53.

Berretta, E.J., Risso, D.F., Levratto, J.C. and Zamit, W.S. (1998) Mejoramiento de campo natural de Basalto fertilizado con nitrógeno y fósforo. In: Berretta, E.J. (ed.) *Seminario de Actualización en Tecnologías para Basalto*. Serie Técnica 102, INIA, Montevideo, pp. 63–73.

Castro, E. and Ganzábal, A. (1988) *Sistemas Lanares Intensivos*. Miscelánea 66, MGAP–CIAAB, La Estanzuela, Colonia, 34 pp.

Comision de Inversiones y Desarrollo Económico (CIDE) (1962) *Los suelos del Uruguay, su uso y manejo*. Montevideo, 68 pp.

Corsi, W. (1978) Clima. In: *Pasturas IV*, 2nd edn. MGAP–CIAAB, Montevideo, pp. 255–256.

Ferreira, G. and Pittaluga, O. (1998) Propuestas tecnológicas para la mejora en la producción sobre suelos de Basalto Superficial y su evaluación económica. In: Berretta, E.J. (ed.) *Seminario de Actualización en Tecnologías para Basalto*. Serie Técnica 102, INIA, Montevideo, pp. 341–353.

Formoso, D. (1990) Pasturas naturales: componentes de la vegetación, producción y manejo de diferentes tipos de campos. In: *III Seminario Técnico de Producción Ovina*. SUL, Paysandú, pp. 225–237.

Ganzábal, A. (1996) *Presentación de resultados en producción de leche ovina: jornada de presentación de leche ovina*. Serie Actividades de Difusión 97, INIA, Las Brujas, Canelones, 15 pp.

Ganzábal, A., Montossi, F., Banchero, G. and San Julián, R. (1999) Sistemas ovinos intensivos del Uruguay. In: *4° Congreso Mundial y 3° Sudamericano de la raza Polwarth-Ideal*. Expo-Prado, Montevideo, pp. 97–118.

Montossi, F., San Julián, R., Ayala, W., Bermúdez, R. and Ferreira, G. (1997) Alternativas de Intensificación de la Producción de Carne Ovina en Sistemas Ganaderos de Uruguay. In: *XXV Jornadas Uruguayas Buiatría, IX Congreso Latinoamericano*. Centro Médico Veterinario, Paysandú, pp. 23–32.

Montossi, F., San Julián, R., de Mattos, D., Berretta, E.J., Ríos, M., Zamit, W. and Levratto, J. (1998a) Alimentación y manejo de la oveja de cría durante el último tercio de gestación en la región de Basalto. In: Berretta, E.J. (ed.) *Seminario de Actualización en Tecnologías para Basalto*. Serie Técnica 102, INIA, Montevideo, pp. 195–208.

Montossi, F., San Julián, R., Risso, D.F., Berretta, E.J., Ríos, M., Frugoni, J.C., Zamit, W. and Levratto, J. (1998b) Alternativas tecnológicas para la intensificación de carne ovina en sistemas ganaderos de Basalto. II Producción de Corderos Pesados. In: Berretta, E.J. (ed.) *Seminario de Actualización en Tecnologías para Basalto*. Serie Técnica 102, INIA, Montevideo, pp. 243–256.

Montossi, F., Berretta, E.J., Pigurina, G., Santamarina, I., Bemhaja, M., San Julián, R., Risso, D.F. and Mieres, J. (1998c) Estudio de la selectividad de ovinos y vacunos en diferentes comunidades vegetales de la región de Basalto. In: Berretta, E.J. (ed.) *Seminario de Actualización en Tecnologías para Basalto*. Serie Técnica 102, INIA, Montevideo, pp. 257–285.

Montossi, F., San Julián, R., de Mattos, D., Ferreira, G. and Pérez Jones, J. (1998d)

Producción de lana fina: una alternativa de valorización de la producción ovina sobre suelos superficiales del Uruguay con escasas posibilidades de divesificación. In: Berretta, E.J. (ed.) *Seminario de Actualización en Tecnologías para Basalto*. Serie Técnica 102, INIA, Montevideo, pp. 307–315.

Oficialdegui, R. and Gaggero, C. (1990) Metodología, estructura, funcionamiento y resultados físicos. In: *III Seminario Técnico de Producción Ovina*. SUL, Paysandú, pp. 11–48.

Olmos, F. (1992) *Aportes para el manejo del campo natural: efecto de la carga animal y el período de descanso en la producción y evolución de un campo natural de Caraguatá (Tacuarembó)*. Serie Técnica 20, INIA, Montevideo, 20 pp.

Pigurina, G., Soares de Lima, J.M. and Berretta, E.J. (1998a) Tecnologías para la cría vacuna en el Basalto. In: Berretta, E.J. (ed.) *Seminario de Actualización en Tecnologías para Basalto*. Serie Técnica 102, INIA, Montevideo, pp. 125–136.

Pigurina, G., Soares de Lima, J.M., Berretta, E.J., Montossi, F., Pittaluga, O., Ferreira, G. and Silva, J.A. (1998b) Características del engorde a campo natural. In: Berretta, E.J. (ed.) *Seminario de Actualización en Tecnologías para Basalto*. Serie Técnica 102, INIA, Montevideo, pp. 137–151.

Risso, D.F. (1991) Siembras en el tapiz: consideraciones generales y estado actual de la información en la zona de suelos sobre Cristalino. In: Carámbula, M., Vaz Martins, D. and Indarte, E. (eds) *Pasturas y Producción Animal en Areas de Ganadería Extensiva*. Serie Técnica 13, INIA, Montevideo, pp. 71–82.

Risso, D.F. (1997) Producción de carne sobre pasturas. In: Vaz Martins, D. (ed.) *Suplementación Estratégica para el Engorde de Ganando*. Serie Técnica 83, INIA, Montevideo, pp. 1–6.

Risso, D.F. and Berretta, E.J. (1996) Mejoramientos de campo en suelos sobre Cristalino. In: Risso, D.F., Berretta, E.J. and Morón, A. (eds) *Producción y Manejo de Pasturas*. Serie Técnica 80, INIA, Montevideo, pp. 193–211.

Risso, D.F. and Scavino, J. (1978) Región Centro Sur. In: *Pasturas IV*, 2nd edn. MGAP–CIAAB, Montevideo, pp. 25–36.

Risso, D.F., Ahunchaín, M., Cibils, R. and Zarza, A. (1991) Suplementación en invernadas del Litoral. In: Restaino, E. and Indarte, E. (eds) *Pasturas y Producción Animal en Areas de Ganadería Intensiva*. Serie Técnica 15, INIA, Montevideo, pp. 51–65.

Risso, D.F., Berretta, E.J. and Bemhaja, M. (1997) Avances tegnológicos para la región basáltica: I Pasturas. In: *Tecnologías de Producción Ganadera para Basalto*. Serie Actividades de Difusión 45, INIA, Tacuarembó, pp. I1–I6.

Risso, D.F., Pittaluga, O., Berretta, E.J., Zamit, W., Levratto, J., Carracelas, G. and Pigurina, G. (1998) Intensificación del engorde en la región Basáltica: I) Integración de campo natural y mejorado para la producción de novillos jóvenes. In: Berretta, E.J. (ed.) *Seminario de Actualización en Tecnologías para Basalto*. Serie Técnica 102, INIA, Montevideo, pp. 153–163.

Rosengurtt, B. (1979) *Tabla de comportamiento de las especies de plantas de campos naturales en el Uruguay*. Departamento de Publicaciones y Ediciones, Universidad de la República, Montevideo, 86 pp.

San Julián, R., Montossi, F., Berretta, E.J., Levratto, J., Zamit, W. and Ríos, M. (1998) Alternativas de manejo y alimentación invernal de la recría ovina en la región de Basalto. In: Berretta, E.J. (ed.) *Seminario de Actualización en Tecnologías para Basalto*. Serie Técnica 102, INIA, Montevideo, pp. 209–228.

Scaglia, G., San Julián, R., Bermúdez, R., Carámbula, M., Castro, L., Robaina, R. and Cánepa., G. (1998) *Engorde de corderos pesados y livianos sobre mejoramientos de campo*. Serie Actividades de Difusión 172, INIA, Treinta y Tres, pp. 39–48.

Veira da Silva, J. (1979) *Introduction à la théorie écologique*. Masson, Paris, 122 pp.

20

Argentina's Humid Grazing Lands

V.A. Deregibus

Faculty of Agronomy, University of Buenos Aires, 1417 Buenos Aires, Argentina

---

The Argentine territory runs from the Tropic of Capricorn (21° 46′ S lat.) to the southern tip of the continent (55° 58′ S lat.), covers almost 4 million km^2 and extends for 3700 km from north to south. Limited to the west by the high Andes, most of this territory is remarkably flat (less than 500 m a.s.l).

Moisture and Thermal Environment

A large portion of the arid diagonal of South America, which crosses the continent from north-west to south-east, covers the Argentine territory (Fig. 20.1), determining shrubby steppes in the north-western and central regions and a cool semi-desert in Patagonia. Influenced by dominant winds coming from the Atlantic Ocean, the northern (subtropical) and central (pampean) regions of the territory of Argentina increase humidity towards the east, with environments ranging from subhumid with a dry winter (annual rainfall = 600 mm) to humid (900 mm) and superhumid (1200 mm) (Fig. 20.2).

The subtropical and temperate humid zones occupy almost 0.5 million km^2 in Argentina, with high precipitation and a mild thermal environment. Such mildness is distinctive, the amplitude between mean temperature peaks being lower than that of North American sites located at similar latitudes and receiving the same rainfall (Table 20.1, Fig. 20.3). Such a mild climate allows year-long grass growth and cattle grazing.

 Grassland Ecophysiology and Grazing Ecology (eds G. Lemaire, J. Hodgson, A. de Moraes, C. Nabinger and P.C. de F. Carvalho)

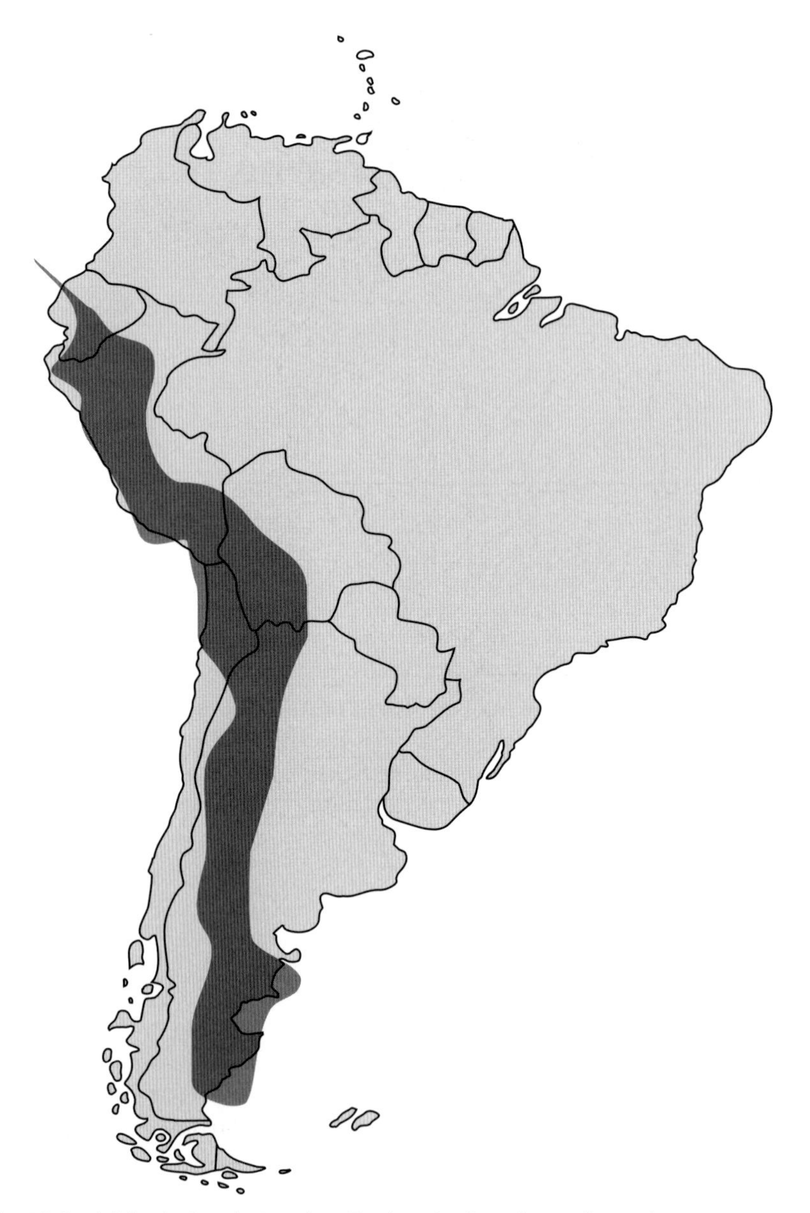

Fig. 20.1. Aridity in South America. Darker shading shows the arid environment, known as the 'arid diagonal'.

Grazing Land Types and Plant Species

Most of the humid subtropical and pampean regions of Argentina have a grassland physiognomy. Even where trees grow, their low density allows grass establishment and cattle grazing. The superhumid north-east features marshes or

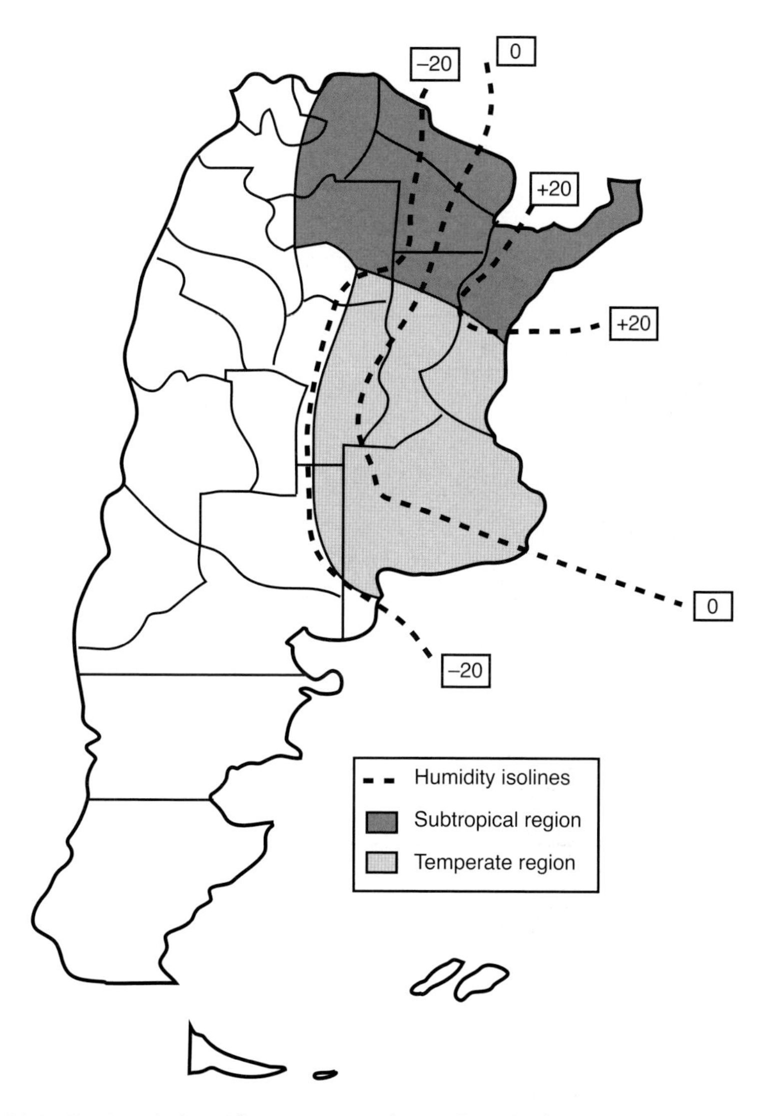

Fig. 20.2. Regions in humid Argentina. Isolines of similar humidity regimes.

humid pastures over acid sandy (deep) and rocky (shallow) soils, while the rest of the subtropical region is covered by a savannah-type vegetation growing on well-developed soils. The characteristics of the pampean region are its lack of native trees, flat terrain, fertile soils, extensive croplands and cultivated pastures. Only the flooding *pampa* is vegetated by native humid grasslands.

Megathermic grazing lands of the subtropical region are dominated by warm-season grasses and have a low density of cool-season grasses. In the mesothermic grazing lands of the pampean region the vegetation alternates between warm- and cool-season grasses (Deregibus, 1988). The warm-season components of these

Table 20.1. Daily mean and minimum temperature and ratio between summer and winter (S/W) daily mean temperatures for places with similar precipitation regime and at similar latitude, in North America and humid Argentina.

	Precipitation (mm)	Mean daily temperature (°C)	Mean daily minimum (°C)	S/W thermal amplitude
Rosario 32° 55′S–60° 47′W	968	16.7	11.7	2.3
Dallas (Texas) 32° 51′N–96° 51′W	879	18.8	13.3	3.8
Buenos Aires 34° 35′S–58° 29′W	1027	18.9	11.1	2.3
Memphis (Tennessee) 35° 03′N–89° 59′W	1262	16.7	10.9	4.9
Mar del Plata 38° 08′S–57° 33′W	768	13.4	8.7	2.5
Washington, DC 38° 51′N–77° 03′W	1036	13.9	8.3	9.5

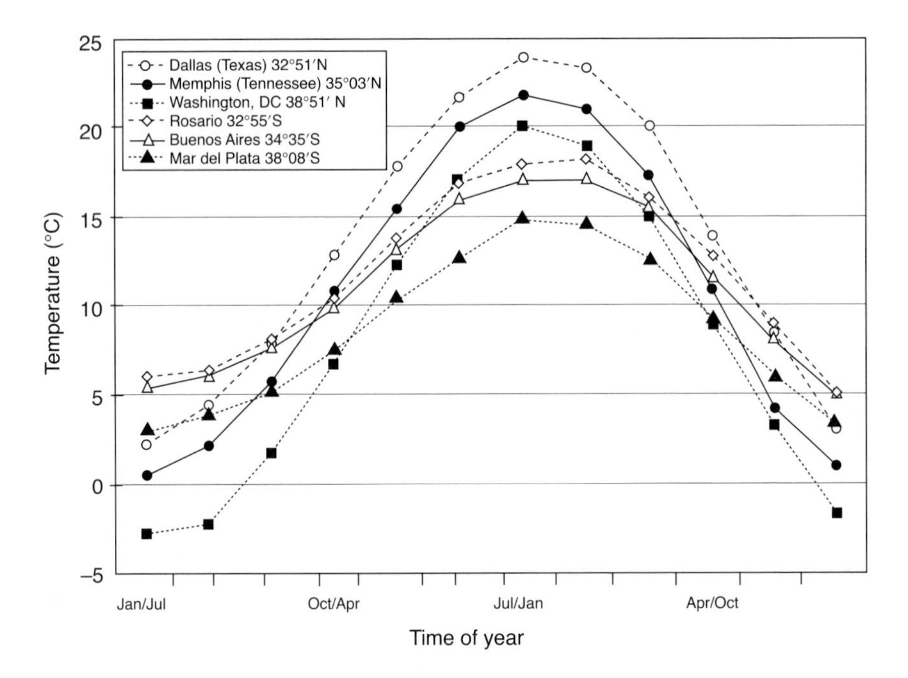

Fig. 20.3. Mean daily minimum temperature for places with similar precipitation regime and at similar latitude, in North America and humid Argentina.

grazing lands are C_4 grasses of the *Panicoideae*, *Chlorideae*, *Andropogoneae* and *Oryzeae* tribes, which are water-efficient, nutrient-thrifty and low-quality forage species. Alternating seasonally with the warm-season grasses, C_3 grasses of the *Agrosteae*, *Aveneae*, *Festuceae*, *Phalarideae* and *Stipeae* tribes thrive in both regions. West of the Paraná River and south of the Río de la Plata, soil fertility increases and a myriad of herbaceous legumes grow (*Captosena* spp., *Cassia* spp., *Crotalaria* spp., *Desmanthus* spp., *Phaseolus* spp., *Vicia* spp., etc.). In the shallow and phosphorus (P)-deficient soils (less than 3 p.p.m.) of the north-east, legumes are sparse, with few individuals of *Adesmia* spp., *Desmodium* spp. and *Rinchosia* spp.

In the north and north east, where soil fertility is very low and the climate warm and superhumid, the dominant, tall, warm-season grasses impede shorter, cool-season grasses and legumes. These characteristics of the megathermic grasslands determine massive production of low-quality forage biomass (60% digestibility or less) during the long warm season. The coarser biomass that is produced from midsummer onwards accumulates, senesces and loses quality (up to 3% protein content) during winter, and often requires to be burned before the onset of the following growing season. This accumulation overtops cool-season grasses and prevents their growth, resulting in scarce forage production during the winter.

Fertile soils and/or shorter and milder summers are essential for C_3 grasses and temperate legumes species to grow during cooler seasons and to alternate with C_4 species. Such characteristics increase towards the pampean region and determine the seasonal combination of species in the mesothermic grasslands. This seasonal combination of grass species maintains the greenness of grasslands year-long and is ideal for resource utilization in a variable climatic environment, with mild water deficits during the summer. From the forage point of view, temperate grasses and legume species of good forage quality (above 15% protein and 70% digestibility) allow utilization during winter of the remnant biomass of summer grasses.

Herbage Productivity and Carrying Capacity

Annual above-ground net primary production (ANPP) of grazing lands is strongly related to mean annual precipitation (ppt). Sala *et al.* (1988) reported an equation that explains 90% of ANPP for US rangelands:

$$\text{ANPP (g m}^{-2}) = -34 + 0.6 \times \text{ppt (mm)} \qquad (1)$$

This equation has been confirmed with the few data available for native grasslands in Argentina. Following this formula, we may estimate that native grasslands in humid Argentina produce annually a biomass that ranges from 4 to 7 t ha^{-1}. With satellite images and using the existing correlation between the integral of the annual normalized difference vegetation index and ANPP (Paruelo *et al.*, 1997), similar estimates were made.

The utilization of improved varieties of forage species and higher nutrient availability increase the ANPP of cultivated pastures significantly. In Argentina's croplands, cultivated pastures alternate cyclically with cash crops, in a rotation aimed to maintain soil fertility. Forage legumes, such as lucerne and clovers, are sown mixed with grasses, such as fescue, phalaris, brome and orchard or ryegrasses, and are directly grazed for 4 or 5 years before the soil is cultivated again. ANPP of these non-fertilized tamed pastures during the first and second year is higher than those given above (8–10 t ha^{-1}), although plant death and depletion in soil nutrient availability significantly reduce such productivity during the following years (Oesterheld and León, 1987). When these pastures are adequately fertilized (principally with P), primary production may achieve 12 to 15 t ha^{-1} or even more.

Annual ANPP estimates are based on mean annual precipitation and say little about intra- and interyear variations. Variation between years is low in humid areas (Paruelo and Lauenroth, 1998), ANPP differences for the flooding *pampa* native grasslands deviating by a maximum or a minimum of ±20% from the mean, mainly due to rainfall variations (Paruelo *et al.*, 1998). Within years, NDVI studies (correlated with primary production) show a year-round and peakless pattern, with smooth drops to both sides, a consequence of the mild weather mentioned above (Paruelo *et al.*, 1999).

Knowledge of the amount of grass biomass produced and how its productivity varies seasonally allows us to make rough estimates of the carrying capacity of the native grasslands. Uncertainties in the estimation of ANPP characterize these extensive animal production systems, based almost exclusively on grass feeding, with little or no supplementation and/or forage transfer. Conservative estimates determine that a high proportion of the biomass produced during peak months remains ungrazed, growing old and losing quality. As mentioned before, such accumulation of low-quality fodder limits the forage harvested by grazing animals and prevents seasonal grass regrowth. Because of this, harvest efficiency is low in Argentina's grazing lands, estimated to be 50% or less of the forage biomass produced. In the cropland region, grazers are usually supplemented with grain, hay and silage, while annual forage crops allow the grazing area to increase. However, the determination of a carrying capacity for cultivated grasslands is a sensitive issue, as the amount of inexpensive forage that is directly grazed defines the profit of these production activities.

Farmers are conservative in deciding the amount of stock to be placed on a given pasture, not only to reduce any climatic risk but as a way of evading some subclinical nutrient deficiencies, which frequently hamper animal condition and production. Because of this, farmers decide animal numbers in an empirical way. It is true that, over a wide range of precipitation regimes (500 to 1200 mm), ANPP explains 80% of the stocking rate observed in different administrative departments of Argentina (Oesterheld *et al.*, 1992). However, in the part of that regression line that corresponds to these more humid grazing lands, the correlation weakens as 200 kg ha^{-1} of animal live-weight corresponds to a range of ANPP from 4 to 7 t ha^{-1}. This suggests the existence of other determinants than ANPP to justify stocking rate, such as variations in the quality of the forage produced and the existence of some winter productivity, provided cool-season grasses are present.

Product Crop and Costs

Meat- and milk-producing cattle in Argentina graze the forage resources extensively year-round. Cow–calf operations occur mainly in areas where cropping is limited. Yearlings are transported to cropland areas where tamed pastures or forage crops are grazed, achieving slaughter weight at 24 months old. These same pastures are grazed by dairy cattle. Although the availability of cheap grain in the area provides an important source of supplementation, cattle fattening in confinement is rare.

Cow–calf operations stock 0.5–0.7 cows ha^{-1} on native grasslands, all over the humid region. Because of sanitary and mainly nitrogen (N) and P deficiencies, cow reconception rates and calf growth are constrained, limiting live-weight production to an average of 40–70 kg ha^{-1} throughout the entire year. On some top farms, beef productivity exceeds 100 kg ha^{-1}, with adequate management and some area of cultivated pastures. The cost of production is estimated at \$0.6 kg^{-1} live-weight, although large operations have reported costs of only half of this value.

Yearling raising and steer fattening occurs on lucerne, clover and grass pastures, which may carry two to three head ha^{-1} year-round. Weight gain averages 100–200 kg $head^{-1}$ and ranges from 250 to 400 kg ha^{-1} annually. Top producers report higher values, but it is difficult to determine how much is due to grass feeding and how much to grain supplementation. Estimated cost of production is also \$0.6 kg^{-1} live-weight.

Dairy farms milk cows during the entire year and receive a premium price for winter milk. Holstein cows are able to process important amounts of dry forage and produce accordingly. For this reason, just one-third of their total diet comes from pastures or cultivated forages, the rest of their feed being based on corn grain and maize silage. The production per hectare ranges from three to five t fluid milk ha^{-1}, with a production cost of \$0.12–\$0.16 l^{-1}, which is now becoming non-competitive.

Nutrient (N, P) Limitations

The soils of the *pampa* and the Chaco Plains (subtropical region) developed mainly from loess material and, to a lesser degree, from fluvial sediments. The soil mineralogy varies over the regions, but usually illite is the dominant clay mineral. The main soil taxonomy orders found are Mollisols in the *pampa*, Alfisols to the west and north-east of the subtropical region and Vertisols in the wet borders of the *pampa* (Alvarez and Lavado, 1998).

Adequate plant and animal nutrition may be the clue for improving production in humid Argentine grazing lands and expressing their potential. Nutrient content in soils varies from severe deficiency to richness. Native vegetation occupies the poorer soils, which are deficient in P and N, as well in some other minerals (e.g. Cu, Se). Tamed pastures are sown on the fertile crop soils of the *pampas*

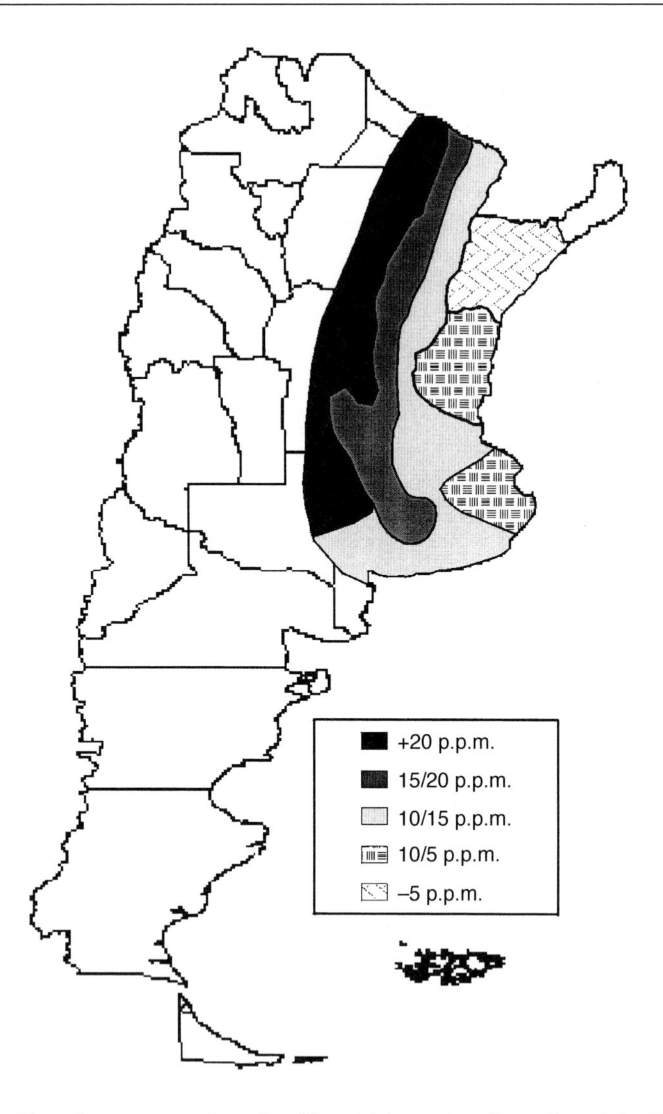

Fig. 20.4. Phosphate content in soils of humid Argentina (from Darwich, 1990).

after 4–6 years of agriculture have thoroughly depleted their fertility. Since agriculture is a more profitable activity, cultivated pastures are intended to recuperate soil fertility, rather than achieving sound animal production. Less than a century of tillage has depleted soils significantly and caused farmers to abandon their land in the northern subtropics. Adequate fertilization is still a missing subject for Argentina's pasture men, although many producers use fertilizer when pastures are sown.

P availability varies from severe deficiency to good availability (Fig. 20.4). In the north-east of the subtropical region, the majority of soils are extremely deficient (less than 5 p.p.m. in soils and less than 0.1% in plant tissue). Such low P availability reduces plant growth, causing the native grasslands of the area to be almost legume-free (except for some *Lotus* and *Adesmia* spp.). The low P content in forages affects cattle performance directly, causing subclinical deficiencies. In the east of Argentina, it is usual to observe bone-chewing animals, low pregnancy percentages and reduced milk production in cows, as well as reduced growth rate in yearlings. This problem is moving west in the *pampas* region, as P and Ca availability are significantly reduced by tilling and extensive cattle husbandry, through acidification, exportation and redistribution of these nutrients.

Year-long evaluation of ANPP (Sala *et al.*, 1981) shows that winter forage production is the weak link in the cattle-ranching chain, affecting both stocking and animal growth rate. The lack of response to N and P fertilization during this season (Rubio *et al.*, 1997) also suggests that the productivity of winter-growing grasses of the flooding *pampa* is controlled by climatic variables, such as temperature. Fortunately, other experiments (Fernandez Grecco *et al.*, 1995; Marino *et al.*, 1995a, b) reported significant winter and early spring growth response to NO_3^- addition to soils, in both native and cultivated grazing lands. NO_3^- concentration during winter in the most humid soils is fairly low (5 p.p.m.), while in the cropland soils it is higher (60 p.p.m.). The difference is possibly caused by nitrification activities during the cool season, highly affected by soil porosity and water content.

Product Quality and Improvement Possibilities

Beef quality depends on the feeding system and diet, in addition to the breed, sex, age, live-weight, etc., of the animals concerned. This has become very important for consumers, who are concerned about healthy diets and avoiding animal fats. The decrease in beef consumption occurring in the developed countries in recent years is explained because high levels of saturated fat and cholesterol in beef increase the risk of developing of certain coronary heart diseases. But animal fat varies in content and composition. Several studies have shown that beef produced on forage diets is leaner than beef produced on grain-fed systems (Crouse *et al.*, 1984; Marmer *et al.*, 1984). In addition, the dietary lipid metabolism of cattle grazing high-quality temperate pasture is different from that occurring in feed-lot cattle. Grass-fed steers are shown to have lower carcass fat content and total cholesterol (Table 20.2), as well as a lower content of saturated fatty acid (García and Casal, 1992).

Humid Argentina is a climatically wealthy region, hosting almost 40 million cattle and whose extensive pastures are grazed year-long. There is no climatic constraint for grass growth or animal comfort, as rainfall is high and fairly even, while temperature is mild and peakless. Grass is green year-round whenever forage biomass is kept well grazed. Although the efficiency of beef cattle production is low, it is increased sharply in top farms by grain supplementation, yearling raising and grassland cultivation. Extensive farms are able to reduce the costs of production.

Table 20.2. Content of intramuscular fat (%) and cholesterol (mg %) in the longissimus dorsi muscle of cattle grown with different diets (García and Casal, 1992).

	Grazing steers	Grainfed steers
Intramuscular fat (%)	2.9 ± 0.94*	3.9 ± 1.1
Cholesterol (mg %)	66.6 ± 8.80*	72.2 ± 13.8

* $P < 0.05$.

Limitations are merely nutritional and an adequate fertilization technology should be developed for preventing environmental pollution. Grass-fed animals provide a healthy product and are produced in ecologically sound systems. With these conditions, the possibilities of improving grassland productivity through adequate fertilization, the combination of selected forage species varieties and controlling the way of grazing offer a very promising future.

References

Alvarez, R. and Lavado, R.S. (1998) Climate, organic matter and clay content relationships in the *Pampa* and Chaco soils, Argentina. *Geoderma* 83, 127–141.

Crouse, J.D., Cross, H.R. and Seideman, J. (1984) Effects of a grass or grain diet on the quality of three beef muscles. *Journal of Dairy Science* 66, 1881–1890.

Darwich, N.A. (1990) Fertilizantes: nuevo balance de requerimientos. Juicio a nuestra agricultura. *INTA* (Argentina) Special Publication, pp. 1–10.

Deregibus, V.A. (1988) Importancia de los pastizales naturales en la República Argentina: situación presente y futura. *Revista Argentina de Producción Animal* 8(1), 67–68.

Fernandez Grecco, R., Mazzanti, A. and Echeverría, H.E. (1995) Efecto de la fertilización nitrogenada sobre el crecimiento de forraje de un pastizal natural de la *Pampa* Deprimida Bonaerense (Argentina). *Revista Argentina de Producción Animal* 15(1), 179–182.

García, P.T. and Casal, J.J. (1992) Lipids in longissimus muscles from grass or grain fed steers. In: *Proceedings 38th International Congress of Meat Science and Technology*, Vol. 2. Clermont-Ferrand, France, pp. 53–56.

Marino, M.A., Mazzanti, A. and Echeverria, H.E (1995a) Fertilización nitrogenada en cultivos forrajeros anuales de invierno en el sudeste bonaerense. 1. Crecimiento y acumulación de forraje. *Revista Argentina de Producción Animal* 15(1), 179–181.

Marino, M.A., Mazzanti, A., Echeverria, H.E. and Andrade, F. (1995b) Fertilización nitrogenada de cultivos forrajeros anuales de invierno en el sudeste bonaerense: 2 Concentración de nitrógeno en el forraje durante el crecimiento inverno-primaveral. *Revista Argentina de Producción Animal* 15(1), 182–185.

Marmer, W.N., Maxwell, R.J. and Williams, E.J. (1984) Effects of dietary regimen and tissue site on bovine fatty acid profiles. *Journal of Animal Science* 59, 109–121.

Oesterheld, M., and León, R.J.C. (1987) El envejecimiento de las pasturas implantadas. Su efecto sobre la productividad primaria. *Turrialba* 37, 29–36.

Oesterheld, M., Sala, O.E. and McNaughton, S.J. (1992) Effect of animal husbandry on herbivore-carrying capacity at a regional scale. *Nature* 356, 234–236.

Paruelo, J.M. and Lauenroth, W.K. (1998) Interannual variability of NDVI and its relationship to climate for North American shrublands and grasslands. *Journal of Biogeography* 25, 721–733.

Paruelo, J.M., Epstein, H.E., Lauenroth, W.K. and Burke, I.C. (1997) ANNP estimates from NDVI for the central grassland region of the US. *Ecology* 78, 953–958.

Paruelo, J.M., Oesterheld, M. and Lafontaine, J. (1998) Evaluación de recursos forrajeros. Utilización de modelos e información satelital. *Revista* CREA 208, 36–40.

Paruelo, J.M., Garbulsky, M.F., Guerschman, J.P. and Oesterheld, M. (1999) Caracterización regional de los recursos forrajeros de las zonas templadas de Argentina mediante imágenes satelitarias. *Revista Argentina de Producción Animal* 19(1), 121–125.

Rubio, G., Taboada, M.A., Lavado, R.S., Rimski-Korsakov, I.I. and Zubillaga, M.S. (1997) Acumulación de biomasa, nitrógeno y fósforo en un pastizal natural fertilizado de la *Pampa* Deprimida, Argentina. *Ciencia del Suelo* 15, 48–50.

Sala, O.E., Deregibus, V.A., Schliter, T. and Alippe, H. (1981) Productivity dynamics of a native temperate grassland in Argentina. *Journal of Range Management* 34, 48–51.

Sala, O.E., Parton, W.J., Joyce, L.A. and Lauenroth, W.K. (1988) Primary production of the central grassland region of the United States. *Ecology* 69, 40–45.

The Final Resolution

Given the main objectives of this symposium and because of the quality of the scientific exchanges, the Scientific Committee was asked by the group of invited speakers coming from Europe, North America and New Zealand, and also by the scientific leaders of the three countries, Argentina, Brazil and Uruguay, to form a Continuing Committee for organizing further symposia every 2 or 3 years in the different countries of this large subtropical and temperate area of South America.

This Committee was asked by the participants of the symposium to write a final resolution giving the objectives for further action.

On behalf of the members of the symposium, the Committee declares:

1. The necessity for grassland sciences to be organized at the world level to face the challenges of:

(a) understanding the functioning of complex pastoral ecosystems;
(b) determining procedures for sustainable management of these large areas in the world for maintaining or enhancing the viability of rural activities, and protecting over the long term the soil, water, air and life resources;
(c) evaluating the contribution of these grassland areas at a global scale to mitigating the effects of climate change and to the preservation of the biodiversity of our planet.

2. The necessity for recognition at the world level of the importance of the temperate and subtropical grassland areas as functional ecosystems, as is the case for boreal, temperate or tropical forest ecosystems, or for tropical savannah and semi-desert vegetation. Research on temperate grasslands has for a long time been dominated by the needs of animal production and efforts have been split in several parts of the world according to national interests. The concepts and methods

(eds G. Lemaire, J. Hodgson, A. de Moraes, C. Nabinger and P.C. de F. Carvalho)

developed at a world level have been mainly dominated by the agronomic and economic objectives of the intensive grassland areas in Europe, New Zealand and North America. But now, facing the environmental limits of these systems, we have to reconsider grassland ecosystems from a more ecological point of view, including a more appropriate use of extensive systems within the context of an ever-increasing range of environmental constraints. Within such a perspective, the scientific challenge will be to determine whether and how the concepts, methods and models developed for intensive grassland management in a relatively small area of the world can be adapted to conform with the diversity of conditions of the large, extensive grassland areas throughout the world.

3. The recognition that the temperate grassland area of South America, which is composed of the *pampa* region of Argentina, and the *campos* in Uruguay and South Brazil, covering an area of about 700,000 km^2, represents an important extensive grazing ecosystem of world interest. Due to its size, this vegetation system has an important role in regulating fluxes of nutrients, carbon and water on a continental scale and with an impact on oceanic and atmospheric quality. The vegetation is a complex mixture of natural species (both C_3 and C_4) and naturalized species introduced since colonization. This vegetation has evolved under a very extensive grazing system, and its equilibrium will be dependent upon the impact of economic constraints on future grazing pressures.

4. The urgent requirement for this area is to have a clear indication of the dynamics of the ecosystem and its evolution under different scenarios of agronomic development. The ever-reducing economic returns of the extensive meat production systems in the region will encourage farmers and landowners to develop cropping systems in order to sustain profits, with possible high risks for the degradation of the unique natural grassland communities, the rapid loss of soil resources and the loss of biodiversity and the very rich wildlife.

5. The necessity is for intensive studies of the dynamics of soil resources and interactions with vegetation, the role of herbivores in the evolution of mixed plant communities and the role and impact of grazing management in the dynamics of the system. Such research carried on in this region should not only have an impact on its own rural development, but should also be valuable at a world level for advancing the science of grazing ecology, with important implications for other similar areas of the world.

As a consequence, the Committee wishes to encourage and promote the organization of future symposia and future scientific collaboration between all the various international institutions interested in the management of pastoral ecosystems, and will seek ways to give maximum international publicity to this declaration.

Membership of the Continuing Committee

- Dr G. Lemaire, INRA, France (Chair)
- Dr E. Berretta, INIA, Uruguay
- Prof. A. De Moraes, University of Parana, Brazil
- Prof. A Deregibus, University of Buenos Aires, Argentina
- Prof. J. Hodgson, Massey University, New Zealand
- Prof. C.J. Nelson, University of Missouri, USA

Index